Mass Spectrometry of Proteins and Peptides

METHODS IN MOLECULAR BIOLOGY™

John M. Walker, SERIES EDITOR

146. **Mass Spectrometry of Proteins and Peptides**, edited by *John R. Chapman, 2000*
145. **Bacterial Toxins:** *Methods and Protocols*, edited by *Otto Holst, 2000*
144. **Calpain Methods and Protocols**, edited by *John S. Elce, 2000*
143. **Protein Structure Prediction:** *Methods and Protocols*, edited by *David Webster, 2000*
142. **Transforming Growth Factor-Beta Protocols**, edited by *Philip H. Howe, 2000*
141. **Plant Hormone Protocols**, edited by *Gregory A. Tucker and Jeremy A. Roberts, 2000*
140. **Chaperonin Protocols**, edited by *Christine Schneider, 2000*
139. **Extracellular Matrix Protocols**, edited by *Charles Streuli and Michael Grant, 2000*
138. **Chemokine Protocols**, edited by *Amanda E. I. Proudfoot, Timothy N. C. Wells, and Christine Power, 2000*
137. **Developmental Biology Protocols, Volume III**, edited by *Rocky S. Tuan and Cecilia W. Lo, 2000*
136. **Developmental Biology Protocols, Volume II**, edited by *Rocky S. Tuan and Cecilia W. Lo, 2000*
135. **Developmental Biology Protocols, Volume I**, edited by *Rocky S. Tuan and Cecilia W. Lo, 2000*
134. **T Cell Protocols:** *Development and Activation*, edited by *Kelly P. Kearse, 2000*
133. **Gene Targeting Protocols**, edited by *Eric B. Kmiec, 2000*
132. **Bioinformatics Methods and Protocols**, edited by *Stephen Misener and Stephen A. Krawetz, 2000*
131. **Flavoprotein Protocols**, edited by *S. K. Chapman and G. A. Reid, 1999*
130. **Transcription Factor Protocols**, edited by *Martin J. Tymms, 2000*
129. **Integrin Protocols**, edited by *Anthony Howlett, 1999*
128. **NMDA Protocols**, edited by *Min Li, 1999*
127. **Molecular Methods in Developmental Biology:** Xenopus *and* Zebrafish, edited by *Matthew Guille, 1999*
126. **Adrenergic Receptor Protocols**, edited by *Curtis A. Machida, 2000*
125. **Glycoprotein Methods and Protocols:** *The Mucins*, edited by *Anthony P. Corfield, 2000*
124. **Protein Kinase Protocols**, edited by *Alastair D. Reith, 2000*
123. **In Situ Hybridization Protocols (2nd ed.)**, edited by *Ian A. Darby, 2000*
122. **Confocal Microscopy Methods and Protocols**, edited by *Stephen W. Paddock, 1999*
121. **Natural Killer Cell Protocols:** *Cellular and Molecular Methods*, edited by *Kerry S. Campbell and Marco Colonna, 2000*
120. **Eicosanoid Protocols**, edited by *Elias A. Lianos, 1999*
119. **Chromatin Protocols**, edited by *Peter B. Becker, 1999*
118. **RNA–Protein Interaction Protocols**, edited by *Susan R. Haynes, 1999*
117. **Electron Microscopy Methods and Protocols**, edited by *M. A. Nasser Hajibagheri, 1999*
116. **Protein Lipidation Protocols**, edited by *Michael H. Gelb, 1999*
115. **Immunocytochemical Methods and Protocols (2nd ed.)**, edited by *Lorette C. Javois, 1999*
114. **Calcium Signaling Protocols**, edited by *David G. Lambert, 1999*
113. **DNA Repair Protocols:** *Eukaryotic Systems*, edited by *Daryl S. Henderson, 1999*
112. **2-D Proteome Analysis Protocols**, edited by *Andrew J. Link, 1999*
111. **Plant Cell Culture Protocols**, edited by *Robert D. Hall, 1999*
110. **Lipoprotein Protocols**, edited by *Jose M. Ordovas, 1998*
109. **Lipase and Phospholipase Protocols**, edited by *Mark H. Doolittle and Karen Reue, 1999*
108. **Free Radical and Antioxidant Protocols**, edited by *Donald Armstrong, 1998*
107. **Cytochrome P450 Protocols**, edited by *Ian R. Phillips and Elizabeth A. Shephard, 1998*
106. **Receptor Binding Techniques**, edited by *Mary Keen, 1999*
105. **Phospholipid Signaling Protocols**, edited by *Ian M. Bird, 1998*
104. **Mycoplasma Protocols**, edited by *Roger J. Miles and Robin A. J. Nicholas, 1998*
103. *Pichia* **Protocols**, edited by *David R. Higgins and James M. Cregg, 1998*
102. **Bioluminescence Methods and Protocols**, edited by *Robert A. LaRossa, 1998*
101. **Mycobacteria Protocols**, edited by *Tanya Parish and Neil G. Stoker, 1998*
100. **Nitric Oxide Protocols**, edited by *Michael A. Titheradge, 1998*
99. **Stress Response:** *Methods and Protocols*, edited by *Stephen M. Keyse, 2000*
98. **Forensic DNA Profiling Protocols**, edited by *Patrick J. Lincoln and James M. Thomson, 1998*
97. **Molecular Embryology:** *Methods and Protocols*, edited by *Paul T. Sharpe and Ivor Mason, 1999*
96. **Adhesion Protein Protocols**, edited by *Elisabetta Dejana and Monica Corada, 1999*
95. **DNA Topoisomerases Protocols:** *II. Enzymology and Drugs*, edited by *Mary-Ann Bjornsti and Neil Osheroff, 2000*
94. **DNA Topoisomerases Protocols:** *I. DNA Topology and Enzymes*, edited by *Mary-Ann Bjornsti and Neil Osheroff, 1999*
93. **Protein Phosphatase Protocols**, edited by *John W. Ludlow, 1998*
92. **PCR in Bioanalysis**, edited by *Stephen J. Meltzer, 1998*
91. **Flow Cytometry Protocols**, edited by *Mark J. Jaroszeski, Richard Heller, and Richard Gilbert, 1998*
90. **Drug–DNA Interaction Protocols**, edited by *Keith R. Fox, 1998*
89. **Retinoid Protocols**, edited by *Christopher Redfern, 1998*
88. **Protein Targeting Protocols**, edited by *Roger A. Clegg, 1998*
87. **Combinatorial Peptide Library Protocols**, edited by *Shmuel Cabilly, 1998*
86. **RNA Isolation and Characterization Protocols**, edited by *Ralph Rapley and David L. Manning, 1998*
85. **Differential Display Methods and Protocols**, edited by *Peng Liang and Arthur B. Pardee, 1997*
84. **Transmembrane Signaling Protocols**, edited by *Dafna Bar-Sagi, 1998*
83. **Receptor Signal Transduction Protocols**, edited by *R. A. John Challiss, 1997*
82. **Arabidopsis Protocols**, edited by *José M Martinez-Zapater and Julio Salinas, 1998*
81. **Plant Virology Protocols:** *From Virus Isolation to Transgenic Resistance*, edited by *Gary D. Foster and Sally Taylor, 1998*

METHODS IN MOLECULAR BIOLOGY™

Mass Spectrometry of Proteins and Peptides

Edited by

John R. Chapman

Sale, Manchester, UK

Humana Press ✳ Totowa, New Jersey

© 2000 Humana Press Inc.
999 Riverview Drive, Suite 208
Totowa, New Jersey 07512

All rights reserved. No part of this book may be reproduced, stored in a retrieval system, or transmitted in any form or by any means, electronic, mechanical, photocopying, microfilming, recording, or otherwise without written permission from the Publisher. Methods in Molecular Biology™ is a trademark of The Humana Press Inc.

The content and opinions expressed in this book are the sole work of the authors and editors, who have warranted due diligence in the creation and issuance of their work. The publisher, editors, and authors are not responsible for errors or omissions or for any consequences arising from the information or opinions presented in this book and make no warranty, express or implied, with respect to its contents.

This publication is printed on acid-free paper. ∞
ANSI Z39.48-1984 (American Standards Institute) Permanence of Paper for Printed Library Materials.

Cover design by Patricia F. Cleary.

Cover art from: Figure 2, Chapter 3, *Characterization of a Mutant Recombinant S100 Protein Using Electrospray Ionization Mass Spectrometry*, by M. J. Raftery.

For additional copies, pricing for bulk purchases, and/or information about other Humana titles, contact Humana at the above address or at any of the following numbers: Tel: 973-256-1699; Fax: 973-256-8341; E-mail: humana@humanapr.com, or visit our Website at www.humanapress.com

Photocopy Authorization Policy:
Authorization to photocopy items for internal or personal use, or the internal or personal use of specific clients, is granted by Humana Press Inc., provided that the base fee of US $10.00 per copy, plus US $00.25 per page, is paid directly to the Copyright Clearance Center at 222 Rosewood Drive, Danvers, MA 01923. For those organizations that have been granted a photocopy license from the CCC, a separate system of payment has been arranged and is acceptable to Humana Press Inc. The fee code for users of the Transactional Reporting Service is: [0-89603-609-x/00 $10.00 + $00.25].

Printed in the United States of America. 10 9 8 7 6 5 4 3 2 1

Library of Congress Cataloging-in-Publication Data

Mass Spectrometry of Proteins and Peptides / edited by John R. Chapman.
     p. cm. -- (Methods in molecular biology ; 146)
     Includes bibliographical references and index.
     ISBN 0-89603-609-x (alk. paper)
     1. Proteins--Analysis. 2. Peptides--Analysis. 3. Peptide sequence. 4. Mass spectrometry. I. Chapman, J. R. II. Methods in molecular biology (Totowa, N.J.) ; v. 146.

QP551.P69565 2000
572'.633--dc21
                                                      99-41621

Preface

Little more than three years down the line and I am already writing the Preface to a second volume to follow *Protein and Peptide Analysis by Mass Spectrometry*. What has happened in between these times to make this second venture worthwhile?

New types of mass spectrometric instrumentation have appeared so that new techniques have become possible and existing techniques have become much more feasible. More particularly, however, the newer ionization techniques, introduced for the analysis of high molecular weight materials, have now been thoroughly used and studied. As a result, there has been an enormous improvement in the associated sample handling technology so that these methods are now routinely applied to much smaller sample amounts as well as to more intractable samples. Again, this particular community of mass spectrometry users has both increased in number and diversified. And, riding this wave of acceptance, leaders in the field have set their sights on more complex problems: molecular interaction, ion structures, quantitation, and kinetics are just a few of the newer areas reported in *Mass Spectrometry of Proteins and Peptides*.

As with the first volume, one purpose of this collection, *Mass Spectrometry of Proteins and Peptides*, is to show the reader what can be done by the application of mass spectrometry, and perhaps even to encourage the reader to venture down new paths. More important, another purpose is to demonstrate how these analyses are carried out in practice by guiding the reader, in a step-by-step manner, through the pitfalls and nuances of apparently straightforward techniques. It is the earnest hope of the editor that the reader, as a user of these techniques, will profit from the concise details that each of the authors has striven to provide. To parody Dr. Johnson: "what is written with effort is, in general, read with pleasure."

It has been suggested that "Writers, like teeth, are divided into incisors and grinders."* So, to spirit you from the Preface to the pleasures of the book, away with the preface grinder and on with the incisors and their chapters.

John R. Chapman

*Walter Bagehot, *Estimates of Some Englishmen and Scotchmen*.

Contents

Preface v
Contributors xi

1 *De Novo* Peptide Sequencing by Nanoelectrospray Tandem Mass Spectrometry Using Triple Quadrupole and Quadrupole/ Time-of-Flight Instruments
   Andrej Shevchenko, Igor Chernushevich, Matthias Wilm, and Matthias Mann 1
2 Direct Analysis of Proteins in Mixtures: *Application to Protein Complexes*
   John R. Yates, III, Andrew J. Link, and David Schieltz 17
3 Characterization of a Mutant Recombinant S100 Protein Using Electrospray Ionization Mass Spectrometry
   Mark J. Raftery 27
4 Searching Sequence Databases via *De Novo* Peptide Sequencing by Tandem Mass Spectrometry
   Richard S. Johnson and J. Alex Taylor 41
5 Signature Peptides: *From Analytical Chemistry to Functional Genomics*
   Haydar Karaoglu and Ian Humphery-Smith 63
6 Investigating the Higher Order Structure of Proteins: *Hydrogen Exchange, Proteolytic Fragmentation, and Mass Spectrometry*
   John R. Engen and David L. Smith 95
7 Probing Protein Surface Topology by Chemical Surface Labeling, Crosslinking, and Mass Spectrometry
   Keiryn L. Bennett, Thomas Matthiesen, and Peter Roepstorff 113
8 Secondary Structure of Peptide Ions in the Gas Phase Evaluated by MIKE Spectrometry: *Relevance to Native Conformations*
   Igor A. Kaltashov, Aiqun Li, Zoltán Szilágyi, Károly Vékey, and Catherine Fenselau 133

9   Preparation and Mass Spectrometric Analysis of
    S-Nitrosohemoglobin
    *Pasquale Ferranti, Gianfranco Mamone,
    and Antonio Malorni* 147
10 Multiple and Subsequent MALDI-MS On-Target Chemical
    Reactions for the Characterization of Disulfide Bonds and
    Primary Structures of Proteins
    *H. Peter Happersberger, Marcus Bantscheff, Stefanie Barbirz,
    and Michael O. Glocker* 167
11 Epitope Mapping by a Combination of Epitope Excision
    and MALDI-MS
    Carol E. Parker and Kenneth B. Tomer 185
12 Identification of Active Site Residues in Glycosidases by Use
    of Tandem Mass Spectrometry
    David J. Vocadlo and Stephen G. Withers 203
13 Probing Protein–Protein Interactions with Mass Spectrometry
    Richard W. Kriwacki and Gary Siuzdak 223
14 Studies of Noncovalent Complexes in an Electrospray
    Ionization/Time-of-Flight Mass Spectrometer
    *Andrew N. Krutchinsky, Ayeda Ayed, Lynda J. Donald,
    Werner Ens, Harry W. Duckworth,
    and Kenneth G. Standing* 239
15 Kinetic Analysis of Enzymatic and Nonenzymatic Degradation
    of Peptides by MALDI-TOFMS
    *Fred Rosche, Jörn Schmidt, Torsten Hoffmann,
    Robert P. Pauly, Christopher H. S. McIntosh,
    Raymond A. Pederson, and Hans-Ulrich Demuth* 251
16 Characterization of Protein Glycosylation by MALDI-TOFMS
    *Ekaterina Mirgorodskaya, Thomas N. Krogh,
    and Peter Roepstorff* 273
17 Positive and Negative Labeling of Human Proinsulin, Insulin
    and C-Peptide with Stable Isotopes: *New Tools for In Vivo
    Pharmacokinetic and Metabolic Studies*
    *Reto Stöcklin, Jean-François Arrighi, Khan Hoang-Van,
    Lan Vu, Fabrice Cerini, Nicolas Gilles, Roger Genet,
    Jan Markusssen, Robin E. Offord, and Keith Rose* 293

Contents ix

18 Identification of Snake Species by Toxin Mass Fingerprinting
of Their Venoms
*Reto Stöcklin, Dietrich Mebs, Jean-Claude Boulain,
Pierre-Alain Panchaud, Henri Virelizier,
and Cécile Gillard-Factor* 317
19 Mass Spectrometric Characterization of the β-Subunit of Human
Chorionic Gonadotropin
Roderick S. Black and Larry D. Bowers 337
20 Analysis of Gluten in Foods by MALDI-TOFMS
Enrique Méndez, Israel Valdés, and Emilio Camafeita 355
21 Quantitation of Nucleotidyl Cyclase and Cyclic
Nucleotide-Sensitive Protein Kinase Activities by Fast-Atom
Bombardment Mass Spectrometry: *A Paradigm for Multiple
Component Monitoring in Enzyme Incubations by Quantitative
Mass Spectrometry*
Russell P. Newton 369
22 Influence of Salts, Buffers, Detergents, Solvents, and Matrices
on MALDI-MS Protein Analysis in Complex Mixtures
K. Olaf Börnsen 387
23 Sample Preparation Techniques for Peptides and Proteins
Analyzed by MALDI-MS
Martin Kussmann and Peter Roepstorff 405
24 Analysis of Hydrophobic Proteins and Peptides by Mass
Spectrometry
Johann Schaller 425
25 Analysis of Proteins and Peptides Directly from Biological Fluids
by Immunoprecipitation/Mass Spectrometry
*Sacha N. Uljon, Louis Mazzarelli, Brian T. Chait,
and Rong Wang* 439
26 Detection of Molecular Determinants in Complex Biological
Systems Using MALDI-TOF Affinity Mass Spectrometry
Judy Van de Water and M. Eric Gershwin 453
27 Rapid Identification of Bacteria Based on Spectral Patterns
Using MALDI-TOFMS
Jackson O. Lay, Jr. and Ricky D. Holland 461
28 Appendices
John R. Chapman 489
Index 527

Contributors

JEAN-FRANÇOIS ARRIGHI • *Department of Medical Biochemistry, University Medical Centre, Geneva, Switzerland*
AYEDA AYED • *Department of Chemistry, University of Manitoba, Winnipeg, Canada*
MARCUS BANTSCHEFF • *Faculty of Chemistry, University of Konstanz, Konstanz, Germany*
STEPHANIE BARBIRZ • *Faculty of Chemistry, University of Konstanz, Konstanz, Germany*
KEIRYN L. BENNETT • *Protein Research Group, Department of Molecular Biology, Odense University, Odense, Denmark*
RODERICK S. BLACK • *Athletic Drug Testing and Toxicology Laboratory, Department of Pathology and Laboratory Medicine, Indiana University School of Medicine, Indianapolis, IN*
K. OLAF BÖRNSEN • *Drug Metabolism and Pharmacokinetics, Novartis Pharma AG, Basel, Switzerland*
JEAN-CLAUDE BOULAIN • *Département d'Ingénierie et d'Etude des Protéines, CEA-Saclay, France*
LARRY D. BOWERS • *Athletic Drug Testing and Toxicology Laboratory, Department of Pathology and Laboratory Medicine, Indiana University School of Medicine, Indianapolis, IN*
EMILIO CAMAFEITA • *Unidad de Análisis Estructural de Proteínas, Centro Nacional de Biotecnología, Cantoblanco, Madrid, Spain*
FABRICE CERINI • *Department of Medical Biochemistry, University Medical Centre, Geneva, Switzerland*
BRIAN T. CHAIT • *Laboratory for Mass Spectrometry, The Rockefeller University, New York, NY*
JOHN R. CHAPMAN • *Sale, Manchester, UK*
IGOR CHERNUSHEVICH • *PE Sciex, Concord, Canada*
HANS-ULRICH DEMUTH • *Probiodrug Research, Halle, Germany*
LYNDA J. DONALD • *Department of Chemistry, University of Manitoba, Winnipeg, Canada*
HARRY W. DUCKWORTH • *Department of Chemistry, University of Manitoba, Winnipeg, Canada*

JOHN R. ENGEN • *Department of Chemistry, University of Nebraska-Lincoln, Lincoln, NE*
WERNER ENS • *Department of Physics, University of Manitoba, Winnipeg, Canada*
CATHERINE FENSELAU • *Department of Chemistry and Biochemistry, University of Maryland, College Park, MD*
PASQUALE FERRANTI • *International Mass Spectrometry Facility Centre and Mass Spectrometry Network-CNR, Naples, Italy*
ROGER GENET • *Département d'Ingénierie et d'Etude des Protéines, CEA-Saclay, France*
M. ERIC GERSHWIN, MD • *Division of Rheumatology, Allergy, and Clinical Immunology, School of Medicine, University of California, Davis, CA*
CÉCILE GILLARD-FACTOR • *Département des Procédés d'Enrichissement, DCC/DPE/SPCP/LASO, CEA-Saclay, France*
NICOLAS GILLES • *Département d'Ingénierie et d'Etude des Protéines, CEA-Saclay, France*
MICHAEL O. GLOCKER • *Faculty of Chemistry, University of Konstanz, Konstanz, Germany*
H. PETER HAPPERSBERGER • *Faculty of Chemistry, University of Konstanz, Konstanz, Germany*
KHAN HOANG-VAN • *Department of Medical Biochemistry, University Medical Centre, Geneva, Switzerland*
TORSTEN HOFFMANN • *Probiodrug Research, Halle, Germany*
RICKY D. HOLLAND • *Division of Chemistry, National Center for Toxicological Research, Food and Drug Administration, Jefferson, AR*
IAN HUMPHERY-SMITH • *Department of Pharmaceutical Proteomics, Universiteit Utrecht, Utrecht, The Netherlands*
RICHARD S. JOHNSON • *Immunex Corporation, Seattle, WA*
IGOR A. KALTASHOV • *Polymer Science Department, University of Massachusetts, Amherst, MA*
HAYDAR KARAOGLU • *Department of Microbiology, University of Sydney, Sydney, Australia*
RICHARD W. KRIWACKI • *Department of Structural Biology, St. Jude Children's Research Hospital, Memphis, TN*
THOMAS N. KROGH • *Department of Molecular Biology, Odense University, Odense, Denmark*
ANDREW N. KRUTCHINSKY • *Department of Physics, University of Manitoba, Winnipeg, Canada*
MARTIN KUSSMANN • *Cerbios-Pharma SA, Barbengo, Switzerland*
JACKSON O. LAY, JR. • *Division of Chemistry, National Center for Toxicological Research, Food and Drug Administration, Jefferson, AR*

Contributors

AIQUN LI • *Department of Chemistry and Biochemistry, University of Maryland, College Park, MD*
ANDREW J. LINK • *Department of Microbiology and Immunology, Vanderbilt University Medical Center, Nashville, TN*
ANTONIO MALORNI • *International Mass Spectrometry Facility Centre and Mass Spectrometry Network-CNR, Naples, Italy*
GIANFRANCO MAMONE • *International Mass Spectrometry Facility Centre and Mass Spectrometry Network-CNR, Naples, Italy*
MATTHIAS MANN • *Protein Interaction Laboratory, University of Southern Denmark, Odense, Denmark*
JAN MARKUSSEN • *Novo Nordisk, Bagsvaerd, Denmark*
THOMAS MATTHIESEN • *Protein Research Group, Department of Molecular Biology, Odense University, Odense, Denmark*
LOUIS MAZZARELLI • *Laboratory for Mass Spectrometry, The Rockefeller University, New York, NY*
CHRISTOPHER H. S. MCINTOSH • *Department of Physiology, University of British Columbia, Vancouver, Canada*
DIETRICH MEBS • *Zentrum der Rechtsmedizin, Frankfurt, Germany*
ENRIQUE MÉNDEZ • *Unidad de Análisis Estructural de Proteínas, Centro Nacional de Biotecnología, Cantoblanco, Madrid, Spain*
EKATERINA MIRGORODSKAYA • *Department of Molecular Biology, Odense University, Odense, Denmark*
RUSSELL P. NEWTON • *Biochemistry Group, School of Biological Sciences, University of Wales, Swansea, UK*
ROBIN E. OFFORD • *Department of Medical Biochemistry, University Medical Centre, Geneva, Switzerland*
PIERRE-ALAIN PANCHAUD • *Reptiles du Monde, Grandvaux, Switzerland*
CAROL E. PARKER • *Laboratory of Structural Biology, National Institute of Environmental Health Sciences, Research Triangle Park, NC*
ROBERT P. PAULY • *Department of Physiology, University of British Columbia, Vancouver, Canada*
RAYMOND A. PEDERSON • *Department of Physiology, University of British Columbia, Vancouver, Canada*
MARK J. RAFTERY • *Cytokine Research Unit, School of Pathology, University of New South Wales, Kensington, Australia*
PETER ROEPSTORFF • *Department of Molecular Biology, Odense University, Odense, Denmark*
FRED ROSCHE • *Probiodrug Research, Halle, Germany*
KEITH ROSE • *Department of Medical Biochemistry, University Medical Centre, Geneva, Switzerland*

JOHANN SCHALLER • *Department für Chemie und Biochemie, Universität Bern, Freiestrasse, Bern, Switzerland*
DAVID SCHIELTZ • *Department of Molecular Biotechnology, University of Washington, Seattle, WA*
JÖRN SCHMIDT • *Probiodrug Research, Halle, Germany*
ANDREJ SHEVCHENKO • *European Molecular Biology Laboratory, Heidelberg, Germany*
GARY SIUZDAK • *Beckman Center for Chemical Sciences, The Scripps Research Institute, La Jolla, CA*
DAVID L. SMITH • *Department of Chemistry, University of Nebraska-Lincoln, Lincoln, NE*
KENNETH G. STANDING • *Department of Physics, University of Manitoba, Winnipeg, Canada*
RETO STÖCKLIN • *Atheris Laboratories, Research and Development, Geneva, Switzerland*
ZOLTÁN SZILÁGYI • *Institute of Chemistry, Chemical Research Center, Hungarian Academy of Sciences, Budapest, Hungary*
J. ALEX TAYLOR • *Immunex Corporation, Seattle, WA*
KENNETH B. TOMER • *Laboratory of Structural Biology, National Institute of Environmental Health Sciences, Research Triangle Park, NC*
SACHA N. ULJON • *Laboratory for Mass Spectometry, The Rockefeller University, New York, NY*
ISRAEL VALDÉS • *Unidad de Análisis Estructural de Proteínas, Centro Nacional de Biotecnología, Madrid, Spain*
KÁROLY VÉKEY • *Institute of Chemistry, Chemical Research Center, Hungarian Academy of Sciences, Budapest, Hungary*
HENRI VIRELIZIER • *CEA Saclay, SPEA-SAIS, Cedex, France*
DAVID J. VOCADLO • *Department of Chemistry, University of British Columbia, Vancouver, Canada*
LAN VU • *Department of Medical Biochemistry, University Medical Centre, Geneva, Switzerland*
RONG WANG • *Laboratory for Mass Spectometry, The Rockefeller University, New York, NY*
JUDY VAN DE WATER • *Division of Rheumatology, Allergy and Clinical Immunology, School of Medicine, University of California, Davis, CA*
MATTHIAS WILM • *European Molecular Biology Laboratory, Heidelberg, Germany*
STEPHEN G. WITHERS • *Department of Chemistry, University of British Columbia, Vancouver, Canada*
JOHN R. YATES, III • *Department of Molecular Biotechnology, University of Washington, Seattle, WA*

1

De Novo Peptide Sequencing by Nanoelectrospray Tandem Mass Spectrometry Using Triple Quadrupole and Quadrupole/Time-of-Flight Instruments

Andrej Shevchenko, Igor Chernushevich, Matthias Wilm, and Matthias Mann

1. Introduction

Recent developments in technology and instrumentation have made mass spectrometry the method of choice for the identification of gel-separated proteins using rapidly growing sequence databases (1). Proteins with a full-length sequence present in a database can be identified with high certainty and high throughput using the accurate masses obtained by matrix-assisted laser desorption/ionization (MALDI) mass spectrometry peptide mapping (2). Simple protein mixtures can also be deciphered by MALDI peptide mapping (3) and the entire identification process, starting from in-gel digestion (4) and finishing with acquisition of mass spectra and database search, can be automated (5). Only 1–3% of a total digest are consumed for MALDI analysis even if the protein of interest is present on a gel in a subpicomole amount. If no conclusive identification is achieved by MALDI peptide mapping, the remaining protein digest can be analyzed by nanoelectrospray tandem mass spectrometry (Nano ESI-MS/MS) (6). Nano ESI-MS/MS produces data that allow highly specific database searches so that proteins that are only partially present in a database, or relevant clones in an EST database, can be identified (7). It is important to point out that there is no need to determine the complete sequence of peptides in order to search a database—a short sequence stretch consisting of three to four amino acid residues provides enough search specificity when combined with the mass of the intact peptide and the masses of corresponding fragment

From: Methods in Molecular Biology, vol. 146:
Protein and Peptide Analysis: New Mass Spectrometric Applications
Edited by: J. R. Chapman © Humana Press Inc., Totowa, NJ

ions in a peptide sequence tag *(8)* (*see* **Subheading 3.4.**). Furthermore, proteins not present in a database that are, however, strongly homologous to a known protein can be identified by an error-tolerant search *(9)*.

Despite the success of ongoing genomic sequencing projects, the demand for *de novo* peptide sequencing has not been eliminated. Long and accurate peptide sequences are required for protein identification by homology search and for the cloning of new genes. Degenerate oligonucleotide probes are designed on the basis of peptide sequences obtained in this way, and subsequently used in polymerase chain reaction-based cloning strategies.

The presence of a continuous series of mass spectrometric fragment ions containing the C terminus (y″ ions) *(10)* has been successfully used to determine *de novo* sequences using fragment ion spectra of peptides from a tryptic digest *(11)*. The peptide sequence can be deduced by considering precise mass differences between adjacent y″ ions. However, it is necessary to obtain additional evidence that the particular fragment ion does indeed belong to the y″ series. To this end, a separate portion of the unseparated digest is esterified using 2 M HCl in anhydrous methanol (**Fig. 1A**) (*see* **Subheading 3.2.**). Upon esterification, a methyl group is attached to the C-terminal carboxyl group of each peptide, as well as to the carboxyl group in the side chain of aspartic and glutamic acid residues. Therefore the *m/z* value of each peptide ion is shifted by $14(n + 1)/z$, where n is the number of aspartic and glutamic acid residues in the peptide, and z is the charge of the peptide ion. The derivatized digest is then also analyzed by Nano ESI-MS/MS, and, for each peptide, fragment ion spectra acquired from underivatized and derivatized forms are matched. An accurate peptide sequence is determined by software-assisted comparison of these two fragment spectra by considering precise mass differences between the adjacent y″ ions as well as characteristic mass shifts induced by esterification (*see* **Subheading 3.4.1.**) (**Fig. 2**). Since esterification with methanol significantly shifts the masses of y″ ions (by 14, 28, 42, ... mass units), it is possible to use low-resolution settings when sequencing is performed on a triple quadrupole mass spectrometer, thus attaining high sensitivity on the instrument. This sequencing approach employing esterification is laborious and time consuming and requires much expertise in the interpretation of tandem mass spec-

Fig. 1. Chemical derivatization for mass spectrometric *de novo* sequencing of peptides recovered from digests of gel separated proteins. (**A**) A protein is digested in-gel (*see* **Subheading 3.1.**) with trypsin and a portion of the unseparated digest is esterified by 2 M HCl in anhydrous methanol (*see* **Subheading 3.2.**). (**B**) A protein is digested in-gel with trypsin in a buffer containing 50% (v/v) $H_2^{18}O$ and 50% (v/v) $H_2^{16}O$ (*see* **Subheading 3.1.**). (**C**) A protein is digested in-gel with trypsin, and the digest is esterified and subsequently treated with trypsin in the buffer containing 50% (v/v)

$H_2^{18}O$ and 50% (v/v) $H_2^{16}O$ (*see* **Note 22**). Here, R_1 repesents the side chain of arginine or lysine amino acid residues (these are trypsin cleavage sites) whereas R_x represents the side chain of any other amino acid residue except for proline.

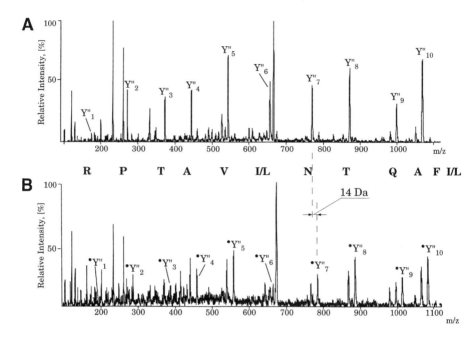

Fig. 2. Peptide *de novo* sequencing by comparison of tandem mass spectra acquired from intact and esterified peptide. A 120-kDa protein from *E. aediculatis* was purified by one-dimensional gel electrophoresis *(24)* and digested in-gel with trypsin; a part of the digest was analyzed by Nano ESI-MS/MS on an API III triple quadrupole mass spectrometer (PE Sciex, Ontario, Canada). A separate part of the digest was esterified and then also analyzed by Nano ESI-MS/MS. **(A)** Tandem (fragment-ion) mass spectrum recorded from the doubly charged ion with *m/z* 666.0 observed in the conventional (Q1) spectrum of the original digest. **(B)** Matching tandem spectrum acquired from the ion with *m/z* 673.0 (Δ mass = [673–666] × 2 = 14) in the conventional (Q1) spectrum of the esterified digest. The peptide sequence was determined by software-assisted comparison of spectra **A** and **B**. The only methyl group was attached to the C-terminal carboxyl of the peptide (designated by a filled circle) and therefore the masses of the singly charged y″ ions in spectrum **B** are shifted by 14 mass units compared with the corresponding y″ ions in spectrum **A**.

tra. However, it allows the determination of accurate peptide sequences even from protein spots that can only be visualized by staining with silver *(12,13)*.

An alternative approach to *de novo* sequencing became feasible after a novel type of mass spectrometer—a hybrid quadrupole/time-of-flight instrument (Q/TOF *[14]* or QqTOF *[15]*) was introduced. QqTOF instruments allow the acquisition of tandem mass spectra with very high mass resolution (>8000 full-width at half-maximum height [FWHM]) without compromising sensitivity. These instruments also benefit from the use of a nonscanning TOF analyzer that records all ions simultaneously in both conventional and MS/MS modes and therefore increases sensitivity. These features make it possible and practical to apply selective isotopic labeling of the peptide C-terminal carboxyl group in order to distinguish y″ ions from other fragment ions in tandem mass spectra (*see* **Subheading 3.4.2.**). Proteins are digested with trypsin in a buffer containing 50% $H_2^{16}O$ and 50% $H_2^{18}O$ (v/v) (*see* **Subheading 3.1.**) so that half of the resulting tryptic peptide molecules incorporate ^{18}O atoms in their C-terminal carboxyl group, whereas the other half incorporate ^{16}O atoms (**Fig. 1B**). During subsequent sequencing by MS/MS, the entire isotopic cluster of each peptide ion, in turn, is selected by the quadrupole mass filter (Q) and fragmented in the collision cell *(9)*. Since only the fragments containing the C-terminal carboxyl group of the peptide appear to be partially (50%) isotopically labeled, y″ ions are distinguished by a characteristic isotopic pattern, viz. doublet peaks split by 2 mass units (*see* **Subheading 3.4.2.**) (**Fig. 3**); other fragment ions have a normal isotopic distribution. Thus, only a single analysis is required, peptide sequence readout is much faster and the approach lends itself to automation *(15)*.

2. Materials

For general instructions, *see* **Note 1**.

2.1. In-Gel Digestion

For contamination precautions, *see* **Note 2**.

1. 100 m*M* ammonium bicarbonate in water (high-performance liquid chromatography [HPLC] grade [LabScan, Dublin, Ireland]).
2. Acetonitrile (HPLC grade [LabScan]).
3. 10 m*M* dithiothreitol in 100 m*M* ammonium bicarbonate.
4. 55 m*M* iodoacetamide in 100 m*M* ammonium bicarbonate.
5. 100 m*M* $CaCl_2$ in water.
6. 15 µL aliquots of trypsin, unmodified, sequencing grade (Boerhringer Mannheim, Germany) in 1 m*M* HCl (*see* **Note 3**).
7. 5% (v/v) formic acid in water.
8. Heating blocks at 56°C and at 37°C.
9. Ice bucket.
10. Laminar flow hood (optional) (*see* **Note 2**).

2.2. Esterification with Methanol

1. Methanol (HPLC grade), distilled shortly before the derivatization process.
2. Acetyl chloride (reagent grade), distilled shortly before the derivatization (*see* **Note 4**).

2.3. Isotopic Labeling Using $H_2^{18}O$

1. Reagents as in **Subheading 2.1.**
2. $H_2^{18}O$ (Cambridge Isotopic Laboratories, Cambridge, MA), distilled (*see* **Note 5**).

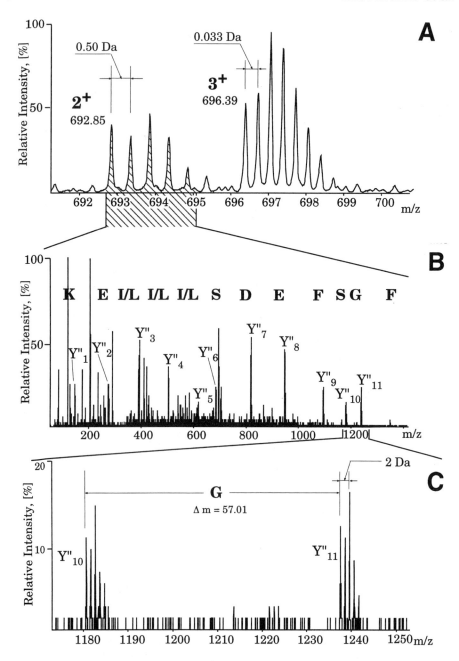

Fig. 3. Sequencing of ^{18}O C-terminally labeled tryptic peptides by Nano ESI-MS/MS. A 35-kDa protein from *Drosophila* was purified by gel electrophoresis, digested in-gel in a buffer containing 50% (v/v) H$_2$^{18}O, and analyzed using a QqTOF mass

2.4. Desalting and Concentrating In-Gel Tryptic Digests Prior to Analysis by Nano ESI-MS/MS

1. 5% (v/v) formic acid in water.
2. 60% methanol in 5% aqueous formic acid (both v/v).
3. Perfusion sorbent POROS 50 R2 (PerSeptive Biosystems, Framingham MA) *(see* **Note 6**).
4. Borosilicate glass capillaries GC120F-10 (1.2-mm OD × 0.69-mm ID) (Clark Electromedical Instruments, Pangbourne, UK) *(see* **Note 7**).
5. Purification needle holder, made as described in **ref. 16** or purchased from Protana (Odense, Denmark).
6. Benchtop minicentrifuge (e.g., PicoFuge, Stratagene, Palo Alto, CA).

3. Methods
3.1. In-Gel Digestion (see Notes 8 and 9)
3.1.1. Excision of Protein Bands (spots) from Gels

1. Rinse the entire gel with water and excise bands of interest with a clean scalpel, cutting as close to the edge of the band as possible.
2. Chop the excised bands into cubes (≈ 1 × 1 mm).
3. Transfer gel pieces into a microcentrifuge tube (0.5- or 1.5-mL Eppendorf test tube).

3.1.2. In-gel Reduction and Alkylation (see Note 10)

1. Wash gel pieces with 100–150 µL of water for 5 min.
2. Spin down and remove all liquid.
3. Add acetonitrile (the volume of acetonitrile should be at least twice the volume of the gel pieces) and wait for 10–15 min until the gel pieces have shrunk. (They become white and stick together.)
4. Spin gel pieces down, removing all liquid, and dry in a vacuum centrifuge.
5. Swell gel pieces in 10 mM dithiothreitol in 100 mM NH$_4$HCO$_3$ (adding enough reducing buffer to cover the gel pieces completely) and incubate (30 min at 56°C) to effect reduction of the protein.
6. Spin gel pieces down and remove excess liquid.

---

spectrometer (PE Sciex). **(A)** Part of the conventional spectrum of the unseparated digest. Although the isotopic pattern of labeled peptides is relatively complex, the high resolution of the QqTOF instrument allows a determination of the charge on the ions. **(B)** The entire isotopic cluster, which contains the doubly charged ion with m/z 692.85, was isolated by the quadrupole mass analyzer and transmitted to the collision cell, and its fragment ion spectrum was acquired. **(C)** Zoom of the region close to m/z 1200 of the fragment ion spectrum in **B**. Isotopically labeled y″ ions are observed as doublets split by 2 mass units. The peptide sequence was determined by considering the mass differences between adjacent labeled y″ ions.

7. Shrink gel pieces with acetonitrile, as in **step 3**. Replace acetonitrile with 55 mM iodoacetamide in 100 mM NH$_4$HCO$_3$ and incubate (20 min, room temperature, in the dark).
8. Remove iodoacetamide solution and wash gel pieces with 150–200 µL of 100 mM NH$_4$HCO$_3$ for 15 min.
9. Spin gel pieces down and remove all liquid.
10. Shrink gel pieces with acetonitrile as before, remove all liquid, and dry gel pieces in a vacuum centrifuge.

3.1.3. Additional Washing of Gel Pieces (for Coomassie-Stained Gels Only) (see **Note 11**)

1. Rehydrate gel pieces in 100–150 µL of 100 mM NH$_4$HCO$_3$ and after 10–15 min add an equal volume of acetonitrile.
2. Vortex the tube contents for 15–20 min, spin gel pieces down, and remove all liquid.
3. Shrink gel pieces with acetonitrile (see **Subsection 3.1.2.**) and remove all liquid.
4. Dry gel pieces in a vacuum centrifuge .

3.1.4. Application of Trypsin (see **Note 12**)

1. Rehydrate gel pieces in the digestion buffer containing 50 mM NH$_4$HCO$_3$, 5 mM CaCl$_2$, and 12.5 ng/µL of trypsin at 4°C (use ice bucket) for 30–45 min. After 15–20 min, check the samples and add more buffer if all the liquid has been absorbed by the gel pieces. For ^{18}O isotopic labeling of C-terminal carboxyl groups of tryptic peptides, prepare the buffer for this step and for **step 2** in 50:50 (v/v) H$_2$^{16}O + H$_2$^{18}O (see **Note 12**).
2. Remove remaining buffer. Add 10–20 µL of the same buffer, but prepared without trypsin, to cover gel pieces and keep them wet during enzymatic digestion. Leave samples in a heating block at 37°C overnight.

3.1.5. Extraction of Peptides

1. Add 10–15 µL of water to the digest, spin gel pieces down, and incubate at 37°C for 15 min on a shaking platform.
2. Spin gel pieces down, add acetonitrile (add a volume that is two times the volume of the gel pieces), and incubate at 37°C for 15 min with shaking.
3. Spin gel pieces down and collect the supernatant into a separate Eppendorf test tube.
4. Add 40–50 µL of 5% formic acid to the gel pieces.
5. Vortex mix and incubate for 15 min at 37°C with shaking.
6. Spin gel pieces down, add an equal volume of acetonitrile, and incubate at 37°C for 15 min with shaking.
7. Spin gel pieces down, collect the supernatant, and pool the extracts.
8. Dry down the pooled extracts using a vacuum centrifuge.

3.2. Esterification of In-Gel Digests with Methanol

1. Put 1 mL of methanol (for the preparation of reagents, see **Subheading 2.2.**) into a 1.5-mL Eppendorf test tube. Place the tube in a freezer at –20°C (or lower) for 15 min.

2. Take the tube from the freezer and immediately add 150 μL of acetyl chloride (*Caution!* *Put on safety goggles and gloves. The mixture may boil up instantly!*). Leave the tube to warm up to room temperature and use this reagent 10 min later.
3. Add 10–15 μL of the reagent *(see* **Note 13**), prepared as in **step 2**, to a dried portion of the peptide pool recovered after in-gel digestion of the protein *(see* **Subsection 3.1.5.**).
4. Incubate for 45 min at room temperature.
5. Dry down the reaction mixture using a vacuum centrifuge.

3.3. Desalting and Concentration of In-Gel Digest prior to Nano ESI-MS/MS Sequencing

1. Pipette ≈ 5 μL of POROS R2 slurry, prepared in methanol, into the pulled glass capillary (here and in subsequent steps now referred to as a "column"). Spin the beads down and then open the pulled end of the column by gently touching against a bench top. Wash the beads with 5 μL of 5% formic acid and then make sure the liquid can easily be spun out of the column by gentle centrifuging. Open the column end wider if necessary. Mount the column into the micropurification holder *(see* **Subheading 2.4.**).
2. Dissolve the dried digest *(see* **Subheading 3.1.5.**) or the esterified portion of the digest *(see* **Subheading 3.2.**) in 10 μL of 5% formic acid and load onto the column. Pass the sample through the bead layer by centrifuging.
3. Wash the adsorbed peptides with another 5 μL of 5% formic acid.
4. Align the column and the nanoelectrospray needle in the micropurification holder and elute peptides directly into the needle with 1 μL of 60% of methanol in 5% formic acid by gentle centrifuging.
5. Mount the spraying needle together with the sample into the nanoelectrospray ion source and acquire mass spectra *(see* **Note 14** and **Subheading 3.4.**).

3.4. Acquisition of Mass Spectra and Data Interpretation

Before the analysis, the tandem mass spectrometer—triple quadrupole or quadrupole/time-of-flight—should be tuned as discussed in **Notes 15** and **16**, respectively. Since in-gel digestion using unmodified trypsin is accompanied by trypsin autolysis, it is necessary to acquire the spectrum of a control sample (blank gel pieces processed as described in **Subheading 3.1.**) in advance. Spectra should be acquired in both conventional scanning (Q1) and precursor-ion detection modes (as in **Subheading 3.4.1., step 1**).

3.4.1. Sequencing on a Triple Quadrupole Mass Spectrometer

1. After desalting and concentration *(see* **Subheading 3.3.**), initiate spraying and acquire a conventional (Q1 scan) spectrum of the peptide mixture from digestion. Introduce collision gas into the instrument and acquire a spectrum in the precursor-scan mode (e.g., scanning to record only ions that are precursors to *m/z* 86 fragment ions on collisional fragmentation) *(17) (see* **Note 17**).

2. Stop spraying by dropping the spraying voltage to zero. Drop the air pressure applied to the spraying capillary. Move the spraying capillary away from the inlet of the mass spectrometer.
3. Examine the acquired spectra and compare them with the spectra acquired from the control sample. Select precursor ions for subsequent tandem mass spectrometric sequencing.
4. Add 0.3–0.5 µL of 60% of methanol in 5% formic acid directly to the spraying capillary if the remaining sample volume is less than ≈ 0.5 µL. Reestablish spraying and acquire tandem (fragment-ion) mass spectra from selected precursor ions.
5. Interpret the acquired spectra. An m/z region above the multiply charged precursor ion is usually free from chemical noise in tandem mass spectra of tryptic peptides and is dominated by y″ ions. Therefore in this region it is relatively easy to retrieve short amino acid sequences by considering the masses of fragment ions. Assemble peptide sequence tags and perform a database search using PeptideSearch software installed on a Macintosh computer or via the Internet (*see* **Note 18**).
6. If the protein proves to be unknown (i.e., not present in a sequence database) take the remaining portion of the digest, esterify with methanol (*see* **Subheading 3.2.**), redissolve in 10 µL of 5% formic acid, perform desalting and concentration (*see* **Subheading 3.3.**), and acquire spectra by nanoelectrospray as described above.
7. Correlate peptide molecular ions in the unmodified and derivatized digests (*see* **Note 19**). Deduce peptide sequences by comparison of the tandem (fragment-ion) spectra from each pair of derivatized and unmodified peptides (**Fig. 2**).

3.4.2. Sequencing of ^{18}O-Labeled Peptides on a Quadrupole/Time-of-Flight Mass Spectrometer

1. Perform nanoelectrospray analysis of in-gel digests, including acquisition of tandem (fragment-ion) spectra, just as described for a triple quadrupole instrument in **Subheading 3.4.1., steps 1–4**, but using a quadrupole/time-of-flight instrument (*see* also **Note 20**).
2. Interpret the fragment spectra and deduce the corresponding peptide sequences (*see* **Note 21**) (**Fig. 3**).
3. Note that in principle only one set of acquired data is required to deduce the peptide sequence. However, if necessary, the remaining portion of a digest could be esterified (*see* **Subheading 3.2.**) and analyzed separately to generate an independent set of peptide sequence data (*see also* **Note 22**)

4. Notes

1. All chemicals should be of the highest degree of purity available. Solutions of dithiothreitol and iodoacetamide should be freshly prepared. It is recommended to use 50–100 mL stocks of water, ammonium bicarbonate buffer, and acetonitrile and to discard old solvents before starting the preparation of a new series of samples. In our experience, stock solutions rapidly accumulate dust, pieces of

Peptide Sequencing by Tandem MS   11

hair, threads, etc. from the laboratory environment. Plastic ware (pipette tips, gloves, dishes, and so on) may acquire a static charge and attract dust. Accumulation of even a minute amount of dust in solutions and reagents results in massive contamination of samples with human and sheep keratins and makes sequencing exceedingly difficult if not impossible. Polymeric detergents (Tween, Triton, etc.) should not be used for cleaning the laboratory dishes and tools.

2. All possible precautions should be taken to avoid the contamination of samples with keratins and polymeric detergents (*see* **Note 1**). Gloves should be worn at all times during operations with gels (staining, documenting, excision of bands or spots of interest) and sample preparation. It is necessary to rinse new gloves with water to wash away talcum powder and it is recommended to rinse them again with water occasionally during sample preparation since gloves with a static charge attract dust. In our experience, it is advisable to perform all operations in a laminar flow hood, which helps to preserve a dust-free environment.

3. Add 250 µL of 1 m*M* HCl to the commercially available vial containing 25 µg of trypsin. Vortex the vial and aliquot the trypsin stock solution in 0.5-mL Eppendorf test tubes (15 µL per tube). Freeze the aliquots and store at –20°C before use. Unfreeze the aliquot shortly before preparation of the digestion buffer. Discard the rest of the aliquot if it is not totally used. Surplus digestion buffer containing trypsin (*see* **Subheading 3.1.4.** and *also* **Note 12**) should also be discarded.

4. A glass tube filled with calcium chloride or molecular sieve should be used to protect acetyl chloride during distillation.

5. Commercially available $H_2^{18}O$ has a chemical purity of ≈ 95% and is unsuitable for protein sequencing by mass spectrometry. Therefore, a 0.5-mL portion of water is purified by microdistillation in a sealed glass apparatus and stored at –20°C in 15-µL aliquots until use. Each aliquot is used only once.

6. Methanol (1 mL) is added to ≈ 30 µL of POROS R2 resin to prepare a slurry. A fraction of the resin beads of submicrometer size, whose presence increases the resistance to liquid flow, can be efficiently removed by repetitive sedimentation. Vortex the test tube containing the slurry and then let it stay in a rack until the major part of the resin reaches the bottom of the tube. Aspirate the supernatant with a pipette and discard it. Repeat the procedure 3–5 times if necessary.

7. Capillaries for micropurification are manufactured in the same way as capillaries for nanoelectrospray *(18)* but are not coated with a metal film.

8. The procedure described in **Subheading 3.1.** *(19)* is applicable, with no modifications, to spots (bands) excised from one- or two-dimensional polyacrylamide gels stained with Coomassie brilliant blue R 250 or G 250, as well as to silver-stained (*see* **Note 9**) or negatively stained gels *(20)*.

9. Any convenient protocol for silver staining can be employed to visualize proteins present on a gel in a subpicomole amount. However, the reagents used to improve the sensitivity and the contrast of staining must not modify proteins covalently. Thus, treatment of gels with the crosslinking reagent glutaraldehyde or with strong oxidizing agents, such as chromates and permanganates, should be avoided.

10. In-gel reduction and subsequent alkylation of free SH groups in cysteine residues is recommended even if the proteins have been reduced prior to electrophoresis. Note that alkylation of free cysteine residues by acrylamide sometimes occurs during electrophoretic separation. Treatment with dithiothreitol does not cleave these acrylamide residues. Thus, possible acrylamidation of cysteines should be taken into consideration when interpreting the spectra or searching a database with peptide sequence tags.
11. This step of the protocol is applied only when Coomassie-stained gel pieces still look blue after reduction and alkylation of the protein are complete. This usually occurs when intense bands (spots) containing picomoles of protein material are being analyzed. If a single washing cycle does not remove the residual staining, the procedure is repeated.
12. To prepare the digestion buffer, add 50 µL of 100 mM NH$_4$HCO$_3$, 50 µL of water, and 5 µL of 100 mM CaCl$_2$ to a 15-µL aliquot of trypsin stock solution (see **Note 3**). Keep the test tube containing digestion buffer on ice before use. To prepare the buffer for ^{18}O labeling use H$_2$^{18}O water instead of H$_2$^{16}O water with the same stock solution of 100 mM NH$_4$HCO$_3$.
13. The added volume of reagent should just cover the solid residue at the bottom of the tube. Avoid an excessive volume since this increases chemical background in the mass spectra.
14. For detailed instructions on the manufacture of the nanoelectrospray needles and on the operation of the nanoelectrospray ion source, see **ref. 18**. The theoretical background of the nanoelectrospray is discussed in **ref. 21**.
15. The calibration of a triple quadrupole mass spectrometer is performed in accordance with the manufacturer's instructions. However, for sequencing of proteins present at the low picomole level, several settings should be specially tuned. Make sure that the settings controlling resolution of the first quadrupole (Q1) allow good transmission of precursor ions. On the other hand, unnecessarily low resolution of Q1 results in the transmission of too many background ions, which may densely populate the low m/z region of the fragment-ion spectra. The third quadrupole (Q3) should likewise be operated at a low resolution setting in order to improve its transmission and to achieve acceptable ion statistics in the fragment-ion spectra. In our experience, a resolution in Q3 as low as 250 (FWHM) still allows accurate readout of peptide sequences. The Q1 and Q3 resolution settings can be tuned in a tandem mass spectrometric experiment using synthetic peptides.
16. Calibration of a QqTOF instrument is performed by acquiring the spectrum of a mixture of synthetic peptides. External calibration with two peptide masses allows 10-ppm mass accuracy for both conventional and tandem mass spectra, if calibration and sequencing experiments are performed within approximately 2 h. A calibration acquired in the mode that records conventional mass spectra does not change when the instrument is switched to tandem mode. The resolution of the first quadrupole (Q1) should be set in a similar way to that described for a triple quadrupole mass spectrometer (see **Note 15**).

17. The conventional (Q1) spectrum ideally contains only peptide molecular ions. However, impurity ions may be present or the peptide ions may be weak and therefore difficult to distinguish from noise. The use of a specific scan for precursor ions that produce m/z 86 fragment ions (immonium ion of leucine or isoleucine) helps to distinguish genuine peptide ions from chemical noise and is therefore indispensable for sequencing at low levels. It is also helpful to acquire precursor-ion spectra even if a somewhat larger (picomole) amount of protein is present on the gel. For example, precursor-ion scanning facilitates the rejection of polyethyleneglycol-like contamination, which is often seen in the low m/z region of conventional (Q1) spectra as series of intense peaks at 44-mass unit intervals.
18. PeptideSearch ver. 3.0 software can be downloaded from the EMBL Peptide & Protein Group WWW-page (http://www.mann.embl-heidelberg.de/). For detailed information on PeptideSearch software *see* ref. 22. Searching a nonredundant protein database can also be performed at the same server via the Internet.
19. The number of residues of aspartic and glutamic acids present in any particular peptide is not known. Therefore, to identify the matching peptide ion in the spectrum of the esterified digest, it is necessary to consider all ions shifted from the mass of the ion in the unmodified peptide by $14(n + 1)/z$ (where $n = 0, 1, 2, 3...$); *see* **Subheading 1.**) and to fragment all of them.
20. Because of limited efficiency of ion transmission from the collision cell to the time-of-flight analyzer in QqTOF instruments, the precursor-ion scan mode is far less sensitive than with triple quadrupole machines. In this mode of operation, the second mass analyzer (TOF or Q3, respectively) is used in a nonscanning mode (e.g., recording ions with $m/z = 86$ only) on both instruments. For this reason, the advantage of the TOF analyser, i.e., that it can record all fragment ions without scanning, is not of value and the precursor ion scan mode on the QqTOF instrument is therefore not useful for sequencing at low levels.

    It is, however, relatively easy to distinguish precursor ions from chemical background by taking advantage of the high resolution of the QqTOF instrument. Isotopically labeled peptide ions are detected as sharp, characteristic isotopic patterns superimposed on a broad, irregularly shaped, background *(23)*. Isotopic peaks of multiply charged ions are very well resolved, and the charge of the precursor ion can be instantly calculated from the mass difference between the isotopic peaks. If a conventional mass spectrum of the digest is noisy, it is not always straightforward to recognize the peak of the first isotope in the complex isotopic pattern of a multiply charged ^{18}O-labeled peptide ion. In this case, the isotopic pattern of singly charged fragment ions produced by collisional fragmentation has to be rapidly examined. If the isotopic pattern of fragment ions is disturbed (for example, there is only one isotopic peak for unlabeled ions, or the second isotopic peaks of the ^{18}O-labeled fragments are missing) then the selection of the precursor ion has to be corrected.
21. y'' ions are distinguished from other fragment ions by their characteristic isotopic profile (*see* **Subheading 1.**). It is easier to start the interpretation in the m/z region above the precursor ion, where fragment spectra usually contain less back-

ground ions and isotopic profiles of labeled ions are clearly visible. The series of y″ ions is followed downward in mass and should terminate at the labeled y″ ion of arginine or lysine. Upward in mass, the y″ series can be extended to the mass of the singly protonated ion of an intact peptide.

The high resolution of a QqTOF instrument greatly assists in spectrum interpretation and allows one to obtain additional pieces of information that are not available in low-resolution tandem mass spectra acquired on triple quadrupole instruments. Thus, fragmentation of doubly charged precursor ions mainly results in a series of singly charged fragments whereas the series of doubly charged fragments usually has a much lower intensity. However, the high resolution of the QqTOF instrument enables them to be identified and used as independent verification of the sequence determined from the series of singly charged fragment ions. Since only the C-terminal carboxyl group of peptides is labeled during tryptic digestion, the N-terminal series of fragment ions (b-series) appear to be unlabeled. Although these ions often have low intensity, they can be recognized in the fragment spectrum and are useful for data interpretation. Again, the high resolution of QqTOF instruments makes it possible to determine the masses of fragment ions very accurately. Thus it is possible to distinguish phenylalanine from methionine-sulfoxide (their masses differ by 0.033 Daltons) as well as glutamine from lysine (mass difference 0.037 Daltons).

22. If the protein was in-gel digested with trypsin in a buffer that did not contain $H_2^{18}O$, selective C-terminal isotopic labeling can still be performed. The digest should be esterified with methanol (*see* **Subheading 3.2.**), dissolved in a buffer containing 50% (v/v) $H_2^{18}O$, treated with trypsin for 30 min, and dried in a vacuum centrifuge. Treatment with trypsin efficiently removes the ester group from the C-terminal carboxyl group of tryptic peptides. At the same time, the C-terminal carboxyl group of peptides incorporates ^{18}O or ^{16}O atoms from the buffer (**Fig. 1C**). Carboxyl groups in the side chains of aspartic and glutamic acid residues remain esterified. However, the procedure results in a much higher chemical noise and in an increased level of keratin peptides. Therefore it can be used only for sequencing of peptides from chromatographically isolated fractions that contain only a small number of peptides.

References

1. Shevchenko, A., Jensen, O. N., Podtelejnikov, A. V., Sagliocco, F., Wilm, M., Vorm, O., et al. (1996) Linking genome and proteome by mass spectrometry: large scale identification of yeast proteins from two dimensional gels. *Proc. Natl. Acad. Sci. USA* **98,** 14,440–14,445.
2. Jensen, O. N., Podtelejnikov, P., and Mann, M. (1996) Delayed extraction improves specificity in database searches by MALDI peptide maps. *Rapid Commun. Mass Spectrom.* **10,** 1371–1378.
3. Jensen, O. N., Podtelejnikov, A. V., and Mann, M. (1997) Identification of the components of simple protein mixtures by high-accuracy peptide mass mapping and database searching. *Anal. Chem.* **69,** 4741–4750.

4. Houthaeve, T., Gausepohl, H., Mann, M., and Ashman, K. (1995) Automation of micro-preparation and enzymatic cleavage of gel electrophoretically separated proteins. *FEBS Lett.* **376,** 91–94.
5. Jensen, O. N., Mortensen, P., Vorm, O., and Mann, M. (1997) Automatic acquisition of MALDI spectra using fuzzy logic control. *Anal. Chem.* **69,** 1706–1714.
6. Wilm, M., Shevchenko, A., Houthaeve, T., Breit, S., Schweigerer, L., Fotsis T., et al. (1996) Femtomole sequencing of proteins from polyacrylamide gels by nanoelectrospray mass spectrometry. *Nature* **379,** 466–469.
7. Lamond, A. and Mann M. (1997) Cell biology and the genome projects—a concerted strategy for characterizing multiprotein complexes by using mass spectrometry. *Trends Cell Biol.* **7,** 139–142.
8. Mann, M. and Wilm, M. (1994) Error tolerant identification of peptides in sequence databases by peptide sequence tags. *Anal. Chem.* **86,** 4390–4399.
9. Shevchenko, A., Keller, P., Scheiffele P., Mann M., and Simons, K. (1997) Identification of components of trans-Golgi network-derived transport vesicles and detergent-insoluble complexes by nanoelectrospray tandem mass spectrometry. *Electrophoresis* **18,** 2591–2600.
10. Roepstorff, P. and Fohlman, J. (1984) Proposed nomenclature for sequence ions. *Biomed. Mass Spectrom.* **11,** 601.
11. Shevchenko, A., Wilm, M., and Mann, M. (1997) Peptide sequencing by mass spectrometry for homology searches and cloning of genes. *J. Protein Chem.* **16,** 481–490.
12. Muzio, M., Chinnaiyan, A. M., Kischkel, F. C., Rourke, K. O., Shevchenko, A., Ni, J., et al. (1996) FLICE, a novel FADD-homologous ICE/CED-3-like protease, is recruited to the CD95 (Fas/APO-1) death-inducing signaling complex. *Cell* **85,** 817–827.
13. McNagny, K. M., Petterson, I., Rossi, F., Flamme, I., Shevchenko, A., Mann, M., et al. (1997) Thrombomucin, a novel cell surface protein that defines thrombocytes and multipotent hematopoetic progenitors. *J. Cell Biol.* **138,** 1395–1407.
14. Morris, H. R., Paxton, T., Dell, A., Langhorn, J., Berg, M., Bordoli,R. S., et al. (1996) High sensitivity collisionally-activated decomposition tandem mass spectrometry on a novel quadrupole/orthogonal-acceleration time-of-flight mass spectrometer. *Rapid Commun. Mass Spectrom.* **10,** 889–896.
15. Shevchenko, A., Chernushevich, I., Ens, W, Standing, K. G, Thomson, B., Wilm, M., et al. (1997) Rapid 'de novo' peptide sequencing by a combination of nanoelectrospray, isotopic labeling and a quadrupole/time-of-flight mass spectrometer. *Rapid Commun. Mass Spectrom.* **11,** 1015–1024.
16. Shevchenko, A., Jensen, O. N., Wilm, M., and Mann, M. (1996) Sample preparation techniques for femtomole sequencing of proteins from polyarylamide gels, in *Proceedings of the 44th ASMS Conference on Mass Spectrometry and Allied Topics*, Portland, OR, p. 331.
17. Wilm, M., Neubauer, G., and Mann, M. (1996) Parent ion scans of unseparated peptide mixtures. *Anal. Chem.* **68,** 527–533.
18. Wilm, M. and Mann, M. (1996) Analytical properties of the nano electrospray ion source. *Anal. Chem.* **66,** 1–8.

19. Shevchenko, A., Wilm, M., Vorm O., and Mann, M. (1996) Mass spectrometric sequencing of proteins from silver stained polyacrylamide gels. *Anal. Chem.* **68,** 850–858.
20. Fernandez-Patron, C., Calero, M., Collazo, P. R., Garcia, J. R., Madrazo, J., Musacchio, A., et al. (1995) Protein reverse staining: high efficiency microanalysis of unmodified proteins detected on electrophoresis gels. *Anal. Biochem.* **224,** 203–211.
21. Wilm, M. and Mann, M.(1994) Electrospray and Taylor-cone theory, Dole's beam of macromolecules at last? *Int. J. Mass Spectrom. Ion Processes* **136,** 167–180.
22. Mann, M. (1994) Sequence database searching by mass spectrometric data, in *Microcharacterization of Proteins* (Kellner, R., Lottspeich, F., and Meyer, H. E., eds.), VCH, Weinheim, pp. 223–245.
23. Shevchenko, A., Chernushevich, I., and Mann, M. (1998). High sensitivity analysis of gel separated proteins by a quadrupole-TOF tandem mass spectrometer, in *Proceedings 46th ASMS conference on Mass Spectrometry and Allied Topics*, Orlando, FL, p. 237.
24. Lingner, J., Hughes, T. R., Shevchenko, A., Mann, M., Lundblad, V., and Cech, T. R. (1997) Reverse transcriptase motifs in the catalytic subunits of telomerase. *Science* **276,** 561–567.

2

Direct Analysis of Proteins in Mixtures

Application to Protein Complexes

John R. Yates, III, Andrew J. Link, and David Schieltz

1. Introduction

Tandem mass spectrometry is a powerful mixture analysis technique suitable for sequence analysis of peptides (1). A tandem mass spectrometer uses two stages of analysis to generate structurally informative fragmentation. The first stage involves separation of an ion from all the other ions that may be entering the mass spectrometer analyser. Ion isolation can be accomplished by separating an ion in time or in space; these processes are used by ion trap instruments or by triple quadrupole instruments, respectively (2–4). The isolated ion is then subjected to ion activation using energetic gas-phase collisions. In the second analysis stage, the mass-to-charge ratio (m/z) values of the fragmentation products are determined. This method is used for protein sequence analysis by first creating a collection of peptides using site-specific enzymatic or chemical proteolysis (1). The collection of peptides is introduced into the mass spectrometer through a separation technique (liquid chromatography [LC] or capillary electrophoresis [CE]) or by batch infusion and finally ionized using electrospray ionization (5–7). Computer control of the data acquisition process allows highly efficient acquisition of these tandem mass spectra as well as unassisted operation of the mass spectrometer (8,9). The resulting tandem mass spectra can reveal the amino acid sequence of peptides by interpretation, or, with the recent expansion of sequence databases, the tandem mass spectra can be used to search protein and nucleotide sequence databases directly to identify the amino acid sequence represented by the spectrum (1,8,10,11). Because tandem mass spectra can be acquired quickly and selectively on individual peptides present, the identities of proteins in a mixture can be determined *(10,12,13)*.

The process of protein mixture analysis is similar to that used for analysis of homogenous proteins. The protein mixture is digested with a site-specific enzyme to create a complex mixture of peptides. This complex mixture is then separated on a reversed-phase high-performance liquid chromatography (HPLC) column or by two-dimensional HPLC (2D LC) prior to entering the tandem mass spectrometer *(14–20)*. By increasing the resolution of the separation through use of a longer gradient separation, or by 2D LC, more tandem mass spectra are acquired so that acquisition of peptide tandem mass spectra from each protein present is ensured. After the acquisition of tandem mass spectra, these are searched through sequence databases to match individual peptide spectra to the sequences of proteins, thus identifying the proteins present.

One application of protein mixture analysis is the identification of components of protein complexes. In most physiological functions, collections of proteins come together to perform a reaction or a series of reactions. For example, protein translation involves a large complex of proteins collectively called the ribosome. The complex provides the scaffold to bring mRNA and translation enzymes together to synthesize proteins. To understand fully processes performed by complexes, quantitative and posttranslational details of the components need to be determined as a function of cellular state. In other words, how does the composition of the complex change as a function of time? On a practical level, these experiments require analytical technology that is capable of high throughput.

This chapter describes the steps taken to identify proteins in mixtures: proteolytic digestion, tandem mass spectrometry data acquisition, and database searching.

2. Materials

2.1. Instruments

1. An LCQ tandem mass spectrometer (Finnigan MAT, San Jose, CA) is used in our laboratory. Other tandem mass spectrometers capable of automated acquisition of tandem mass spectra should also be suitable for this purpose.
2. A standard system for reversed-phase HPLC is needed. We currently use several different types: Hewlett Packard 1100 (Palo Alto, CA), Thermo Separation Products SpectraSystem P4000 (San Jose, CA), and Applied Biosystems ABI140B (Foster City, CA).
3. A multidimensional HPLC system (Integral Microanalytical System, Perspective Biosystems, Framingham, MA) is currently used in our laboratory. Other systems capable of multidimensional chromatography should be suitable.
4. Laser puller (P250, Sutter Instruments, Novato, CA) is used to create a 2-μm tip for microcolumns (*see* **Note 1**).

2.2. Reduction and Carboxyamidation

1. Nitrogen 99.99%.
2. Dithiothreitol (Pierce, Rockville, IL).
3. Iodoacetamide (Pierce).

2.3. Endoproteinase Lys-C Digestion

1. 100 mM Tris-HCl (pH 8.6) 8 M urea.
2. Endoproteinase Lys-C (sequencing grade, Boehringer Mannheim, Indianapolis, IN).

2.4. Trypsin Digestion

1. 100 mM Tris-HCl (pH 8.6).
2. Trypsin (sequencing grade, Boehringer Mannheim).

2.5. Single-Dimension Reversed-Phase HPLC

1. Buffer A: 0.5% aqueous acetic acid.
2. Buffer B: acetonitrile (HPLC grade) + 0.5% aqueous acetic acid 4:1 (v/v).

2.6. Strong Cation-Exchange/Reversed-Phase Two-Dimensional HPLC

1. Buffer A: 0.5% aqueous acetic acid + acetonitrile (HPLC grade) 95:5 (v/v).
2. Buffer B: acetonitrile (HPLC grade) + 0.5% aqueous acetic acid 4:1 (v/v).
3. Buffer C: 0.5% aqueous acetic acid + acetonitrile (HPLC grade) 95:5 (v/v) containing 250 mM KCl.
4. Buffer D: 0.5% aqueous acetic acid + acetonitrile (HPLC grade) 95:5 (v/v) containing 500 mM KCl.

2.7. HPLC Columns

1. C18 reversed-phase HPLC column 1.0 mm × 25 cm (Vydac, Hesperia, CA).
2. Strong cation-exchange column 1.0 mm × 25 cm (PolyLC, Columbia, MD).

2.8. Micro-HPLC Columns

1. 100-μm ID × 360-μm OD fused-silica capillary (J+W Scientific, Folsom, CA).
2. 50-μm ID × 360-μm OD fused-silica capillary, ≈30 cm length (J+W Scientific).
3. Reversed-phase packing material (POROS R2, Perseptive Biosystems). Other C18 materials are also suitable.
4. PEEK micro cross (Upchurch, Oak Harbor, WA).
5. 0.025-in. diameter gold wire (Scientific Instruments Service, Ringoe, NY).

3. Methods

After isolation of a protein complex, the proteins must be denatured and digested with a protease to create a mixture of peptides for analysis. To denature the complex, it is dissolved in 8 M urea in ammonium bicarbonate. To dissociate the complex, proteins are reduced and alkylated. The complex is incubated with endoproteinase Lys-C followed by dilution of the solution to 2 M urea. The complex is then digested overnight with trypsin.

3.1. Reduction and Carboxyamidation

1. Add 1 µg dithiothreitol (DTT) for every 50 µg of protein.
2. Incubate at 37°C for 1 h under nitrogen.
3. Add 5 µg iodoacetamide for every 50 µg of protein.
4. Incubate at room temperature in the dark for 30 min.

3.2. Endoproteinase Lys-C Digestion

1. Add 1 µg endoproteinase Lys-C for every 50 µg of protein.
2. Incubate at 37°C for 8 h.
3. Dilute the solution to 2 M urea, 50 mM ammonium bicarbonate, pH 8.6.

3.3. Trypsin Digestion

1. Add 1 µg trypsin for every 50 µg of protein.
2. Incubate at 37°C for 4–8 h.
3. Terminate the reaction by the addition of a few microliters of glacial acetic acid.

3.4. Single Dimension HPLC Separation of Complex Peptide Mixtures

The resulting complex mixture of peptides is separated using single-dimension HPLC, typically using microcolumns with diameters of 100 µm or less for high sensitivity analysis. The HPLC flow is thus reduced to 300 nL/min and ions for mass spectrometric analysis are created using a microelectrospray ion source (**Fig. 1**).

1. Load a sample on the column by disconnecting it from the ion source and placing it in a pneumatic "bomb."
2. Dip the entrance to the column into the sample solution and pressurize the "bomb" to force liquid onto the column.
3. Collect the liquid displaced from the column to determine the amount loaded onto the column.
4. Replace the column in the microelectrospray source.
5. Wash salts and other small molecules from the column at 0% B buffer.
6. Use a long gradient (90 min) for the separation (**Fig. 2**) because of the complexity of the peptide mixture (*See* **Notes 2–4**).

3.5. Two-Dimensional HPLC Separation of Complex Peptide Mixtures

To increase separation resolution for complex peptide mixtures, two-dimensional separations can be used. A convenient form of two-dimensional separation, combining two orthogonal methods, is to couple together strong cation-exchange (SCE) and reversed-phase chromatography (**Fig. 3**). To ensure that all peptide

Direct Analysis of Proteins in Mixtures

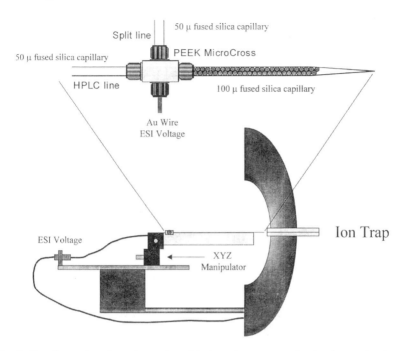

Fig. 1. Configuration for micro-HPLC/microelectrospray ionization. The tip of the column is directed approx 1–2 mm from the opening of the heated capillary.

material enters the mass spectrometer, a step gradient is employed on the ion exchange separation where each salt step transfers material from the SCE column to the reversed-phase column. The separation effects fractionation initially by charge and then by hydrophobicity (see **Notes 5 and 6**).

1. Initially wash the SCE column with 100% buffer A and then apply a linear acetonitrile gradient to separate peptides by hydrophobicity on the reverced-phase column. Peptides initially transferred from the ion exchange column are separated and detected in the mass spectrometer.
2. Use an initial salt step-gradient of 0–5% 250 mM KCl to elute a new fraction of peptides onto the reversed-phase column (**Fig. 4**).
3. Repeat the process of using step gradients to elute peptides sequentially from the ion exchange column onto the reversed-phase column, followed by a reversed-phase gradient, in 5% increments of salt concentration.

3.6. Tandem Mass Spectrometry (LC/MS/MS or LC/LC/MS/MS)

As peptides elute into the mass spectrometer, data-dependent data acquisition is performed. In this experiment, the mass spectrometer is set to acquire a conventional scan over the m/z range 400–1600. Ions detected by the data system, above a preset ion-current threshold, are then automatically selected and a tandem mass

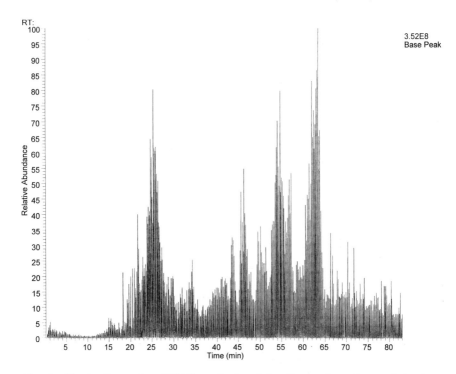

Fig. 2. Single-dimension HPLC analysis of the digested products of the human ribosomal complex. A linear gradient of 90 min from 0–60% buffer B was used to perform the separation. The oscillations of the ion current are generated by the instrument rapidly switching between MS and MS/MS modes.

spectrometry (MS/MS) experiment performed each at corresponding m/z value (see **Notes 7** and **8**). In the analysis of complex peptide mixtures, the three most intense peaks are selected for MS/MS experiments and, once a specific ion has been selected in this way, it is not reselected until after a specified time interval.

3.7. Data Analysis using Database Searching

An enormous amount of data can be generated in both one- and two-dimensional analyses of proteolytically digested protein mixtures. A fast method to analyze mass spectral data is to use a database of protein sequences *(12)*. MS/MS data can be readily and automatically matched to amino acid sequences in the database. By matching to a nonconserved amino acid sequence, at least 7 residues in length, the protein from which the amino acid sequence was obtained can be identified. The data obtained from the one- or two-dimensional chromatographic analysis are analyzed using the appropriate database. If the proteins were derived from an organism whose genome has been com-

Direct Analysis of Proteins in Mixtures

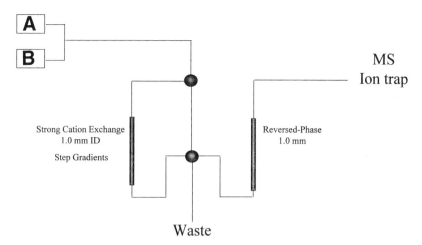

Fig. 3. Configuration for 2D LC. A single set of pumps is used to generate the gradients from a selection of four solvents. The solvent flow is passed through a valve, which directs the flow to the ion-exchange column. The flow exiting the column can be directed onto the reversed-phase column or to waste. When a salt step is performed, the effluent exiting the IEX column is sent to the reversed-phase column. Peptides are bound to the hydrophobic stationary phase, desalted, and then separated with a linear gradient. Solvent flow to the reversed-phase column bypasses the IEX column during this step.

pleted, then it is appropriate to search only that database. The ability to search a database of sequences from an organism whose genome is completed is a strength of the method for the analysis of protein complexes (*see* **Note 9**).

4. Notes

1. Small-diameter tips can also be created by attaching a weight to the fused-silica and heating the capillary with a hot flame. As the fused-silica melts, the pull of gravity on the weight stretches the fused-silica to a fine point. A drawback is that the glass needs to be trimmed. Glass tips produced by this method are frequently not reproducible.
2. The folding and association of proteins in a multimeric protein complex can inhibit complete proteolysis. To achieve complete proteolysis the complex is denatured in 8 M urea, reduced, and alkylated. The complex is digested with endoproteinase Lys-C, which is active in 8 M urea, and then the solution is diluted to 2 M urea. An overnight digestion with trypsin is performed to create peptides suitable for tandem mass spectrometry.
3. An integrated microcolumn/microelectrospray ionization system described by Gatlin et al. (*21*), is used to separate peptides in a single dimension of reversed-phase chromatography. The column is fabricated from 100-µm capillary tubing that has been pulled to a 2–5-µm tip. A 12-cm length of tubing is filled with 10-µm POROS beads. The column is connected to a PEEK micro-cross contain-

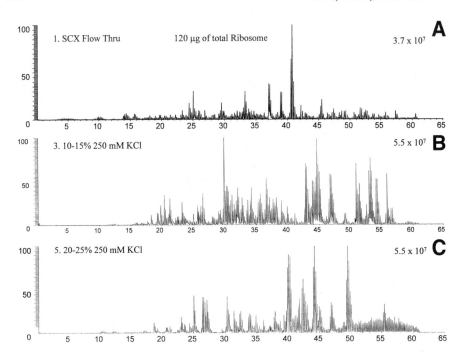

Fig. 4. Ion chromatograms for three different step gradients used in the analysis of the digested products of the *S. cerevisiae* ribosomal complex. **(A)** Ion chromatogram showing the peptides that passed through the ion-exchange column during sample loading and bound to the reversed-phase column. **(B)** Ion chromatogram showing the peptides removed from the ion-exchange column during a 10–15% 250 mM KCl step gradient. Peptides were then desalted on the reversed-phase column and separated by a 60-min linear gradient. **(C)** Ion chromatogram showing the peptides removed from the ion-exchange column during a 20–25% 250 mM KCl step gradient. In all three steps, the ion current for peptides eluting into the mass spectrometer is approximately the same, indicating good fractionation of peptides across the ion-exchange separation.

ing a gold electrode through another side arm and is split through a restriction capillary to send a flow of 200–300 nL/min through the column. A single dimension separation is typically suitable for the analysis of moderately complex peptide mixtures. A preponderance of single tandem mass spectral matches to proteins indicates that the mixture needs a higher resolution separation.

4. Typically, protein mixtures containing up to 30 components can be identified in a single-dimension separation. Two factors must be taken into account to judge the comprehensiveness of the analysis: the relative quantity of the proteins present and the molecular weights of the proteins. In general, proteins can be readily identified when they are within a 30-fold molar ratio of the most abundant component.

5. The pH of the solution is adjusted to 2 using concentrated glacial acetic acid prior to loading on the 2D-LC system.

6. Two-dimensional chromatography employs a strong cation-exchange resin followed by reversed-phase chromatography. Peptides are eluted from the cation-exchange resin using a salt step-gradient. Eluted peptides are transferred to the reversed-phase column, desalted and separated over a 60-min linear gradient. The process is repeated with an increasing concentration of KCl to elute another set of peptides. A Perseptive Biosystems Integral Workstation is used to perform the 2D LC. At the flow rates required for 1-mm ID columns (≈50 μL/min), the limit of detection is approximately one pmole. This experimental system is capable of identifying up to 100 components in a mixture using 120 μg total protein.
7. On the LCQ instrument, three microscans are recorded for both the conventional scan (*m/z* 400–1600) and for the three MS/MS experiments.
8. On the LCQ instrument, the mass range for MS/MS is calculated based on a +2 charge state.
9. Tandem mass spectra are searched against databases using the SEQUEST search algorithm. The database searching method is described in **ref. 10**. All single tandem mass spectral matches to proteins should be validated against the sequence matched in the database. SEQUEST will match a tandem mass spectrum to a similar, mass conserved sequence if the correct sequence is not present in the database. SEQUEST can match tandem mass spectra with relatively poor signal-to-noise ratios to the correct sequence, but the spectrum should be of sufficient quality for validation or the match should be considered tentative. Multiple hits to the same protein sequence with poor quality tandem mass spectra can be considered a valid identification. To summarize the protein identification data, hits to the same protein are collated by gene name and then peptide sequence identified. The ion-exchange fraction in which the peptide was found is listed along with a character abbreviation of the cross-correlation score. SEQUEST software is available commercially from Finnigan MAT.

References

1. Hunt, D. F., Yates III, J. R., Shabanowitz, J., Winston, S., and Hauer, C. R. (1986) Protein sequencing by tandem mass spectrometry. *Proc. Natl. Acad. Sci. USA* **83,** 6233–8238.
2. Yost, R. A. and Boyd, R. K. (1990) Tandem mass spectrometry: quadrupole and hybrid instruments. *Methods Enzymol.* **193,** 154–200.
3. Louris, J. N., Brodbelt Lustig, J. S., Cooks, R. G., Glish, G. L., van Berkel, G. J., and McLuckey, S. A. (1990) Ion isolation and sequential stages of mass spectrometry in a quadrupole ion trap mass spectrometer. *Int. J. Mass Spectrom. Ion Proc.* **96,** 117–137.
4. Jonscher, K. R. and Yates, III., John R. (1997) The quadrupole ion trap mass spectrometer—a small solution to a big challenge. *Anal. Biochem.* **244,** 1–15.
5. Griffin, P. R., Coffman, J. A., Hood, L. E., and Yates, III., J. R. (1991) Structural studies of proteins by capillary HPLC electrospray tandem mass spectrometry. *Int. J. Mass Spectrom. Ion Proc.* **111,** 131–149.
6. Hunt, D. F., Henderson, R. A., Shabanowitz, J., Sakaguchi, K., Michel, H., Sevilir, et al. (1992) Characterization of peptides bound to the class I MHC molecule HLA-A2.1 by mass spectrometry [see comments]. *Science* **255,** 1261–1263.

7. Wilm, M. and Mann, M. (1996) Analytical properties of the nanoelectrospray ion source. *Anal. Chem.* **68,** 1–8.
8. Yates, III, J. R, Eng, J. K., McCormack, A. L., and Schieltz, D. (1995) Method to correlate tandem mass spectra of modified peptides to amino acid sequences in the protein database. *Anal. Chem.* **67,** 1426–1436.
9. Davis, M. T., Stahl, D. C., Hefta, S. A., and Lee, T. D. (1995) A microscale electrospray interface for on-line, capillary liquid chromatography/tandem mass spectrometry of complex peptide mixtures. *Anal. Chem.* **67,** 4549–4556.
10. Eng, J. K., McCormack, A. L., and Yates III, J. R. (1994) An approach to correlate tandem mass spectral data of peptides with amino acid sequences in a protein database. *J. Am. Soc. Mass Spectrom.* **5,** 976–989.
11. Yates III, J. R., Eng, J. K., and McCormack, A. L. (1995) Mining genomes: correlating tandem mass spectra of modified and unmodified peptides to nucleotide sequences. *Anal. Chem.* **67,** 3202–3210.
12. McCormack, A. L., Schieltz, D. M., Goode, B., Yang, S., Barnes, G., Drubin, D., et al. (1997) Direct analysis and identification of proteins in mixtures by LC/MS/MS and database searching at the low-femtomole level. *Anal. Chem.* **69,** 767–776.
13. Link, A. J., Carmack, E., and Yates, III., J. R. (1997) A strategy for the identification of proteins localized to subcellular spaces: application to *E. coli* periplasmic proteins. *Int. J. Mass Spectrom. Ion Proc.* **160,** 303–316.
14. Larmann, J. P., Jr., Lemmo, A. V., Moore, A. W., Jr., and Jorgenson, J. W. (1993) Two-dimensional separations of peptides and proteins by comprehensive liquid chromatography-capillary electrophoresis. *Electrophoresis* **14,** 439–447.
15. Holland, L. A. and Jorgenson, J. W. (1995) Separation of nanoliter samples of biological amines by a comprehensive two-dimensional microcolumn liquid chromatography system. *Anal. Chem.* **67,** 3275–3283.
16. Moore, A. W., Jr. and Jorgenson, J. W. (1995) Comprehensive three-dimensional separation of peptides using size exclusion chromatography/reversed phase liquid chromatography/optically gated capillary zone electrophoresis. *Anal. Chem.* **67,** 3456–3463.
17. Opiteck, G. J., Lewis, K. C., Jorgenson, J. W., and Anderegg, R. J. (1997) Comprehensive on-line LC/LC/MS of proteins. *Anal. Chem.* **69,** 1518–1524.
18. Opiteck, G. J. and Jorgenson, J. W. (1997) Two-dimensional SEC/RPLC coupled to mass spectrometry for the analysis of peptides. *Anal. Chem.* **69,** 2283–2291.
19. Anderegg, R. J., Wagner, D. S., Blackburn, R. K., Opiteck, G. J., and Jorgenson, J. W. (1997) A multidimensional approach to protein characterization. *J. Protein Chem.* **16,** 523–526.
20. Bushey, M. M. and Jorgenson, J. W. (1990) Automated instrumentation for comprehensive two-dimensional high-performance liquid chromatography of proteins. *Anal. Chem.* **62,** 161–167.
21. Gatlin, C. L., Kleemann, G. R., Hays, L. G., Link, A. J., and Yates, III, J. R. (1998) Protein identification at the low femtomole level from silver-stained gels using a new fritless electrospray interface for liquid chromatography-microspray and nanospray mass spectrometry. *Anal. Biochem.* **263,** 93–101.

3

Characterization of a Mutant Recombinant S100 Protein Using Electrospray Ionization Mass Spectrometry

Mark J. Raftery

1. Introduction

Recombinant protein expression is of fundamental importance for the production of small or large quantities of biologically active proteins for laboratory and therapeutic uses. The availability of milligram quantities has made elucidation of biological activity and structural characterization possible, even if only small amounts of native protein are available. Commercial insulin, growth hormone, cytokines, and other therapeutic proteins are now produced in large quantities each year using recombinant technologies (1).

Electrospray ionization mass spectrometry (ESI-MS) is a rapid and precise method for determining masses of proteins and peptides and can be used to validate protein sequences (2,3). Mass accuracy is generally within 0.01% of the calculated mass for proteins with masses < ≈40 kDa (2,4). ESI-MS has been used to characterize many recombinant proteins, including the S100 proteins calvasculin (5), calcyclin (5), and S100A3 (6). No mutant or posttranslationally modified forms were identified after comparison of theoretical and experimental masses. ESI-MS was also used to characterize an unusual posttranslational modification of the murine S100 protein MRP14 (7) and to identify errors in the cDNA sequences of rat MRP8 and 14 (8). Matrix-assisted laser desorption/ionization mass spectrometry (MALDI-TOFMS) has been used as an alternative to sodium dodecyl sulfate polyacrylamide gel electrophoresis (SDS-PAGE), to monitor the extent of factor Xa cleavage of a fusion protein between glutathione-S-transferase (GST) and HIV-1_{IIIB} p26 (9).

CP10 is a potent chemotractant for murine and human polymorphonuclear leukocytes in vivo and in vitro *(10)* with optimum activity at approx 10^{-12} *M* in vitro *(11,12)*. The amino acid sequence was determined biochemically and from the derived complementary DNA (cDNA) sequence *(11,13)*. It is composed of 88 amino acids, contains no posttranslational modifications, and is a member of the S100 Ca^{2+} binding protein family *(14)*. Initial studies indicated that CP10 was produced in small quantities by activated murine spleen cells, but isolation from supernatants was a lengthy and complex procedure *(11)*. To facilitate biochemical and structural characterization, a relatively large-scale source of CP10 was obtained by chemical synthesis *(15)* and as a recombinant protein using the pGEX expression system *(16)*.

The pGEX expression system produces the desired recombinant protein as a fusion with GST, enabling isolation from bacterial lysates by affinity chromatography under nondenaturing conditions. The fusion protein is cleaved at a cloned consensus sequence, located between the two proteins, with either thrombin or factor Xa *(17)*. The GST/CP10 plasmid was produced using a CP10-fragment in which the ATG start codon was mutated to a *Bam*HI restriction site by polymerase chain reaction-mediated mutagenesis *(16)*. After digestion with *Bam*HI, the insert was subcloned into the pGEX-2T expression vector (**Fig. 1**) and transfected into *E. coli;* fusion protein expression was subsequently induced with isopropyl-β-D-thiogalactopyranoside (IPTG). The nucleotide and derived protein sequence of GST-CP10 fusion are shown in **Fig. 1B** *(11,18)*. Fusion protein was isolated from the *E. coli* lysate by affinity chromatography on glutathione-agarose. Recombinant CP10 isolated following thrombin cleavage and C18 RP-HPLC has two additional amino acids (Gly and Ser) at the N terminus and a theoretical mass of 10307.6 Daltons *(16)*.

A detailed account of the expression and characterization of rCP10 has been reported *(16)*. A mutant form of rCP10 separated at a slightly higher apparent molecular weight on SDS-PAGE and as a small, early-eluting shoulder on C18 RP-HPLC *(16)*. In this chapter we fully characterize a mutant form of rCP10, which contains 10 additional C-terminal amino acids, and determine the likely mechanism of its production in *E. coli (19)*.

2. Materials

2.1. Chemicals

1. Glutathione-agarose beads (Sigma, St. Louis, MO).
2. Human thrombin (American Diagnostica, New York, NY).

Fig. 1. (**A**) Schematic of the CP10-expressing plasmid (pCP10-12) showing several genes required for fusion protein expression. (**B**) Coding cDNA and derived protein sequences of GST-CP10 fusion cloned into the pGEX-2T vector. Fusion protein sequence starts at Met1 and ends at Glu314. The thrombin cleavage site is at residue 224.

Characterization of an S100 Protein

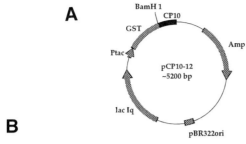

B

```
-61                                      -31
CGG CTC GTA TAA TGT GTG GAA TTG TGA GCG GAT AAC AAT TTC ACA CAG GAA ACA GTA TTC
 .   .   .   .   .   .   .   .   .   .   .   .   .   .   .   .   .   .   .
1/1                                      31/11
ATG TCC CCT ATA CTA GGT TAT TGG AAA ATT AAG GGC CTT GTG CAA CCC ACT CGA CTT CTT
 M   S   P   I   L   G   Y   W   K   I   K   G   L   V   Q   P   T   R   L   L
61/21                                    91/31
TTG GAA TAT CTT GAA GAA AAA TAT GAA GAG CAT TTG TAT GAG CGC GAT GAA GGT GAT AAA
 L   E   Y   L   E   E   K   Y   E   E   H   L   Y   E   R   D   E   G   D   K
121/41                                   151/51
TGG CGA AAC AAA AAG TTT GAA TTG GGT TTG GAG TTT CCC AAT CTT CCT TAT TAT ATT GAT
 W   R   N   K   K   F   E   L   G   L   E   F   P   N   L   P   Y   Y   I   D
181/61                                   211/71
GGT GAT GTT AAA TTA ACA CAG TCT ATG GCC ATC ATA CGT TAT ATA GCT GAC AAG CAC AAC
 G   D   V   K   L   T   Q   S   M   A   I   I   R   Y   I   A   D   K   H   N
241/81                                   271/91
ATG TTG GGT GGT TGT CCA AAA GAG CGT GCA GAG ATT TCA ATG CTT GAA GGA GCG GTT TTG
 M   L   G   G   C   P   K   E   R   A   E   I   S   M   L   E   G   A   V   L
301/101                                  331/111
GAT ATT AGA TAC GGT GTT TCG AGA ATT GCA TAT AGT AAA GAC TTT GAA ACT CTC AAA GTT
 D   I   R   Y   G   V   S   R   I   A   Y   S   K   D   F   E   T   L   K   V
361/121                                  391/131
GAT TTT CTT AGC AAG CTA CCT GAA ATG CTG AAA ATG TTC GAA GAT CGT TTA TGT CAT AAA
 D   F   L   S   K   L   P   E   M   L   K   M   F   E   D   R   L   C   H   K
421/141                                  451/151
ACA TAT TTA AAT GGT GAT CAT GTA ACC CAT CCT GAC TTC ATG TTG TAT GAC GCT CTT GAT
 T   Y   L   N   G   D   H   V   T   H   P   D   F   M   L   Y   D   A   L   D
481/161                                  511/171
GTT GTT TTA TAC ATG GAC CCA ATG TGC CTG GAT GCG TTC CCA AAA TTA GTT TGT TTT AAA
 V   V   L   Y   M   D   P   M   C   L   D   A   F   P   K   L   V   C   F   K
541/181                                  571/191
AAA CGT ATT GAA GCT ATC CCA CAA ATT GAT AAG TAC TTG AAA TCC AGC AAG TAT ATA GCA
 K   R   I   E   A   I   P   Q   I   D   K   Y   L   K   S   S   K   Y   I   A
601/201                                  631/211
TGG CCT TTG CAG GGC TGG CAA GCC ACG TTT GGT GGT GGC GAC CAT CCT CCA AAA TCG GAT
 W   P   L   Q   G   W   Q   A   T   F   G   G   G   D   H   P   P   K   S   D
661/221                                  691/231
CTG GTT CCG CGT GGA TCC CCG TCT GAA CTG GAG AAG GCC TTG AGC AAC CTC ATT GAT GTC
 L   V   P   R   G   S   P   S   E   L   E   K   A   L   S   N   L   I   D   V
721/241                                  751/251
TAC CAC AAT TAT TCC AAT ATA CAA GGA AAT CAC CAT GCC CTC TAC AAG AAT GAC TTC AAG
 Y   H   N   Y   S   N   I   Q   G   N   H   H   A   L   Y   K   N   D   F   K
781/261                                  811/271
AAA ATG GTC ACT ACT GAG TGT CCT CAG TTT GTG CAG AAT ATA AAT ATC GAA AAC TTG TTC
 K   M   V   T   T   E   C   P   Q   F   V   Q   N   I   N   I   E   N   L   F
841/281                                  871/291
AGA GAA TTG GAC ATC AAT AGT GAC AAT GCA ATT AAC TTC GAG GAG TTC CTT GCG ATG GTG
 R   E   L   D   I   N   S   D   N   A   I   N   F   E   E   F   L   A   M   V
901/301                                  931/311
ATA AAA GTG GGT GTG GCA TCT CAC AAA GAC AGC CAC AAG GAG TAG CAG AGG GGG ATC CCC
 I   K   V   G   V   A   S   H   K   D   S   H   K   E   AMB
961/                                     991/
GGG AAT TCA TCG TGA CTG ACT GAC GAT CTG CCT CGC GCG TTT CGG TGA TGA CGG TGA AAA
 .   .   .   .   .   .   .   .   .   .   .   .   .   .   .   .   .   .   .
```

3. Endoprotease Asp-N (sequencing grade from Boehringer Mannheim, Castle Hill, NSW, Australia).
4. Trifluoroacetic acid (Pierce, Rockford, IL).
5. Deionized water (18 MΩ from a Milli-Q system, Millipore, Bedford, MA).
6. Other reagents and chemicals (analytical grade from Sigma and BioRad, Hercules, CA).
7. Solvents (high-performance liquid chromatography [HPLC] grade from Mallinckrodt, Clayton South, Victoria, Australia).

2.2. Electrophoresis

1. Mini Protean II SDS-PAGE apparatus (BioRad) with 15% gels and a Tris/Tricene buffer system *(20)*.
2. Polyvinylidene difluoride membrane (Immobilon-P, Millipore).
3. Mini Trans-blot cell (BioRad).
4. Enhanced chemiluminescence kit (Amersham, Buckinghamshire, UK).

2.3. RP-HPLC

1. Nonmetallic LC626 or LC625 HPLC system (Waters, Milford, MA).
2. 996 photodiode array or 490 UV/visible detector (Waters).
3. Analytical RP-HPLC columns (C4 and C18, 5 µm, 300 Å, 4.6 × 250 mm from Vydac Separations Group, Hesperia, CA).

2.4. Mass Spectrometry

1. Single quadrupole mass spectrometer equipped with an electrospray ion source (Platform, VG-Fisons Instruments, Manchester, UK).
2. HPLC syringe pump (Phoenix 40 from VG-Fisons Instruments).
3. Loop injector (Rheodyne 7125).
4. Fused-silica capillary coupling to the ion source (50 µm × 40 cm).
5. Sample transport solvent (50:50 water + acetonitrile containing 0.05% trifluoroacetic acid [TFA]).

2.5. Sequences

1. Protein and cDNA sequences (Swiss-Prot or Genbank databases via ANGIS [Australian National Genomic Information Service, http://www.angis.org.au; *see* **Notes 1** and **2**]).
2. GCG program package (ver. 8.1, Genetics Computer Group, Madison, WI) (*see* **Note 3**).
3. Applied Biosystems model 473 or 470A automated protein sequencer (Applied Biosystems, Burwood, Victoria, Australia) (*see* **Note 4**).

3. Methods
3.1. Expression and Isolation of Recombinant CP10

Recombinant CP10 was produced in *E. coli* (strain JPA101 *[21]* or BL-21 [Novagen, Madison, WI]) as a fusion protein with glutathione-*S*-transferase *(16)*.

1. Isolate the fusion protein after cell lysis (Tris-buffered saline [TBS], 1% Triton X-100] using glutathione-agarose beads. Cleave with thrombin (30 NIH units; Tris 50 mM, pH 8; ethylenediaminetetraacetic acid [EDTA], 5 mM; 60 min; 37°C).
2. Wash the beads with TBS (2 × 2 mL) and store the combined eluate at –80°C or apply it (100 µL) to a C4 RP-HPLC column and elute the proteins with a linear gradient of 35 to 65% acetonitrile, containing 0.1% TFA, at 1 mL/min over 30 min (*see* **Notes 5–7**).
3. Manually collect peaks with major ultraviolet (UV) absorbances at 214 or 280 nm and concentrate these fractions using a Speedvac (Savant, Farmingdale, NY) to a final concentration of approx 50 ng/µL.
4. Analyze samples (approx 200 ng) by SDS/PAGE.
5. Silver-stain gels or blot these onto PVDF membrane (75 V for 30 min at 4°C).
6. Block membranes with 5% nonfat skim milk and then incubate with a polyclonal antibody to native CP10 *(16)*. Detect with a horseradish peroxidase-conjugated goat anti-rabbit H+L chain (BioRad) using enhanced chemiluminescence.

3.2. Characterization of Recombinant CP10

Two forms of rCP10 readily separated using analytical C4 RP-HPLC with a shallow acetonitrile gradient (*see* **Subheading 3.1., step 2**). The mutant form eluted at 13.6 min as a distinct peak followed by rCP10 at 14.1 min (**Fig. 2A**). The disulphide-linked homodimer of rCP10 eluted at 15.1 min followed by GST at 17.5 min (**Fig. 2A**). The ratio of mutant protein to rCP10 was approx 1:10, and this ratio did not vary over 10 different preparations. The amount of disulphide-linked homodimer varied from batch to batch and was probably formed by oxidation of Cys43 during thrombin cleavage and/or isolation using glutathione-agarose. S100 proteins with free Cys residues readily form disulphide-linked homodimers. S100b, the disulphide-linked dimer of S100β, is a growth factor for glial cells whereas the monomer is inactive *(22)*.

SDS/PAGE followed by silver staining or Western blotting of the proteins isolated from C4 RP-HPLC is shown in **Fig. 2B** (lanes 2–5 and 6–9, respectively). Recombinant CP10 (lane 3) had an apparent mol wt of approx 8000, whereas the mutant protein (lane 2) migrated with a slightly higher mol wt, confirming the previous report *(16)*. The disulphide-linked homodimer (lane 4) had an apparent mol wt of 20,000 and GST had an apparent mol wt of 26,000 (lane 5). The Western blot of the same proteins showed that an anti-CP10 rabbit polyclonal antibody produced using native CP10 *(16)* reacted with both forms of rCP10 (lanes 6 and 7), indicating that they possess common antigen epitopes. The antibody also reacted with the disulphide-linked homodimer of rCP10 (lane 8) and with the GST fusion protein (lane 9).

3.3. Mass Spectrometry

1. Set the electrospray ionization source temperature to 50°C. Adjust the nebulizing nitrogen flow to ≈20 L/min and the drying gas flow to ≈150 L/min. Set the mass spectrometer to give a peak width at half-height of 1 mass unit.

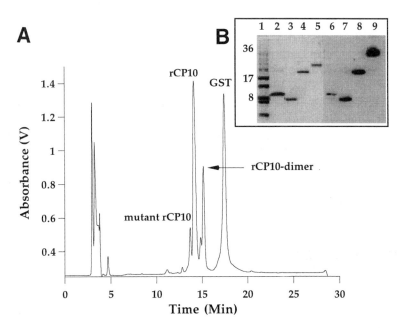

Fig. 2. (**A**) C4 RP-HPLC chromatogram of thrombin cleavage products of GST-CP10 fusion protein. (**B**) SDS-PAGE analysis of isolated proteins. Lane 1, mol wt markers; lane 2, mutant rCP10; lane 3, rCP10; lane 4, CP10-homodimer; lane 5, GST (lanes 1–5 were silver stained); lane 6, mutant rCP10; lane 7, rCP10; lane 8, CP10-homodimer; lane 9, fusion protein (lanes 6–9 were detected with an anti-CP10 antibody using enhanced chemiluminescence).

2. Set the electrospray droplet ionization voltage to ≈3 kV and the cone voltage (the sample cone to skimmer lens voltage controls ion transfer to the mass analyzer) to 50 V.
3. Inject ≈50 pmol, 10-μL samples of protein and peptide sample fractions eluted from C4 RP-HPLC into electrospray transport solvent flowing at 10 μL/min.
4. Acquire electrospray spectra in the multichannel acquisition mode. Use a spectrum scan from m/z 700 to 1800 and a scan time of 5 s (*see* **Note 8**).
5. Carry out a mass calibration with horse heart myoglobin (Sigma).

The mass of each protein isolated from C4 RP-HPLC (*see* **Subheading 3.1.**, **step 2**) was determined using ESI-MS, (**Table 1**), which enabled comparison with the theoretical masses derived from the known cDNA sequences (**Fig. 1B**). The experimental mass of rCP10 was 10,308 (**Table 1**), which compares well with the theoretical mass (10,307.6 Daltons), whereas that of the mutant protein was 11,333 Da, 1025 Da greater than that of rCP10. The experimental masses of rCP10 homodimer and GST were 20,615 and 26,168 Daltons, respectively, which compare well with their theoretical masses (**Table 1**) and confirm their identity.

Table 1
Comparison of the Masses of Proteins Isolated after C4 RP-HPLC and Determined by ESI with the Theoretical Masses Derived from their cDNA Sequences (see text for details)

Protein	Calculated mass (Daltons)	ESI mass (Daltons)
rCP10	10,307.6	10,308
Mutant CP10	11,332.7	11,333
rCP10Dimer	20,614.1	20,615[a]
GST	26,166.6	26,169[b]
Fusion Protein	36,456.1	36,459[b]
Mutant fusion Protein	37,481.2	37,484

[a]A small peak at 21,640 Daltons was also observed, which was attributed to the rCP10-mutant rCP10 disulphide-linked heterodimer (calc. mass 21,639.2 Daltons).

[b]A peak 131 Daltons lower was also observed, probably due to removal of the initiator Met after translation of fusion protein.

One major series and a second minor series of multiply charged ions were observed in the ESI-MS spectrum of the fusion protein fraction (see **Note 3**). Deconvolution over the mass range of 35,000–39,000 Daltons using the maximum entropy software supplied with the data system (MaxEnt *[23]*, VG-Fisons Instruments) gave masses of 36,459 and 37,484 Daltons in an approximate ratio of 10:1 (**Fig. 3**). The mass of the major form corresponded to the theoretical mass of the fusion protein (**Table 1**). The minor form was 1025 Daltons greater than the theoretical mass of the fusion protein and corresponded to the difference in mass observed between the two forms of rCP10, suggesting that they were both derived by thrombin cleavage of two fusion proteins. This indicated that the mutant form of rCP10 was produced as a result of translation rather than by aberrant cleavage of the fusion protein by thrombin.

3.4. N-Terminal Sequencing and Endoprotease Asp-N Digest

The first 10 amino acids obtained after automated N-terminal sequencing (at either Sydney University Macromolecular Analysis Centre [SUMAC] or the School of Biochemistry, La Trobe University, Bundoora, Victoria) of the two forms of rCP10 were GSPSELEKAL, which corresponds to the predicted sequence of rCP10 and indicated an unmodified N terminus. To determine the location and identity of the modification, rCP10 and the mutant form were both treated with endoprotease Asp-N as follows:

1. Digest recombinant proteins (50 µg) isolated from C4 RP-HPLC in ammonium bicarbonate (250 µL, 50 m*M*, pH 8.0) using endoprotease Asp-N at an enzyme to substrate ratio of approx 1:100 at 37°C for 2 h.

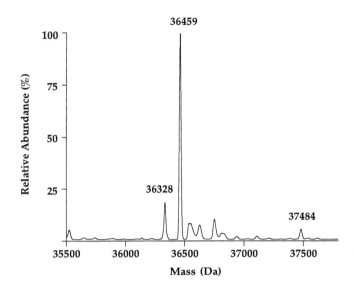

Fig. 3. MaxEnt-transformed mass spectrum of the two fusion proteins isolated after C4 RP-HPLC. The spectrum was acquired over the range m/z 1200–2000 in 5 s, using a cone voltage of 75 V. The peak width at half-height was 1.5 mass units. The two forms of the fusion protein had masses of 36,459 and 37,484 ±5 Daltons (*see text* for details).

2. Lower the pH of the digest to approx 2 (1% TFA) and apply the mixture directly to a C18 RP-column.
3. Elute the peptides with a gradient of 5–75% acetonitrile (containing 0.1% TFA) at 1 mL/min over 30 min (**Fig. 4**). Collect fractions with major A_{214nm} values manually.
4. Determine the mass of each collected fraction by ESI-MS (**Table 2**).

The isolated peptides covered 90% of the rCP10 sequence. Both digests gave exactly the same C18 RP-HPLC trace except for an additional peak at 10.0 min in the digest of the mutant protein (**Fig. 4**). **Table 2** shows that the mass of each co-eluting peptide was identical and that it corresponded to the predicted digestion pattern of rCP10. The mass of peptide 1 was 1639 Daltons (**Table 2**), which did not correspond to any theoretical Asp-N digest product of rCP10, suggesting that this peptide contained the modification. Automated N-terminal sequence analysis indicated the sequence DSHKEQQRGIPGNSS. The calculated mass is 1639.7 Daltons, which compares well with the experimental mass of 1639 Daltons (**Table 2**). The first 5 amino acids, i.e., DSHKE, were identical to the last five C-terminal amino acids of rCP10. The unmodified peptide (DSHKE) from the rCP10 digest was not isolated after C18 RP-HPLC because this peptide is highly hydrophilic and was not retained on the column. Thus, the mutant form of rCP10 contains 100 amino acids and has the same sequence

Table 2
Masses of Asp-N Peptides Isolated by C18 RP-HPLC and Determined by ESI After Digestion of rCP10 and Mutant rCP10, Together with their Theoretical Masses (*see text* for details)

Peptide	Fragment	Calculated mass (Daltons)	ESI mass (Daltons) rCP10/mutant rCP10
1	86–90/86–110	614.6/1639.7	Not isolated/1639
2	15–33	2287.5	2288/2288
3	1–14	1457.7	1457/1457
4	34–59	3157.7	3158/3158
5	64–85	2432.8	2433/2433

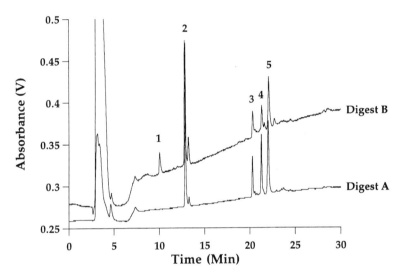

Fig. 4. Comparison of the C18 RP-HPLC chromatograms of the Asp-N digestion of rCP10 (Digest A) and mutant rCP10 (Digest B). All peptides were present in both digests except for one additional peptide (labeled 1) present in the digest of the mutant protein.

as rCP10 but with an additional 10 amino acids (QQRGIPGNSS) at the C terminus. The calculated mass of rCP10 incorporating the additional C-terminal amino acids is 11,332.7 Daltons, in good agreement with the experimental mass (11,333 Daltons).

These data indicate that two forms of fusion protein were translated during protein synthesis, i.e., either two different mRNAs translated two different fusion proteins or a single mRNA translated two distinct fusion proteins. It is unlikely that two fusion protein mRNAs were produced because one plasmid was transfected into *E. coli* and the *E. coli* used to make the recombinant pro-

```
                    CP10                              pGEX-2T
         ------------------------•●------------------------
A   ..GCA TCT CAC AAA GAC AGC CAC AAG GAG TAG CAG AGG GGG ATC CCC GGG AAT TCA TCG TGA..
    -----------------------------  ----------------------------------------------
B   .. A  S  H  K  D  S  H  K  E  *
C   .. A  S  H  K  D  S  H  K  E  Q  Q  R  G  I  P  G  N  S  S  *
```

Fig. 5. Partial nucleotide sequence and derived C-terminal region of the proteins from the expression vector used to produce rCP10. The expression vector contains the cDNA sequence of CP10 and the pGEX-2T fusion vector gene. (**A**) The nucleotide sequence contains two stop codons, TAG and TGA. (**B**) The first stop codon is TAG, which produces rCP10. (**C**) If this codon translates a glutamine (because of the *supE* mutation), then the mutant protein with 10 additional C-terminal amino acids would be expressed.

tein was derived from a single colony. Spontaneous mutation of the plasmid may have occurred, but this cannot account for the constant ratio of mutant protein to rCP10, which remained over 10 preparations, suggesting that the most likely source of mutant rCP10 was an errant translation of a single fusion protein mRNA.

3.5. Recombinant CP10-Plasmid cDNA Sequence

The nucleotide sequence of the plasmid used to produce the recombinant proteins (plasmid pCP10-12), was verified using the chain termination method of DNA sequencing *(16)*. **Figure 1B** shows the coding nucleotide sequence derived from the expression vector used to transfect *E. coli,* and **Fig. 5** shows partial sequences near the C terminus. Two stop codons are in frame, TAG and TGA. If translation proceeded normally it would end at the first stop codon (TAG) to yield the last nine amino acids located at the C terminus of rCP10 (**Fig. 5B**). If translation proceeded through the TAG codon, and a glutamine was inserted as a consequence, then a mutant protein identical to the one described here would be formed. Translation would continue until the second stop codon (TGA), located 27 base pairs downstream from the TAG codon-yielding rCP10 with 10 additional amino acids at the C terminus (**Fig. 5C**). Based on the ratio of the two forms isolated by C4 RP-HPLC, the mutation would occur in approx 10% of transcripts (**Fig. 2A**).

E. coli contains a number of suppressor mutations that allow nonsense codons to code for amino acids *(24)*. In strains of *E. coli* with suppressor mutations in tRNA genes, the three termination codons, TAG, TAA, and TGA, can each encode an amino acid. If *E. coli* has the suppressor mutation *supE*, TAG codes for glutamine in 5–10% of transcripts *(24)*. Incorporation of glutamine at a TAG stop codon is caused by a mutation in a glutamine-tRNA gene, i.e., the anticodon, which base-pairs with the codon on the mRNA, is mutated from

CTG to CTA (G to A), thereby allowing the usual chain termination codon TAG to code for glutamine *(24)*. Suppressor genes are generally produced from tRNA genes that are redundant, resulting in only partial formation of mutant proteins and, because of this redundancy, suppressor mutations are generally not lethal. This mutation occurs in a number of strains of *E. coli* used as hosts for expression vectors for production of recombinant proteins. The JPA101 strain used to express rCP10 was derived from *E. coli* JM109, which has the *sup*E44 gene *(21)*. Eukaryotic mRNAs use the three termination codons with approximately equal preference, whereas prokaryotes use the TAG stop codon 25 times less often than TAA *(25)*. Therefore, strains of *E. coli* with the *sup*E genotype are unsuitable for production of recombinant proteins from cDNAs that use the TAG stop codon. Transfection of plasmid pCP10-12 into a strain of *E. coli* without the *sup*E mutation (BL-21) followed by protein expression produced one fusion protein, and only full-length CP10 (ESI mass 10,308 Daltons) was isolated after thrombin cleavage, further supporting our results.

3.6. Conclusions

We have identified and characterized a mutant form of rCP10 derived from a mutant fusion protein formed as a consequence of a glutamine insertion at the normal stop codon, allowing translation to proceed to the second stop codon 27 base pairs downstream. This occurred in approx 10% of transcripts and was most likely due to the amber mutation *sup*E in the strain of *E. coli* used for protein expression.

4. Notes

1. These databases are also available at other locations (http://www.expasy.ch).
2. GenBank accession numbers for pGEX-2T and CP10 are U13850 and S57123, respectively.
3. Programs within this database are used to manipulate the sequence data.
4. A typical sample level for sequencing of proteins or digest peptides is 250–500 pmol.
5. The elution of the disulphide-linked homodimer was completely eliminated by addition of DTT (1 mM, 30 min, 37°C) before C4 RP-HPLC (not shown).
6. GST, derived from the fusion protein after cleavage with thrombin, was partially washed off the affinity column and subsequently isolated by C4 RP-HPLC. The yield of GST also varied from batch to batch and was dependent on the age of the glutathione-agarose, suggesting some loss of specificity of the beads with use.
7. The fusion protein was isolated directly from the glutathione-agarose beads before thrombin cleavage by incubation with glutathione (20 mM, 5 min). Only one peak corresponding to the fusion protein was separated by C4 RP-HPLC (not shown).
8. Programs used to calculate the mass of protein/peptide and endoprotease digestion products were BioLynx (VG-Fisons Instruments) and MacProMass (Dr. Terry Lee, Division of Immunology, Beckman Research Institute of the City of Hope, Duarte, CA).

Acknowledgments

This work was supported in part by grants from the National Health and Medical Research Council of Australia. Members of the Cytokine Research Unit and Immunology group at the Heart Research Institute, Missenden Rd., Camperdown, NSW, Australia, are acknowledged for rCP10 preparations and helpful discussions.

References

1. Cleland, J. L. (1993) *Impact of protein folding on biotechnology*, in *Protein Folding In Vivo and In Vitro*, vol. 526 (Cleland, J. L., ed.), American Chemical Society, Washington DC, pp. 1–21.
2. Ashton, D. S., Beddell, C. R., Green, B. N., and Oliver, R. W. A. (1994) Rapid validation of molecular structures of biological samples by electrospray-mass spectrometry. *FEBS Lett.* **342**, 1–6.
3. Roepstorff, P. (1997) Mass spectrometry in protein studies from genome to function. *Curr. Opin. Biotechnol.* **8**, 6–13.
4. Chait, B. T. and Kent, B. H. (1992) Weighing naked proteins: practical, high-accuracy mass measurement of peptides and proteins. *Science* **257**, 1885–1894.
5. Pedrocchi, M., Schafer, B. W., Durussel, I., Cox, J. A., and Heizmann, C. W. (1994) Purification and characterization of the recombinant human calcium-binding S100 proteins CAPL and CACY. *Biochemistry* **33**, 6732–6738.
6. Fohr, U. G., Heizmann, C. W., Engelkamp, D., Schafer, B. W., and Cox, J. A. (1995) Purification and cation binding properties of the recombinant human S100 calcium-binding protein A3, an EF-hand motif protein with high affinity for zinc. *J. Biol. Chem.* **270**, 21,056–21,061.
7. Raftery, M. J., Harrison, C. A., Alewood, P., Jones, A., and Geczy, C. L. (1996) Isolation of the murine S100 protein MRP14 from activated spleen cells: characterisation of post-translational modifications and zinc binding. *Biochem. J.* **316**, 285–293.
8. Raftery, M. J. and Geczy, C. L. (1998) Identification of post-translational modifications and cDNA sequencing errors in the rat S100 proteins MRP8 and 14 using electrospray ionization mass spectrometry. *Anal. Biochem.* **258**, 285–292.
9. Parker, C. E., Papac, D. I., and Tomer, K. B. (1996) Monitoring cleavage of fusion proteins by matrix-assisted laser desorption ionization/mass spectrometry: recombinant HIV-1IIIB p26. *Anal. Biochem.* **239**, 25–34.
10. Devery, J. M., King, N. J., and Geczy, C. L. (1994) Acute inflammatory activity of the S100 protein CP-10. Activation of neutrophils in vivo and in vitro. *Immunology* **152**, 1888–1897.
11. Lackmann, M., Rajasekariah, P., Iismaa, S. E., Cornish, C. J., Simpson, R. J., Reid, G., et al. (1992) Identification of a chemotactic domain of the proinflamatory S100 protein CP10. *J. Immunol.* **150**, 2981–2991.
12. Geczy, C. L. (1996) Regulation and proinflammatory properties of the chemotactic protein CP10. *Biochim. Biophys. Acta* **1313**, 246–253.

13. Lackmann, M., Cornish, C. J., Simpson, R. J., Moritz, R. L., and Geczy, C. L. (1992) Purification and structural analysis of a murine chemotactic cytokine (CP-10) with sequence homology to S-100 proteins. *J. Biol. Chem.* **267,** 7499–7504.
14. Fano, G., Biocca, S., Fulle, S., Mariggio, M. A., and Belia, S. (1995) The S-100—a protein family in search of a function. *Prog. Neurobiol.* **46,** 71–82.
15. Alewood, P. F., Alewood, D., Jones, A., Lackmann, M., Cornish, C., and Geczy, C. L. (1993) *Chemical synthesis, purification and characterization of the murine pro-inflammatory protein CP-10,* in Peptides, Escom, Leiden, pp. 359–361.
16. Iismaa, S. E., Hu, S., Kocher, M., Lackmann, M., Harrison, C. A., Thliveris, S., et al. (1994) Recombinant and cellular expression of the murine chemotactic protein. CP-10. *DNA Cell Biol.* **13,** 183–192.
17. Smith, D. B. and Johnson, K. S. (1988) Single-step purification of polypetides expressed in Eschericha coli as fusion proteins with glutathione S-tranferase. *Gene* **67,** 31–40.
18. Smith, D. B., Davern, K. M., Board, P. G., Tiu, W. U., Garcia, E. G., and Mitchell, G. F. (1986) Mr 26,000 antigen of *Schistoma japonicum* recognized by resistant WEHI 129/J mice is a parasite glutathione S-transferase. *Proc. Natl. Acad. Sci. USA* **83,** 8703–8707.
19. Raftery, M. J., Harrison, C. H., and Geczy, C. L. (1997) Characterisation of a mutant recombinant S100 protein using electrospray ionization mass spectrometry. *Rapid Commun. Mass Spectrom.* **11,** 405–409.
20. Schagger, H. and von Jagow, G. (1987) Tricine-sodium dodecyl sulphate-polyacrylamide electrophoresis for the separation of peptides in the range 1 to 100 kDa. *Anal. Biochem.* **166,** 368–379.
21. Iismaa, S. E., Ealing, P. M., Scott, K. F., and Watson, J. M. (1989) Molecular linkage of the nif/fix and nod gene regions in *Rhizobium leguminosarum* biovar *trifolii. Mol. Microbiol.* **3,** 1753–1764.
22. Barger, S. W. and Van Eldik, L. J. (1992) S100 beta stimulates calcium fluxes in glial and neuronal cells. *J. Biol. Chem.* **267,** 9689–9694.
23. Ferrige, A. G., Seddon, M. J., Green, B. N., Jarvis, S. A., and Skilling, J. (1992) Disentangling electrospray spectra with maximum entropy. *Rapid Commun. Mass Spectrom.* **6,** 707–711.
24. Eggertsson, G. and Soll, D. (1988) Transfer ribonucleic acid-mediated suppression of termination codons in *Escherichia coli. Microbiol. Rev.* **52,** 354–374.
25. Kohli, J. and Grosjean, H. (1981) Usage of the three termination codons: compilation and analysis of the known eukaryotic and prokaryotic translation termination sequences. *Mol. Gen. Genet.* **182,** 430–439.

# 4

Searching Sequence Databases via *De Novo* Peptide Sequencing by Tandem Mass Spectrometry

Richard S. Johnson and J. Alex Taylor

1. Introduction

The initial stages of tandem mass spectral data interpretation for a proteomics project usually involve looking for exact matches between the query spectrum and theoretical fragmentation data derived from database sequences *(1–3)*. A number of programs are available that use the observed peptide mass as a filter for identifying portions of database sequences with the same calculated mass prior to evaluating the congruency between the database sequence candidates and the tandem mass spectrum *(4–7)*. These programs generally fall short when analyzing peptides whose sequences differ in a manner that alters its mass compared with the database sequence (*see* **Note 1**). For example, a single conservative substitution of isoleucine for valine is sufficient to thwart algorithms that rely on peptide mass prefilters. We commonly obtain high-quality tandem mass spectral data of peptides for which no exact database match can be made, and the question remains whether these nonmatching spectra are due to novel sequences, or are a result of less interesting possibilities such as interspecies variation, database sequence errors, or unexpected proteolytic cleavages.

We have written three computer programs that together provide an alternative approach to searching sequence databases using tandem mass spectra *(8)*. The first (Lutefisk97) performs a *de novo* sequence interpretation and provides, as output, a short list of candidate sequences. Lutefisk97 uses a graph theory approach similar to those described by Bartels *(9)* and Fernandez-de-Cossjo et al. *(10)*, and the algorithm has been described in detail elsewhere *(8)*. It is important to note that manual or computer interpretations of low-energy collision-induced decomposition (CID) data from tandem mass spectra of peptides are bound to yield multiple sequence candidates, and it is often impossible to distinguish the correct sequence from the incorrect ones. Frequently the variations between sequence possibilities are minor and involve inversions of dipeptides, swapping dipeptides of the same mass, or swapping single amino acids with dipeptides of the same mass (e.g., two glycines have the same mass as a single asparagine). To deal with these multiple yet similar sequence candidates, a second computer program (CIDentify) was written. CIDentify is a version of the FASTA homology-based search program *(11,12)* that was modified to accommodate the ambiguous sequencing results obtained by tandem mass spectrometry. The third program, CIDentify Result Compiler, compiles the CIDentify output for peptides derived from the same protein, and produces a list of database sequences that are ranked according to the number and quality of individual matches found by CIDentify. Here we describe how to obtain, implement, and run these programs.

2. Materials

2.1. Computers, Operating Systems, and Development Environment

Lutefisk97 was written in the ANSI compliant C programming language on an Apple Power Macintosh 7500/100 running system 7.6.1. The Windows version was tested on a Compaq 4000 6200 running the NT version 4.0 operating system. Metrowerks' Codewarrior development environment (Metrowerks, Austin, TX) was used on both the Macintosh and NT. UNIX versions were tested on a Sun Ultra-30 running Solaris 5.5 and a DEC Alpha running OSF 4.0 using the GNU C compiler gcc ver. 2.7. All executables, source code, and accessory files can be downloaded over the World Wide Web (http://www.lsbc.com:70/Lutefisk97.html).

2.2. Instrumentation

Data were obtained using either a Finnigan TSQ 700 triple quadrupole instrument or a Finnigan LCQ ion trap. Ionization was performed using a home-built nanoflow electrospray source using spray tips from Protana (Denmark) *(13)*. In-gel digestion was performed as described previously *(8)*.

3. Methods
3.1. Lutefisk97
3.1.1. General Description

Lutefisk97 *(8)* is a computer program that produces a short list of sequence candidates from a tandem mass spectrum of a peptide *(14)*. The program first

Searching Sequence Databases

reads either unprocessed mass peak profile data, peak centroid data, or manually selected mass-to-charge (m/z) values, and derives a list of 30–70 ions that are considered significant. From these ions, a "sequence spectrum" *(9)*, which is a graph in which all ions are mathematically converted to ions of a single type, is derived. From this sequence spectrum, or graph, thousands of sequence possibilities are generated via a stepwise buildup of subsequences. These subsequences begin at the N terminus, and amino acid residues are added at the C terminus until the subsequences match the observed mass of the peptide. The final step in the algorithm involves assigning scores and ranking the thousands of sequences in order to produce a list of 10 or fewer candidate sequences.

Lutefisk97 works best using CID data obtained from tryptic peptides. Due to the important influence of charged residues on CID fragmentation *(15)* and the fact that tryptic peptides usually have charged residues at the C terminus (lysine or arginine), CID spectra of tryptic peptides produce a very consistent, complete, and more predictable fragmentation pattern *(16)*. Peptides derived from proteolysis using other enzymes tend to be more difficult to interpret, either manually or with Lutefisk97.

Since tandem mass spectra of singly or doubly charged precursor ions contain more complete sequence information than data derived from precursor ions of higher charge states, we have found that Lutefisk97 works best when used with spectra obtained from these lower charge states. However, peptides that produce singly charged precursor ions by electrospray ionization are usually of lower molecular weight and have sequences that are too short to be of value in a database search. Tandem mass spectra of triply charged precursor ions typically produce a series of doubly charged y-type ions that delineate a short sequence of three to five residues, and Lutefisk97 is capable of producing candidate sequences that contain this correct stretch of sequence. Unfortunately, such a short sequence is usually not sufficient for the subsequent homology-based database search. To get more extensive sequence from triply charged ions, we often perform multiple stages of tandem mass spectrometry (MS/MS/MS or MS^3) using nanoelectrospray on an ion trap *(17)*. Thus, the first stage of tandem mass spectrometry produces the series of doubly charged y-type ions, and these y-type ions are subsequently selected for a second stage of tandem mass spectrometry. Since doubly charged y-type ions are thought to have the same structure as doubly charged peptide ions, the process of interpreting such MS^3 spectra is identical to the interpretation of doubly charged peptide ions. Lutefisk97 is usually successful at interpreting such MS^3 spectra.

In general, Lutefisk97 requires data of higher quality than other database search programs *(4–7)*. Whereas these other programs require that a tandem mass spectrum contain only partial sequence information, Lutefisk97 attempts to produce a complete sequence and therefore needs a more complete fragment ion series. Lutefisk97 has the ability to skip two amino acids at a time if a few critical fragment ions are absent; however, we usually limit such skips to one per sequence, in addition to skips at the N and C termini. This limitation does not preclude using a CID spectrum that lacks sequence information for large stretches of the molecule; the program is likely to derive candidate sequences that contain the correct stretch of sequence and then fill in the missing regions with various incorrect sequences derived from background ions. Thus, the data quality influences the length of sequence that can be correctly determined, which in turn influences the success of the subsequent database search.

The mass accuracy of the data has a significant influence on the success of the program. Since there are a number of amino acids that differ in mass by 1 mass unit (1 u), a fragment ion tolerance greater than 0.5 u will introduce a number of sequence possibilities that would be absent with better mass accuracy. For poorly resolved fragment ions, a tolerance of 1 u is reasonable; however, where Asn might be incorporated into a subsequence, two other subsequences that contain Asp and Leu at the same position in the sequence (Asp and Leu both differ from Asn by 1 u) would be generated as well. A much greater number of sequence possibilities are generated, thereby increasing the likelihood that limited computer memory will cause the correct subsequence to be purged before it can be completely sequenced. A tighter fragment ion tolerance reduces the number of sequence possibilities and increases the likelihood of successfully sequencing the peptide. The current implementation of Lutefisk97 converts *m/z* values into nominal (integer) mass values, so tolerances less than 0.5 u have no influence on the outcome. Hence, the greater mass accuracy afforded by newer mass analyzers cannot be exploited in the current version of Lutefisk97 (version 2.0.5).

3.1.2. Implementing Lutefisk97

Lutefisk97 has been successfully implemented under the MacOS, UNIX, and Win32 operating systems. Source code, executables, and accessory files for compiling Lutefisk97 on each platform are provided in the Lutefisk97src.tar.Z archive. The "0_README file" contains descriptions of the archive files and compilation instructions. The Lutefisk97PPC.sea.hqx and Lutefisk97Win.zip archives contain precompiled executable and accessory files for Macintosh PPC and Win32, respectively.

3.1.2.1. MACINTOSH

To run Lutefisk97 on a Macintosh, put the ASCII-formatted CID data file (*see* **Subheading 3.1.4.**) in the same folder as the Lutefisk97 application (*see* **Note 2**). Also include in this folder are the accessory files (*see* **Subheading 3.1.5.**) called Lutefisk.details, Lutefisk.edman, and Lutefisk.params. Once these four

Searching Sequence Databases

files (data plus accessory files) have been appropriately edited, as described below, double click on the Lutefisk97 application icon. A dialog box will appear wherein command-line arguments can be typed (*see* **Subheading 3.1.3.** and **Note 3**) that will, for example, assign a name to the output file. By leaving the argument blank the output file is given a default name, which is the CID data file name with ".lut" appended to it. After the user clicks the "OK" button in the argument dialog box, the program proceeds without any further user intervention. The SIOUX interface provides a text output window, which indicates the various stages of processing that have been achieved. When the program has successfully finished, it produces a list of sequences ranked according to their "intensity score," which is a score that is roughly equivalent to the percentage of fragment ion current that can be accounted for as one of the anticipated ion types (e.g., **b**- or **y**-type ions). Following this list of sequences is a short list containing 10 or fewer sequences, which were selected based on both the "intensity score" and a crosscorrelation score *(5)*. This final list is placed in the output file, which can be directly read by CIDentify.

3.1.2.2. UNIX

The Lutefisk97Src.tar.Z archive contains makefiles for OSF and Solaris, and all necessary source code files for compiling Lutefisk97. Copy the appropriate makefile to "Makefile" and issue the "make all" command. Place the Lutefisk97 executable in the same directory as the ASCII-formatted data file and the accessory files Lutefisk.edman, Lutefisk.params, and Lutefisk.details (*see* **Note 2**). The program can be started and command-line options (*see* **Subheading 3.1.3.**) can be implemented via standard UNIX command-line statements (e.g., "lutefisk -o output.lut" will start the program and place the output into a file called output.lut). Otherwise, the default output file name and other descriptions for the Macintosh (*see* **Subheading 3.1.2.1.**) also apply to the UNIX version.

3.1.2.3. WINDOWS 95/NT

Place the Lutefisk.exe program file into the same directory as the ASCII-formatted data file and the accessory files Lutefisk.edman, Lutefisk.params, and Lutefisk.details (*see* **Note 2**). The program can be started by double-clicking on the Lutefisk.exe icon, and the program will run as described above for the Macintosh (*see* **Note 3**). The default output file name and other descriptions for the Macintosh also apply to the Windows version.

3.1.3. Command-Line Options

To get a usage statement, listing the most important command-line options, invoke Lutefisk97 with the '-h' option (*see* **Note 3** about how to use command-line options in the Macintosh and Windows environment). By using these command-line options, Lutefisk97 can be run in an automatic or batch mode to process multiple CID files at one time (*see* **Notes 4** and **5**).

3.1.4. Data File Structures

Due to instrument vendor secrecy regarding data file structures, we have been unable to read binary data files directly. Therefore, we use a more cumbersome method of converting data files to text or ASCII files, which can then be read by Lutefisk97. For data generated with the Finnigan TSQ triple quadrupole instrument, an ASCII file is created by the "List" program, which is started within the "ICIS Executive" program. From "List", open the data file of interest, and under the "File" menu select "Print..." (*see* **Note 6**). A dialog box appears wherein "ASCII" can be selected as the saved format; also select "Multiple Pages" under "Text Displays". After typing in a file name, click on the "Save to File" button. The ASCII file that is produced can be read by Lutefisk97. For data obtained using the Finnigan LCQ instrument, start the "File Converter" program from "Navigator." From the "Destination" box, select "Text" as the "Format." Select a file from the "Source" box, click on the arrow button, and then click on the "Convert" button. If successfully converted, this file can also be read by Lutefisk. The ".dta" files used by Sequest can also be read by Lutefisk97. Alternatively, a tab-delimited ASCII file can be read where the first column is m/z values, and the second column, separated by a tab space, lists the corresponding unitless intensity. This latter file format is sufficiently generic to allow the reading of data from instruments from a variety of vendors. Whatever the format, data can be either read as raw profile data, centroid data, or selected peaks.

3.1.5. Accessory Files

Three accessory files are required for executing Lutefisk97—Lutefisk.details, Lutefisk.edman, and Lutefisk.params.

3.1.5.1. LUTEFISK.DETAILS

The Lutefisk.details file contains the "ion probabilities" *(9)* for each type of ion (**Fig. 1**). Ion probabilities can be thought of as the significance of any particular ion type for delineating the peptide sequence. The **b**-type and **y**-type ions have the highest values (lines 1 and 7) and are of paramount importance in low-energy CID spectra of peptides (**Fig. 1**). Three columns contain the ion probabilities for three types of fragmentation conditions. The first column is currently not used, but the second and third columns contain values for low-energy CID from triple quadrupole and ion trap mass spectrometers, respectively. The column that is utilized is determined by the "Fragmentation pattern" parameter in the Lutefisk.params file.

4	6	8	/ b ion values
2	2	0	/ a ion values
0	0	0	/ c ion values
0	0	0	/ d ion values
2	2	2	/ b-17 or b-18 ion values
2	0	0	/ a-17 or a-18 ion values
4	8	8	/ y ion values
0	0	0	/ y-2 ion values
2	2	2	/ y-17 or y-18 ion values
0	0	0	/ x ion values
0	0	0	/ z+1 ion values
0	0	0	/ w ion values
0	0	0	/ v ion values
0	0	0	/ b-OH ion values
0	0	0	/ b-OH-17 ion values

Fig. 1. Lutefisk.details file. The second and third columns contain the ion probabilities for data obtained from low-energy collision-induced dissociation (CID) of tryptic peptides in a triple quadrupole and ion trap mass spectrometer, respectively. The first column is not currently utilized.

3.1.5.2. LUTEFISK.EDMAN

Lutefisk97 can employ Edman sequencing data as an aid in deducing sequences from CID spectra. The Edman data can be placed into the Lutefisk.edman file (**Fig. 2**), and it need not be unambiguous or complete in order to be useful. Lutefisk97 uses this information as suggestions rather than requirements, and if the Edman data contradict the CID data, the latter take precedence.

3.1.5.3. LUTEFISK.PARAMS

Most of the user-selected parameters can be changed from within the Lutefisk.params file (**Fig. 3**) and are detailed below in the order in which they appear in the file. Although there are numerous parameters, in practice only two of them are typically changed, viz. the CID filename and the peptide molecular weight. The remaining parameters increase the flexibility of the program to accommodate altered conditions or unusual situations. The program reads these parameters line by line starting from the beginning of each line until a space or tab character is reached.

1. *CID filename:* This is the name of the ASCII-formatted CID data file, which can be specified as a full or partial path name.
2. *Peptide molecular weight:* List the peptide molecular weight here including any number of decimal places. Note that this is not the mass of the protonated molecule [M+H]$^+$. If a value of zero is entered here and if a Sequest ".dta" file is used as input, then the program reads the molecular weight directly from the ".dta" file. If a nonzero value is entered, then the information in the ".dta" file is ignored.

```
X      / Cycle 1.
X      / Cycle 2.
C      / Cycle 3.
P      / Cycle 4.
D      / Cycle 5.
A      / Cycle 6.
X      / Cycle 7.
X      / Cycle 8.
MP     / Cycle 9.
DK     / Cycle 10.
A      / Cycle 11.
K      / Cycle 12.
K      / Cycle 13.
       / Cycle 14.
       / Cycle 15.
       / Cycle 16.
       / Cycle 17.
       / Cycle 18.
       / Cycle 19.
       / Cycle 20.
```

Fig. 2. Lutefisk.edman file. Incomplete and ambiguous Edman data can be entered in this file. Use single letter code without spaces; "x" indicates cycles where any residue is possible.

3. *Number of charges on the precursor ion:* This number is an integer. As described above, Lutefisk97 is most frequently used to interpret CID data from doubly charged precursors or doubly charged fragment ions in MS^3 spectra. For Sequest ".dta" files, if a zero is entered then the charge s1tate is read from the ".dta" file.

The following parameters appear under the heading "Mass tolerances":

4. *Peptide molecular weight tolerance:* This is the error in peptide mass measurement.
5. *Fragment ion tolerance:* This is the error in the mass measurement of the fragment ions. For unit-resolved CID spectra, use a value of 0.5 u, and for lower resolution triple quadrupole spectra use a value of 0.75 or 1 u. Since the fragment ion *m/z* values are converted to nominal masses in the current version (2.0.5) of Lutefisk97, there is no advantage in having tolerances less than 0.5 u.

The following parameters appear under the heading "Memory and speed":

6. *Number of final sequences stored:* This is the maximum number of completed sequences that can be stored before discarding low-scoring sequences. The higher the number the better, but for some computers there may be RAM availability limitations (*see below*). A value of 10,000 sequences is sufficient to ensure that the correct completed sequence is not discarded.
7. *Number of subsequences allowed:* This is the maximum number of subsequences, or partial sequences, that can be stored before discarding low scoring subsequences. The largest number possible that is within the limitations of available

Searching Sequence Databases 49

```
98011901d    / CID Filename.
1128.7  /  Peptide molecular weight.
2            / Number of charges on the precursor ion.
/ Mass tolerances ////////////////////////////////////////////////////////////////////
0.75         / Peptide molecular weight tolerance (u).
0.75         / Fragment ion tolerance (u).
/ Memory and speed/////////////////////////////////////////////////////////////////
10000        / Number of final sequences stored.
4000         / Number of subsequence allowed.
/ Manipulating data //////////////////////////////////////////////////////////////
F            / CID file type: F for ICIS text file, T for tab text, L for LCQ "text", D for .dta
D            / Is this CID data in profile or centroid form? P=Profile, C=Centroid, D=Default.
0            / Peak width at about 10%. A value of 0 (zero) activates the auto-peak width mode.
0.1          / Ion threshold. (Ions > average intensity x Ion threshold are utilized.)
0.0          / Mass offset (u).
6            / Ions per input window (windows are 120 u wide).
4            / Number of ions per average residue.
/ Subsequencing //////////////////////////////////////////////////////////////////
1800         / Cutoff for monoisotopic to average mass calculations (u).
D            / Fragmentation pattern (T=triple quad tryptic,L=ion trap tryptic, D=Default)
1            / Maximum number of gaps per subsequence.
0.15         / Extension threshold.
5            / Maximum number of extensions per subsequence.
/ Extras /////////////////////////////////////////////////////////////////////////
160.03  /  Residue mass of cysteine.
Y            / Auto-tag (Y/N).
0            / Sequence tag - low mass y ion
*            / Sequence tag - single letter code, no spaces, from low mass to high mass y ion
0            / Sequence tag - high mass y ion
N            / Are there any Edman data (Y = yes, N = no)?
T            / Type of proteolysis? T=tryptic, K=Lys-C, E=V8, D=AspN, and N=none of the above
*            / Amino acids known to be present in the peptide. * means none.
*            / Amino acids known to be absent from the peptide. * means none.
N            / Modified N-terminus? (N=none, A=acetylated, C=carbamylated, P=pyroglutamic acid)
N            / Modified C-terminus? (N=none, A=amidated)
```

Fig. 3. Lutefisk.params file. For the collision-induced decomposition (CID) spectrum in **Fig. 4A**, this parameter file was used to obtain the results shown in **Fig. 5A**. See text for a detailed description of each parameter.

RAM and CPU speed should be used. For example, 8192 K of allocated memory is sufficient to handle 4000 subsequences, and a 100 MHz Power Macintosh is able to complete the process in a few minutes. You should adjust this value according to your computer's capabilities.

The following parameters appear under the heading "Manipulating data":

8. *CID file type:* Enter "F" if the CID data is derived from the Finnigan "List" program, "T" for a tab-delineated ASCII file, "L" for a "text" file obtained from the LCQ File Converter program or "D" for a Sequest ".dta" file.
9. *Profile or centroid:* Unprocessed profile data are subjected to five-point digital smoothing. By entering "D" for default, the program automatically differentiates between the two data types.
10. *Peak width at 10%:* This resolution parameter for the CID spectrum is used for peak detection. For unit resolved data, a value of 1–1.5 should be used, whereas values of 2–3 are suitable for poorly resolved spectra. When using profile data, a value of zero, which triggers an automatic procedure to determine peak width, should be used.

11. *Ion threshold:* CID data points with an intensity greater than the average intensity times this threshold are used for identifying peaks. The default value is 0.1.
12. *Mass offset (u):* If the fragment ion m/z values are consistently in error by the same amount throughout the spectrum, then a correcting offset can be applied. A negative number would reduce all fragment ion masses by the specified amount (usually only a few tenths of a mass unit [u]), whereas a positive number would raise their values. Normally, a value of zero (no offset) is assigned.
13. *Ions per input window:* The program examines a mass window 120 u wide (the average residue mass weighted for frequency of occurrence in the protein sequence database) around each ion, and, if the number of ions within each window exceeds the set value, those with the lowest intensity are eliminated. For regions of the CID spectrum that could contain multiply charged ions, this window is narrowed accordingly (e.g., 60 u width for regions that could contain doubly charged ions). A value of 6 was empirically derived for this parameter.
14. *Number of ions per average residue:* This value sets an overall limit to the number of ions to be used in determining sequence candidates. A typical value for this parameter is 4, which means that, for a peptide of molecular weight 1218.6 and an average residue mass of 120 u, the program would limit the total number of ions to 40. In some cases, the program might use slightly more or less than 40 ions; for example, if the signal-to-noise ratios of certain ions are too low, then these are eliminated and the final tally of ions is lower than what would otherwise be calculated from this parameter.

The following parameters appear under the heading "Subsequencing":

15. *Cutoff for monoisotopic to average mass calculations:* Below this cutoff, monoisotopic masses are calculated for fragment ion masses whereas above this cutoff average mass values are calculated. The cutoff is not abrupt, rather, a switch occurs linearly over a 400 u range below the specified cutoff mass. For example, if 1800 u is chosen as the cutoff, then fragment ions below 1400 are calculated as monoisotopic masses, whereas above 1800 u the ions are calculated as average masses and, from 1400 to 1800 u, fragment ions that are between the monoisotopic and average values are calculated. If the CID data are less than unit-mass resolved and the peptide molecular weight is given as an average mass, then this cutoff should be around 1800 u. Since ion traps provide unit-mass resolved spectra and the monoisotopic peptide molecular weight can be determined from the high-resolution "zoom scans," this cutoff should be set sufficiently high for ion trap data (e.g., 5,000) so as to avoid the need for average mass calculations.
16. *Fragmentation pattern:* Currently, two different fragmentation patterns are recognized by the program. Enter "T" for data obtained from a triple quadrupole instrument and "L" for data from an ion trap. An entry of "D", for default, will automatically differentiate between the two data types by looking for the presence of data at the low mass end, since ion trap spectra lack data in the low mass region. Although triple quadrupole and ion trap data are quite similar (**b**-type and **y**-type ions feature prominently in data from both instruments), there are subtle

Searching Sequence Databases

differences that, when properly recognized, can improve the odds that the correct sequence will be listed in the output.

17. *Maximum number of gaps per subsequence:* Lutefisk97 can skip two amino acids at a time if certain sequence-specific ions are missing. Such skips introduce "gaps" in the sequence, which are dipeptides of known mass but unknown sequence. For higher quality data, it is best to limit the number of gaps to one, but for poorer quality data allow two gaps. Since it is often the case that the two N-terminal amino acids cannot be sequenced, this "gap" is not counted in the limit. Likewise, the program always assumes that tryptic peptides could have C-terminal arginines or lysines even if there are no fragment ions indicating the presence of these amino acids. This assumption also introduces a "gap" that is not counted toward the limit.

18. *Extension threshold:* For a given subsequence there may be fragment ions indicating several possible amino acid extensions. The score of the highest scoring extension multiplied by this "extension threshold" is used as a threshold that the other extensions must exceed to avoid being discarded. This is an empirically determined value that has been set at 0.15. This threshold should be reduced to 0.1 if larger numbers of subsequences (> 4,000) are allowed.

19. *Maximum number of extensions per subsequence:* This value places a limit on the number of extensions that are allowed for each subsequence. A value of 5 has been empirically derived; however, larger values (6–8) could be used if larger numbers of subsequences (> 4,000) are allowed.

The following parameters appear under the heading "Extras":

20. *Residue mass of cysteine:* This variable is necessary to account for the different ways of alkylating cysteine. Enter the anticipated residue mass of cysteine— 161.01 for carboxymethylated cysteine, 208.07 for pyridylethylated cysteine, or 160.03 for carbamidomethylated cysteine.

21. *Auto-tag:* Auto-tag is a routine that can be implemented by specifying a "Y" for this parameter. It looks for the most abundant ions at m/z values greater than that of the precursor ion. It then tries to find short stretches of sequences (*see* **Note 7**), which are then used to limit the number of sequences generated. This should be used when analyzing doubly charged precursors using a triple quadrupole instrument, since this region of the spectrum usually contains only y-type ions. In contrast, ion traps often produce abundant **b**-type and **y**-type ions in the region above the precursor m/z value, which tend to make the peptide sequence tag approach more difficult. We normally do not use the "Auto-tag" feature for ion trap data.

22. *Sequence tag—low-mass y-ion*: Triple quadrupole CID spectra of tryptic peptides very frequently contain a series of **y**-type ions in the region above the mass of the precursor ion. With minimal effort, partial manual interpretation of this region can often provide a sequence tag (*see* **Note 7**), which can be specified in the parameters "Sequence tag—low-mass **y**-ion", "Sequence tag", and "Sequence tag—high-mass ion". For "Sequence tag—low-mass **y**-ion" enter the m/z value

of the lowest mass y-type ion in the series of y-type ions that delineate the peptide sequence tag. If you do not have a sequence tag, enter zero here.
23. *Sequence tag:* Enter the amino acid sequence using the single-letter code, without spaces, in order from the low-mass y-type ion to the high-mass y-type ion. If you do not have a sequence tag, enter "*" here.
24. *Sequence tag—high-mass y-ion*: Enter the *m/z* value of the highest mass y-type ion in the series of y-type ions that delineate the peptide sequence tag. If you do not have a sequence tag, enter zero here.
25. *Are there any Edman data?:* Enter "Y" if there are Edman data in the Lutefisk.edman file, or "N" if not.
26. *Type of proteolysis:* If tryptic proteolysis ("T") is selected, then both Arg and Lys are forced into the C-terminal position regardless of whether there are any fragment ions indicating their presence (e.g., y_1 = 147 for Lys and 175 for Arg). This does not eliminate other possible C-terminal amino acids, rather, it ensures that the two most likely residues are considered. Although other proteolysis options are available (*see* **Note 8**), Lutefisk97 works best when using data obtained from tryptic digestion.
27. *Amino acids known to be present:* If a complete sequence lacks one of these amino acids it is discarded. Use single-letter codes without spaces.
28. *Amino acids known to be absent:* These amino acids are not considered when generating the sequence candidates. Use single-letter codes without spaces.
29. *Modified N terminus:* Three possibilities are available: Enter "C" for carbamylation, "A" for acetylation, or "P" for pyroglutamylation. For unmodified peptides enter "N".
30. *Modified C terminus:* Enter "A" for amidated, or "N" for unmodified.

3.1.6. Example

Four CID spectra were obtained, using a triple quadrupole mass spectrometer, from an in-gel digested sample obtained from an assay-based protein purification project that used rabbit tissue as the starting material (**Fig. 4**). No Edman data were available, and the Lutefisk.details file shown in **Fig. 1** was used. The Lutefisk.params file shown in **Fig. 3** was used for the spectrum in **Fig. 4A**, and the remaining spectra used the same parameters except for the CID filenames and peptide molecular weights. Using a Power Macintosh 7500/100, the output shown in **Fig. 5** was obtained within a few minutes.

3.2. CIDentify

3.2.1. General Description

Conventional homology search programs such as BLAST *(18)* and FASTA *(11,12)* are not well suited for utilizing the *de novo* sequencing output produced by Lutefisk97. First, it is difficult to use mass spectrometry to differentiate between leucine and isoleucine (identical masses) or glutamine and lysine

Searching Sequence Databases

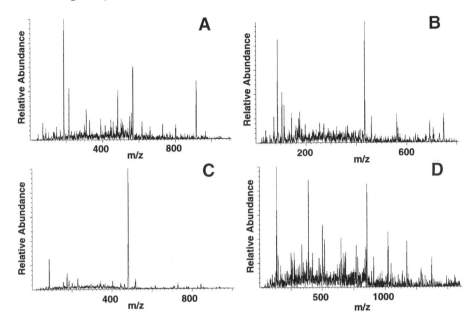

Fig. 4. (A–D) CID spectra of in-gel tryptic peptides obtained from a protein purified from rabbit tissue.

(mass difference of only 0.036 u) (**Fig. 6**). This is only a minor problem as the scoring matrix can be easily altered to accommodate these ambiguities. Second, incomplete CID fragmentation can lead to situations in which a dipeptide of known mass but unknown sequence is present within a sequence (described as "gaps" above) (**Fig. 6**). In addition, certain amino acids have the same residue mass as pairs of amino acids (e.g., tryptophan = glycine + glutamic acid), which can lead to the misinterpretation of a CID spectrum (**Fig. 6**). Finally, BLAST and FASTA homology searches are designed to search with only one query at a time instead of the multiple candidate sequences that are usually produced by Lutefisk97.

To overcome these limitations, we modified the source code to FASTA (*see* **Note 9**), made publicly available by Dr. W.R. Pearson of the University of Virginia, and called the resulting program CIDentify. CIDentify uses the multiple sequences produced by Lutefisk as queries to search a protein database. For each database sequence, the scores for the individual queries are summed to yield the score of that database sequence. No weight is given to the individual query scores based on their ranking in the Lutefisk output. The version of CIDentify that performs six-frame translations for searching nucleotide databases is called CIDentifyX (corresponds to TFASTX). Due to the short query lengths, no gaps are allowed, and the FASTA ktup *(11)* value is set to

A [214]NADLLLEK
[214]NANNLLEK
[214]NANGGLLEK
[214]NADLLNKK
[214]NANNLNKK
[214]NANGGLNKK
[214]NADLLLTR
[214]NANNLLTR

C LDNVL[228]R
LDNVN[227]R

B LGM[201]SPR
LGM[288]PR

D [199]HSNFNEEFLNEK
[199]HSNFNKLYNNEK
[260]YLFDEKFLNEK
[259]YNFDKEFLNEK
[259]YNFNEEFLNEK
[259]YDFNEEFLNEK
[259]YNFDEKFLNEK
[260]YNFNEKMEDEK
[260]YLFDEKFNLEK
[199]HSNFNKLYMPEK

Fig. 5. (A–D) Lutefisk97 output for the CID spectra shown in **Fig. 4**. The numbers in brackets indicate the mass of a dipeptide that could not be sequenced. The letters L and K are used to indicate isoleucine/leucine and glutamine/lysine pairs, respectively.

one. The default scoring matrix is Blosum90MS, which is the Blosum90 table *(19)* that has been modified to use the average values for the isobaric amino acid pairs leucine/isoleucine and lysine/glutamine, and where all scores for X have been set to zero. CIDentify converts ambiguous dipeptides in the query sequence to "XX." When these are encountered in the initial FASTA scoring routine, which looks for identities, the mass of the dipeptide is compared with the combined nominal residue masses of the two database amino acids in the corresponding position. If the dipeptide masses of the query and database are identical, the matrix identities for the two database residues are added to the score.

The secondary scoring routine increases the sensitivity of the search by reexamining the best hit for each query/database sequence pair. This reexamination takes into consideration the possibility of swapping 1 residue for 2 isobaric residues, 2 residues for 1 isobaric residue, and 2 residues for 2 isobaric residues; this is in addition to the usual 1 residue to 1 residue homologous substitutions that are performed by the standard FASTA program. When the mass of one query residue is equal to the mass of two database residues, the matrix identity for the single query residue is added to the score. When the mass of two query residues is equal to the mass of one database residue, the matrix identity for the database residue is added to the score. Where the mass of two query residues is equal to the mass of two database residues, half of the sum of the matrix identities for the database residues is added to the score. If the database residue immediately preceding the alignment is an arginine or a lysine,

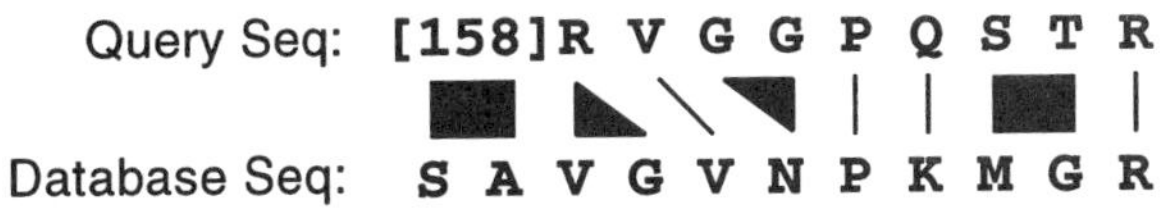

```
Query Seq:   [158] R  V  G  G  P  Q  S  T  R
                  ■ ▲\▼  |  |  ■  |
Database Seq: S  A  V  G  V  N  P  K  M  G  R
```

Fig. 6. An example of the possible isobaric combinations considered by CIdentify. The ambiguous dipeptide mass [158] is isobaric to Ser-Ala in the database sequence. Arg in the query is an isobaric match to Val-Gly in the database sequence. Gly-Gly in the query is an isobaric match to Asn in the database sequence. Gln and Lys are considered isobaric, and the query combination of Ser-Thr is an isobaric match with Met-Gly in the database sequence.

the residue's matrix identity is added to the score as a tryptic cleavage bonus (*see* **Note 10**).

3.2.2. Implementing CIdentify

CIdentify has been successfully implemented under the UNIX, MacOS, and Win32 operating systems. Source code, executables, and accessory files for compiling the CIdentify applications on each platform are provided in the CIdentifySrc.tar.Z archive. The '0_README' file contains descriptions of the archive files and brief compilation instructions. The CIdentifyMac.sea.hqx and CIdentifyWin32.zip archives contain precompiled executable and accessory files for Macintosh PPC and Win32, respectively.

3.2.2.1. MACINTOSH PPC

When executed, CIdentify will prompt the user to locate the Lutefisk97 output file to be used as a query. Next, a list of database options is presented if a FASTLIBS file (*see* **Note 11**) is present. To use a database that is not listed (or if the FASTLIBS file is not present), select an unused letter and then use the file open dialog to locate the database. After the search has been completed, the user is asked for an output filename and the number of hits and alignments to include in the output file. Note that several source files are different between CIdentify and CIdentifyX and that a compiler directive in ffasta.h must be uncommented before compiling CIdentifyX.

3.2.2.2. UNIX

The CIdentifySrc.tar.Z archive contains makefiles for OSF and Solaris, as well as all necessary source code files for compiling CIdentify, CIdentifyX, and CIdentifyRC (the CIdentify result compiler). The source code has been successfully compiled and tested for OSF and Solaris using the GNU C compiler gcc ver. 2.7. The program is run as described for the Macintosh in **Subheading 3.2.2.1.**

3.2.2.3. WINDOWS 95/NT

Note that several source files are different between CIDentify and CIDentifyX and that a compiler directive in ffasta.h must be uncommented before compiling CIDentifyX. The program is run as described for the Macintosh in **Subheading 3.2.2.1.**

3.2.3. Command-Line Options

To get a usage statement listing the most important command-line options, invoke CIDentify (or CIDentifyX) with the '-h' option (*see* **Note 3**). CIDentify runs in user-interactive mode by default, but these command-line options in conjunction with the '-q' quiet mode option can be used to run CIDentify in an automated or background mode (*see* **Notes 5** and **12**).

3.2.4. CIDentify Input

CIDentify/CIDentifyX is designed to take a Lutefisk output file as input. After keying in on the header line "Sequence Rank X-corr IntScr", it simply examines each line that follows for a left-justified sequence (ignoring all scores) and stops on encountering a line that begins with a space or return. Thus, a mock CIDentify input file can be easily created by copying a header line into a file and following it with multiple uppercase sequences to search concurrently. Ambiguous dipeptide masses must be formatted by placing the nominal mass in square brackets.

3.2.5. Databases

CIDentify/CIDentifyX can be used with any FASTA formatted database (*see* **Note 13**). Some publicly available FASTA formatted databases are shown in **Table 1** (*see* **Notes 11** and **14**).

3.2.6. Example

The Lutefisk97 output shown in **Fig. 5** was used as input for CIDentify using the nr.aa protein sequence database. The results for **Fig. 5D** obtained from CIDentify are shown in **Fig. 7**.

3.3. CIDentify Result Compiler

3.3.1. General Description

When examining multiple peptides derived from the same protein, the CIDentify Result Compiler, or CIDentifyRC, can be used to amplify the effect of several individual matches to the same protein by combining the CIDentify/CIDentifyX output files. The results in each CIDentify output file are given a rank score according to their placement. The best hit is given a rank score of 200, the second 199, and so forth. The results from each of the files are then combined into one final list. If a sequence hit is present in more than one output file, its scores are summed.

Table 1
Obtaining FASTA-Formatted Databases

Protein databases (for use with CIdentify)
SwissProt	ftp://www.expasy.ch/databases/sp_tr_nrdb/fasta/sprot.fas.Z
	ftp://ncbi.nlm.nih.gov/blast/db/swissprot.Z
TrEMBL	ftp://www.expasy.ch/databases/sp_tr_nrdb/fasta/trembl.fas.Z
OWL	ftp://ncbi.nlm.nih.gov/repository/OWL/owl.fasta.Z
nr.aa	ftp://ncbi.nlm.nih.gov/blast/db/nr.Z
yeast.aa	ftp://ncbi.nlm.nih.gov/blast/db/yeast.aa.Z
ecoli.aa	ftp://ncbi.nlm.nih.gov/blast/db/ecoli.aa.Z

DNA databases (for use with CIdentifyX)
nr	ftp://ncbi.nlm.nih.gov/blast/db/nt.Z
est	ftp://ncbi.nlm.nih.gov/blast/db/est.Z
yeast	ftp://ncbi.nlm.nih.gov/blast/db/yeast.nt.Z
ecoli	ftp://ncbi.nlm.nih.gov/blast/db/ecoli.nt.Z

A

	initn	initn sum	std. dev.
protein kinase C (EC 2.7.1.-) delta 13	67	518	7.6
protein kinase C delta-type	67	518	7.6
KPCD_HUMAN PROTEIN KINASE C, DELTA	67	518	7.6
protein kinase C (EC 2.7.1.-) theta - hum	64	516	7.6
protein kinase C isoform theta - human gi| 64	516	7.6	
KPCT_HUMAN PROTEIN KINASE C, THETA 64	516	7.6	
KPCT_MOUSE PROTEIN KINASE C, THETA	57	454	6.1
KPCD_RAT PROTEIN KINASE C, DELTA TYPE	56	437	5.7
protein kinase C (EC 2.7.1.-) delta - mouse	56	430	5.5
KPCD_MOUSE PROTEIN KINASE C, DELTA	56	430	5.5

B

```
              XXYNFDKEFLNEK
              ::::::::::::
GNIKIHPFFKTINWTLLEKRRLEPPFRPKVKSPRDYSNFDQEFLNEKARLSYSDKNLIDSMDQSA
F
```

Fig. 7. (**A**) The top 10 database sequences using the Lutefisk97 output shown in **Fig. 5D**. The value "initn" is the score for the best query, "initn sum" is the sum of the values from each query, and "std. dev." is the number of standard deviations of "initn sum" compared with the average "initn sum" across the database. (**B**) Alignment of the top-ranked database sequence with the best query sequence.

3.3.2. CIdentifyRC Input

When executed, CIdentifyRC requests the name of a control file that contains the path names of the CIdentify/CIdentifyX output files to be combined. The control file contains one file name per line (*see* **Note 5**).

The overall best scores are:

	Rank Sum	init1	#
KPCD_RAT PROTEIN KINASE C, DELTA TYPE	548	812	3
KIMSCD protein kinase C (EC 2.7.1.-) delta - mouse	541	795	3
KPCD_MOUSE PROTEIN KINASE C, DELTA TYPE	541	795	3
protein kinase [Mus musculus]	541	795	3
KPC1_CAEEL PROTEIN KINASE C-LIKE 1 (PKC)	540	512	3
alternatively spliced form of Caenorhabditis elegans 538	512		3
TPA-1B [C. elegans]	530	512	3
TPA-1 [C. elegans]	523	512	3
protein kinase C delta-type 398	603		2
protein kinase C (EC 2.7.1.-) delta 13 - h	397	603	2
KPCD_HUMAN PROTEIN KINASE C, DELTA TYPE	396	603	2
KPCT_MOUSE PROTEIN KINASE C, THETA TYPE	385	533	2
eye-specific protein kinase	313	430	2
PslA [Dictyostelium discoideum]	311	432	2
KPCT_HUMAN PROTEIN KINASE C, THETA TYPE	305	589	2
protein kinase C isoform theta - human gi|	304	589	2
protein kinase C (EC 2.7.1.-) theta - hum	304	589	2
KPCL_RAT PROTEIN KINASE C, ETA TYPE	295	135	2
Serine/Threonine protein kin	294	435	2
KPCL_MOUSE PROTEIN KINASE C, ETA TYPE	293	135	2

Fig. 8. Using all of the CIDentify output files from the four tryptic peptides, the Result Compiler identifies database sequences that were found more than once. The column labeled "#" indicates the number of peptides that were matched to the database sequence. The column labeled "init1" is the sum of the "initin sum" values shown in **Fig. 7**. The Result Compiler assigns a "rank value" of 200 to the top ranked sequence, with lower ranked sequences receiving lower "rank values" decremented by one. The column labeled "Rank Sum" is the sum of these "rank values." All of the top twenty database sequences are kinases, except for the protein labeled "PslA [Dictyostelium dicoideum]," which may instead be an ATPase.

3.3.3. Example

The CIDentify outputs for the four peptides in this example (e.g., **Fig. 7**) were used as input for CIDentifyRC, and the output is shown in **Fig. 8**. The alignment of the Lutefisk97 query sequence that best matched rat PKC delta is shown in **Fig. 9**. We did not verify the sequence of this rabbit kinase; however, this sample was later shown to behave as a PKC with respect to both substrate specificity and susceptibility to PKC inhibitors. The isolated protein was either rabbit PKC delta, or a close homolog.

4. Notes

1. All these programs provide some limited means of handling nonexact matches; however, none of them employs a homology-based comparison with the database sequences. Having said this, we also wish to reiterate the great utility of using these programs as an initial screen for identifying simple database matches.

Searching Sequence Databases

```
TFYAAEIMCGLQFLHSKGIIYRDLKLDNVLLDRDGHIKIADFGMCKENIF
        ::::::::
        LDNVL[]R

GESRASTFCGTPDYIAPEILQGLKYTFSVDWWSFGVLLYEMLIGQSPFHG


DDEDELFESIRVDTPHYPRWITKESKDILEKLFEREPTKRLGVTGNIKIH
                                            :::  .
                                            LGM[]SPR

PFFKTINW TLLEKRRLEPPFRPKVKSPRDYSNFDQEFLNEKARLSYSDKN
:::::.::::                    ::::::::::
[]NADLLLEK                    []YNFDKEFLNEK

LIDSMDQSAFAGFSFVNPKFEHLLED
```

Fig. 9. Alignment of the best query sequences from **Fig. 5** with the top-ranked database sequence produced by the Result Compiler (**Fig. 8**). Only the C-terminal region of the protein kinase C is shown. Empty brackets indicate unsequenced dipeptide masses, with the mass numbers left out to preserve the alignments (*see* **Fig. 5** for the nominal dipeptide masses). Note that the nominal mass of W equals the sum of A and D, so for the query sequence [ ]NADLLLEK, the corresponding database sequence includes a space after the W to maintain the alignment. Colons indicate exact matches, and periods indicate conserved matches.

2. The easiest way to use the program is to place all the files into the same directory or folder. However, it is possible to specify a full or partial path to the CID data file. See **Note 5** for details of the differences among UNIX, MacOS, and Win32 path specifications.
3. On the Macintosh, the first dialog box provides a text box for entering command-line options. To use command-line options on a PC, the program must be invoked from a DOS command line by using the Command Prompt program supplied with the Windows operating system. Lutefisk97 options are as follows:
lutefisk [param file]
   -o   output file pathname
   -q   quiet mode (default OFF)
   -f   CID file pathname
   -m   precursor ion mass
   -d   details file pathname
   -v   verbose mode (default OFF)
   -h   print this help text

   CIDentify options not present in FASTA are shown as follows:
   -j   path to the input file (Lutefisk output file)
   -p   path to the FASTA formatted database
   -C   nominal mass of the modified cysteine (e.g., '-C 161' for carbamidomethyl Cys)

4. An example perl script for running Lutefisk97 in batch mode, batch_lutefisk.pl, can be found in the Lutefisk97src.tar.Z archive.
5. When specifying paths, either in command-line options or in the fastgbs file, remember that UNIX uses '/' as a directory separator, the Macintosh uses ':', and Win32 uses '\'. Also, directory names containing spaces or other special characters will probably not work properly.
6. Be careful not to select "Print". By doing so you will generate reams of paper containing the printed list of m/z and intensity values.
7. A peptide sequence tag is defined as a short stretch of sequence that is surrounded by regions of unknown sequence but known mass *(4)*.
8. Besides trypsin, other proteolysis options are Lys-C, Glu-C, and Asp-N.
9. The program FASTA was chosen over BLAST because it was easier to understand and modify the source code. In addition, it seems that BLAST often fails to find suitable database matches when using short query sequences.
10. The matrix identity for lysine is added as a bonus if the alignment starts at the N terminus of the database sequence.
11. A FASTLIBS file, such as the "fastgbs" file included with the CIDentify distribution, can be used to build a menu of available databases to choose from (for more information see the FASTA documentation provided in the "FASTA_doc" directory of all of the archives). Under UNIX the pathname of the FASTLIBS file is specified by the FASTLIBS environment variable. Since the Macintosh and Windows operating systems do not utilize environment variables, a file named "environment" is used, which is located in the same directory as the executable. The line "FASTLIBS = fastgbs" specifies that the FASTLIBS file is a file from the same directory named "fastgbs".
12. We have written a perl module wrapper for automating CIDentify that we use for web-based searches on our corporate intranet. A version of this perl cgi is available from the Lutefisk website.
13. The FASTA format is where each sequence entry has a single header line that begins with a ">" followed by one or more lines containing the sequence, for example: >gi|486479 (Z28264) ORF YKR039w similar to subliminal message binding prot ELVISISKINGELVISISKINGELVISISKINGELVISISSCAN DINAVIANSEATSMELLYFISHKINGELVISISKINGELVISISKING.
14. When working under Win32, it is not necessary to convert databases to "DOS format with linebreaks" before using them with CIDentify. CIDentify can process databases generated under the UNIX, Macintosh, or Win32 operating systems.

Acknowledgments

The authors wish to acknowledge Martin Baker for his debugging efforts and his contributions to the code.

References

1. Humphery-Smith, I., Cordwell, S. J., and Blackstock, W. P. (1997) Proteome research: complementarity and limitations with respect to the RNA and DNA worlds. *Electrophoresis* **18,** 1217–1242.

2. Pennington, S. R., Wilkins, M. R., Hochstrasser, D. F., and Dunn, M. J. (1997) Proteome analysis: from protein characterization to biological function. *Trends Cell Biol.* **7**, 168–173.
3. Roepstorff, P. (1997) Mass spectrometry in protein studies from genome to function. *Curr. Opin. Biotechnol.* **8**, 6–13.
4. Mann, M. and Wilm, M. (1994) Error-tolerant identification of peptides in sequence databases by peptide sequence tags. *Anal. Chem.* **66**, 4390–4399.
5. Eng, J. K., McCormack, A. L., and Yates III, J. R. (1994) An approach to correlate tandem mass spectra data of peptides with amino acid sequences in a protein database. *J. Am. Soc. Mass Spectrom.* **5**, 976–989.
6. Clauser, K. R., Baker, P., and Burlingame, A. L., in The 44th ASMS Conference on Mass Spectrometry and Allied Topics, Portland, OR, 1996, p. 365.
7. Fenyo, D., Qin, J., and Chait, B. T. (1998) Protein indentification using mass spectrometric information. *Electrophoresis* **19**, 998–1005.
8. Taylor, J. A. and Johnson, R. S. (1997) Sequence database searches via de novo peptide sequencing by tandem mass spectrometry. *Rapid Commun. Mass Spectrom.* **11**, 1067–1075.
9. Bartels, C. (1990) Fast algorithm for peptide sequencing by mass spectroscopy. *Biomed. Environ. Mass Spectrom.* **19**, 363–368.
10. Fernandez-de-Cossjo, J., Gonzalez, J., and Besada, V. (1995) A computer program to aid the sequencing of peptides in collision-activated decomposition experiments. *CABIOS* **11**, 427–434.
11. Pearson, W. R. and Lipman, D. J. (1988) Improved tools for biological sequence analysis. *Proc. Natl. Acad. Sci. USA* **85**, 2444–2448.
12. Pearson, W. R. (1990) Rapid and sensitive sequence comparison with FASTP and FASTA. *Methods Enzymol.* **183**, 63–98.
13. Wilm, M. and Mann, M. (1996) Analytical properties of the nanoelectrospray ion source. *Anal. Chem.* **68**, 1–8.
14. Papayannopoulos, I. A. (1995) The interpretation of collision-induced dissociation tandem mass spectra of peptides. *Mass Spectrom. Rev.* **14**, 49–73.
15. Hunt, D. F., Yates, J. R. I., Shabanowitz, J., Winston, S., and Hauer, C. R. (1986) Protein sequencing by tandem mass spectrometry. *Proc. Natl. Acad. Sci. USA* **83**, 6233–6237.
16. Tang, X. and Boyd, R. K. (1992) An investigation of fragmentation mechanisms of doubly protonated tryptic peptides. *Rapid Commun. Mass Spectrom.* **6**, 651–657.
17. Korner, R., Wilm, M., Morand, K., Schubert, M., and Mann, M. (1996) Nano electrospray combined with a quadrupole ion trap for the analysis of peptides and protein digests. *J. Am. Soc. Mass Spectrom.* **7**, 150–156.
18. Altschul, S. F., Gish, W., Miller, W., Myers, E. W., and Lipman, D. J. (1990) Basic local alignment search tool. *J. Mol. Biol.* **215**, 403–410.
19. Henikoff, S. and Henikoff, J. G. (1992) Amino acid substitution matrices from protein blocks. *Proc. Natl. Acad. Sci. USA* **89**, 10,915–10,919.

5

Signature Peptides

From Analytical Chemistry to Functional Genomics

Haydar Karaoglu and Ian Humphery-Smith

1. Introduction

For almost half a century protein chemists have depended heavily on Edman degradation chemistry for protein microsequencing and characterization. Over the last decade, however, mass spectrometry has played an increasingly important role in both processes, to a point that this analytical platform is now, in conjunction with DNA sequence information, the preferred option in most protein chemistry and proteomics laboratories globally. In all cases, the initial process involves the use of endoproteinases to generate peptide fragments.

Measurement of the molecular masses of these peptide fragments is then used to produce a peptide mass fingerprint (PMF). In turn, this information is used alone, or in conjunction with additional orthogonal attributes, to characterize purified proteins, the latter having been obtained by either one- or two-dimensional gel electrophoresis, column chromatography, capillary zone electrophoresis, or a combination of these separation techniques prior to mass spectrometry. The principal methods used are those combining PMF data with (1) predicted fragmentation patterns *(1)*, (2) a peptide sequence tag *(2)*, or (3) the PMF provided by the use of at least one other endoproteinase *(3)*—all of which depend on replacing the protein chemical code with unique numerical parameters for the interrogation of modified protein databases to find like entries.

The pivotal importance of peptide mass spectrometry to proteomics (the protein output of a genome) and indeed to protein chemistry, led Wise et al. *(4)* to investigate the inherent virtues and limitations of PMF as a tool for protein characterization. These authors examined some 51 million peptides generated by perfect digestion using 23 different enzymes (three multiple cutters and the remainder cutting at the carboxy terminus of each amino acid residue) by computer simulation in an attempt to quantify the bounds of reliability and efficiency achievable by this approach. Obviously, in an experimental setting, such quality of data is never achieved consistently, but then neither would such a large-scale analysis have ever been possible in an experimental setting. The challenge for experimental findings is clearly to attempt to emulate the quality of the computer-generated data, while taking into account the findings of this analysis, which served to highlight the following points:

1. Some proteins could not be unequivocally identified using this approach, depending on the cut-site employed (8.4–18.0% of database entries).
2. Some proteins could not be cut with endoproteinases due to an absence of the cut-site, i.e., absence of specific amino acid residue(s). Examples of the largest uncut proteins included 60,906 Daltons for trypsin (R & K) and 258,314 Daltons for cysteine.
3. Peptide fragments of >5000 Daltons had little effect on the outcome of PMF, whereas those of lower mass were of greatest utility, e.g., peptides near to 350 Daltons in the case of trypsin.
4. On average, for both common and rare cutters, the combination of approximately two fragments (1.91–2.34) was sufficient to identify most database entries able to be uniquely identified by PMF.
5. Rare cutters left most database entries unidentified.
6. Any combination of either one or up to 23 different cutters could not identify all database entries.
7. On average, between 5.09 (W cut-site) and 30.88 *(4,5)* peptide fragments were generated per database entry.
8. The maximum number of peptides sharing a similar molecular mass and generated by a single enzymatic cutter was 84,888. (This number will have already increased significantly since the study was conducted.)
9. The most statistically useful molecular masses within the peptide databases were found alongside the most useless, i.e., in the lower molecular mass range <1,000 Daltons).

Independent of the frequency of cut-site and the number of peptides generated, if the protein was able to be uniquely identified, then this could be achieved with a mean of near to just two peptide fragments. This allowed the discovery of the concept of "combinational sieving" so as to facilitate the generation of statistical population outliers corresponding to the best available answer from within a set of either theoretical or experimentally generated peptide masses, that is, faced with a set of 15 peptide masses, plus or minus some chemical noise, one could interrogate peptide databases not with just one query event, but rather with interrogation events corresponding to all the possible pairs of peptides. Since the work of Cordwell et al. *(3)*, we have been aware that the quality of data used when searching peptide mass databases dramatically affects the quality of the data output. Particularly, matrix-assisted laser desorption/ionization–time-of-flight (MALDI-TOF) mass spectrometric data, but also other mass spectrometric platforms, regularly generate erroneous data (incorrect mass-to-charge ratios assigned to peaks) due to impurities in the protein sample, the desorption matrix, chemical adducts, column elutant, or digestion buffer. Posttranslational modifications and incomplete digestion can also further complicate the spectra. In addition, one is continually confronted with a deteriorating signal-to-chemical noise ratio as the quantity of protein analyte available for analysis is reduced. This is most often the case for proteins resolved in a single two-dimensional electrophoresis gel. Thus, in all the instances mentioned above, it would be highly desirable to employ only the "good" or statistically "most useful" data, while conducting protein characterization/identification, i.e., effectively filter out chemical noise and less statistically valuable information. By such means, the correct or best answer would be extracted from noisy data, effectively decreasing the detection threshold and improving the quality of experimentally acquired data. This computational approach has yet to be implemented as an automated tool for use in conjunction with mass spectrometry; however, the mathematics underwriting the approach have been detailed in the published record *(4,5)*.

From the aforementioned work, it became obvious that not all peptides shared equal statistical weight when it came to protein characterization. Elsewhere *(6–9)*, it became obvious that PMF was not merely a "dumb" analytical procedure and, in fact, it was seen that some peptide masses corresponded to portions of, or contained entire, protein "motifs" of functional or structural significance, which had been conserved within protein molecules and/or used as building blocks in larger molecules. Thus, mass spectrometric data were also shown to be capable of highlighting functional attributes within protein molecules, in addition to the generation of data useful in gene-product characterization. It was, indeed, some of these peptides that were responsible for identifying a particular molecule or family of molecules. Such observations incited the development of a graphic user interface (GUI) designed to highlight regions within proteins that were statistically more significant than others (color coding from blue through hot red colors to indicate increased statistical significance). The methodology employed will be outlined later in this chapter (*see* **Subheading 4.**). In this manner, if such peptides were also encountered experimentally, one could readily access the value-added information corresponding to the particular peptide in relation to its statistical weight and any known PROSITE entries (http:/www.expasy.ch/prosite/ is a database of protein families and domains), also included in the GUI along with their corresponding statistical significance. Such a bioinformatic tool allows the display of an automated protein profile with respect to potential functionally significant attributes.

It is worthy of note that molecular "mimicry" and/or sequence similarity can vary from 0–100% for amino acids strings of length one to the largest known proteins. Here, this study was initially limited to 100% similarity, which, apart from whole gene duplications, tended to limit the length of peptides perfectly preserved or re-evolved within Nature to a manageable length, usually of <10–12 amino acids and occasionally up to twice that size, a size also compatible with peptides generated by endoproteinase digestion. However, work has been initiated to accommodate amino acid substitutions at >80% sequence similarity, in an attempt to grow these small words out to a size more familiar with motifs regularly encountered in PROSITE. Such measures should allow strings of small words to form larger words, while maintaining or even increasing their inherent statistical significance.

Furthermore, and on a similar plane to the lessons learnt in genomics and bioinformatics, it has become apparent that for proteomics also: "It is not merely a matter of amassing great quantities of data, but rather exploiting that data maximally." If proteomics is to become the mainstay of functional genomics, it must go further than merely attributing novel or previously sequenced status to gene products. It is important to note that approximately one-third of all genes/open reading frames (ORFs) currently being identified as part of genomic sequencing initiatives remain uncharacterized and still more bear the tag of hypothetical. If the value-added information contained within mass spectra can be further exploited, real insight into protein function and evolutionary affinities are likely to become evident.

With this objective in view, we have initiated efforts to transform an analytical science based on the characterization of protein samples into one that is also directed towards providing value-added information in the form of highlighted regions of relevance to protein function. By using previously transformed relational databases containing all known proteins and theoretical ORFs, one can be immediately aware of the existence of such value-added information when examining *de novo* protein sequences obtained by mass spectrometry. Site-directed mutagenesis can then be used instead of, or in conjunction with, "gene knock-out" in an attempt to dissect gene function. The co-occurrence of a number of motifs linked with a known functional family of proteins can be used to construct an impression of function. The relevant functional or structural motifs can correspond to pieces of genes with quite difference phylogenetic ancestries and thus be obscured by more traditional whole gene comparisons. A variety of other bioinformatic approaches have been developed to examine affinities of portions of genes or protein sequence (reviewed in **ref. *10***); however, these have not been devel-

Signature Peptides

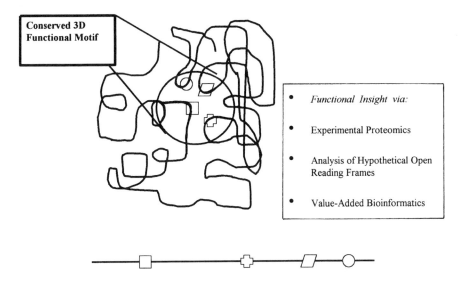

Fig. 1. Theoretical projection for the use of permutations and combinations of signature peptides to detect structural motifs at the lower end of sequence similarity.

oped with a view to profiling the entire protein sequence with respect to its multiple affinities.

When genes resemble one another greatly or result from gene duplication, then their affinities become obvious by any gene comparison tools currently employed within bioinformatics. However, once this level of sequence similarity falls below 30–35%, these supposed functional similarities based on sequence similarity become far more difficult to detect. At the lower end of sequence similarity between proteins, it is feasible that the use of permutations and combinations of signature peptides may allow hypothetical proteins to be afforded partial functionality, even for motifs only present in tertiary molecular space (**Fig. 1**). Information provided from the analytical laboratory or by processing known sequence information by alternative means based on data similar to that obtained from PMF can provide protein scientists with information relevant to both characterization and attribution of function. Over the next few years, there will exist a window of opportunity for many genes/proteins to be detected first by proteomics, rather than by genomics and DNA sequencing. The latter is more probable, if high throughput proteomics rapidly becomes a reality *(11)*.

Here, we review aspects of the origin of information content within genes and thereafter provide concrete examples of statistically significant strings of amino acids (signature peptides) detected within the genome of *Haemophilus influenzae*. The utility of such signature peptides for intra- and inter-genomics will also be demonstrated.

2. Origin of Information Content Within Genes

The past few years have been marked by the determination of the complete sequences of nine bacterial, three archaeal, and one eukaryotic genome *(12–30)*. This has ushered in novel approaches to the way we can now perform biological studies. A major challenge is, however, to elucidate the function of the multitude of uncharacterized genes or genomes; namely, function with respect to active sites, molecular interactions, and evolutionary affinities. A common theme is that genome evolution is perhaps far more complex then most of us could have imagined prior to the onset of the genomic era. Genes possessing a significant similarity to each other are presumed to have evolved from a single ancestral gene *(31–33)*, and one of the main challenges in the understanding of sequence data of complete genomes is to go from the analysis of the evolutionary processes at the level of a single gene to those at the level of gene families and complete genomes. Many questions still remain unanswered concerning the molecular evolution of genes and proteins. Clearly, there are many theories, such as gene duplication, horizontal gene transfer, random genetic drift, and exon and domain shuffling, which have been proposed to explain some aspects of gene evolution. Variations in bacterial genomes can be further mediated by any number of mechanisms, such as recombination, conjugation, transduction, and transformation *(34–39)*, which are frequently employed to transfer or gain genes that give rise to genetic diversity. Plasmids *(40)*, bacteriophages *(41)*, transposable elements *(42)*, and pathogenicity islands *(43)* provide further genomic additions that contribute to the evolution of bacterial pathogens.

2.1. Ancient Conserved Regions

Ancient conserved regions (ACRs) *(44–49)* are sequences contained within genes and proteins that originated early in evolution and have remained much the same throughout phylogeny in bacteria, eukaryotes, and archaea. Over long evolutionary periods, the less constrained portions of the sequences will have significantly diverged. Consequently, the regions of similarity are usually those of greatest structural or functional importance. More precisely, ACRs often correspond to specific motifs present in a variety of proteins, such as the zinc finger DNA binding domain or active sites of enzymes *(50,51)*. ACRs can also comprise most or all of the sequence of a single highly conserved protein or protein family, such as actins and histones *(49)*. The very high sequence similarity in these gene families is probably due to a continuous process of genetic transfer by unequal crossover, gene conversion, and transposition. Preliminary analysis has indicated that the number of ACRs is about 1000 *(48)*. Not all proteins contain ACRs, but it has been estimated that about 40% of all genes contain at least one ACR *(45)*. A recent detailed analysis *(52)* revealed very similar distributions of protein sequence conservation in bacterial and archaeal genomes. Biologically relevant sequence similarity on the basis of statistically significant alignments and/or motif conservation showed that the fraction of proteins containing regions conserved over evolutionary distances (ACRs) appears to be nearly constant at 70% *(52)*. Additional genome sequences are likely to follow this rule. This is seen in highly conserved and more variable proteins with conserved housekeeping roles and those that define the biological uniqueness of each species. The fraction of proteins possessing ACRs may further increase as detailed sequence analysis is conducted on "weak" similarities, and this information is complemented by genomic sequences of other eukaryotes.

2.2. Orthologous and Paralogous Genes

Genes and proteins change during the course of evolution, forming families of related molecules that have similar primary, secondary, and tertiary structures, but that have divergent functions. Homologous genes are classified into orthologs (genes related by vertical descent or horizontal transfer) and paralogs (genes related by duplication within the same organism) *(53–55)*. Estimates of the fraction of gene products in sequenced genomes that have detectable homologues in current databases, range from 44% for *Methanococcus jannaschii* to 80% for *Haemophilus influenzae (12,14)*. Obviously, these percentages will increase as the number of fully sequenced genomes also increases. Even in the case of phylogenetically distant bacteria, for instance *Mycoplasma genitalium,* about half of the genes possess orthologs in both *Escherichia coli* and *H. influenzae (56)*. A considerable fraction of genes in bacteria and archaea, namely 30–55%, belongs to families of paralogs *(47,57)*. This fraction is greater in species with larger genomes. Furthermore, one-half of *E. coli, H. influenzae,* and *Synechocystis* sp. genomes contain important proportions of paralogous genes *(15,58,59)*. Many of these paralogs group into families of various sizes *(60)*. Remarkably, the largest classes of paralogs are the same in each of the sequenced genomes, even though there are notable exceptions, such as the helix-helix DNA-binding proteins in *Mycoplasma genitalium* and the abundant Fe-S oxidoreductase in *M. jannaschii. H. influenzae* and *M. jannaschii* each contain a similar number of genes and proteins in the three largest classes of paralogs, namely, ATPases and GTPases, NAD(FAD)-utilizing enzymes and helix-turn-helix DNA binding proteins *(12,14,61,62)*.

2.3. Domains and Modules

Recent advances in molecular biology have revealed the hidden features within gene and protein structures. A eukaryotic gene generally is not continuous, but is divided into subregions, introns and exons *(63–67)*, whereas proteins are composed of domains and modules *(68–74)*. These subregions, rather than the entire ORF, are known to contain and essentially determine the biological functions of the gene or protein. Although the concept of protein domains as structurally independent units dates back several decades, the idea that similar structural units might be shuffled around as building blocks (modules) to form proteins is relatively new *(75–78)*. Modules are usually recognized as contiguous sequence motifs defined by a common subset of conserved residues that are necessary to form the domain core and preserve a given fold. Modules usually consist of between 40 and 100 amino acid residues and maintain a stable structure when isolated from the native protein. Although, in principle, the terms "domain" and "module" are conceptually quite distinct, they have often been used quite synonymously.

Some domains appear to be narrowly distributed *(78)*, whereas others appear to be universal *(79,80)*. It is presumed that such domains were in existence before the last common ancestor of prokaryotes and eukaryotes, approximately two to three billion years ago *(80)*. Most of these domains are now locked into position in proteins at the core of metabolism. Still, prokaryotes and eukaryotes may each possess some unique domains. Certainly, this would seem to be true for fungi, plants, and animals, which have many sequences that appear not to have counterparts elsewhere. For example, the widespread WD-40 domain, with a characteristic sequence motif that has so far been found only in eukaryotes, was first identified in animals and yeast but is now known to occur in upwards of 50 different proteins *(81)*. This domain has been conserved over a billion years since fungi, animals, and plants last had a common ancestor. It has not been found in bacteria, and its evolutionary origin remains mysterious. Other apparently critical modules are known to be common to eukaryotes but have not yet been traced to prokaryotic origin. For example, the leucine-zipper domain, the zinc-binding domain called the RING finger *(82)*, the pleckstrin homology (PH) domain *(83)*, and the *v-src* homology region 2 (SH2) domain *(84)*. In contrast, many domains or modules still appear to have moved around in prokaryotes and eukaryotes with equal alacrity. Thus, they are often called evolutionarily mobile modules. The evidence of their ancient mobility is apparent only from three-dimensional comparisons *(85)*, all sequence resemblance having been eroded long ago. However, occasionally a weak sequence similarity remains at the level of motifs. This movement is not restricted to single species, and can occur from animals to bacteria or vice versa via horizontal transfer.

Approximately 50 different domains have been identified that are shuffled about in various animal extracellular proteins. These include the epidermal growth factor (EGF)-like domain, and the immunoglobulin and fibronectin type III domains *(86–89)*. Some modules occur in long tandem repeats. Titin, a muscle protein, contains about 300 modules with repeated units of 100 amino acids *(90)*, whereas many proteins contain only two or more domains, although some are restricted to single domains *(91–93)*. The function of many modules is not clear, but many of these evolutionarily mobile units are versatile in their binding properties. They form the packing units that bind to themselves as well as to other modules. Some modules serve as recognition tags, many bind ligands, many appear to be linkers or spacers, and some may have no function at all *(94)*. The amino acid sequence of modules contains the information required for folding of the protein into a stable and functional three-dimensional structure. Not surprisingly, recent evidence indicates that an altered amino acid sequence changes protein folding and is the molecular basis of a growing list of human diseases, such as Marfan's syndrome and cystic fibrosis *(95,96)*. A number of prokaryotic structural proteins are also made up in an obvious domain and modular fashion. Some of these are long multidomain proteins that contain coiled segments *(97)*. Others, like the wall-associated proteins *(98)*, are highly repetitive and exhibit sequence motifs seen in other microbial proteins *(99,100)*.

2.4. Motifs

Motifs are small conserved signature regions within domains or modules *(101,102)*. Motifs have been regarded as playing important roles in the function of the protein in which they reside *(103)*. Motifs are characterized by a specific arrangement of evolutionary conserved amino acids in a protein. Motifs are expected to be related to biological function, as the neutral theory of molecular evolution *(104)* states that, as the biological importance of a region in a gene or protein increases, the evolutionary pressure on the region becomes higher, making it more highly conserved. Examples include DNA/RNA binding motifs *(105)*, calcium binding sites *(106)*, and structures encoded by the exons associated with shuffling and duplication. The fibronectin type III module has a characteristic motif of approximately 90–100 amino acid residues typified by a pattern of well-defined structural features *(107)*. Certain structural motifs are widespread among proteins, but whether they have resulted from a common ancestor or whether there has been a convergence to a common structure remains open *(46)*. For example, a conserved amino acid sequence motif was identified in four different groups of enzymes, namely, GMP synthetase, asparagine synthetase, arginosuccinate synthetase, and ATP sulfurylases *(58)*. Another example is provided by the transcription factors *Fos* and *Jun*, which contain unique conserved motifs. In turn, these motifs were used to identify other transcription factors within other proteins with similar modular characteristics *(108)*. Detection and alignment of these locally conserved motifs among closely as well as distantly related sequences can be highly informative in providing insight into protein structure, function, and evolution *(109–111)*.

3. Evolution of Genes and Proteins

Duplication of ancestral genes, or regions therein, is thought to play an important role in the evolution of genes, because it facilitates the differentiation of one copy, while retaining function in the original copy. Many prokaryotic and eukaryotic genes are thought to have been duplicated and then to have subsequently diverged functionally during the evolution of organisms *(112–117)*, as typically exemplified by four clusters of serum albumin gene families *(118)*. Most duplications are of just a single gene, but recent studies have shown that multiple gene duplications in *H. influenzae (57)*, or even whole-genome duplication, as has been found in *Streptomyces cerevisiae* genome *(119)*, are also important evolutionary mechanisms. Evolution in daughter genes after a large-scale gene duplication event follows a two-step mechanism. At first, selectively neutral changes dominate, and then the coding regions of duplicated genes generally accumulate missense mutations. Sometimes, new genes can be generated by an overprinting mechanism, i.e., new genetic sequences are produced from unused reading frames or antisense strands *(120,121)*. This mechanism seems promising for newly created bacterial genes, since the new genes produced by overprinting could have a novel function quite different from that of the ancestral genes. The overlapping genes have been found in bacteriophage ΦX174 and in animal virus SV40 *(122)*. Variation in eukaryotic genes can be mediated by any number of mechanisms, but the resulting changes fall into three general classes: point mutations, homologous exchange, and chromosomal rearrangements, involving acquisition, deletion, and reorganization of segments of the genome. Mutations first arise in the germline, then spread through the population as polymorphisms, and are fixed in the species as an evolutionary change.

An important conceptual consideration is the statistical likelihood of random point mutations combining with evolutionary forces to produce novel function, as opposed to recombination, duplication, and other errors that occur during replication (e.g., deletion or expansion via slippage events, but arising from information already contained within self [a given genome]). Here, the statistical likelihood must be higher for a chance production of meaningful change by the rearrangement of building blocks contained within the existing genomic complexity. In addition, horizontal transfer of genomic information by a variety of mechanisms (see above) can further increase the rate of evolution of novel functional units within genes and, indeed, genes themselves or groups of genes. One must then ask whether the complexity observed today within biological systems is the result of predominantly lengthy evolutionary processes due to random drift away from progenitor sequence information or, most likely and superimposed upon the former, more rapid evolution due to genetic rearrangement from within already acquired complexity and functional utility found within self. Whereas both are known to be active, the latter has the greatest potential to produce biological diversity most rapidly.

Until the early seventies, the accepted theory about how proteins evolved centered mostly on duplication and modification. Over the last 25 yr, a better understanding of protein evolution has arisen. A knowledge of the three-dimensional structure of lactate dehydrogenase was determined by X-ray diffraction *(123,124)* and thereby allowed the realization that some parts of the molecule closely resembled features seen in other proteins. This implied that a part of the enzyme obviously had counterparts in other dehydrogenases. Because these structural similarities did not occur in the same parts of the molecule, it was concluded that proteins were constructed of domains or modules *(125)*. The existence of these repeated sequences or domains in vertebrate proteins is now well recognized *(126)*. It appears that in the process of evolution, these structural units are used over and over again. The same domain is sometimes used to perform different tasks *(80,127)*. The common ancestry of eukaryotes, archaea, and eubacteria is well demonstrated by amino acid sequence comparisons of numerous proteins that are common to all three groups. On the other hand, some proteins like ubiquitin are common to eukaryotes and archaea but have yet to be observed in eubacteria *(128)*, whereas others appear to be restricted to eukaryotes, e.g., the cytoskeletal proteins *(129)*. In theory, sequence similarity could be due to genetic drift, convergence, or common ancestry *(130,131)*.

Convergence differs from chance similarity in that it depends on adaptive replacements that are positively selected. Sequence convergence involves the parallel evolution of sequences that are similar to begin with. This is quite different from evolving similar sequences from completely unrelated ancestors. However, convergence is not the major force, because many three-dimensional structures of proteins are known to possess similar folds that must have descended from a common ancestor *(132)* and yet they may have no sequence resemblance. It is also common for similar three-dimensional structures to be formed from totally dissimilar sequences *(133)*. Such examples involve common ancestry, but many of the three-dimensional barrel folds are so simple and common that they may have evolved independently. Evidence for domain shuffling *(134)* has been gaining support as more protein sequences have been compiled. As proteins of related function are often similar in sequence, this also reflects a common phylogenetic origin. Proteins with no known homology are probably divergent proteins too distantly related to known sequences in databases to have retained similarity. All proteins, however, probably share common ancestries if one goes back far enough in evolution. Therefore, given the huge accumulation of protein sequences in current databases, it could be expected that some proteins, with no obvious sequence resemblance to any other, share some residues that could represent footprints of ancient common ancestries. To trace protein evolution, it is easier to follow the distribution of protein domains from different phyla by comparing these with the most widespread modules, such as extracellular EGF-like domains *(135)*. In animals, most of the extracellular proteins are composed of multiple domains that are found to be functional in diverse proteins *(67)*. The spread of these domains is thought to be the result of genetic shuffling mechanisms. Sequence comparison suggests that the spread of these domains in diverse organisms may have occurred via horizontal gene transfer *(136–144)*.

The occurrence of these similar domains in different proteins supports the existence of divergent evolution from a common ancestral gene that has been duplicated and fused with a variety of different genes. This suggests that large proteins have been constructed using a modular mechanism in which domains are the building blocks. The appearance of proteins with new or diverged functions could then arise by shuffling of these domains as facilitated by exon shuffling. The shuffling of domains in recently evolved proteins has been greatly promoted by introns *(64,145)*. However, this does not imply that all domain rearrangements involve introns. Only a small fraction of known exons show evidence of having been shuffled *(125,146)*, i.e., some genes fit well with the concept of exon shuffling, whereas others do not. Homologous segments, which contain highly conserved amino acid sequence motifs for a protein, appear to play a very important role in the evolution of novel function *(103,147)*. Evolution of a new protein function can apparently be accompanied by a shift of the dominant selection pressure from one conserved motif to another. For example, the pattern of motif conservation in the ATP pyrophosphatases and phosphoadenosine-phosphosulfate (PAPS) reductases exposes several general aspects of enzyme evolution in which the different folds in these enzymes converge toward a similar function *(148,149)*. Once a functional motif is established, gene duplication, subsequent mutational modifications, and gene fusion provide the basis for the spread of domains.

Phylogenetic analysis based on alignment studies indicates that the ancestral extradiol dioxygenase was a one-domain enzyme and that the two-domain enzymes arose from a single genetic duplication event *(150)*. Subsequent divergence among the two-domain dioxygenases has resulted in several families, two of which are based on substrate selection *(151)*. In several other cases, the two domains of a given enzyme express different phylogenies, suggesting the possibility that such enzymes arose from the recombination of genes encoding different dioxygenases *(152)*. Nonetheless, studies on *E. coli* genes have shown that the process of duplication and divergence are not the only mechanism of evolution. Relationships among the sequences of multiple enzymes indicate that nearly half could have arisen either by convergent evolution or by lateral transfer *(153,154)*.

4. Significant Short Words—Methodology

The incidence of peptide "words" of length 3 to 5 amino acid residues was counted. Significantly overrepresented words were detected using the formula described by Karlin et al. *(155)* but previously employed only on DNA sequence data. The use of Karlin's method tests each peptide word for a significantly more frequent occurrence against the assumptions of a Poisson distribution and a random sequence by using a known abundance for each amino acid letter (f_i) and a known total length for the ORFs.

In practice, Karlin's method determines the copy number threshold, r_w, as the smallest integer satisfying the inequality:

$$\exp(-p_w L)\, (p_w L)^{r_w}/r_w! \leq 1/L \tag{1}$$

where $p_w = f_{i1} f_{i2} \ldots f_{is}$, with f_i defined as above. L represents the sequence length and is the word length given by inequality (2) which applies in the case of an amino acid alphabet of size A containing letters with uniform frequency $1/A$.

$$A^{s-1} \leq L < A^s \tag{2}$$

The short words thus discovered were used as input for a program based on the "word-up" algorithm *(156)*, in which overlapping words were replaced with their common superstring, if it produced a higher level of significance than that obtained for the initial string (using the same formula, as outlined above, to calculate significance).

Since the "word-up" output still contained families of similar words, a "word-up" step was added, in which words were eliminated from the set, if the set also included a sub-word that was more significant than the word in question.

4.1. Occurrence of Frequent Amino Acid Strings ("Words") Within the Genome of Haemophilus influenzae

The frequent word list of *H. influenzae* contained 169 sequences ranging from 3 to 18 amino acid residues in length and varying in statistical significance (p value) from 1.41e-233 to 1.78e-06 (data not shown). A search revealed that most of the words (99%) occurred more than once in the *H. influenzae* genome (**Table 1**). The missing 1% was accounted for by the fact that only the top 100 hits were picked from the genome. A further 80% of the words were also found to reside within the proteins of other organisms, most prominent being the ATP/GTP binding proteins. This observation alone suggests that these words are likely to be important within proteins. Their higher than expected frequency is indicative of occurrence due to more than chance and probably demonstrative of building blocks with inherent structural and/or functional significance within protein molecules. They recur in multiple numbers within the genome of other diverse organisms from prokaryotes to archaea,

Table 1
Occurrence of *Haemophilus influenzae* Frequent Words Found Elsewhere in the NR Protein Database, as Determined by FASTA Analysis

Sequence length (aa)	No. of sequences	Frequency of paralogy (in *H. influenzae*) (100% identity)	Frequency of orthology (elsewhere) (100 % identity)	Frequency of orthology with >70% identity
18	1	2	—	2
17	2	4	—	2–6
16	1	2	—	—
14	5	2	12	7
13	2	2	—	—
12	1	2	—	—
11	2	2	—	—
10	6	2–6	1–4	2–10
9	7	2–3	6–76	7–57
8	17	2–5	96	2–72
7	12	2–9	28–100	2–68
6	4	5–22	1–66	41–67
5	5	4–31	1–57	1–4
4	18	68–100	31	4–31
3	86	>100	>100	>100

and hence one can imply a special relationship between proteins of these organisms. This wide distribution may be the result of widespread horizontal gene transfer, as evidenced by the use of similar building blocks in all living organisms.

Eighteen of the words (greater than 5 residues long) were found in proteins of *H. influenzae* as well as in other organisms (**Table 2**). Close analysis of PROSITE patterns revealed them to occur within, or in close association with, important protein family signatures. For example, a sequence "AHVDCPGHA," which is part of the elongation factor family of proteins, was found to reside 100 amino acids upstream of a previously known motif, the casein kinase II phosphorylation signature. In some cases, the words were found to flank already characterized motifs, whereas in others, they were found to be on either side of these motifs. An example is the short sequence "DTPGHG," which is found in the ATP/GTP binding proteins, resides 6 amino acids upstream of the ATP transporters family signature motif, and is also about 100 amino acids upstream of ATP/GTP-binding site motif A (P-loop).

Sixteen of the words were found to occur only in *H. influenzae* proteins (**Table 3**). Understandably, these are scattered within a wide range of proteins ranging from proteins associated with nitrogen fixation to cell division. Associated PROSITE patterns were also diverse by comparison with words occur-

Table 2
Significant Orthologous Words Greater than Five Amino Acid Residues in Length Occurring in *H. influenzae* and Elsewhere

Sequence	Distribution/ protein(s) function	Associated PROSITE patterns
ELGWGGWWFWDPVE	Cytochrome c protein	FKBP-type peptidyl-prolyl cis-trans isomerase signature 1
QRVALAR	ATP binding proteins	ATP-GTP-binding site motif A (P-loop)
SGGQQQR	Transport ATP binding	ABC transporters family signature
LLDEPFSALD	ATP/GTP binding protein	100 aa upstream of ATP-GTP-binding site motif A (P-loop); 6 aa upstreams of ATP transporters family signature
ILLDEP	Cytochrome c protein	ATP-GTP-binding site motif A (P-loop) ATP transporters family signature
DGPMPQTR	ATP-binding protein	ATP-GTP-binding site motif A (P-loop) ATP transporters family signature
VATDVAARG	RNA helicase protein	DEAD-box subfamily ATP-dependent helicases signature
QRVAIARAL	ATP-binding protein	ATP-GTP-binding site motif A (P-loop)
DEATSALD	ATP-binding protein	ATP-GTP-binding site motif A (P-loop) ABC transporters family signature
LLIADEPT	Peptide transport ATP-binding protein	100 aa upstreat of ATP/GTP-binding site motif A (P-loop)
TCIDCHKGI	Cytochrome c protein	Within the cytochrome c family heme-binding site signature
REKADPY	Transferrin-binding protein	14–15 aa of tyrosine kinase phosphorylation site
AHVDCPGHA	Elongation factor (EF-Tu)	100 aa upstream of casein kinase II phosphorylation signature
SGGQRQR	Leukotoxin secretion protein	100 aa upstream of ATP-GTP-binding site motif A (P-loop); GTP-binding elongation factor signature
SGYAVR	Pyruvate formate lyase protein	Within glycine radical signature
GPSGAGK	Arginine transport protein	ATP-GTP-binding site motif A (P-loop)
TFKHPDF	Elongation factor protein	ATP-GTP-binding site motif A (P-loop); 8 aa upstream of GTP-binding elongation factor signature
DTPGH	GTP-binding, initiation factor, EP-Tu factor	ATP-GTP-binding site motif A (P-loop); GTP-binding elongation signature (21 aa upstream of)

Table 3
Significant Paralogous Words Greater than Five Amino Acid Residues in Length Occurring in *H. influenzae*

Sequence	Protein/function	Associated PROSITE patterns
NRFYAPGRNY	Iron binding protein	Occurs within Ton B-dependent protein signature
RSDEYCHQSTCNG	Transferrin ginding protein	Ton B-dependent protein signature
VRNRFDCWISVD	Dihydropteroate synthase	21 aa upstream of dihydropterone synthase signater 2
AIPKGQWE	Arginine and glutamine transport synthase permease	Within the binding protein-dependent transport systems' inner membrane composition signature
LPVCIMHMQGQ	Dihydropteroate synthase	Upstream of dihydropteroate synthase signature 1 and 2
VTWDMFQMHA	Xanthine guanine phosphoribosyl transferase	Downstream of purine/pyrimidine phosphoribosyl transferase signature
TWDSARS	Lactoferrin ginding and transferrin binding proteins	18 aa downstream of Ton B-dependent receptor protein signature 2
MSVQFEF	Lactoferrin binding protein	Ton B-dependent receptor protein signature 2
YLQMSAWC	Ferritin-like protein	Adjacent to leucine zipper pattern
WHETAKKC	Nitrogen fixation protein	Downstream of aminotransferases class-V pyridoxal-phosphate attachment site
CIAASYHDH	Xanthine-guanine phosphoribosyl transferase	21 aa downstream of purine/pyrimidine phosphoribosyl transferases signature
MHMQGQPR	Folate protein	Upstream of dihydropterone synthase signature 1 and 2
NWNTVKWN	Nodulation, hypothetical proteins	Next to tyrosine kinase phosphorylation site
EWAWRIPF	Hypothetical	Downstream of sugar transport signature
DYKHHSYN	Transferrin binding protein 1 and 2 proteins	Downstream of Ton B-dependent receptor protein signature 2
ISQKQLE	Cell protein division protein	With protein kinase c phosphorylation site

ring in different organisms. One interesting example is the leucine zipper pattern, which is accommodated in the ferritin-like protein. This well-known motif was found adjacent to "YLQMSAWC," an important motif involved in DNA binding. Thus, the likelihood of this or other relationships that appear to exist within the proteins studied suggests that the relationships are unlikely to be due to chance alone. By contrast, the most revealing finding was the 20 words that appeared to be novel in *H. influenzae* and other organisms (**Table 4**). Of these, six are made up of repeating residues, the longest being 17 amino acids long. These repeats are currently unique to *H. influenzae* and may contribute to important functions, for example, replication and repair processes *(155)*. Other proteins harboring these novel words are mostly hypothetical but also include transport proteins.

In general, the analysis confirmed significant relationships between these short sequences and existing motifs. However, in many cases their function remains to be clarified. Moreover, the findings presented here indicate that the appearance of these short words shared by nonrelated proteins could be due to either sequence convergence or descent from a common ancestor. Conversely, the related proteins sharing these sequences may be the result of gene duplication. For the purpose of this study, it would make no difference how their evolutionary fates were determined. In both cases, the result is the identification of relationships with patterns of probable biological significance. The patterns detected may help cast light on the origin and/or function of proteins, including those with no known homologs or motifs. Once established as being of significance, such peptide sequences take on greater value in that they can be linked to probable partial protein function, independent of often misleading whole-gene nomenclature, and can further direct experimental design to refine knowledge of protein function. Because individual proteins can contain several such "building blocks," each conceivably possessing different ancestral origins, this information can be used to create an image of the functional potential of a given gene product.

4.2. Comparative Genomics Using Signature Peptides ("Significant Short Words")

Just as statistically significant signature peptides have virtues for automatic protein profiling with respect to small amino acid strings of both known and unknown functional significance, they also appear to present a novel approach for both intra- and intergenomic analysis. As shown above, signature peptides occur with different frequencies within and between genes and genomes. Their incidence in a particular genome has probably resulted principally from internal duplication events or origins shared with a particular biological lineage and/or most living organisms.

Table 4
Larger Frequent Words from *H. influenzae* of Unknown Function

Sequence	Annotation of host gene
PTNQPTNQPTNQPTNQP	Tranferrin-binding protein
SQSVSQSVSQSVSQSVSQ	Type 3 restriction modification enzyme
TDRQTDRQTDRQTD	Glycosyl transferase
QSINQSINQSINQS	Lipooligosaccharide biosynthesis protein
QPTNQPTNQPTNQPTNQ	Hypothetical, transferrin-binding protein
LTSELTSEL	Type 1 restriction enzyme specificity protein
VGSSSSP	Hypothetical
KVDGDQI	Cytochrome c-type protein
SSPISGI	Disulfide interchange, hypothetical
DQGWQYQM	DNA transposase
EPKFEPKFEP	Hypothetical
PILHSDQGW	Hypothetical
FDTFHFKCDTCRL	Hypothetical
WQENWQACDQ	Hypothetical
FETKDART	Transferrin-binding, heme-utilization protein
MVERCRQH	Hypothetical
WQYQMIGY	Hypothetical
ISQKQLE	Transferrin-binding 1 and 2 proteins
DHDHDHKHEHKHDH	Adhesin B precursor protein
LTSELTSEL	Transport, mercuric transport proteins

First, if the incidence of signature peptides in all genes within the *E. coli* genome is examined, it is notable that each gene can be ranked with respect to the preponderance of such peptides, i.e., from those containing the most shared attributes through to those containing the least (**Fig. 2**). Far from being simply of academic interest, this information may be of real relevance to selecting genes for inclusion in multiple subunit vaccines or for selecting targets for intervention strategies. Traditionally, such approaches have been directed toward one gene at a time. These genes are subsequently regrouped into a subunit vaccine. However, targeting genes rich in signature peptides offers a means of simultaneously targeting multiple genes. Genes with multiple shared peptides suggest that the likelihood is greatest that an intervention directed against such genes is also likely to have an effect on genes sharing similar vulnerable regions within other gene products, i.e., drug targets or the site being targeted by the immune defenses of a suitably primed host. At the other extreme are those genes that would appear to be most unique, as determined by their lack of shared signature peptides (not necessarily the same genes, as determined by whole gene comparison using bioinformatic tools based on FASTA or BLAST,

E. coli

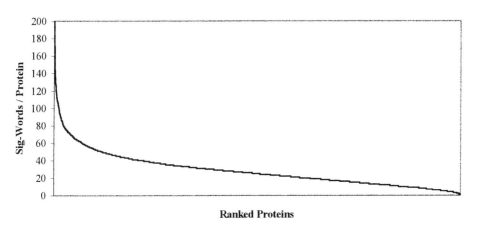

Fig. 2. Abundance of significant words ("signature peptides") per protein encoded by the genome of *Escherichia coli*. Each protein/open reading frame is positioned in order of decreasing rank from left to right across the x-axis.

for example). Here, if the object of an intervention strategy is not a broad-spectrum antibiotic, but rather a highly specific intervention, then the latter class of genes could provide ideal targets. An example here would be genes of interest in controlling *Helicobacter pylori*, the causative agent of stomach ulcers, as opposed to those genes that are likely to be shared by the normal flora, i.e., bacteria possessing highly desirable attributes for their hosts.

An analysis of signature peptides was conducted on 14 fully sequenced bacterial genomes (**Table 5**). Such significant words did not occur in all bacterial genomes with equal incidence (**Table 5**) and showed a decreasing incidence of shared occurrence between genomes (**Fig. 3**). Indeed, no such signature peptides were found to occur across all bacterial genomes. However, some were shared by up to 11 of 14 examined (**Fig. 3**), whereas the incidence of paralogous words across fully sequenced bacterial genomes is presented in **Fig. 4**. Variability was observed in the ratio of paralogous words not significant in other genomes and those not present in other genomes. The latter can be interpreted as having arisen entirely by intragenomic mechanisms, for example, duplication, error in replication, internal recombination, and transposition of genetic information.

In addition, the number of significant words unique to a particular genome was plotted against the number of significant words per genome and in turn as a function of the number of proteins/ORFs per genome (**Fig. 5**). This clearly demonstrated that some genomes possess an incidence that is disparate from that expected and is indicative, therefore, of varying degrees of peptide duplication

Table 5
Fully Sequenced Bacterial Genomes Examined for the Presence of Significant Words

Abbreviation	Bacterium	No. of proteins/ ORFs	No. of significant words	Paralogous words not significant in other genomes	Paralogous words not present in other genomes
Aa	Aguifex aeolicus	1522	240	66	0
Af	Archeaeglobus fulgidus	2409	365	174	36
Bs	Bacillus subtilis	4367	495	236	17
Bb	Borrelia burgdorferi	850	69	18	5
Ca	Clostridium acetobutylicum	4018	289	132	34
Ec	Escherichia coli	4289	537	285	67
Hi	Haemophilus influenzae	1717	140	35	6
Hp	Helicobacter pylori	1577	423	264	62
Mth	Methanobacterium thermoautotrophicum	1871	295	128	17
Mj	Methanococcus jannaschii	1735	286	121	13
Mtb	Mycobacterium tuberculosis	3918	538	413	149
Mg	Mycoplasma genitalium	467	81	37	6
Mp	Mycoplasma pneumoniae	677	300	271	147
Sx	Synechocystis species	3169	398	236	64

within different genomes. It can be speculated that the functional utility of this is to assist in confounding host defenses and may be linked to the nature of the host/parasite interface and the associated need for antigenic variation. Gene duplication is a good example, but it can confound analyses based on signature peptides, as highly similar genes produce large numbers of apparently similar peptides as a direct result of their sequence similarity. Signature peptides are most relevant as shared regions within genes with sequence similarity below 30–35%, where other homology detection methodologies tend to work less efficiently. Although not instigated in the present study, analysis of the *Mycobacterium tuberculosis* genome, rich in whole gene duplications and

Signature Peptides

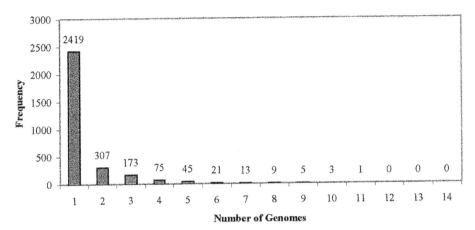

Fig. 3. Maximum number of significant words common to one or more bacterial genomes.

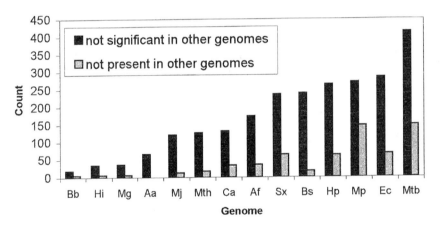

Fig. 4. Paralogous significant words per protein/open reading frame encoded by each of 14 bacterial genomes.

therefore also in significant words (**Fig. 4**), may demand that a cutoff measure for the exclusion of gene duplications be introduced, so as to exclude these events from paralogous word detection.

5. Conclusions

Just as the evolution of genomics as a respectable scientific discipline has been intimately linked with the evolution of bioinformatics for initially stitching together contiguous DNA sequence data and subsequently for providing gene detection and annotation, so too the evolution of proteomics as a respectable

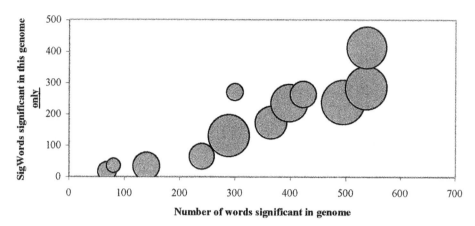

Fig. 5. Number of significant words unique to a particular genome plotted against the number of significant words per genome. The diameter of each circle represents the number of proteins/open reading frames per genome. Note that there are a number of population outliers distinct from the expected linear regression.

scientific discipline will be associated with the development of dedicated bioinformatic tools. Here the task will be to link data obtained experimentally with that indicative of function so as to facilitate improved protein annotation. At present, most signature peptides (small amino acid words) detected remain associated with unknown structure and function. There are nonetheless many similarities between this approach and that of chemometrics. An example of the latter would be the analysis of petroleum from Bahrain, Nigeria, California, and Colombia. Comparison of the chromatograms obtained from each, in conjunction with prior knowledge, can be used to infer that those with a particular profile or possessing one or two particular peaks in the profile give rise to excellent carburants after refining. These peaks may or may not be associated with known fractions, but they provide the purchaser of crude oil with some value-added information of commercial worth. Functional genomics is the seeking out that value-added information, as an add-on to DNA sequence information, which then needs to be confirmed experimentally. The data presented here have been chosen with a view to demonstrating the considerable potential for proteomics, and appropriately structured databases, to play an increasingly important role in functional genomics and most certainly going beyond that first step of linking protein sequence to a previously sequenced piece of DNA.

Acknowledgments

We would like to thank Peter Maxwell for computational assistance and indispensable help in data handling and analysis. This work was supported by grants from Glaxo Wellcome, Australia.

References

1. Eng, J. K., McCormack, A. L., and Yates, III, J. R. (1994) An approach to correlate tandem mass spectral data of peptides with amino acid sequences in a protein database. *J. Am. Soc. Mass Spectrom* **5**, 976–989.
2. Mann, M. and Wilm, M. (1994) Error-tolerant identification of peptides in sequence databases by peptide sequence tags. *Anal. Chem.* **66**, 4390–4399.
3. Cordwell, S. J., Wilkins, M. R., Cerpa-Poljak, A., Gooley, A. A., Duncan, M., Williams, K. L., et al. (1995) Cross-species identification of proteins separated by two-dimensional gel electrophoresis using matrix-assisted laser desorption ionisation/time-of-flight mass spectrometry and amino acid composition. *Electrophoresis* **16**, 438–443.
4. Wise, M. J., Littlejohn, T., and Humphery-Smith, I. (1997) Peptide-mass fingerprinting and the ideal covering set for protein characterisation. *Electrophoresis* **18**, 1399–1409.
5. Wise, M. J., Littlejohn, T., and Humphery-Smith, I. (1997) Better cutters for protein mass fingerprinting: preliminary findings. *ISMB* **5**, 340–343.
6. Humphery-Smith, I., and Blackstock, W. P. (1997) Proteome analysis: genomics via the output rather than input code. *J. Protein Chem.* **16**, 537–544.
7. Cordwell, S. J., Wasinger, V. C., Cerpa-Poljak, A., Duncan, M., and Humphery-Smith, I. (1997) Conserved domains as the basis of recognition for homologous proteins identified across species boundaries using peptide-mass fingerprinting. *J. Mass Spectrom.* **32**, 373–378.
8. Humphery-Smith, I., Cordwell, S. J., and Blackstock, W. P. (1997) Proteome research, complementarity and limitations with respect to the DNA and RNA worlds. *Electrophoresis* **18**, 1217–1242.
9. Cordwell, S. J., Basseal, D., Cerpa-Poljak, A., and Humphery-Smith, I. (1997) Malate/lactate dehydrogenase in mollicutes: evidence for a multienzyme proteins. *Gene* **195**, 113–120.
10. Kanehesa, M. (1998) Grand challenges in bioinformatics. *Bioinformatics* **14**, 309.
11. O'Brien, C. (1996) Protein fingerprints, proteome projects and implications for drug discovery. *Mol. Med. Today* **28**, 316.
12. Fleischmann, R. D., Adams, M. D., White, O., Clayton, R. A., Kirkness, E. F., Kerlavage, A. R., et al. (1995) Whole-genome random sequencing and assembly of *Haemophilus influenzae* Rd. *Science* **269**, 496–512.
13. Blattner, F. R., Plunkett, G., II, Bloch, C. A., Perna, N. T., Burland, V., Riley, M., et al. (1997) The complete genome sequence of *Escherichia coli* K-12. *Science* **277**, 1453–1462.
14. Bult, C. B., White, O., Olsen, G. J., Zhou, L., Fleischmann, R. D., Sutto, G. G., et al. (1996) Complete Genome Sequence of the Methanogenic Archeon, *Methanococcus jannaschii*. *Science* **273**, 1058–1073.
15. Kaneko, T., Sato, S., Kotani, H., Tanaka, A., Asamizu, E., Nakamura, Y., et al. (1996) Sequence analysis of the genome of the unicellular *Cyanobacterium synechocystis sp.* strain PCC6803. I. Sequence determination of the entire genome and assignment of potential protein-coding regions. *DNA Research* **3**, 109–136.

16. Fraser, C. M., Casjens, S., Huang, W. M., Sutten, G. G., Clayton, R., Lathigra, R., et al. (1997) Genomic sequence of a Lyme disease spirochaete, *Borrelia burgdorferi*. *Nature* **390**, 580–586.
17. Fraser, C. M., Gocayne, J. D., White, O., Adams, M. D., Clayton, R. A., Fleiscmann, R. D., et al. (1995) The Minimal gene complement of *Mycoplasma genitalium*. *Science* **270**, 397–403.
18. Himmelreich, R., Hilbert, H., Plagens, H., Pirkl, E., Li, B. C., and Herrmann, R., et al. (1996) Complete sequence analysis of the genome of the bacterium *Mycoplasma pneumoniae*. *Nucleic Acids Res.* **24**, 4420–4451.
19. Tomb, J., White, O., Kerlevage, A., Clayton, R., Sutten, G., Fleishmann, R., et al. (1997) The complete genome sequence of the gastric pathogen *Helicobacter pylori*. *Nature* **388**, 539–547.
20. Goffeau, A., Barrell, B. G., Bussey, H., Davis, R. W., Dujon, B., Feldmann, H., et al. (1996) Life with 6000 genes. *Science,* **274**, 546–567.
21. Klenk, H. P., White, O., Tomb, J. F., Clayton, R. A., Nelson, K. E., Ketchum, K. E., et al. (1997) The complete genome sequence of the hyperthermophilic sulfate-reducing archeon *Archaeoglobus fulgidus*. *Nature* **390**, 364–370.
22. Smith, D. R., Doucette-Stamm, L. A., Deloughery, C., Dubous, J., Aldredge, T., Bashirzadeh R. *et al.* (1997) Complete genome sequence of *Methanobacterium thermoautrophicum* deltaH: functional analysis and comparative genomics. *J. Bacteriol.* **179**, 7135–7155.
23. Kunst, F., Ogasawara, N., Moszer, I., Albertini, A. M., Alloni, G., Azevedo, V., et al. (1997) The complete genome sequence of the gram-positive bacterium *Bacillus subtilis*. *Nature* **390**, 249–256.
24. Gardner, M. J., Tettelin, H., Carucci, D. J., Cummings, L. M., Aravind, L., Koonin, E. V., et al. (1998) Chromosome 2 sequence of the human malaria parasite *Plasmodium falciparum*. *Science* **282**, 1126–1132.
25. Andersson, S. G., Zomorodipour, A., Andersson, J. O., et al. (1998) The genomesequence of *Rickettsia prowazekii* and the origin of mitochondria. *Nature* **396**, 133–140.
26. Fraser, C. M., Norris, S. J., Weinstock, G. M., White, O., Sutton, G. G., Dodson, R., et al. (1998) Complete genome sequence of *Treponema pallidum*, the syphilis spirochete. *Science* **281**, 375–388.
27. Cole, S. T., Brosch, R., Parkhill, J., Garnier, T., Churcher, C., Harris, D., et al. (1998) Deciphering the biology of *Mycobacterium tuberculosis* from the complete genome sequence. *Nature* **393**, 537–544.
28. Kawarabayasi, Y., Sawada, M., Horikawa, H., Haikawa, Y., Hino, Y., Yamamoto, S., et al. (1998) Complete sequence and gene organization of the genome of a hyperthermophilicarchae bacterium *Pyrococcus horikoshii* OT3 (supplement*). DNA Res* **5**, 147–155.
29. Stephens, R. S., Kalman, S., Lammel, C., Fan, J., Marathe, R., Aravind, L., et al. (1998) Genome sequence of an obligate intracellular pathogen of humans: *Chlamydia trachomatis*. *Science* **282**, 754–759.

30. Deckert, G., Warren, P. V., Gaasterland, T., Young, W. G., Lenox, A. L., Graham, D. E., et al. (1998) The complete genome of the hyperthermophilic bacterium *Aquifex aeolicus*. *Nature* **392**, 353–355
31. Huynen, M. A., and Nimwegen, E. V. (1998) The frequency distribution of gene family sizes in complete genomes. *Mol Biol Evol* **15**, 583–589.
32. Flores, A. I. and Cuezva, J. M. (1997) Identification of sequence similarity between 60 kDa and 70 kDa molecular chaperones: evidence for a common evolutionary background? *Biochem. J.* **2**, 641–647.
33. Doolittle, R. F., Feng Da-Fei, Tsang, S., Cho, G., and Little, E. (1996) Determining divergence times of the major kingdoms of living organisms with a protein clock. *Science* **271**, 70–477.
34. Arber, W. (1995) The generation of variation in bacterial genomes. *J. Mol. Evol.* **40**, 7–12.
35. Tsinoremas, N. F., Kutach, A. K., Strayer, C. A., and Golden, S. S. (1994) Efficient gene transfer in *Synechococcus sp.* strain PCC 7942 and PCC 6301 by interspecies conjugation and chromosomal recombination. *J. Bacteriol.* **176**, 6764–6768.
36. Lan, R. and Reeves, P. R. (1996) Gene transfer is a major factor in bacterial evolution. *Mol. Biol. Evol.* **13**, 47–55.
37. Guttman, D. S., and Dykhuizen, D. E. (1994) Clonal divergence in *Escherichia coli* as a result of recombination. *Science* **266**, 1380–1383.
38. Voelker, L. L. and Dybvig, K. (1996) Gene transfer in *Mycoplasma arthritidis*: transformation, conjugal transfer of Tn916, and evidence for a restriction system recognizing AGCT. *J. Bacteriol.* **178**, 6078–6081.
39. Radicella, J. P., Park, P. U., and Fox, M. S. (1995) Adaptive mutation in *Escherichia coli*: a role for conjugation. *Science* **268**, 418–420.
40. Cohen, S. N. (1993) Bacterial plasmids: their extraordinary contribution to molecular genetics. *Gene* **135**, 67–76.
41. Cheetham, B. F. and Katz, M. E. (1995) A role for bacteriophages in the evolution and transfer of bacterial virulence determinants. *Mol. Microbiol.* **18**, 201–208.
42. Kim, J. M., Vanguri, S., Boeke, J. D., Gabriel, A., and Voytas, D. F. (1998) Transposable elements and genome organization: a comprehensive survey of retrotransposons revealed by the complete *Saccharomyces cerevisiae* genome sequence. *Genome Res.* **8**, 464–478.
43. Maurelli, A. T., Frenemdez, R. E., Bloch, C. A., Rode, C. K., and Fasano, A. (1998) "Black holes" and bacterial pathogenicity: a large genomic deletion that enhances the virulence of *Shigella* spp. and enteroinvasive *Escherichia coli*. *Proc. Natl. Acad. Sci. USA* **95**, 3943–3948.
44. Aoyama, Y., Noshiro, M., Gotoh, O., Imaoka, S., Funae, Y., Kurosawa, N., et al. (1996) Sterol 14-demethylase (P45014DM) is one of the most ancient and conserved P450 species. *J. Biochem.*, **119**, 926–933
45. Green, P., Lipman, D., Hillier, L., Waterston, R., States, D., and Claverie, J. M. (1993) Ancient conserved regions in new gene sequences and the protein databases. *Science* **259**, 1711–1716.

46. Thode, G., Garcia-Ranea, J. A., and Jimenez, J. (1996) Search for ancient patterns in protein sequences. *J. Mol. Evol.* **42,** 224–233.
47. Koonin, E. V., Tatusov, R. L., and Rudd, K. E. (1995) Sequence similarity analysis of *Escherichia coli* proteins: functional and evolutionary implications. *Proc. Natl. Acad. Sci. USA* **92,** 11921–11925.
48. Aitken, A., Collinge, D. B., van Heusden B. P. H., Isobe, T., Roseboom, P. H., Rosenfeld, G., and Soll, J. (1992) 14-3-3 proteins: a highly conserved, widespread family of eukaryotic proteins. *Trends Biochem. Sci.* **17,** 498–501.
49. Baxevanis, A. D. and Landsman, D. (1996) Histone Sequence Database: a compilation of highly-conserved nucleoprotein sequences. *Nucleic Acid Res.* **24,** 245–247.
50. Keeling, P. J. and Doolittle, W. F. (1995) Archaea: narrowing the gap between prokaryotes and eukaryotes. *Proc. Natl. Acad. Sci. USA* **92,** 5761–5764.
51. Henikoff, S. and Henikoff, J. G. (1991) Automated assembly of protein blocks for database searching. *Nucleic Acids Res.* **19,** 6565–6572.
52. Koonin, E. V., Mushegian, A. R., Galperin, M. Y., and Walker, D. R. (1997) Comparison of archaeal and bacterial genomes: computer analysis of protein sequences predicts novel functions and suggests a chimeric origin for the archaea. *Mol. Biol.* **25,** 619–937.
53. Labedan, B. and Riley, M. (1995) Gene products of *Escherichia coli*: sequence comparisons and common ancestries. *Mol. Biol. Evol.* **12,** 980–987.
54. Labedan, B. and Riley, M. (1995) Widespread protein sequence similarities: origin of *Escherichia coli* genes. *J. Bacteriol.* **177,** 1585–1588.
55. Mushegian, A. R. and Koonin, E. (1996) A minimal gene set for cellular life derived by comparison of complete bacterial genomes. *Proc. Natl. Acad. Sci. USA* **93,** 10,268–10,273.
56. Renaud de, R., and Labedan, B. (1998) The evolutionary relationships between the two bacteria *Escherichia coli* and *Haemophilus influenzae* and their putative last common ancestor. *Mol. Biol. Evol.* **15,** 17–27.
57. Brenner, S. E., Hubbard, T., Murzin, A., and Chothia, C. (1995) Gene duplications in *H. influenzae*. *Nature* **378,** 140.
58. Koonin, V. K., and Galperin, Y. (1997) Prokaryotic genomes: the emerging paradigm of genome-based microbiology. *Curr. Biol.* **7,** 757–763.
59. Koonin, E. V., Tatusov, R. L., and Rudd, K. E. (1996) Protein sequence comparison at genome scale. *Methods Enzymol.* **266,** 295–322.
60. Koonin, E. V. and Mushegian, A. R. (1996) Complete genome sequences of cellular life forms: glimpses of theoretical evolutionary genomics. *Curr. Opin. Genet. Dev.* **6,** 757–762.
61. Koonin, E. V. (1997) Evidence for a family of archaeal ATPases. *Science* **275,** 1489–1490.
62. Koonin, E. V., Mushegian, A. R., and Rudd, K. E. (1996) Sequencing and analysis of bacterial genomes. *Curr. Biol.* **6,** 404–416.
63. Gilbert, W., de Souza, S. J., and Long, M. (1997) Origin of genes. *Proc. Natl. Acad. Sci. USA* **94,** 7698–7703.

64. Long, M., de Souza, S. J., and Gilbert, W. (1995) Evolution of the intron-exon structure of eukaryotic genes. *Curr. Opin. Genet. Dev.* **5,** 774–778.
65. Long, M., Rosenberg, C., and Gilbert, W. (1995) Intron phase correlations and the evolution of intron/exon structure of genes. *Proc. Natl. Acad. Sci. USA* **92,** 12,495–12,499.
66. Patthy, L. (1991) Exons—original building blocks of proteins? *Bioessays* **13,** 187–192.
67. Patthy, L. (1994) Introns and exons. *Curr. Opin. Struct. Biol.* **222,** 10–21.
68. Doolittle, R. F. (1995) The multiplicity of domains. *Annu. Rev. Biochem.* **64,** 287–314.
69. Baron, M., Norman, D. G., and Campbell, I. D. (1991) Protein modules. *Trends Biochem. Sci.* **16,** 13–17.
70. Doolittle, R. F., and Bork, P. (1993) Evolutionarily mobile modules in proteins. *Sci. Am.* **269(4),** 50–56.
71. Bork, P., Downing, A. K., Kieffer, B., and Campbell, I. D. (1996) Structure and distribution of modules in extracellular proteins. *Q. Rev. Biophys.* **29,** 119–167.
72. Henrissat, B. and Bork, P. (1996) On the classification of modular proteins. *Protein Eng.* **9,** 725–726.
73. Sonnhammer, Eric L. L. and Khan, D. (1994) Modular arrangement of proteins as inferred from analysis of homology. *Protein Sci.* **3,** 482–492.
74. Pfuhl, M., Improta, S., Politou, A. S., and Pastore, A. (1997) When a module is also a domain: the role of the N-terminus in the stability and the dynamics of immunoglobulin domains from titin. *J. Mol. Biol.* **265,** 242–256.
75. Bork, P. (1991) Shuffled domains in extracellular proteins. *FEBS Lett.* **286,** 47–54.
76. Bork, P. (1992) Mobile modules and motifs. *Curr. Opin. Struct. Biol.* **2,** 413–421.
77. Pao, G. M. and Saier, G. M., Jr (1995) Response regulators of bacterial signal transduction systems: selective domain shuffling during evolution. *J. Mol. Evol.* **40,** 136–154.
78. Bork, P. and Bairoch, A. (1995) A proposed nomenclature for the extracellular protein modules of animals. *Trends Biochem. Sci.* **20,** poster C02.
79. Bork, P., and Doolittle, R. F. (1992) Proposed Acquisition of an Animal Protein Domain by Bacteria. *Proc. Natl. Acad. Sci. USA* **89,** 8990–8994.
80. Ponting, C. P. (1997) Evidence for PDZ domains in bacteria, yeast, and plants. *Protein Sci.* **6,** 464–468.
81. Neer, E. J., Schmidt, C. J., Nambudripad, R., and Smith, T. F. (1994) The ancient regulatory-protein family of WD-repeat proteins. *Nature* **371,** 297–300.
82. Boddy, M. N., Freemont, P. S., and Borden K.L.B. (1994) The p53-associated protein MDM2 contains a newly characterized zinc-binding domain called the RING finger. *Trends Biochem. Sci.* **19,** 198–199.
83. Musacchio, A., Gibson, T., Rice, P., Thompson, J., and Saraste, M. (1993) The PH domain: a common piece in the structural patchwork of signalling proteins. *Trends Biochem. Sci.* **18,** 343–348.
84. Waksman, G., Kominos, D., Robertson, S. C., Pant, N., Baltimore, D., Birge, R. B., et al. (1992) Crystal structure of the phosphotyrosine recognition domain SH2 of v-src complexed with tyrosine-phosphorylated peptides. *Nature* **358,** 646–653.

85. Rudd, P. M. and Dwek, R. A. (1997) Glycosylation: heterogeneity and the 3D structure of proteins. *Crit. Rev. Biochem. Mol. Biol.* **32,** 1–100.
86. Mendoza, L. M., Nishioka, D., and Vacquier, V. D. (1993) A GPI-anchored sea urchin membrane protein containing EGF domains is related to uromodulin. *J. Cell Biol.* **121,** 1291–1297.
87. Derrick, J. P. and Wigley, D. B. (1994) The third IgG-binding domain from streptococcal protein, G., An analysis by X-ray crystallography of the structure alone and in a complex with Fab. *J. Mol. Biol.* **243,** 906–918.
88. Hughes, A. L. (1997) Rapid evolution of immunoglobulin superfamily C2 domains expressed in immune system cells. *Mol. Biol. Evol.* **14,** 1–5.
89. Spitzfaden, C., Grant, R. P., Mardon, H. J., and Campbell, I. D. (1997) Module-module interactions in the cell binding regions of fibronectin: stability, flexibility and specificity. *J. Mol. Biol.* **265,** 565–579.
90. Higgins, D. G., Labeit, S., Gautel, M., and Gibson, T. J. (1994) The evolution of titin and related giant muscle proteins. J Mol Evol **38,** 395–404.
91. Foster, S. J. (1993) Molecular analysis of three major wall-associated proteins of *Bacillus subtilis* 168: evidence for processing of the product of a gene encoding a 258 kDa precursor two-domain ligand-binding protein. *Mol. Microbiol.* **8,** 299–310.
92. Riley, M. and Labedan, B. (1997) Protein evolution viewed through *Escherichia coli* protein sequences: introducing the notion of a structural segment of homology, the module. *J. Mol. Biol.* **268,** 857–868.
93. Hawkins, A. R., and Lamb, H. K. (1995) The molecular biology of multidomain proteins, selected examples. *Eur. J. Biochem.* **232,** 7–17.
94. Dejgaard, K., and Leffers, H. (1996) Characterization of the nucleic-acid-binding activity of KH domains. Different properties of different domains. *Eur. J. Biochem.* **241,** 425–431.
95. Savill, J. (1997) Molecular genetic approaches to understanding disease. *BMJ* **314,** 126–129.
96. Thomas, P. J., Qu, B. H., and Pedersen, P. L. (1995) Defective protein folding as a basis of human disease. *Trends Biochem. Sci.* **20,** 456–460.
97. Engel, A. M., Cejka, Z., Lupaz, L., Lottspeich, F., and Baumeister, W. (1992) Isolation and cloning of Omp alpha, a coiled-coil protein spanning the periplasmic space of the ancestral eubacterium *Thermotoga maritima*. *EMBO J.* **11,** 4319–4378.
98. Lupaz, A., Engelhardt, H., Peters, J., Santarius, U., Volker, S., and Baumeister, W. (1994) Domain structure of the *Acetogenium kivui* surface layer revealed by electron crystallography and sequence analysis. *J. Bacteriol.* **176,** 1224–1233.
99. Bork, P., Ouzous, C., and McEntyre, J. (1995) Ready for a motif submission? A proposed checklist. *Trends Biochem. Sci.* **20,** 104.
100. Koonin, E. V., and Van der Vies, S. M. (1995) Conserved sequence motifs in bacterial and bacteriophage chaperonins. *Trends Biochem. Sci.* **20,** 14–15.
101. Bork, P., and Koonin, E. V. (1996) Protein sequence motifs. *Curr. Opin. Struct. Biol.* **6,** 366–376.

102. Bork, P., and Koonin, E. V. (1994) A P-loop-like motif in a widespread ATP pyrophosphatase domain: implications for the evolution of sequence motifs and enzyme activity. *Proteins* **20,** 347–355.
103. Koonin, E. V. (1995) A protein splice-junction motif in hedgehog family proteins. *Trends Biochem. Sci.* **20,** 141–142.
104. Takahata, N. (1996) Neutral theory of molecular evolution. *Curr. Opin. Genet. Dev.* **6,** 767–772.
105. Heinrichs, V. and Baker, B. S. (1997) *In vivo* analysis of the function of the Drosophila splicing regulator RBP1. *Proc. Natl. Acad. Sci. USA* **94,** 115–120.
106. Kelly, C. R., Dickinson, C. D., and Ruf, W. (1997) Calcium binding to the first epidermal growth factor module of coagulation factor VIIa is important for cofactor interaction and proteolytic function. *J. Biol. Chem.* **272,** 17,467–17,472.
107. Sticht, H., Pickford, A. R., Potts, J. R., and Campbell, I. D. (1998) Solution structure of the glycosylated second type-2 module of fibronectin. *J. Mol. Biol.* **276,** 177–187.
108. Sutherland, J. A., Cook, A., Bannister, A. J., and Kouzarides, T. (1992) Conserved motifs in Fos and Jun define a new class of activation domain. *Genes Dev.* **6,** 1810–1819.
109. Neuwald, A. F. and Green, P. (1994) Detecting patterns in protein sequences. *J. Mol. Biol.* **239,** 698–712.
110. Neuwald, A. F., Liu, J. S., and Lawrence, C. E. (1995) Gibbs motif sampling: detection of bacterial outer membrane protein repeats. *Protein Sci.* **4,** 1618–1632.
111. Rost, B. and Sander, C. (1994) Combining evolutionary information and neural networks to predict protein structure. *Proteins* **19,** 55–72.
112. Sogin, M. L. (1991) Early evolution and the origin of eukaryotes. *Curr. Opin. Genet. Dev.* **1,** 457–463.
113. Clark, A. G. (1994) Invasion and maintenance of a gene duplication. *Proc. Natl. Acad. Sci. USA* **91,** 2950–2954.
114. Haack, K. R. and Roth, J. R. (1995) Recombination between chromosomal IS200 elements supports frequent duplication formation in *Salmonella typhimurium*. *Genetics* **141,** 1245–1252.
115. Holland, P. W., Garcia-Fernandez, J., Williams, N. A., and Sidow, A. (1994) Gene duplications and the origins of vertebrate development. *Dev. Supp.* 125–133.
116. Koonin, E. V. and Bork P (1996) Ancient duplication of DNA polymerase inferred from analysis of complete bacterial genomes. *Trends Biochem. Sci.* **21,** 128–129.
117. Sidow, A. (1996) Gen(om)e duplications in the evolution of early vertebrates. *Curr. Opin. Genet. Dev.* **6,** 715–722.
118. Gibbs P., E. M., Witke, W. F., and Dugaiczyk, A. (1998) The molecular clock runs at different rates among closely related members of a gene family. *J. Mol. Evol.* **46,** 552–561.
119. Wolfe, K. H. and Shields, D. C. (1992) Molecular evidence for an ancient duplication of the entire yeast genome. *Nature,* **387,** 708–713.
120. Keese, P. K. and Gibbs, A. (1992) Origins of genes: "big bang" or continuous creation? *Proc. Natl. Acad. Sci. USA* **89,** 9489–9493.

121. Ikehara, K. and Okazawa, E. (1993) Unusually biased nucleotide sequence on sense strands of Flavobacterium sp. genes produce nonstop frames on the corresponding antisense strands. *Nucleic Acids Res.* **21,** 2193–2199.
122. Ikehara, K., Amada, F., Yoshida, S., Mikata, Y., and Tanaka, A. (1996) A possible origin of newly-born genes: significance of GC-rich nonstop frame on antisense strands. *Nucleic Acids Res.* **24,** 4249–4255.
123. Tsuji, S., Qureshi, M. A., Hou, E. W., Fitch, W. M., and Li, S. S. (1994) Evolutionary relationships of lactate dehydrogenases (LDHs) from mammals, birds, an amphibian, fish, barley, and bacteria: LDH cDNA sequences from Xenopus, pig, and rat. *Proc. Natl. Acad. Sci. USA* **91,** 9392–9396.
124. Morton, B. R., Gaut, B. S., and Clegg, M. T. (1996) Evolution of alcohol dehydrogenase genes in the palm and grass families. *Proc. Natl. Acad. Sci. USA* **93,** 11,735–11,739.
125. Ikeo, K., Takahashi, K., and Gojobori, T. (1995) Different evolutionary histories of kringle and protease domains in serine proteases: a typical example of domain evolution. *J. Mol. Evol.* **40,** 331–336.
126. Beckmann, G. and Bork, P. (1993) An adhesive domain detected in functionally diverse receptors. *Trends Biochem. Sci.* **18,** 40–41.
127. Keeling, P. J., Charlebois, R. L., and Doolittle, W. F. (1994) Archaebacterial genomes: eubacterial form and eukaryotic content. *Curr. Opin. Genet. Dev.* **4,** 816–822.
128. Vrana, P. B. and Wheeler, W. C. (1996) Molecular evolution and phylogenetic utility of the polyubiquitin locus in mammals and higher vertebrates. *Mol. Phylogenet. Evol.* **6,** 259–269.
129. Weber, K., Schneider, A., Muller, N., and Plessmann, U. (1996) Polyglycylation of tubulin in the diplomonad *Giardia lamblia*, one of the oldest eukaryotes. *FEBS Lett.* **393,** 27–30.
130. Doolittle, R. F. (1994) Convergent evolution: the need to be explicit. *Trends Biochem. Sci.* **19,** 15–18.
131. Bork, P., Sander, C., and Valencia, A. (1993) Convergent evolution of similar enzymatic function on different protein folds: the hexokinase, ribokinase, and galactokinase families of sugar kinases. *Protein Sci.* **2,** 31–41.
132. Hubbard, T., Park, J., Leplae, R., and Tramontano, A. (1996) Protein structure prediction: playing the fold. *Trends Biochem. Sci.* **21,** 279–281.
133. Graumann, O. and Marahiel, M. A. (1996) A case of convergent evolution of nucleic acid binding modules. *Bioessays* **18,** 309–315.
134. Patthy, L. (1996) Exon shuffling and other ways of module exchange. *Matrix Biol.* **15,** 301–310.
135. Bork, P. (1993) Hundreds of ankyrin-like repeats in functionally diverse proteins: mobile modules that cross phyla horizontally. *Proteins* **17,** 363–374.
136. Baumler, A. J., Gilde, A. J., Tsolis, R. M., Van der Velden, A. W., Ahmer, B. M., and Heffron, F. (1997) Contribution of horizontal gene transfer and deletion events to development of distinctive patterns of fimbrial operons during evolution of *Salmonella* serotypes. *J. Bacteriol.* **179,** 317–322.

137. Delwiche, C. F. and Palmer, J. D. (1996) Rampant horizontal transfer and duplication of rubisco genes in eubacteria and plastids. *Mol. Biol. Evol.* **13,** 873–882.
138. Kehoe, M. A., Kapur, V., Whatmore, A. M., and Musser, J. K. (1996) Horizontal gene transfer among group A streptococci: implications for pathogenesis and epidemiology. *Trends Microbiol.* **4,** 436–443.
139. Moens, L., Vanflateren, J., Van de Peer, Y., Peters, K., Kapp, O., Czeluzniak, J., et al. (1996) Globins in nonvertebrate species: dispersed by horizontal gene transfer and evolution of the structure-function relationships. *Mol. Biol. Evol.* **13,** 324–333.
140. Quillet, L., Barray, S., Labedan, B., Peyit, F., and Guespin-Michel, J. (1995) The gene encoding the beta-1,4-endoglucanase (CelA) from *Myxococcus xanthus*: evidence for independent acquisition by horizontal transfer of binding catalytic domains from actinomycetes. *Gene* **158,** 23–29.
141. Robertson, H. M. and Lampe, D. J. (1995) Recent horizontal transfer of a mariner transposable element among and between Diptera and Neuroptera. *Mol. Biol. Evol.* **12,** 850–862.
142. Smith, M. W., Feng, D. F., and Doolittle, R. F. (1992) Evolution by acquisition: the case for horizontal gene transfers. *Trends Biochem. Sci.* **17,** 489–493.
143. Stratz, M., Mau, M., and Timmis, K. N. (1996) System to study horizontal gene exchange among microorganisms without cultivation of recipients. *Mol. Microbiol.* **22,** 207–215.
144. Syvanen, M. (1994) Horizontal gene transfer: evidence and possible consequences. *Annu. Rev. Genet.* **28,** 237–261.
145. de Souza, S. J., Long, M., Schoenbach, L., Roy, L., and Gilbert, W. (1996) Intron positions correlate with module boundaries in ancient proteins. *Proc. Natl. Acad. Sci. USA* **93,** 14,632–14,636.
146. Opavsky, R., Pastorekova, S., Zelnic, V., Gibadulinova, A., Stanbridge, E. J., Zavada, J., et al. (1996) Human MN?CA9 gene, a novel member of the carbonic anhydrase family: structyre and exon to protein domain relationships. *Genomics* **33,** 480–487.
147. Traut, W. (1994) The functions and consensus motifs of nine types of peptide segments that form different types of nucleotide binding sites. *Eur. J. Biochem.* **222,** 9–19.
148. Bork, P. and Koonin, E. V. (1994) A P-loop-like motif in a widespread ATP pyrophosphatase domain: implications for the evolution of sequence motifs and enzyme activity. *Proteins* **20,** 347–355.
149. Bork, P. and Doolitle, R. F. (1994) *Drosophila* kelch motif is derived from a common enzyme fold. *J. Mol. Biol.* **236,** 1277–1282.
150. Guigo, R., Mushnic, I., and Smith, T. F. (1996) Reconstruction of ancient molecular phylogeny. *Mol. Phylogenet. Evol.* **6,** 189–213.
151. Fryxell, K. J. (1996) The coevolution of gene family trees. *Trends Genet.* **12,** 365–369.
152. Eltis, L. D. and Bolin, J. T. (1996) Evolutionary relationships among extradiol dioxygenases. *J. Bacteriol.* **178,** 5930–5937.
153. Sakellaris, H., Balding, D. P., and Scott, J. R. (1996) Assembly proteins of CS1 pili of enterotoxigenic *Escherichia coli. Mol. Microbiol.* **21,** 529–541.

154. Aoyama, K., Haase, A. M., and Reeves, P. R. (1994) Evidence for effect of random genetic drift on G+C content after lateral transfer of fucose pathway genes to *Escherichia coli* K-12. *Mol. Biol. Evol.* **11,** 829–838.
155. Karlin, S., Mrazek, J., and Campbell, A. M. (1996) Frequent oligonucleotides and peptides of *Haemophilus influenzae* genome. *Nucleic Acids Res.* **24,** 4263–4272.
156. Pesole, G., Attimonelli, M., and Saccone, C. (1996) Linguistic analysis of nucleotide sequences: algorithms for pattern recognition and analysis of codon strategy. *Methods Enzymol.* **266,** 281–294.

6

Investigating the Higher Order Structure of Proteins

Hydrogen Exchange, Proteolytic Fragmentation, and Mass Spectrometry

John R. Engen and David L. Smith

1. Introduction

1.1. Exploring the Dynamics of Proteins

It has been apparent for some years that the structures of proteins are dynamic rather than static. For some proteins, dynamics is essential to function (e.g., refs. *1–7*). These structural changes have been detected for more than 30 yr by observing hydrogen exchange between peptide amide hydrogens and solvent containing the hydrogen isotopes tritium or deuterium *(8–10)*. Although tritium is no longer used extensively for this purpose, deuterium is widely used in hydrogen exchange studies, especially in multidimensional nuclear magnetic resonance (NMR), in which amide hydrogen signals disappear on deuteration. Since deuterium weighs 1 Dalton more than protium, hydrogen exchange in proteins can also be detected by mass spectrometry. This approach is complementary to NMR in some respects and clearly advantageous in others.

1.2. Mass Spectrometry Can Do more than Weigh Proteins

By coupling hydrogen exchange (HX) with mass spectrometry of the combined technique (HX/MS) can be used to learn more about proteins than their molecular mass. For example, HX/MS has been used to investigate folding or unfolding, conformational changes, structural heterogeneity, and effects of binding or aggregation (for a review, *see* **ref. *11***). Hydrogen exchange within

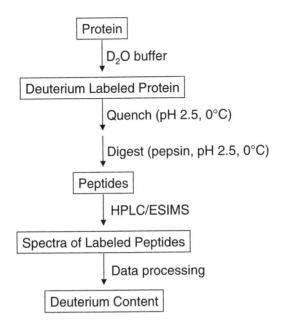

Fig. 1. Flow chart for hydrogen exchange and proteolytic fragmentation mass spectrometry.

entire proteins can be followed with mass spectrometry. Some examples in which only the intact protein was analyzed include conformational studies of ubiquitin *(12)*, apo-myoglobin *(13)* and recombinant insulins *(14)*, unfolding studies of lysozyme *(15)* and an SH3 domain *(16)*, peptide binding studies *(17,18)*, and HX in proteins folded by chaperones *(19,20)*. Analysis of the intact protein following HX is not sufficient to answer some questions, however, necessitating the development of methods to localize exchange measurements to smaller and smaller pieces of the polypeptide backbone.

Zhang and Smith *(21)* developed a method of deuterium labeling followed by pepsin digestion to localize exchange to short fragments of the protein backbone. This approach is based on methods described previously for tritium exchange *(22–24)*. The general method, illustrated in **Fig. 1** and described in detail below, has been used to investigate many structural features of proteins including (1) structural heterogeneity in oxidized cytochrome c *(25)*; (2) the effects of mutagenesis on ferredoxin *(26)*, ferrocytochrome c2 *(27)* and cytochrome c553 *(28,29)*; (3) small molecule (NADPH) binding-induced conformational changes in diaminopimelate dehydrogenase *(30)* and dihydrodipicolinate reductase *(31)* and changes in protein structure caused by rhodopsin binding to arrestin *(32)*; (4) exchange rates in portions of apo- and holomyoglobin *(13,33)*; (5) slow protein unfolding in Hck SH3 and the effects of peptide binding *(16)*;

(6) protein kinase activation in wild-type and mutant MAP kinase kinase-1 *(34)*; (7) the exchange of amide hydrogens on the surface of proteins *(35)*; (8) millisecond folding of cytochrome c *(36)*; (9) exchange rates in rabbit muscle aldolase *(37)* and αA crystallin *(38)*; (10) identification of unfolding domains in large proteins *(39,40)*, and other applications *(41,42)*.

Successful application of HX/MS to a particular protein or protein system requires careful consideration of several technical aspects, such as pH, temperature, digestion, separation, and processing the large quantities of data. Although such experiments are technically challenging, the rewards far outweigh any experimental obstacles. Before delving into an HX experiment, one should understand the current models used to link HX to protein structure and dynamics.

1.3. Hydrogen Exchange Models

The details of HX have been reviewed elsewhere *(11,43–46)*. Protons located at peptide amide linkages in peptides incubated in D_2O at pD 7 are replaced with deuterons within 1–10 s. In folded proteins, some amide hydrogens may exchange within seconds, whereas others exchange only after months, with rates of the most slowly exchanging amide hydrogens reduced to as much as 10^{-8} of their rates in unfolded forms of the same protein. Nearly all peptide amide hydrogens in folded proteins are hydrogen bonded, either intramolecularly to another part of the protein or to water. Base-catalyzed isotope exchange can occur only when a hydrogen bond is severed in the presence of the catalyst (hydroxide) and the source of the new proton (water). The large reduction of amide HX rates in folded proteins is due primarily to intramolecular hydrogen bonding and restricted access to the solvent.

Mechanisms for isotope exchange of peptide amide hydrogens are illustrated in **Fig. 2**. HX can be summarized by a two-process model in which exchange can occur directly in a folded protein (**Fig. 2A**) or through the competing process of exchange in momentarily unfolded forms (**Fig. 2B**). Exchange from the folded form probably dominates for amide hydrogens that are not participating in intermolecular hydrogen bonding and are located near the surface. Exchange in unfolded forms requires substantial movement of the backbone. Unfolding to expose amide hydrogen positions to deuterium can be isolated to small regions (localized unfolding) or may involve the entire protein (global unfolding).

The observed rate of exchange is described by the rate constant k_{obs}, which can be expressed as the sum of the contributions of exchange from folded and unfolded forms (**Eq. 1**) *(46)*.

$$k_{obs} = k_f + k_u \qquad (1)$$

The rate constant for exchange from the folded state, k_f, is described by **Eq. (2)**:

$$k_f = \beta k_2 \qquad (2)$$

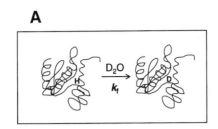

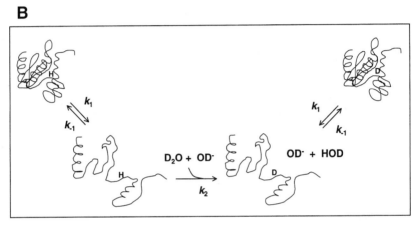

Fig. 2. General mechanisms for isotope exchange at peptide amide hydrogens. Exchange can occur directly into folded forms (**A**), indirectly into unfolded forms (**B**), or through a combination of the two processes.

where β is the probability that a particular amide hydrogen is exposed to water and catalyst at the same time that it is exchange competent, and k_2 is the rate constant for amide hydrogen exchange in an unfolded peptide *(47)*. The proportionality factor, β, is a function of several parameters including solvent accessibility and intramolecular hydrogen bonding. The rate constants for exchange from unfolded forms of proteins depend on the rate constant for exchange from an unfolded peptide (k_2) as well as the unfolding dynamics described by k_1 and k_{-1}. When $k_2 \gg k_{-1}$ (termed EX1 kinetics), the observed hydrogen exchange rate constant is given by the unfolding rate constant k_1 (**Eq. [3]**):

$$k_u = k_1 \quad (3)$$

However, under physiological conditions, it is more common for $k_{-1} \gg k_2$. In this case (EX2 kinetics), the hydrogen exchange rate constant is given by **Eq. (4)**:

$$k_u = (k_1/k_{-1})k_2 = K_{unf} k_2 \quad (4)$$

where K_{unf} is the equilibrium constant describing the unfolding process. All proteins undergo exchange via EX2 kinetics, which may be envisioned as many visits to a state capable of exchange. Since the probability of isotope exchange during one unfolding event is small, individual molecules must unfold many times before they are completely deuterated. Some proteins exhibit EX1 kinetics, or can be induced to exhibit EX1 kinetics with denaturant *(39)* or by increasing pH *(48)*. EX1 kinetics are characterized by a correlated isotope exchange event that accompanies protein unfolding. Since isotope exchange is fast relative to refolding, exchange occurs at each linkage before refolding occurs (**Eq. [3]**). Some proteins may contain regions that undergo EX1 and EX2 kinetics simultaneously. Regions in which exchange occurs primarily by EX1 and EX2 kinetics can be identified by characteristic isotope patterns in mass spectra.

2. Materials

2.1. Buffers

Phosphate buffers provide excellent buffering capacity for exchange-in at pH 7 and quenching at pH 2.5 (*see* **Subheading 3.2.**).

1. A weak phosphate buffer (5 mM, H_2O) for equilibration.
2. A weak phosphate buffer (5 mM, D_2O) for the labeling step.
3. A stronger phosphate buffer (100 mM, H_2O) for the quenching step.

2.2. Materials for D_2O Introduction

1. Microcentrifuge filter units (Alltech, Deerfield, IL #2440 or similar) packed with G-25 superfine gel filtration media and equilibrated with D_2O buffer (*see* **Subheading 3.1.**).

2.3. Materials for Chromatography

1. Empty capillary columns (254 µm ID × 10 cm) for perfusion chromatography.
2. POROS 10-R2 packing (PerSeptive Biosystems, Foster City, CA) for perfusion chromatography.
3. Silica-based columns (1 mm × 5 cm) containing Vydac C18 material (MicroTech, Sunnyvale, CA).
4. Acetonitrile, trifluoroacetic acid (TFA).
5. Stainless steel (rather than PEEK) transfer lines for effective cooling.
6. Ice bath containing injector, sample loop, column, and all transfer lines.

2.4. Software for Data Processing

1. PeakFit (SPSS, Chicago, IL) can be used to estimate the relative intensities of incompletely resolved envelopes.
2. Kaleidograph (Synergy Software, Reading, PA) or SigmaPlot (SPSS) can be used for data fitting.

3. Methods

3.1. Introduction of Deuterium

Two general approaches have been used to expose proteins to D_2O to initiate hydrogen exchange: (1) dilution, and (2) spin-column solvent exchange. Manual dilution is easiest to perform, but requires careful planning to avoid final concentrations of protein that are too dilute to be analyzed by mass spectrometry. In contrast, the spin-column method of introducing deuterium relies on gel-filtration chromatography, a method commonly used for changing solvents. The dilution method is used when short labeling times are required. For example, manual dilution labeling times can be as short as 10 s, whereas automated dilution labeling times can be reduced to 5 ms *(36)*. The shortest labeling time possible with spin-column solvent exchange is approx 1–2 min. Despite its limitations, the spin-column method is preferred when minimal dilution of the sample is essential.

1. To introduce deuterium by the dilution method, we normally dilute a protein solution (H_2O/buffer) 20-fold with D_2O/buffer. This makes a solution of 95% D_2O.
2. Microcentrifuge filter units (Alltech #2440 or similar) are packed with G-25 superfine gel-filtration medium and equilibrated with D_2O buffer. The sample (in H_2O) is placed on the column and spun until the same volume that was placed on the column has passed through the medium and is collected in the receiving tube. The H_2O is trapped in the G-25 and the protein elutes in D_2O.
3. For both methods, it is prudent to equilibrate the protein in solution prior to the start of any labeling. Addition of D_2O to dry protein is not advised because the dissolution process may lead to erroneous isotopic exchange results. A weak buffer (5 mM, H_2O) is usually adequate for equilibration (5 mM, H_2O) and labeling (5 mM, D_2O) steps, whereas a stronger buffer (100 mM, H_2O) is used for the quenching step.

3.2. Control of Exchange Rate by Buffering

Perhaps the most important factor that facilitates measurements of hydrogen exchange with mass spectrometry is the relationship of exchange rate and pH. Hydrogen exchange is both acid and base catalyzed with an exchange minimum occurring at pH 2.5–3.0 (**Fig. 3**). Note that changing the pH by 1 unit changes the HX rate 10-fold. By performing a labeling experiment at pH 7.0 and then decreasing the pH to 2.5, the average exchange rate of amide hydrogens is reduced by 4 orders of magnitude. For reproducible exchange and for effective quenching, the pH must controlled very precisely.

Buffers are essential for precise pH control. Phosphate buffers are used because phosphate has a pK_1 of 2.15 and a pK_2 of 6.82, providing excellent buffering capacity for exchange-in at pH 7 and quenching at pH 2.5. A weak

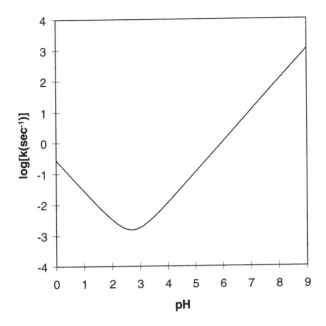

Fig. 3. The pH dependence of amide hydrogen exchange in unfolded polypeptides. Calculated based on parameters derived from NMR analysis of model peptides *(47)*.

buffer is usually adequate for equilibration (5 m*M* H$_2$O) and labeling (5 m*M*, D$_2$O), whereas a stronger buffer (100 m*M*, H$_2$O) is used for quenching. This combination is advantageous because it ensures that the pH is decreased reproducibly to 2.5.

1. A stronger buffer (100 m*M*, H$_2$O) is used for the quenching step. For this reason, only a small volume of quench solution is required to change the pH from 7 to 2.5, thereby minimizing dilution of the sample. A 100-m*M* quench buffer (pH 2.5) can be used in a 1:1 volume ratio with 5 m*M* buffer to effectively reduce the pH to 2.5 (*see* **Note 1**).

3.3. Control of Exchange Rate by Temperature

Although reducing the pH to 2.5 decreases the average HX rate by 4 orders of magnitude, exchange is still too fast for analysis. To decrease the exchange rate another order of magnitude, the temperature of the quenched samples is maintained at 0°C for the remainder of the experiment. The exchange rate changes threefold for each 10°C change *(47)*. Under quench conditions of 0°C and pH 2.5, the loss of deuterium label during analysis occurs with a half-life of 30–120 min, thereby providing enough time to carry out the analysis without substantial loss of deuterium from the peptide amide linkages.

3.4. Digestion and Peptide Separation

3.4.1. Nonspecific Partial Digestion of Labeled Proteins Using Pepsin

Pepsin is a low-specificity acid protease that is active under quench conditions of pH 2.5. Other acid proteases with different specificities have been investigated *(24)*. Although the specificity of pepsin is low, it is highly reproducible for similar experimental conditions. Autodigestion of pepsin is generally minimal. To localize HX to specific regions of a protein, peptic fragments must be identified. This step can be accomplished with off-line tandem mass spectrometry (MS/MS) experiments and/or carboxy-terminal sequencing *(21,49–52)*, provided that off-line pepsin digestion conditions are identical to those used for the actual exchange experiment. Alternatively, fragments can be identified by on-line liquid chromatography (LC)/MS/MS.

1. Prepare the pepsin solution in pure water (pH approx 5.7) and place at 0°C.
2. Add pepsin to the quenched sample at pH 2.5 so that it becomes active.
3. Digest for 5 min at 0°C with a weight ratio of 1:1 pepsin/protein. These conditions usually fragment the protein into peptides with an average length of 10 residues (*see* **Note 2**).

3.4.2. Peptide Separation and the Importance of Speed

Under quench conditions of 0°C and pH 2.5, some deuterium label reverts to hydrogen during analysis. Back-exchange occurs in the quench buffer (H_2O), and the digestion step (pepsin in H_2O) and during fragment separation (H_2O-based solvents). The half-life for back-exchange is 30–120 min. It is, therefore, critical to complete the analysis as quickly as possible to minimize back-exchange. The pepsin digestion step typically accounts for approximately half the total analysis time and cannot be reduced significantly without sacrificing effective digestion. The chromatography step for fragment separation is usually optimized for speed rather than resolution. Because the mass spectrometer is able to resolve fragments from one another based on mass, complete chromatographic separation is not essential. Separations achieved in 5–7 min for 5–60% acetonitrile gradients are generally adequate for proteins with fewer than 500 residues.

Because the mass spectrometer can be used to analyze peptide mixtures, one might question the need to separate the peptic fragments with chromatography at all. It would be much simpler if spectra for all the fragments of the protein were acquired simultaneously. Unfortunately, this is not generally advisable for several reasons. First, electrospray ionization (ESI) mass spectra may be very complicated, especially with larger peptides that may have up to four charge states. Second, data processing is complicated by the presence of overlapping peaks, a problem that can be reduced with chromatography. Third, signal suppression in complex mixtures can mask peptides that give only weak signals or that are present in small amounts. Finally, the separation step serves to wash away deuterium from side chains. Labile hydrogens on amino acid side chains (such as serine, lysine, and so forth) exchange within seconds. Since the separation step occurs in H_2O-based solvents, any deuterium that exchanged into side chain positions is exchanged back to hydrogen during separation. Consequently, mass changes detected for peptic fragments are due only to deuterium located at peptide amide linkages.

The column used for the separation can be either perfusion- or silica-based. We have used both and find advantages with both depending on the type of analysis. Experiments involving denaturants, such as urea or guanidine hydrochloride, are better done with a perfusion column because all the denaturant is effectively washed away after each sample due to the high linear velocity of the mobile phase. For silica-based packing material, it takes a much longer time to wash the denaturant out of the column. Set up the appropriate column as follows:

1. For perfusion chromatography, we pack our own capillary columns (254 μm ID × 10 cm) with POROS 10-R2 (PerSeptive Biosystems) (*see* **Notes 3** and **4**).
2. Our silica-based columns (1 mm × 5 cm) are purchased from MicroTek and contain Vydac C18 material (*see* **Note 4**).
3. For either column, the flow rate is 40–50 μL/min, all of which is directed into the mass spectrometer (*see* **Notes 5** and **6**).
4. To minimize loss of label at the backbone amide positions, the injector, sample loop, column, and all transfer lines are placed in an ice bath. Stainless steel, rather than PEEK, is used for transfer lines for effective cooling.
5. Separations are based on a gradient of 5–60% acetonitrile containing TFA at a concentration of 0.05% *(53)* (*see* **Note 7**). Adjust the gradient to achieve a partial separation in 5–7 min.

3.5. Mass Spectrometry

Although continuous-flow fast-atom bombardment mass spectrometry was used for the earliest HX/MS measurements *(21)*, electrospray ionization has become the preferred ionization method because of its high sensitivity and facile coupling to high-performance liquid chromatography (HPLC). Several mass analyzers, including magnetic sector, quadrupole, ion cyclotron resonance, and ion trap, have been used successfully for HX/MS. The instrumentation used to acquire the information presented here was as follows:

1. Mass spectra of intact proteins were acquired on a Micromass (Altrincham, UK) Platform quadrupole mass spectrometer equipped with a standard source operated with a cone voltage of approx 50 V.

2. Analysis of peptide fragments was accomplished using a Micromass Autospec magnetic sector mass spectrometer equipped with a focal place detector and a standard ESI source. The source was operated at 60°C.

3.5.1. Adjustment for Back-Exchange

Although some loss of deuterium label from the peptic fragments during peptic digestion and HPLC is unavoidable, simple adjustments can be made for these losses. These adjustments are based on analysis of appropriate controls, a 0% deuterium control, and a 100% deuterium control *(21,53)*. Control samples representing 0% and 100% exchanged protein should be analyzed with each set of samples, as conditions may vary from day to day. The 0% control is protein that has never been exposed to deuterium, and the 100% control is protein in which all exchangeable hydrogens have been replaced with deuterium. The following formula can be used to adjust for any loss or gain during analysis:

$$D = (m - m_{0\%})/(m_{100\%} - m_{0\%}) \times N \qquad (5)$$

where D is the adjusted deuterium level, m is the experimentally observed mass, $m_{0\%}$ is the 0% or undeuterated control mass, $m_{100\%}$ is the totally deuterated control mass and N is the total number of amide hydrogens in the fragment. Since this adjustment is based on the assumption that loss of deuterium from a totally deuterated fragment is proportional to the loss from the same fragment when partially deuterated, the accuracy of the adjustment depends on the sequence of the peptide *(21)*. The average error that results from **Eq. (5)** is about 5%, with 92% of fragments having an error <10%, 3% having an error >15%, and 0.5% having an error >25%. Adjustment is most useful when determining the amount of deuterium incorporated in a fragment. However, when the levels of deuterium in the same fragment are compared (such as comparing the amount of deuterium in a given region in the free or ligand bound state), the correction is not required as long as the analysis conditions are identical.

Totally deuterated protein, and hence totally deuterated fragments, can usually be prepared as follows:

1. Incubate the protein in a solution of 100% deuterium, pH approx 7, at elevated temperature for a prolonged period (4–24 h). The optimum temperature depends on the protein and should be experimentally determined for each protein to avoid irreversible denaturation or precipitation.
2. Alternatively, or in combination with **step 1**, complete exchange may be achieved by incubating a protein in 100% D_2O at pD 2.4–3.4 for several hours. Although exchange at this pH is slow, most proteins are denatured and therefore allow exchange into portions of the protein that are protected in the native state (*see* **Note 8**).

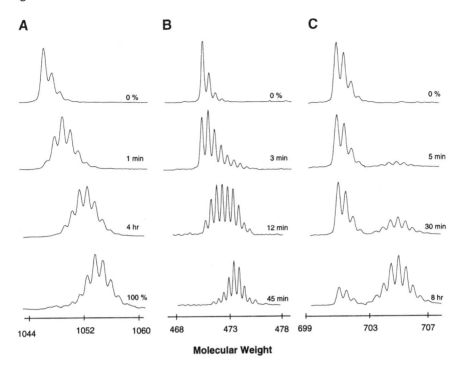

Fig. 4. Distinctive isotope patterns characteristic of EX2 and EX1 kinetics. (**A**) A fragment of a protein exhibiting EX2 kinetics. A single isotopic envelope is obvious. (**B**) A fragment exhibiting EX1 kinetics; the overlap between the two envelopes is severe. (**C**) A fragment exhibiting EX1 kinetics and clear separation of the two envelopes.

3. Analyze a control sample from **steps 1** and/or **2**, representing 100% exchanged protein, together with an undeuterated protein control, with each set of samples, as conditions may vary from day to day.

3.5.2. Interpretation of Mass Spectra

3.5.2.1. AVERAGE MASS

When spectra of a deuterated fragment are acquired, the most notable observation is a distribution of isotopes (*see* **Fig. 4A** as an example). Because loss of deuterium label during analysis is a random process, the molecules comprising a sample have a distribution of deuterium levels. The increase in the average mass of a fragment is equal to the average amount of deuterium in the entire population of molecules. For some experiments, the average mass of the fragment at a given D_2O incubation time is all that is desired. Increase in average mass plotted against exchange-in or D_2O incubation time, yields an exchange-in curve for a given fragment. This plot can be used to calculate the exchange rates of amide hydrogens within the fragment (*see* **Subheading 3.5.2.3.**) and for comparison with exchange-in curves when the protein is labeled under different conditions (such as bound vs free).

3.5.2.2. Isotope Patterns: ex2/ex1

Before determining the average mass of a fragment, it is important to determine which type of exchange kinetics are occurring for that region of the protein because this will determine how the data are processed. Whether isotope exchange follows the EX2 or EX1 kinetic models discussed above can usually be determined from isotope patterns characteristic of these models. **Figure 4A and C** shows characteristic EX2 and EX1 isotope patterns, respectively. HX via EX2 kinetics gives a binomial distribution of isotope peaks, from which the centroid or average mass can be determined. HX via EX1 kinetics gives a bimodal isotope pattern, as illustrated in **Fig. 4C**. Two envelopes of mass appear: (1) a low-mass envelope, which indicates the non-exchanged population; and (2) a high-mass envelope, which represents the exchanged population. The distance between the two envelopes indicates the number of amide hydrogens involved in cooperative unfolding; the relative intensity (area) of the two envelopes is directly related to the relative population of folded and unfolded forms of the protein. This information can be used to determine the unfolding rate constant for the cooperative transition. The average mass of each envelope can be determined separately or the average mass of both peaks together can be determined as representative of the entire population.

The ability to distinguish between EX1 and EX2 kinetics depends on the distance that separates the two envelopes and the widths of the two peaks. Dominance of EX1 kinetics is easily recognized when the envelopes are completely separated, as illustrated in **Fig. 4C**. However, when the size of the unfolding region is small, the envelopes may be close together, as illustrated in **Fig. 4B**. Factors that affect the separation of envelopes of isotope peaks have been discussed elsewhere *(37,54)*. Various computer programs (such as PeakFit, SPSS Software) can be used with the appropriate reference peaks to estimate the relative intensities of incompletely resolved envelopes. For example, the width of the low-mass envelope at 3 min in **Fig. 4B** is estimated from spectra taken prior to the EX1 unfolding. The width of the high-mass portion of the bimodal isotope pattern is estimated from spectra taken after the EX1 event, such as at the 45-min time point in **Fig. 4B**. With these two widths fixed, the intensities of the two envelopes are varied to obtain the best fit between the model and the experimental data.

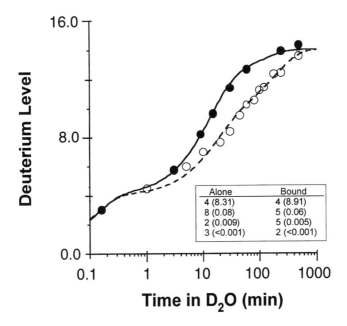

Fig. 5. Example of determining the distribution of hydrogen exchange rate constants within a segment of a protein using **Eq. (6)**. Hck SH3 was labeled alone (solid symbols) or in the presence of a high-affinity ligand (open symbols). Experimental data were fit to a three-term exponential with **Eq. (6)** and displayed as solid (SH3 alone) or dashed (SH3 bound) lines. The results of fitting are tabulated in the inset, where the number of amide hydrogens is shown with the average rate constant (in parentheses, expressed as min^{-1}) for those hydrogens.

3.5.2.3. Calculating Exchange Rates

From exchange-in time-course data, one can estimate the distribution of rate constants describing isotope exchange at each linkage within a peptide. Although single amino acid resolution is not generally possible, exchange rates describing groups of hydrogens with similar exchange rates can be calculated. Rate constants in **Eq. (6)** are fit to the experimental data with a series of first-order rate expressions where D is the number of deuterium atoms present in a peptide, N is the number of peptide amide linkages in a segment, and k_i are the hydrogen-deuterium exchange rate constants for each peptide linkage *(37)*. **Figure 5** illustrates typical fitting of a three-term exponential version of **Eq. (6)** to HX results obtained for residues 119–136 of the Hck SH3 domain *(16)*. The number of amide hydrogens that exchange at similar exchange rates is calculated for each term of **Eq. (6)**. At the same time the average rate constant for each group is calculated. The results for fitting the experimental data in **Fig. 5** are shown in the inset, where the number of amide hydrogens that exchange at a given rate are tabulated along with the average rate constant calculated for each group. Software packages such as Kaleidograph (Synergy Software) or SigmaPlot (SPSS) can be used for fitting.

$$D = N - \sum_{i=1}^{N} \exp(-k_i t) \qquad (6)$$

These isotope exchange rate constants can be compared with results obtained by NMR or used to describe exchange behaviors in different forms of a protein. For example, HX results for a fragment of SH3 when the intact domain is alone or bound to a ligand are compared in **Fig. 5** *(16)*. It is apparent from the raw data that peptide binding causes a significant reduction in the rate of HX for amide hydrogens in this region of the protein. What becomes apparent only after fitting the data with **Eq. (6)** is the alteration in the population of amide hydrogens and their respective rates. In the free state, eight NH of groups exchange at a rate of 0.08 min^{-1}. In the bound state, however, both the number of NH of groups and the average rate decrease.

3.6. Conclusions

Amide hydrogen exchange combined with proteolytic fragmentation and mass spectrometry introduces the possibility of learning about the higher order structure and the dynamics of proteins. The technique promises to increase our understanding of the link between protein structure and function. The method is particularly attractive since it can be used to localize HX to small regions of proteins that are too large to be studied with multidimensional NMR. Mass spectrometry also has advantages over NMR in the area of sensitivity and protein solubility *(11)*. The most sensitive NMR with ^{15}N-labeled protein requires approximately 5 µmol of protein, whereas conventional NMR requires at least 50 µmol. Mass spectrometry requires only a few nanomoles. Mass spectrometry requires only micromolar concentrations of protein, whereas NMR requires millimolar concentrations. Mass spectrometry can also detect cooperative unfolding events based on EX1 kinetics, which are difficult to determine with other methods.

4. Notes

1. Minimal dilution is important when the quantity of protein is limited or when the concentration of protein must be low during isotopic exchange. The strength of buffer required also depends on the concentrations of proteins and salts that may be a major constituent of the protein sample. Testing the ability of a buffer to maintain the pH when protein is added is recommended, as is testing effective quench ratios and quench buffer strengths.
2. Although these conditions are a good starting point, they may be modified to alter the extent of digestion. Increasing the pepsin/protein ratio typically has little effect, but

digestion time can be shortened or lengthened somewhat with some success. Some groups have also achieved good results using pepsin immobilized on beads *(26)*.

3. Packing the columns is easy and inexpensive (approx $20/column), and directions are provided with the packing material.
4. The choice of column diameter depends on the desired flow rate for ESI-MS (and *see* **Note 6***)*.
5. Flow rates in the 25–75-µL/min range are a good compromise between sensitivity and residence time of the peptides in the transfer lines.
6. When sample consumption is not a concern, effluent from a large column may be split prior to analysis by ESI-MS.
7. Although the optimum pH for quenching HX depends somewhat on the amino acid sequence of the peptides, a TFA concentration of 0.05% in the mobile phase is a good compromise for most peptides.
8. Denaturants may also be used to prepare totally exchanged reference samples.

References

1. Nimmesgern, E., Fox, T., Fleming, M. A., and Thomson, J. A. (1996) Conformational changes and stabilization of inosine 5'-monophosphate dehydrogenase associated with ligand binding and inhibition by mycophenolic acid. *J. Biol. Chem.* **271**, 19,421–19,427.
2. Creighton, T. E. (1992) Protein Folding. W. H. Freeman, New York.
3. Wilson, I. A. and Stanfield, R. L. (1994) Antibody-antigen interactions: New structures and new conformational changes. *Curr. Opin. Struct. Biol.* **4**, 857–867.
4. Spolar, R. S. and Record, M. T., Jr. (1994) Coupling of local folding to site-specific binding of proteins to DNA. *Science* **263**, 777–784.
5. Kriwacki, R. W., Hengst, L., Tennent, L., Reed, S. I., and Wright, P. E. (1996) Structural studies of p21$^{Waf1/Cip1/Sdi1}$ in the free and Cdk-2 bound state: conformational disorder mediates binding diversity. *Proc. Natl. Acad. Sci. USA* **93**, 11,504–11,509.
6. Shen, F., Triezenberg, S. J., Hensley, P., Porter, D., and Knutson, J. (1996) Transcriptional activation domain of the herpesvirus protein VP16 becomes conformationally constrained upon interaction with basal transcription factors. *J. Biol. Chem.* **271**, 4827–4837.
7. Alexandrescu, A. T., Abeygunawardana, C., and Shortle, D. (1994) Structure and dynamics of a denatured 131-residue fragment of Staphlococcal nuclease: a heteronuclear study. *Biochemistry* **33**, 1063–1072.
8. Hvidt, A. and Nielsen, S. O. (1966) Hydrogen exchange in proteins. *Adv. Protein Chem.* **21**, 287–385.
9. Woodward, C., Simon, I., and Tuchsen, E. (1982) Hydrogen exchange and the dynamic structure of proteins. *Mol. Cell. Biochem.* **48**, 135–160.
10. Englander, S. W. and Kallenbach, N. R. (1984) Hydrogen exchange and structural dynamics of proteins and nucleic acids. *Q. Rev. Biophys.* **16**, 521–655.
11. Smith, D. L., Deng, Y., and Zhang, Z. (1997) Probing the non-covalent structure of proteins by amide hydrogen exchange and mass spectrometry. *J. Mass Spectrom.* **32**, 135–146.

12. Katta, V. and Chait, B. T. (1991) Conformational changes in proteins probed by hydrogen-exchange electrospray-ionization mass spectrometry. *Rapid Commun. Mass Spectrom.* **5,** 214–217.
13. Wang, F. and Tang, X.-J. (1996) Conformational heterogeneity and stability of apomyoglobin studied by hydrogen-deuterium exchange and electrospray ionization mass spectrometry. *Biochemistry* **35,** 4069–4078.
14. Ramanathan, R., Gross, M. L., Zielinski, W. L., and Layloff, T. P. (1997) Monitoring recombinant protein drugs: a study of insulin by H/D exchange and electrospray ionization mass spectrometry. *Anal. Chem.* **69,** 5142–5145.
15. Miranker, A., Robinson, C. V., Radford, S. E., Aplin, R. T., and Dobson, C. M. (1993) Detection of transient protein folding populations by mass spectrometry. *Science* **262,** 896–900.
16. Engen, J. R., Smithgall, T. E., Gmeiner, W. H., and Smith, D. L. (1997) Identification and localization of slow, natural, cooperative unfolding in the hematopoietic cell kinase SH3 domain by amide hydrogen exchange and mass spectrometry. *Biochemistry* **36,** 14,384–14,391.
17. Kragelund, B. B., Knudsen, J., and Poulsen, F. M. (1995) Local perturbations by ligand binding of hydrogen deuterium exchange kinetics in a four-helix bundle protein, acyl coenzyme A binding protein (ACBP). *J. Mol. Biol* **250,** 695–706.
18. Anderegg, R. J. and Wagner, D. S. (1995) Mass spectrometric characterization of a protein ligand interaction. *J. Am. Chem. Soc.* **117,** 1374–1377.
19. Gross, M., Robinson, C. V., Mayhew, H., Hartl, F. U., and Radford, S. E. (1996) Significant hydrogen exchange protection in GroEL-bound DHFR is maintained during iterative rounds of substrate cycling. *Protein Sci.* **5,** 2506–2513.
20. Robinson, C. V., Gross, M., Eyles, S. J., Ewbank, J. J., Mayhew, M., Hartl, F. U., et al. (1994) Conformation of GroEL-bound alpha-lactalbumin probed by mass spectrometry. *Nature* **372,** 646–651.
21. Zhang, Z. and Smith, D. L. (1993) Determination of amide hydrogen exchange by mass spectrometry: a new tool for protein structure elucidation. *Protein Sci.* **2,** 522–531.
22. Rosa, J. J. and Richards, F. M. (1979) An experimental procedure for increasing the structural resolution of chemical hydrogen-exchange measurements on proteins: application to ribonuclease S peptide. *J. Mol. Biol.* **133,** 399–416.
23. Rosa, J. J. and Richards, F. M. (1981) Hydrogen exchange from identified regions of the S-protein component of ribonuclease as a function of temperature, pH, and the binding of S-peptide. *J. Mol. Biol.* **145,** 835–851.
24. Englander, J. J., Rogero, J. R., and Englander, S. W. (1985) Protein hydrogen exchange studied by the fragment separation method. *Anal. Biochem.* **147,** 234–244.
25. Dharmasiri, K. and Smith, D. L. (1997) Regional stability changes in oxidized and reduced cytochrome c located by hydrogen exchange and mass spectrometry. *J. Am. Soc. Mass. Spectrom.* **8,** 1039–1045.
26. Remigy, H., Jaquinod, M., Petillot, Y., Gagnon, J., Cheng, H., Xia, B., et al. (1997) Probing the influence of mutation on the stability of a ferredoxin by mass spectrometry. *J. Protein Chem.* **16,** 527–532.

27. Jaquinod, M., Guy, P., Halgand, F., Caffrey, M., Fitch, J., Cusanovich, M., et al. (1996) Stability of *Rhodobacter capsulatus* ferrocytochrome c2 wild-type and site-directed mutants using hydrogen/deuterium exchange monitored by electrospray ionization mass spectrometry. *FEBS Lett.* **380,** 44–48.
28. Guy, P., Remigy, H., Jaquinod, M., Bersch, B., Blanchard, L., Dolla, A., et al. (1996) Study of the new stability properties induced by amino acid replacement of tyrosine 64 in cytochrome C553 from *Desulfobibrio vulgaris* Hildenborough using electrospray ionization mass spectrometry. *Biochem. Biophys. Res. Commun.* **218,** 97–103.
29. Guy, P., Jaquinod, M., Remigy, H., Andrieu, J. P., Gagnon, J., Bersch, B., et al. (1996) New conformational properties induced by the replacement of Tyr-64 in *Desulfovibrio vulgaris* Hildenborough ferricytochrome d553 using isotopic exchange monitored by mass spectrometry. *FEBS Lett.* **395,** 53–57.
30. Wang, F., Scapin, G., Blanchard, J. S., and Angeletti, R. H. (1998) Substrate binding and conformational changes of *Clostridium glutamicum* diaminopimelate dehydrogenase revealed by hydrogen/deuterium, exchange and electrospray mass spectrometry. *Protein Sci.* **7,** 293–299.
31. Wang, F., Blanchard, J. S., and Tang, X. J. (1997) Hydrogen exchange/electrospray ionization mass spectrometry studies of substrate and inhibitor binding and conformational changes of *Escherichia coli* dihydrodipicolinate reductase. *Biochemistry* **36,** 3755–3759.
32. Ohguro, H., Palczewski, K., Walsh, K. A., and Johnson, R. S. (1994) Topographic study of arrestin using differential chemical modifications and hydrogen/deuterium exchange. *Protein Sci.* **3,** 2428–2434.
33. Johnson, R. S. and Walsh, K. A. (1994) Mass spectrometric measurement of protein amide hydrogen exchange rates of apo- and holo-myoglobin. *Protein Sci.* **3,** 2411–2418.
34. Resing, K. A. and Ahn, N. G. (1998) Deuterium exchange mass spectrometry as a probe of protein kinase activation. Analysis of wild-type and constitutively active mutants of MAP kinase kinase-1. *Biochemistry* **37,** 463–475.
35. Dharmasiri, K. and Smith, D. L. (1996) Mass spectrometric determination of isotopic exchange rates of amide hydrogens located on the surfaces of proteins. *Anal. Chem.* **68,** 2340–2344.
36. Yang, H., and Smith, D. L. (1997) Kinetics of cytochrome c folding examined by hydrogen exchange and mass spectrometry. *Biochemistry* **36,** 14,992–14,999.
37. Zhang, Z., Post, C. B., and Smith, D. L. (1996) Amide hydrogen exchange determined by mass spectrometry: application to rabbit muscle aldolase. *Biochemistry* **35,** 779–791.
38. Liu, Y. and Smith, D. L. (1994) Probing high order structure of proteins by fast-atom bombardment mass spectrometry. *J. Am. Soc. Mass Spectrom.* **5,** 19–28.
39. Deng, Y. and Smith, D. L. (1998) Identification of unfolding domains in large proteins by their unfolding rates. *Biochemistry* **37,** 6256–6262.
40. Zhang, Z. and Smith, D. L. (1996) Thermal-induced unfolding domains in aldolase by amide hydrogen exchange and mass spectrometry. *Protein Sci.* **5,** 1282–1289.

41. Maier, C. S., Kim, O. H., and Deinzer, M. L. (1997) Conformational properties of the A-state of cytochrome s studied by hydrogen/deuterium exchange and electrospray mass spectrometry. *Anal. Biochem.* **252**, 127–135.
42. Zhang, Z., Li, W., Logan, T. M., Li, M., and Marshall, A. G. (1997) Human recombinant [C22A] FK506-binding protein amide hydrogen exchange rates from mass spectrometry match and extend those from NMR. *Protein Sci.* **6**, 2203–2217.
43. Loh, S. N., Rohl, C. A., Kiefhaber, T., and Baldwin, R. L. (1996) A general two-process model describes the hydrogen exchange behavior of RNase A in unfolding conditions. *Proc. Natl. Acad. Sci. USA* **93**, 1982–1987.
44. Bai, Y., Sosnick, T. R., Mayne, L., and Englander, S. W. (1995) Protein folding intermediates: native-state hydrogen exchange. *Science* **269**, 192–197.
45. Mayo, S. L. and Baldwin, R. L. (1993) Guanidinium chloride induction of partial unfolding in amide proton exchange in RNase A. *Science* **262**, 873–876.
46. Kim, K.-S. and Woodward, C. (1993) Protein internal flexibility and global stability: effect of urea on hydrogen exchange rates of bovine pancreatic trypsin inhibitor. *Biochemistry* **32**, 9609–9613.
47. Bai, Y., Milne, J. S., Mayne, L., and Englander, S. W. (1993) Primary structure effects on peptide group hydrogen exchange. *Proteins* **17**, 75–86.
48. Swint-Kruse, L. and Robertson, A. D. (1996) Temperature and pH dependence of hydrogen exchange and global stability for ovomucoid third domain. *Biochemistry* **35**, 171–180.
49. Zhou, Z. and Smith, D. L. (1990) Assignment of disulfide bonds in proteins by partial acid hydrolysis and mass spectrometry. *J. Protein Chem.* **9**, 523–532.
50. Biemann, K. (1990) Sequencing of peptides by tandem mass spectrometry and high-energy collision-induced dissociation. *Methods Enzymol.* **193**, 455–479.
51. Caprioli, R. M. and Fan, T. (1986) Peptide sequence analysis using exopeptidases with molecular analysis of the truncated polypeptides by mass spectrometry. *Anal. Biochem.* **154**, 596–603.
52. Smith, J. B., Sun, Y., Smith, D. L., and Green, B. (1992) Identification of the posttranslational modifications of bovine lens alpha-B-crystallins by mass spectrometry. *Protein Sci.* **1**, 601–608.
53. Zhang, Z. (1995) Protein hydrogen exchange determined by mass spectrometry: a new tool for probing protein high-order structure and structural changes. *Doctoral Dissertation, Purdue University.*
54. Deng, Y., Zhang, Z., and Smith, D. L. (1998) Comparison of continuous and pulsed labeling amide hydrogen exchange/mass spectrometry for studies of protein dynamics. *J. Am. Soc. Mass Spec.* **10(8)**, 675–684.

7

Probing Protein Surface Topology by Chemical Surface Labeling, Crosslinking, and Mass Spectrometry

Keiryn L. Bennett, Thomas Matthiesen, and Peter Roepstorff

1. Introduction

The higher order structural properties of peptides and proteins are commonly investigated in the solid state by X-ray diffraction crystallography *(1)* or in solution by multidimensional nuclear magnetic resonance (NMR) spectroscopy *(2)*. These techniques can provide detailed structural information; however, they are severely limited by the quantity of protein required for the analyses, difficulties with crystallization of certain proteins (e.g., glycoproteins), and the poor solubility of many macromolecules. Moreover, these methods are generally time consuming and tedious. Computer-based molecular modeling offers an alternative method to obtain information on the three-dimensional spatial arrangement of a protein. The success of this approach, however, is dependent on prior knowledge of the tertiary structure of an homologous protein *(3–5)*. Due to these limitations, there is an increasing demand for an independent method to probe protein structure and/or confirm information obtained by the techniques mentioned above.

The higher order structure of proteins has rarely been correlated with the chemical properties of the constituent amino acid residues. Nevertheless, it is a reasonable assumption that the relative reactivity of amino acid residues in a native protein will reflect the surface accessibility of the residues. This property can be monitored by selective surface labeling of specific amino acid side chains *(6,7)*. The chemical properties of amino acid residues can also be exploited to determine the contact regions between subunits of protein complexes. Chemical crosslinks formed between amino acid side chains of different protein molecules stabilize noncovalently associated protein complexes, thereby enabling identification of interacting regions of the complex.

The introduction of exogenous chemical groups into a protein or protein complex makes it necessary to have a reliable method for measuring the degree of incorporation (stoichiometry), for characterizing the site(s) of modification, and for determining the relative reactivity of amino acid residues in the native configuration of the macromolecule. Classical techniques have included analytical procedures that involve the synthesis and use of specifically labeled reagents (e.g., radioactively or spectrally labeled reagents) followed by laborious isolation and characterization of the modified peptides *(8)*. Direct spectrometric measurements or amino acid analyses have also been extensively used to characterize protein conjugates.

With the development of soft ionization techniques in biological mass spectrometry *(9,10)*, a novel and simple method for the analysis of chemically modified and crosslinked peptides and proteins is offered. Proteolytic digestion of derivatized proteins followed by mass spectrometric peptide mapping provides information on the location of the modified amino acid residues *(11,12)* and identification of crosslinked sites in complex protein systems. Systematic modification of a protein with various chemical reagents, followed by mass spectrometric identification of the modified amino acids, provides information concerning the exposed regions of the protein and the side chains involved in intramolecular interactions. In addition, the analytical power of mass spectrometry enables evaluation of the selectivity of chemical modification agents *(13)* and determination of the relative reactivities of specific amino acid residues *(6)* and provides confirmatory information concerning the chemistry of the modification reactions *(12)*.

The applications of protein surface labeling and/or protein crosslinking are diverse. When combined with mass spectrometry, these techniques are used for probing protein surface topology *(6,11,14)* and protein function *(15,16)*, as well as for determining protein-peptide *(17)*, protein-protein *(18)*, protein-cofactor *(7,19)*, protein-drug *(20)*, and protein-DNA *(21,22)* interactions.

The purpose of this chapter is to outline the strategies of protein surface labeling and crosslinking and to provide examples of protocols used in our laboratory to modify specific amino acid side chains, and/or to crosslink protein subunits. Intact protein conjugates are characterized by mass spectrometry to determine the number of groups introduced into the protein. Modified peptides are located by mass spectrometric mapping of the digested conjugates, and modified residues are determined by tandem mass spectrometry (MS/MS).

A wide range of reagents is available for selectively modifying specific side chains of amino acid residues and for crosslinking protein subunits. The reagents used in the following protocols are intended as a guide only. Ideally,

Protein Chemical Modification and Mass Spectrometry 115

investigating the three-dimensional structure of a protein or protein complex should involve a range of modifying reagents to target different amino acid side chains. We have attempted to outline some of the pitfalls and problems associated with these protocols and the numerous factors that one must consider before embarking on such studies.

2. Materials

2.1. Recombinant Protein

1. ParR dimers were a gift from Marie Godtfredsen (Department of Molecular Biology, Odense University, Odense M, Denmark).

2.2. Solvents and Buffers

1. Ultra-high-quality (UHQ) water (Millipore, Molsheim, France).
2. 20 m*M* sodium phosphate buffer, 0.15 *M* NaCl (phosphate-buffered saline [PBS], pH 7.5) (Merck, Darmstadt, Germany).
3. 100 m*M* ammonium bicarbonate (NH_4HCO_3), 2 *M* urea, pH 8.3 (Sigma, St Louis, MO and Merck).
4. High-performance liquid chromatography (HPLC)-grade trifluoroacetic acid (TFA), acetonitrile (CH_3CN), formic acid (HCOOH), and methanol (MeOH) (Rathburn, Walkerburn, Scotland).

2.3. Chemicals

1. *N*-hydroxysuccinimidobiotin (NHS-biotin) (Sigma). Manufacturer's recommendation: store desiccated at –20°C and prepare a fresh solution in dimethylsulphoxide (DMSO) as required (*see* **Note 1**).
2. DMSO (Sigma). Store over a drying agent, e.g., 3–4 Å molecular sieves (*see* **Note 2**).
3. Iodogen iodination reagent (1,3,4,6-tetrachloro-3α,6α-diphenylglycouril) (Pierce, Rockford, IL). Manufacturer's recommendation: store sealed under nitrogen in an amber screw cap vial (as supplied). Desiccate at 4°C and prepare a fresh solution in dichloromethane as required (*see* **Note 3**).
4. Dichloromethane (DCM) (Sigma).
5. Sodium iodide ($Na^{127}I$) (Merck). Dissolve $Na^{127}I$ in PBS (pH 7.5). It is possible to store aliquots at –20°C, but we prepare fresh solutions as required.
6. β-Mercaptoethanol (Merck).
7. 3,3'-Dithiobis[sulfosuccinimidyl propionate] (DTSSP) (Pierce). Manufacturer's recommendation: store desiccated under nitrogen at –4°C and prepare a fresh solution in an appropriate buffer as required (*see* **Notes 1** and **4**).
8. 1,4-Dithiothreitol (DTT; Sigma). Prepare as required in UHQ water or an appropriate buffer, or store aliquots at –20°C.
9. Poroszyme immobilized trypsin (PerSeptive Biosystems, Cambridge, MA). Store as supplied in small aliquots at –4°C.
10. Modified porcine trypsin (Promega, Madison, WI). Store in small aliquots (0.5 µg lyophilized) at –20°C.

11. POROS R2 50 and R2 20 reversed-phase (RP) media (PerSeptive Biosystems).
12. Peptide standard calibration kit (HP G2052A; Hewlett-Packard, Palo Alto, CA). Store as supplied at –20°C.

2.4. Matrix-Assisted Laser Desorption/Ionization Mass Spectrometry (MALDI-MS) Matrices (see Note 5)

1. Sinapinic acid (SA) (Fluka, Buchs, Switzerland).
2. 4-Hydroxy-α-cyanocinnamic acid (HCCA; Sigma).
3. 2,5-Dihydroxybenzoic acid (DHB) (Hewlett-Packard, Palo Alto, CA).

3. Methods and Discussion

3.1. MALDI-MS

3.1.1. Instrumentation

Mass spectra were recorded on a Voyager Elite MALDI reflectron time-of-flight (MALDI-TOF) mass spectrometer (PerSeptive Biosystems) equipped with a 337-nm nitrogen laser and a delayed-extraction ion source.

3.1.2. Preparation of MALDI-MS Matrix Solutions

Matrix solutions were prepared as follows (see **Note 6**):

1. Dissolve SA (20 µg/mL) in 70% aqueous acetonitrile + 0.1% TFA.
2. Dissolve HCCA (10 µg/mL) in 70% aqueous acetonitrile + 0.1% TFA.
3. Lyophilize 100 µL DHB in a Speed Vac (Savant, Farmingdale, NY) to remove all traces of organic solvent and redissolve in an equal volume of 30% aqueous acetonitrile + 0.1% TFA.

3.1.3. Sample Preparation

Prior to MALDI-MS analysis, all protein and peptide samples were desalted and concentrated on disposable microcolumns packed with POROS R2 50 (proteins) or POROS R2 20 (peptides) medium (see **Note 7**).

3.1.4. MALDI-TOFMS Analyses

In brief, samples were prepared and analyzed *(23)* by MALDI-MS as follows (see **Note 8**):

1. Mix 0.5 µL of the desalted sample with an equal volume of an appropriate matrix solution.
2. Deposit the sample + matrix solution onto the stainless steel MALDI target.
3. Co-crystallize the sample and matrix at ambient temperature.
4. Measure the intact modified proteins in the positive-ion linear mode and the peptides in the positive-ion reflectron mode at an accelerating voltage of 20–25 keV.
5. Acquire data from 150 to 250 laser pulses to obtain a representative spectrum.
6. External calibration for proteins is accomplished using the protonated average mass of the unmodified protein.

7. External calibration for peptide digests is achieved by using the HP peptide calibration standards.
8. Internal calibration for peptide digests is achieved using the known protonated monoisotopic masses of specified tryptic fragments of the protein in question.

3.1.5. MALDI-MS Spectra Interpretation

Peptide mapping data were processed with the General Peptide Mass Analysis for Windows (GPMAW) software, Version 3.12 (Lighthouse Data, Odense, Denmark). The introduction of user-defined modifications and crosslinks aided analysis of complex MALDI-MS digest spectra.

3.2. Electrospray Ionization Tandem Mass Spectrometry (ESI-MS/MS)

3.2.1. Instrumentation

Modified peptides were fragmented in the positive-ion mode using an Esquire ion trap mass spectrometer (Bruker-Franzen, Bremen, Germany) equipped with a nanoelectrospray ion source (flow rate ≈ 30 nL/min).

3.2.2. ESI-Ion Trap MS/MS Analyses

To confirm the site of modification in a derivatized peptide, the procedure outlined below was followed:

1. Separate the tryptic digest of the modified protein on a C_{18} column using standard HPLC conditions (*see* **Note 9**).
2. Locate the modified peptide by MALDI-MS analysis of the HPLC fractions.
3. Desalt and concentrate the modified peptide on a POROS R2 20 microcolumn (*see* **Note 10**).
4. Load the desalted sample into a nanoelectrospray needle *(24)*.
5. Isolate the charged precursor ion with the ion trap instrument.
6. Fragment the modified peptide precursor ion in positive-ion mode by collisional activation.
7. Sum data from approx 50 scans to obtain a representative spectrum.

3.3. Preparation of Surface-Labeled Proteins

3.3.1. General Considerations

The correlation of selective chemical modification of amino acid residues with the three-dimensional structure of a protein is valid only if several criteria are fulfilled:

1. The native conformation of the protein is maintained under the conditions implemented during the course of the reaction (i.e., nondenaturing) *(25,26)*. The subtle and fragile network of interactions that stabilize the native conformation of a protein must not be perturbed, as partial or local unfolding of the protein chain(s) will inevitably lead to misinterpretation of the data.

2. The relative reactivity of individual amino acids should be solely influenced by the three-dimensional structure of the protein. The introduction of modifications can cause local conformational changes and/or elimination of favorable interactions. A low molar ratio of protein to derivatizing reagent limits the extent of protein modification, thereby enhancing the relative differences in chemical reactivity between the amino acids.
3. The protein derivative must be sufficiently stable to enable identification of the modified residue(s) *(11)*.

3.3.2. Modification of Tyrosine Residues

The available methods for direct attachment of the naturally occurring iodine-127 isotope to a protein use oxidative agents to produce an electrophilic iodine species (I^+) *(27)*. Iodide (I^-) is converted to an I^+ species by oxidation using the water-insoluble compound 1,3,4,6-tetrachloro-3α,6α-diphenylglycouril (Iodogen). The I^+ species undergoes electrophilic addition to the phenolic ring of tyrosine residues. Tyrosine residues were modified using the following protocol:

1. Dissolve the Iodogen reagent in dichloromethane (DCM) at a concentration of 20 µg/mL (*see* **Note 11**).
2. Add 10-µL aliquots to reaction vessels (*see* **Note 12**).
3. Coat the Iodogen on the surface of the reaction vial by slowly evaporating the DCM (*see* **Note 13**).
4. Dissolve the protein in 20 mM PBS (pH 7.5) at a concentration of 1 mg/mL.
5. Rinse the coated reaction vial with 20 mM PBS (pH 7.5) to remove loose particles of the iodination reagent (*see* **Note 14**).
6. Add the protein solution to the coated reaction vessel.
7. Add the appropriate quantity of $Na^{127}I$ dissolved in 20 mM PBS (pH 7.5) (*see* **Note 15**).
8. Incubate the sample at ambient temperature for 10 min.
9. Terminate the reaction by decanting the reaction mixture into a clean glass test tube or Eppendorf tube.
10. Analyze the desalted iodinated protein by MALDI-MS.
11. If the acquired mass spectrum indicates that excessive oxidation of the protein has occurred (mass increments of 15.999 Daltons), incubate the iodinated protein with 25% (w/w) β-mercaptoethanol for 18 h at ambient temperature (*see* **Note 16**).
12. Reanalyze the reduced iodinated protein by MALDI-MS (*see* **Note 17**).
13. Digest the modified protein according to standard procedures or on a column of immobilized trypsin (*see* **Note 18**) *(28)*.
14. Analyze the desalted peptides by MALDI-MS.

3.3.3. Modification of Lysine Residues

N-hydroxysuccinimide (NHS) ester derivatives have the ability to introduce structural probes and other unique functional groups into proteins via lysine residues *(29)*. NHS-biotin reacts specifically with primary amines (generally lysine ε-groups) on peptides and proteins to form stable covalent bonds, with the concomitant release of *N*-hydroxysuccinimide. Each addition of a biotin molecule increases the average mass of the protein by 226.30 Daltons. A distinct advantage of modifying a protein with a group such as biotin is that the derivatized protein can be readily separated from the underivatized protein by affinity isolation with streptavidin-coated Dynabeads (Dynal, Oslo, Norway). Lysine residues were modified using the following protocol:

1. Dissolve the protein in 20 m*M* PBS (pH 7.5) at a concentration of 1 mg/mL (*see* **Notes 19** and **20**).
2. Dissolve NHS-biotin in anhydrous DMSO (*see* **Note 21**) at a concentration of 10 mg/mL.
3. Add the appropriate molar ratio of NHS-biotin to the protein sample such that the total quantity of DMSO does not exceed 10% (v/v) (*see* **Note 22**).
4. Incubate the sample at ambient temperature for 30 min (*see* **Notes 23** and **24**).
5. Analyze the desalted sample by MALDI-MS.
6. Digest the biotinylated protein on a miniature column of immobilized trypsin (*see* **Note 18**) *(28)*.
7. Analyze the peptides by MALDI-MS.

The ParR protein is a 13.3-kDa DNA-binding protein that exists exclusively in solution as a dimer (M. Dam and K. Gerdes, unpublished results) and plays an integral role in the segregation (or partitioning) of bacterial plasmids during cell division *(30)*. No X-ray crystallography or molecular modelling data are available for the ParR dimer and NMR spectroscopy of highly purified ParR preparations indicate that the protein is unstructured in solution (M. Dam and K. Gerdes, unpublished results). The ParR dimer was therefore chosen as an model for assessing the feasibility of combining limited surface labeling and chemical crosslinking with MALDI-MS to investigate the three-dimensional structure of a protein complex.

Figure 1 shows the MALDI-TOF mass spectra of the ParR dimer following biotinylation with NHS-biotin in a molar ratio of protein to reagent of (A) 1:0.5 and (B) 1:1 (*see* **Notes 25** and **26**). The spectrum in **Fig. 1A** shows a peak corresponding to the [ParR+H]$^+$ ion, which is the most abundant species under these reaction conditions, and a peak corresponding to the [ParR+biotin+H]$^+$ ion. No further additions of biotin were observed. Increasing the molar excess of protein to reagent, as shown in **Fig. 1B**, resulted in the addition of a maximum of two biotin moieties to the ParR monomer. The major species is again the protonated native protein [ParR+H]$^+$, but, in addition, the spectrum shows peaks corresponding to [ParR+biotin+H]$^+$ and [ParR+2biotin+H]$^+$ ions.

Tryptic digestion of the biotinylated ParR dimers indicated which lysine residues are preferentially modified under conditions of limited surface label-

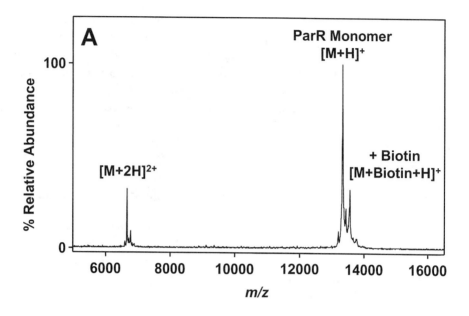

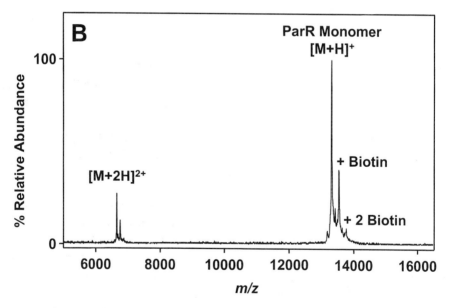

Fig. 1. MALDI-TOF mass spectrum (matrix, SA; accelerating voltage, 25 keV) of biotinylated ParR *monomers*. ParR *dimers* were modified in solution at a molar ratio of protein to NHS-biotin of (**A**) 1:0.5 and (**B**) 1:1. The major species in the spectrum, under the reaction conditions employed, is the unmodified ParR monomer. In addition, the ParR monomer modified with one biotin moiety (**A**) and two biotin moieties (**B**) are evident.

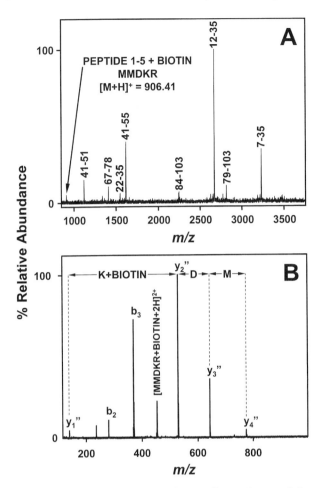

Fig. 2. Mass spectra of biotinylated ParR *dimers* digested on a miniature column of immobilized trypsin. (**A**) MALDI-MS peptide map of the protein modified at a molar ratio of protein to NHS-biotin of 1:0.5 (matrix, HCCA; accelerating voltage, 20 keV). (**B**) MS/MS spectrum of the biotinylated peptide [MMDKR+biotin+2H]$^{2+}$. The biotin label is incorporated exclusively at Lys4.

ing. **Figure 2A** is the MALDI-MS peptide map of the biotinylated ParR monomers (cf. **Fig. 1A**) following tryptic digestion. Assignment of the signals in the spectrum identified peptide 1–5 (MMDKR) as the only modified peptide. The N-terminal peptide has two possible sites for covalent attachment of a biotin group, namely, the α-amino group of Met1 and the ε-amino group of Lys4. To determine whether incorporation of biotin was favored at one of these sites, or whether there is a distribution of biotin between both sites, the peptide was analyzed by ESI-ion trap tandem mass spectrometry (MS/MS). The data

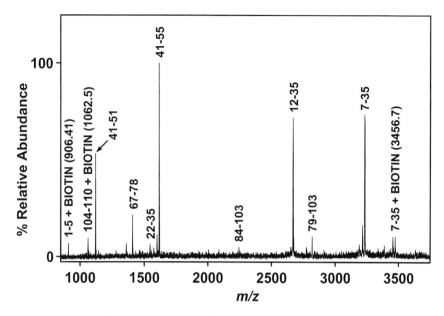

Fig. 3. MALDI-TOF peptide map of biotinylated ParR *dimers* modified at a molar ratio of protein to NHS-biotin of 1:1 (cf. **Fig. 1B**) and then digested on a miniature column of immobilized trypsin. The biotin labels are incorporated at Lys4, Lys109 and Lys11 (matrix, HCCA; accelerating voltage, 20 keV).

obtained (**Fig. 2B**) show that the only site of biotin incorporation was at Lys4. No evidence of the introduction of a biotin group at the N terminus was observed. Based on this information, Lys4 was assigned as the most reactive (or accessible) lysine residue in the ParR monomer.

A maximum of two biotin groups was incorporated into the ParR *monomer* at a protein to NHS-biotin ratio of 1:1 (**Fig. 1B**). Interestingly, the MALDI-MS peptide map of the modified protein (**Fig. 3**) revealed the presence of three peptides, each containing a single biotin modification. The peptide at mass-to-change ratio (m/z) 906.47 is attributed to peptide 1–5 (MMDKR), with the addition of biotin to the side chain of Lys4. In addition, biotinylated peptides were observed at m/z 1062.5 and m/z 3456.7. Three possibilities exist for the peptide at m/z 1062.5. Peptides 1–6 (MMDKRR) and 104–110 (SDEETKK), each with an attached biotin, have calculated monoisotopic $[M+H]^+$ values of 1062.50 and 1062.48, respectively. On this basis, it is not possible to determine whether the α-amino group of Met1 or the ε-amino group of Lys4 from peptide 1–6 or Lys109 from peptide 104–110 bears the modification. It is presumed that Lys110 (peptide 104–110) is not modified since it is improbable that trypsin would cleave at a modified residue. MS/MS analysis of the peptide at

Protein Chemical Modification and Mass Spectrometry 123

m/z 1062.5 (data not shown) revealed that the biotin moiety is, in fact, exclusively incorporated at Lys109. The peptide at m/z 3456.7 corresponds in mass to the protonated ion from peptide 7–35, which contains two lysine residues, Lys11 and Lys21, plus one biotin group. It is not possible to perform MS/MS analysis on a peptide of this size using an ion trap mass spectrometer; therefore digestion of the biotinylated ParR with Asp-N was necessary (data not shown). MALDI-MS analysis of the digest confirmed that Lys11 modification was favored over Lys21 modification. The assumption that can be made from these experiments is that Lys109 and Lys11 have similar reactivities so that the second biotin observed in the intact modified protein in **Fig. 1B** is equally distributed between these two residues. This is not unreasonable, as many investigators have reported that the incorporation of a modifying group into a protein can result in a population of different molecular species with an equal number of modifying groups but with these groups located at different positions *(11)*.

In summary, increasing the molar ratio of NHS-biotin to protein results in a gradual increase in the number of groups incorporated into the protein. The order of amino acid reactivity toward a specific reagent can be determined by MALDI-MS analysis of the derivatized proteins following proteolytic digestion and/or MS/MS of modified peptides. It is then possible to correlate the information obtained from these studies with the location, or more specifically the accessibility, of an amino acid residue within the three-dimensional structure of the protein.

3.4. Preparation of Crosslinked Proteins

3.4.1. General Considerations

The chemical crosslinking of protein components following exposure to a bifunctional reagent implies that only the polypeptides involved in the interaction were contiguous during the course of the reaction *(25)*. Contact relationships established by crosslinking methods are biologically valid if the experiment conforms to two important criteria:

1. The experimental conditions are optimal for maintenance of the structure of the native macromolecular assembly.
2. The introduction of crosslinks does not perturb the organization of the system.

3.4.2. Protein Crosslinking via Lysine Residues

DTSSP is a water-soluble, homobifunctional, thiol-cleavable, *N*-hydroxysuccinimide ester that reacts with primary amines to form stable covalent crosslinks with the concomitant release of *N*-hydroxysuccinimide. The successful incorporation of a DTSSP-generated crosslink increases the average mass of the protein (or protein complex) by 174.24 Daltons. If an appropriate residue is not located in the vicinity of the free end of a conjugated crosslinker, hydrolysis of the second moiety occurs. Such a modification increases the average mass of the protein by 192.26 Daltons. Crosslinked samples were prepared using the following protocol:

1. Dissolve the protein(s) in 20 mM PBS (pH 7.5) at a concentration of 0.1 mg/mL.
2. Dissolve DTSSP in 20 mM PBS (pH 7.5) at a concentration of 10 mg/mL (*see* **Note 27**).
3. Add the appropriate quantity of DTSSP to the protein sample (*see* **Note 28**).
4. Incubate the sample at ambient temperature for 30 min.
5. Analyze the desalted samples by MALDI-MS (*see* **Note 29**).
6. Add 20 µL 2 M urea, 0.1 mM NH_4HCO_3 (pH 8.3) to the crosslinked protein.
7. Add trypsin to 4% (w/w).
8. Digest overnight at 37°C.
9. Divide the digested sample into two equal portions.
10. Incubate one-half of the digest in 50 mM DTT for 30 min at 37°C (*see* **Note 30**).
11. Analyze samples by MALDI-MS (*see* **Note 31**).

ParR dimers crosslinked with a 50-fold excess of DTSSP were digested in solution by trypsin, and the resultant peptides were analyzed by MALDI-MS prior to and following reduction by DTT (**Figs. 4A** and **B**, respectively) *(31)*. The data obtained from differential peptide mapping revealed the presence of a single intermolecular crosslink formed between monomers of ParR. In addition, surface modification by hydrolyzed and reduced DTSSP moieties (monoisotopic mass increments of 191.99 and 87.998 Daltons, respectively) were also evident. The molecular ion at *m/z* 3509.4 corresponds in mass to an intact crosslink between peptide 6–21 from one monomer and to peptide 104–114 from the second monomer, plus the addition of a hydrolyzed DTSSP moiety. Peptide 6–21 has one lysine residue (Lys11), whereas peptide 104–114 has two lysine residues (Lys109 and Lys110). Following reduction (**Fig. 4B**), however, the peak at *m/z* 3509.40 was no longer evident, but a new peptide at *m/z* 1947.8, representing peptide 6–21 with a reduced crosslinker modification, was apparent. The peptide corresponding to residues 104–114 with two additions of the reduced crosslinker (predicted *m/z* 1368.6) was not observed, possibly due to ion suppression, which is a limiting factor in these analyses. These data demonstrate that by combining chemical protein crosslinking with differential MALDI-MS peptide mapping, the site of interaction between monomers of the DNA-binding protein ParR can be ascertained.

3.5. Concluding Remarks

The protocols detailed in this chapter demonstrate the utility of combining classical protein chemistry with biological mass spectrometry. Amalgamation of selective chemical modification procedures with standard mass spectromet-

Protein Chemical Modification and Mass Spectrometry 125

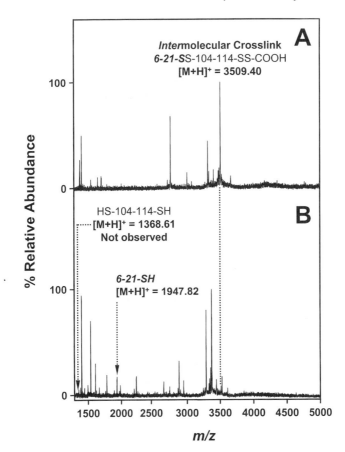

Fig. 4. MALDI-MS peptide map (matrix, HCCA; accelerating voltage, 20 keV) of ParR *dimers* crosslinked in solution by DTSSP at a molar ratio of protein to DTSSP of 1:50. Samples were digested overnight at 37°C with 4% (w/w) trypsin. Peptide maps (**A**) prior to and (**B**) following DTT reduction of the thiol linker. Note in particular the disappearance of the protonated peak m/z 3509.4 (corresponds in mass to peptide 6–21 crosslinked to peptide 104–114 with an added hydrolyzed DTSSP moiety) and the appearance of the protonated peak at m/z 1947.8 (peptide 6–21 with an added reduced DTSSP moiety). This suggests that peptides 6–21 and 104–114 were joined by the crosslinker prior to DTT reduction. The ion corresponding in mass to peptide 104–114 plus two additions of the reduced form of the crosslinker (predicted m/z 1368.6) was not detected.

ric techniques can therefore be used to determine the position of amino acid side chains in the overall configuration of the protein and to localize functional regions of the protein, which are protected on binding to other proteins, peptides, ligands, or cofactors *(11)*.

The correlation of static amino acid residue accessibility with side chain reactivity is often consistent with surface accessibilities established by other techniques *(6)*. Mass spectrometry thereby provides complementary information to supplement dynamic methods such as two-dimensional NMR spectroscopy *(32)*, X-ray diffraction analysis *(1)*, and computer-generated simulation of three-dimensional protein structure *(6)*. Thus, mass spectrometry is an indispensable analytical tool in protein biochemistry and molecular biology *(33)* and is clearly the most convenient and effective method for characterizing chemically-derivatized proteins *(34)* and for correlating protein structure with protein function *(13)*.

4. Notes

1. Hydrolysis of NHS esters is the major competing reaction in aqueous solution. Therefore, for the stability of these reagents, it is essential to avoid exposure to moisture.
2. DMSO is hygroscopic; therefore, co-solvents used in surface labeling or crosslinking experiments must be thoroughly dry. As in **Note 1**, this precaution diminishes competitive hydrolysis of the reactive modification reagent.
3. Due to the inherent stability of Iodogen, reaction vessels can be prepared in advance and stored indefinitely in a desiccator (over a drying agent at ambient temperature).
4. DTSSP is a water-soluble crosslinker; therefore, co-solvents are not required. The manufacturer suggests preparation of a stock solution of the crosslinker in 5 mM sodium citrate buffer (pH 5.0), but we did not find this step necessary.
5. In general, SA is used for MALDI-MS analyses of intact proteins and HCCA or DHB for peptide mapping of protein digests.
6. Ideally, prepare new matrix solutions each day. If stored at 4°C, away from direct light, however, the solutions are usually stable for 1 wk.
7. For details of microcolumn preparation and usage, the reader is referred to Chapter 23 of this volume.
8. For details of MALDI-MS sample preparation and analysis, the reader is referred to Chapter 23 of this volume.
9. The HPLC step is not always necessary. This is primarily dependent on the complexity of the digest, i.e., the number of peptides that are produced. It is possible to desalt entire protein digests and isolate the ion of interest in the mass spectrometer.
10. TFA is incompatible with ESI-MS. The desalting procedure, however, is identical to desalting for MALDI-MS except that 5% HCOOH is substituted for 0.1% TFA, and 50% aqueous CH_3CN + 0.1% HCOOH is substituted for 50% aqueous MeOH + 5% HCOOH.
11. The manufacturer's recommendation is that 10 µg or less of Iodogen reagent should be used per 100 µg of protein. The concentration of Iodogen will vary depending on the molecular mass of the protein and the quantity of protein available for iodination.
12. The manufacturer recommends glass vials, but the reagent can be successfully coated on clean dry plastic 500-µL Eppendorf tubes.

13. The manufacturer recommends slow evaporation of the solvent using dry nitrogen or another inert gas rather than leaving vials to dry in a fume-hood without nitrogen. We have found, however, that the latter procedure is effective for iodination of our proteins. The most important point to note is that failure to plate the Iodogen correctly on the surface of the reaction vessel can result in some of the oxidizing agent remaining in solution with the protein. This results in continuous exposure of the protein to the oxidative agent and eventually leads to substantial oxidative damage.
14. The manufacturer recommends that if visible flakes of Iodogen are present, then coating of new reaction vials is necessary.
15. As for the biotinylation reaction, the molar ratio of protein to Na^{127}I is kept to a minimum. This emphasizes the differences in amino acid reactivity toward the modification reagent. For 10 µL Iodogen in DCM added to the Eppendorf tube, we keep the total volume of protein and Na^{127}I added to the Iodogen-coated Eppendorf tube at 10 µL.
16. Oxidation of methionine residues is unavoidable when a protein is exposed to strong oxidizing agents; however, this step may be unnecessary if excessive oxidation of the protein is not evident in the mass spectrum.
17. If the iodinated protein contains intra-disulphide bridges, then one must be aware that the reduction step will increase the mass of the protein by 2.016 Daltons per disulphide bridge.
18. For details of trypsin microcolumn preparation and usage, the reader is referred to Chapter 23 of this volume.
19. Conjugation reactions are performed in buffered systems that not only maintain the three-dimensional structure of the protein but also allow reaction with a chemical reagent to occur. Modification of lysine ε-groups and/or α-amino groups requires that they be deprotonated. It is our experience that the addition of an exogenous moiety to a protein occurs successfully in buffers such as sodium bicarbonate (pH 8.3–8.5) in which most of the lysine residues are deprotonated. Nonetheless, the possibility of increased local or partial unfolding increases with increasingly basic pH. Therefore, we modify lysine residues in 20 mM PBS (pH 7.5). The degree of conjugation is reduced, but, for the purposes of probing protein surface topology for the most reactive residues, it is an appropriate buffer for conjugation reactions. Additionally, the possibility of polypeptide unfolding is diminished. The use of ammonium bicarbonate buffers (although compatible with mass spectrometric analyses) is not advisable as local pH changes diminish the number of lysine residues that are modified.
20. The concentration of the protein to be modified is crucial. The hydrolysis of modification reagent occurs more readily in dilute protein solutions, whereas the acylation reaction is favored in more concentrated protein solutions. A protein concentration of 1 mg/mL is sufficient for biotinylation reactions.
21. Aqueous systems are often incompatible with many chemical reagents used for modification purposes, as these are insoluble in aqueous buffers. This is overcome by dissolving the reagent in a water-miscible co-solvent. The choice of

co-solvent is important in that the reagent must not decompose in the co-solvent and the co-solvent should not cause irreversible denaturation and precipitation of the protein. DMSO and dimethylformamide (DMF) are the most commonly used co-solvents in protein chemistry.

22. For example, to biotinylate 1 mg of a protein in 1 mL 20 mM PBS (pH 7.5), the volume of DMSO added to the protein solution should not exceed approx 100 µL.
23. It is possible to incubate samples at a lower temperature for longer periods; however, we have found that the reaction is rapid and specific at ambient temperature.
24. Many procedures recommend quenching the reaction with a buffer containing primary amines, e.g., Tris or glycine buffers, but we omit a quenching step as removal of excess reagent is achieved by desalting the sample after modification
25. It is important to note that although the ParR *dimer* was modified in solution, the MALDI-TOF mass spectrum shows the mass of the ParR *monomer*. Noncovalent protein associations are frequently disrupted during sample preparation for MALDI-MS analysis.
26. It is inadvisable to increase the molar ratio of NHS-biotin to protein to greater than an approx 2- to 5-fold excess. Although a gradual increase in the number of biotin moieties incorporated into the protein is observed, it becomes exceedingly difficult to assign relative reactivities to specific lysine residues. The introduction of an increasing number of modifying groups into a protein leads to local and/or partial unfolding of the native structure. Thus, the relative reactivity of the residues is no longer represented by the degree that exogenous moieties are incorporated into the protein. Nonetheless, structural information concerning the three-dimensional spatial arrangement of the protein can be obtained from limited chemical modification of a protein in solution. For this reason, modification reactions were kept at low NHS-biotin to protein ratios.
27. The manufacturer's recommendation is to use the crosslinker at a concentration of 0.25–5 mM.
28. The manufacturer's recommendation is to add a 20– to 50-fold molar excess of the crosslinker over the protein when the protein concentration is below 5 mg/mL. Add a 10-fold molar excess for concentrations above 5 mg/mL. Determining the appropriate quantity of crosslinker to add to a protein solution requires several trial experiments. Monitor crosslinked samples by using non-reducing one- or two-dimensional sodium dodecyl sulfate-polyacrylamide gel electrophoresis to determine the proportion of crosslinked versus monomeric species.
29. MALDI-MS analysis of protein subunits linked by a reagent containing a disulphide bridge must be performed using DHB as a matrix. Matrix-induced fragmentation is reduced with DHB and, consequently, the disulphide link is not cleaved during the MALDI process.
30. The addition of DTT to the protein digest reduces the disulphide bridge in the spacer region of the crosslinker. Therefore, crosslinked peptides are no longer covalently linked. Reduction increases the monoisotopic mass of each of the peptides by 88.998 Daltons.

31. The disulphide bond located in the spacer region of the crosslinker in the nonreduced sample can be cleaved during the MALDI process. This must be taken into consideration when analyzing the MALDI-MS peptide maps.

References

1. Weber, P. C. (1991) Physical principles of protein crystallization. *Adv. Protein Chem.* **41**, 1–36.
2. Wüthrich, K. (1989) The development of nuclear magnetic resonance spectroscopy as a technique for protein structure determination. *Acct. Chem. Res.* **22**, 36–44.
3. Chothia, C. and Lesk, A. M. (1986) The relation between the divergence of sequence and structure in proteins. *EMBO J.* **5**, 823–826.
4. Blundell, T. L., Sibanda, B. L., Sternberg, M. J. E., and Thornton, J. M. (1987) Knowledge-based prediction of protein structures and the design of novel molecules. *Nature* **326**, 347–352.
5. Lesk, A. M. and Tramontano, A. (1989) The computational analysis of protein structures: sources, methods, systems and results. *J. Res. Natl. Intit. Standards Technol.* **94**, 85–93.
6. Suckau, D., Mak, M., and Przybylski, M. (1992) Protein surface topology-probing by selective chemical modification and mass spectrometric peptide mapping. *Proc. Natl. Acad. Sci. USA* **89**, 5630–5634.
7. Ohguro, H., Palczewski, K., Walsh, K. A., and Johnson, R. S. (1994) Topographic study of arrestin using differential chemical modifications and hydrogen deuterium exchange. *Protein Sci.* **3**, 2428–2434.
8. Kaplan, H., Stevenson, K. J., and Hartley, B. S. (1971) Competitive labelling, a method for determining the reactivity of individual groups in proteins. The amino groups of porcine elastase. *Biochem. J.* **124**, 289–299.
9. Karas, M., Bachmann, D., Bahr, U., and Hillenkamp, F. (1987) Matrix-assisted ultraviolet laser desorption of non-volatile compounds. *Int. J. Mass Spectrom. Ion Processes* **78**, 53–68.
10. Fenn, J. B., Mann, M., Meng, C. K., Wong, S. F., and Whitehouse, C. M. (1989) Electrospray ionisation for mass spectrometry of large biomolecules. *Science* **246**, 64–71.
11. Zappacosta, F., Ingallinella, P., Scaloni, A., Pessi, A., Bianchi, E., Sollazzo, M., et al. (1997) Surface topology of minibody by selective chemical modifications and mass spectrometry. *Protein Sci.* **6**, 1901–1909.
12. Krell, T., Chackrewarthy, S., Pitt, A. R., Elwell, A., and Coggins, J. R. (1998) Chemical modification monitored by electrospray mass spectrometry: a rapid and simple method for identifying and studying functional residues in enzymes. *J. Peptide Res.* **51**, 201–209.
13. Glazer, A. N. (1976) The chemical modification of proteins by group-specific and site-specific reagents, in *The Proteins* (Neurath, H. and Hill, R. L., eds.), Academic, New York, pp. 1–103.
14. Glocker, M. O., Borchers, C., Fiedler, W., Suckau, D., and Przybylski, M. (1994) Molecular characterisation of surface topology in protein tertiary structures by aminoacylation and mass spectrometric peptide mapping. *Bioconjugate Chem.* **5**, 583–590.

15. Przybylski, M., Glocker, M. O., Nestel, U., Schnaible, V., Bluggel, M., Diederichs, K., et al. (1996) X-ray crystallographic and mass spectrometric structure determination and functional characterisation of succinylated porin from *Rhodobacter capsulatus*: implications for ion selectivity and single-channel conductance. *Protein Sci.* **5,** 1477–1489.
16. Glocker, M. O., Kalkum, M., Yamamoto, R., and Schreurs, J. (1996) Selective biochemical modification of functional residues in recombinant human macrophage colony-stimulating factor β (rhM-CSF β): identification by mass spectrometry. *Biochemistry* **35,** 14,625–14,633.
17. Steiner, R. F., Albaugh, S., Fenselau, C., Murphy, C., and Vestling, M. (1991) A mass spectrometric method for mapping the interface topography of interacting proteins, illustrated by the melittin-calmodulin system. *Anal. Biochem.* **196,** 120–125.
18. Yu, Z. H., Friso, G., Miranda, J. J., Patel, M. J., Lotseng, T., Moore, E. G., et al. (1997) Structural characterisation of human haemoglobin crosslinked by bis(3,5-dibromosalicyl) fumarate using mass spectrometric techniques. *Protein Sci.* **6,** 2568–2577.
19. Glocker, M. O., Nock, S., Sprinzl, M., and Przybylski, M. (1998) Characterisation of surface topology and binding area in complexes of the elongation factor proteins EF-Ts and EF-Tu·GDP from *Thermus thermophilus*: a study by protein chemical modification and mass spectrometry. *Chem. Eur. J.* **4,** 707–715.
20. Ganguly, A. K., Pramanik, B. N., Huang, E. C., Liberles, S., Heimark, L., Liu, Y. H., et al. (1997) Detection and structural characterisation of Ras oncoprotein-inhibitors complexes by electrospray mass spectrometry. *Bioorg. Med. Chem.* **5,** 817–820.
21. Farrow, M. A., Aboulela, F., Owen, D., Karpeisky, A., Beigelman, L., and Gait, M. J. (1998) Site-specific crosslinking of amino acids in the basic region of human immunodeficiency virus type 1 tat peptide to chemically modified TAR RNA duplexes. *Biochemistry* **37,** 3096–3108.
22. Wong, D. L., Pavlovich, J. G., and Reich, N. O. (1998) Electrospray ionisation mass spectrometric characterisation of photocrosslinked DNA-*Eco*RI DNA methyltransferase complexes. *Nucleic Acids Res.* **26,** 645–649.
23. Kussmann, M., Nordhoff, E., Rahbek-Nielsen, H., Haebel, S., Rossel-Larsen, M., Jakobsen, L., et al. (1997) Matrix-assisted laser desorption/ionisation mass spectrometry sample preparation techniques designed for various peptide and protein analytes. *J. Mass Spectrom.* **32,** 593–601.
24. Wilm, M. and Mann, M. (1996) Analytical properties of the nanoelectrospray ion source. *Anal. Chem.* **68,** 1–8.
25. Lomant, A. J. and Fairbanks, G. (1976) Chemical probes of extended biological structures: synthesis and properties of the cleavable protein crosslinking reagent (35S)dithiobis(succinimidyl proprionate). *J. Mol. Biol.* **104,** 243–261.
26. Zappacosta, F., Pessi, A., Bianchi, E., Venturini, S., Sollazzo, M., Tramontano, A., et al. (1996) Probing the tertiary structure of proteins by limited proteolysis and mass spectrometry—the case of minibody. *Protein Sci.* **5,** 802–813.
27. Seevers, R. H. and Counsell, R. E. (1982) Radioiodination techniques for small organic molecules. *Chem. Rev.* **82,** 575–590.

28. Gobom, J., Nordhoff, E., Ekman, R., and Roepstorff, P. (1997) Rapid micro-scale proteolysis of proteins for MALDI-MS peptide mapping using immobilised trypsin. *Int. J. Mass Spectrom. Ion Processes* **169/170,** 153–163.
29. Lundblad, R. L. (1995) *Techniques in Protein Modification.*, CRC Press, Boca Raton, FL.
30. Dam, M. and Gerdes, K. (1994) Partitioning of plasmid R1. Ten direct repeats flanking the *parA* promoter constitute a centromere-like partition site *parC*, that expresses incompatibility. *J. Mol. Biol.* **236,** 1289–1298.
31. Bennett, K. L., Kussmann, M., Björk, P., Godtfredsen, M., Godzwon, M., Roepstorff, P., et al. (1999) Chemical crosslinking with thiol-cleavable reagents combined with differential mass spectrometric peptide mapping: a novel approach to assess intermolecular protein contacts, *Protein Sci.*, submitted.
32. Miranker, A., Radford, S. E., Karplus, M., and Dobson, C. M. (1991) Demonstration by NMR of folding domains in lysozyme. *Nature* (Lond.) **349,** 633–636.
33. Carr, S. A., Roberts, G. D., and Hemling, M. E. (1990) Structural analysis of posttranslationally modified proteins by mass spectrometry, in *Mass Spectrometry of Biological Materials* (McEwen, C. N. and Larsen, B. S., eds.), Marcel Dekker, New York, pp. 87–136.
34. Burlingame, A. L., Baillie, T. A., and Russell, D. H. (1992) Mass spectrometry. *Anal. Chem.* **64,** 467R-502R.

# 8

Secondary Structure of Peptide Ions in the Gas Phase Evaluated by MIKE Spectrometry

Relevance to Native Conformations

Igor A. Kaltashov, Aiqun Li, Zoltán Szilágyi, Károly Vékey, and Catherine Fenselau

1. Introduction

The recent explosive growth of studies of noncovalent interactions of biomolecules using mass spectrometry *(1)* has greatly stimulated interest in biomolecular conformations in the absence of solvent *(2)*. The two most widely used approaches to study the three-dimensional structure of biomolecules in the gas phase are hydrogen/deuterium exchange *(2,3)* and measurement of collisional cross-sections of bio-ions *(4)*. These experiments characterize biomolecular conformations in the gas phase on the global level, with little or no detail at the local level. In this chapter we describe the application of mass-analyzed ion kinetic energy (MIKE) spectrometry *(5)* to produce local information on the gas phase conformations of peptide ions, i.e., to evaluate secondary structures of metastable peptide ions produced by electrospray ionization. Secondary structure is evaluated based on the intercharge distances for metastable, multiply charged, peptide ions. Unimolecular dissociation of such ions results in significant increases of fragment ion kinetic energy if the fragmentation proceeds via charge-separation channels as in **Eq. (1)**.

$$m_1^{x+} \rightarrow m_2^{y+} + m_3^{z+} \qquad (1)$$

The kinetic energy release (KER) arises due to Coulombic repulsion between the like charges of the separating fragment ions m_2^{y+} and m_3^{z+} and may be used to evaluate the intercharge distances in the transition state of the m_1^{x+} ions (6–10). These distances are then compared with the results of molecular modeling of the peptide ions using various native and nonnative conformations.

1.1. Principle of KER Measurements with MIKE Spectrometry

MIKE spectrometry utilizes the ability of the electrostatic analyzer of a sector mass spectrometer to measure the energy spread of ions (5). Fragment ions produced in the unimolecular dissociation reaction with no interconversion between internal energy and kinetic energy will result in a MIKE peak whose shape is determined by the energy spread of the precursor ions and the ratio of the masses of the fragment and the precursor ions. However, conversion of internal energy to kinetic energy upon the dissociation process (e.g., due to coulombic repulsion in the process [1]) will result in significant broadening of the peak. The shape of the ion peak in the MIKE spectrum will depend on the distribution of KER in the dissociation process. A narrow KER distribution with a maximum at a nonzero energy value would produce a broad dish-topped peak with relatively sharp edges (5). This dish-topped appearance is caused by the so-called "z-discrimination" of the ion signal, since electrostatic analyzers do not provide ion focusing in the vertical direction (z-axis of the mass spectrometer). The average KER value for the process (1) may be simply estimated using the width of the fragment ion peak ΔE in the MIKE spectrum as follows:

$$T_{av} = [(y^2 \cdot m^2_1 \cdot e \cdot U)/(16 \cdot m_2 \cdot m_3)] (\Delta E^2/E^2) \tag{2}$$

where e is the elementary charge, U is the acceleration voltage, and E is the position energy of the center of the precursor ion peak in the MIKE spectrum (5). In most cases KER distribution may be deduced from the fragment ion peak shape using various deconvolution algorithms (7–11).

1.2. Secondary Structure of Peptide Ions

The two most common elements of peptide secondary structure are α-helices and β-pleated sheets. Survival of these structural elements on transition from solution to the gas phase is extremely crucial for the stability of the higher order biomolecular architecture and, therefore, the credibility of studies of noncovalent interactions using mass spectrometry. Since the secondary structure of any given peptide is a result of the cumulative action of several different interactions (e.g., hydrogen bonding, electrostatic interactions, and hydrophobic effects), we have attempted to evaluate the importance of each such interaction for the stability of the secondary structure by selecting appropriate model systems.

1.2.1. Factors Affecting Higher Order Structure of Bio-ions in the Gas Phase

Stability of specific regular patterns of intramolecular hydrogen bonds is essential for both α-helix and β-pleated sheet formation. Hydrophobic effects are also known to play a very important role in determining the secondary structure of polypeptides in solution, although this factor is all but eliminated in the gas phase. Electrostatic repulsion was believed to be a major force determining the peptide ion conformation in the gas phase *(12)*, since the absence of dielectric screening by polar solvent molecules would significantly reinforce both long range (charge–charge) and short range (dipole–dipole) *(13)* repulsive interactions. This might lead to a situation in which all polypeptides and proteins (and especially those with high β-sheet propensities) unfold in the gas phase to form extended conformations (i.e., incipient β-strands) regardless of their native conformation in solution. In our work, we have attempted to study gaseous conformations of peptides whose native secondary structures are well characterized and are maintained largely by hydrogen bonding (a small α-helix and a model β-hairpin peptide, as well as a model three-strand β-sheet former). Two other systems represent peptides whose secondary structure in solution is determined mostly by hydrophobic interactions (bombesin and polyalanine). We have also evaluated solvent-free conformations of short peptides with very high β-sheet propensities (e.g., fragments of dynorphin A) that form random coils in solution. Evaluation of the stability of such "incipient" β-strands in the absence of solvent is expected to provide useful insight into the role of electrostatic repulsion in the gas phase conformations of bio-ions.

2. Materials

The following commercially available peptides were used in these studies:

1. Dynorphin A fragment [1–17] YGGFLRRIRPKLKWDNQ, (Sigma, St. Louis, MO).
2. Dynorphin A fragment [7–15] RIRPKLKWD, (Sigma).
3. Melittin GIGAVLKVLTTGLPALISWIKRKRQQ-NH$_2$, (Sigma).
4. Bombesin Glp-Gln-Arg-Leu-Gly-Asn-Gln-Trp-Ala-Val-Gly-His-Leu-Met-NH$_2$, (BACHEM Bioscience, King of Prussia, PA).

Three other peptides had been synthesized by the BioPolymer Lab at the University of Maryland, Baltimore:

5. Polyalanine peptide, Ac-KA$_{14}$K-NH$_2$ *(14)*.
6. A model β-hairpin peptide, RGITVNGKTYGR *(15)*.
7. A model three-strand β-sheet former, βpep4, SIQDLNVSMKLFRKQAKWKII-VKLNDGRELSLD *(16)*.

Purity and structure of all peptides were determined by high-performance liquid chromatography (HPLC) assays and tandem mass spectrometry (using high-energy collision-induced dissociation [CID]). All polypeptides were used without further purification.

3. Methods

3.1. Mass Spectrometry Instrumentation and Operation

All experiments were performed on a high-performance four-sector (EBEB geometry) mass spectrometer (JEOL HX110/HX110, Tokyo, Japan) equipped with an electrospray source. Peptide solutions (0.25–1.2 mM in 49:49:2 v/v/v water+methanol+acetic acid) were normally electrosprayed at a flow rate of 0.4–1.2 µL/min. Although this solvent composition can not qualify as "native-like" for most complex biological systems, low pH does not necessarily degrade the secondary structure of peptides, and, therefore, is suitable for the purposes of our work. The temperature of a glass capillary at the electrospray interface was maintained at 120°C. Ion acceleration voltage was set at 5 kV (doubly charged precursor ions) or 3 kV (triply charged precursor ions). Precursor ions were mass-selected with the first two sectors (EB) of the mass spectrometer and introduced into the grounded collision cell. No collision gas was supplied to the cell, so that only unimolecular dissociation processes could occur. MIKE spectra of the resulting fragment ions were recorded by scanning the second electrostatic sector of the instrument. Normally 200–500 scans were acquired to record each spectrum.

3.2. Analysis of MIKE Spectra

3.2.1. Description of the Deconvolution Algorithm

The algorithm developed is based on the calculation of the kinetic energy (velocity) of product ions of the metastable decomposition investigated. The kinetic energy released in a metastable decomposition results in a slight change of kinetic energy of the product ions. This effect can be observed experimentally as broadening of metastable peaks; thus the KER distribution can be determined by analysis of the experimental metastable peak shape.

It is well known that the ideal peak profile corresponding to a metastable decomposition characterized by a single KER value is approximately rectangular *(17)*. However, rectangular peak profiles are never observed, showing the importance of using KER *distributions* instead of a single KER value. In our numerical approach, the KER distribution is represented by a series of KER values and the corresponding probabilities. Using a simple geometric model, described in detail elsewhere *(18)*, the rectangular metastable peak profile for each KER and its probability can be calculated, and the real peak profile can be obtained by adding up the rectangular parts. Conversely, the KER distribution can be determined from an experimental peak profile with a "reversed" algorithm.

This simple approach, however, can be applied only if the largest energy release of the KER distribution is relatively small. If the maximum KER exceeds 200–300 meV, discrimination effects (some of the product ions do not reach the detector, due to their large angular divergence) have to be taken into account. Larger angular divergence involves a larger degree of discrimination, and this causes the so-called dish-topped peak shapes. This kind of peak shape is characteristic of metastable decompositions accompanied by large KER, e.g., charge separation processes. **Figure 1** shows an example of the calculation of the metastable peak profile. If the KER distribution is represented by a series of discrete KER values and the corresponding probabilities, the peak profile can be calculated by adding up the distorted rectangular parts, even in the case of significant discrimination. The smooth peak profile shown in **Fig. 1** was calculated as the sum of 500 rectangular parts.

3.2.2. Equations Used in the Calculations

In our numerical method, the velocity of the product ion can be converted into kinetic energy (laboratory frame) to generate a series of kinetic energies of the metastable peak profile $E_2^i (1 \leq i \leq n)$, as in **Eq. (3)**:

$$E_2^i = (1/2) \cdot m_2 \cdot [V_1 - u_2 + i \cdot ([2 \cdot u_2]/[n])]^2 \quad (3)$$

where n is the number of points of the metastable peak profile, V_1 is the velocity of the precursor ion (laboratory frame); and u_2 is the velocity of the product ion (center-of-mass frame), which can be calculated by **Eq. (4)**:

$$u_2 = \sqrt{2T \, [(m_1 - m_2)/(m_1 \cdot m_2)]} \quad (4)$$

The intensity of the metastable peak corresponding to a given kinetic energy can be expressed as in Equation *(5)*:

$$I^i = 1/[E_2^i - E_2^{(i-1)}] \cdot C_{detected}/C_{total} \quad (5)$$

where the denominator of the first factor is the kinetic energy difference between the two consecutive abscissa points calculated by **Eq. (4)** and the second factor ($C_{detected}/C_{total}$) takes into account the discrimination effects (discussed above).

As has been noted, this simple method is valid only in the case of small KER values. If the KER is larger than a few hundred meV, it is necessary to take into account the discrimination effects. In our approach, the focusing properties of the magnetic and electrostatic sectors of the mass spectrometer are neglected. When a single KER value is considered, the product ions appear along a circle in the plane of the collector slit (**Fig. 2**). Using this model, the degree of discrimination can be calculated as the ratio of the detected and total ion current. The detected fraction of the ions is represented by the circumference of the

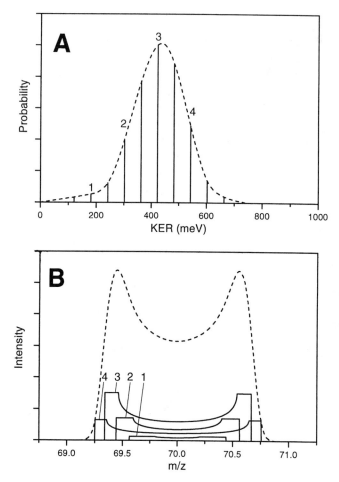

Fig. 1. (**A**) KER distribution of a hypothetical $100^{2+} \rightarrow 70^+ + 30^+$ charge separation metastable decomposition represented by the continuous curve and some discrete KER values. (**B**) Metastable peak shape of the same process calculated from the KER distribution represented by 500 data points (**A**, dashed line), and the peak shapes calculated using the single KER values shown in **A** by numbered bars.

circle falling inside a rectangular area ($C_{detected}$ in **Eq. [5]**), whereas the total-ion current is the whole circumference of the circle (C_{total} in **Eq. [5]**). The rectangular area is specified by the collector slit and a virtual slit, whose position is calculated from the acceptance angle of the electrostatic analyzer.

3.2.3. Practical Considerations

One of the most important steps of the calculation is the determination of the acceptance angle of the instrument. This defines the maximum deviation of the

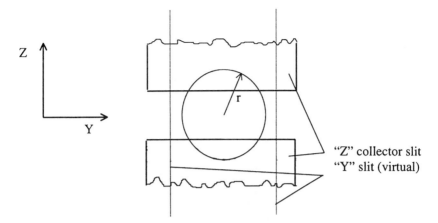

Fig. 2. Taking into account discrimination effects using the real size of the collector slit (z-direction) and a virtual slit that is specified by the acceptance angle of the electrostatic analyzer (y-direction). Both y and z directions are perpendicular to the ion trajectories.

product ion beam from the original flight direction at which the product ions can be detected. In some cases this parameter is specified by the manufacturer of the instrument. Otherwise it can be determined by a trial-and-error procedure based on the analysis and recalculation of peak shapes of a charge separation process.

The method is very sensitive to the signal-to-noise (S/N) ratio of the peak profile. This means that smoothing of the experimental data set is necessary in most cases. This can be easily performed by the data processing software of most mass spectrometers (e.g., Savitzky-Golay or Fast fourier transform algorithms). However, this has to be done with particular care, because smoothing may lead to strong distortion of the peak profile, resulting in inaccurate KER distributions. If the peak profile consists of a large number of data points, this pitfall can be avoided. Our method for determination of KER distribution requires only one-half of the fragment ion peak profile. This is particularly useful if the fragment ion peak profile is distorted by interfering peaks.

3.3. Molecular Modeling

A CHARMm22/QUANTA 4.0 (MSI, Palo Alto, CA) macromolecular mechanics modeling package was used to evaluate conformations of model peptides. Native conformations of polypeptides possessing stable secondary structure were either imported from the Protein Data Bank (PDB; Brookhaven National Laboratory, http://www.resb.org/pdb) or constructed *de novo* based on the available nuclear magnetic resonance (NMR) constraints *(15,16)*. In this latter case the dihedral angles for the segments forming β-pleated sheets were set to the standard "antiparallel strand" values, whereas the angles for the turn regions were set according to the original NMR/molecular modeling work *(15,16)*. The incipient β-strands were constructed as extended chains (dihedral angles corresponding to antiparallel β-sheets). All initial conformations (both exported and constructed *de novo*) were energy minimized using the method of steepest descents (5000 steps) to avoid possible steric hindrance. To evaluate the stability of certain conformations of model peptide ions in the gas phase, the following cycle of molecular dynamics simulation was performed. First, the relaxed (energy-minimized) molecule was heated from 0 to 500 K in 1 ps (through 1000 steps) and equilibrated at this temperature for another 1 ps (also 1000 steps). The dynamics simulation was normally run at constant temperature for 30 ps. Every 10th conformation throughout the entire simulation cycle was saved in a trajectory file that was later analyzed to evaluate the conformational space. It must be noted that the simulation time used in these studies (<50 ps) is inadequate to attempt full-scale modeling of metastable ions on the time scale of the sector instrument (μs). The major goal of the molecular modeling studies was to evaluate the stability of certain conformations of polypeptide ions in the absence of solvent and to aid the interpretation of the experimental data.

3.4. Experimental Data and Molecular Modeling: Implications for Conformational Studies

3.4.1. Melittin: Stability of Helical Conformation in the Gas Phase

Melittin (GIGAVLKVLTTGLPALISWIKRKRQQ-NH$_2$) is a 26-residue amphipathic polypeptide found in bee venom. Numerous biophysical studies have indicated that its native conformation is an α-helix (*see* **ref. *19*** and references therein for a full account). More specifically, this polypeptide consists of two helices, connected together via a hinge at a Pro14 residue. Unimolecular dissociation of the [MH$_3$]$^{3+}$ ion of melittin gives rise to a set of rather abundant y_n^{2+} fragment ion peaks (n = 18–22). Each of them is clearly dish-topped and has rather sharp edges (**Fig. 3A**), suggesting that the dissociation process is accompanied by significant KER. Estimation of the average KER based on the width of the fragment ion peak provides KER_{av}= 1.25 ± 0.12 eV. Deconvolution of the fragment ion peak shape provides a KER distribution (**Fig. 3B**) for this dissociation process. To compare these data with the results of the molecular modeling studies, the charge sites of the peptide ion have to be assigned. Altogether, melittin has six basic sites (two Arg residues, three Lys residues, and an N terminus). Arginine is the most basic natural amino acid its proton affinity exceeds that of the next closest, lysine, by more than 20 kcal/mol *(20)*; therefore two of the three protonation sites may be assigned as the guanidinium

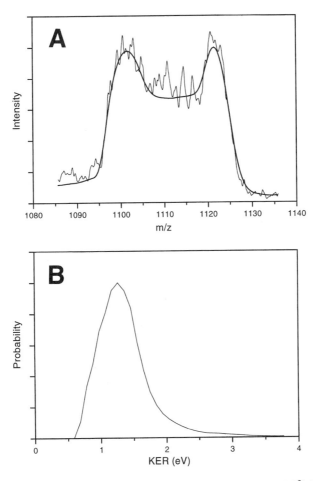

Fig. 3. MIKE spectrum of unimolecular dissociation of an $[MH_3]^{3+}$ ion of melittin, y_{19}^{2+} fragment ion region (**A**), and (**B**) a KER distribution curve deconvoluted from this spectrum.

groups of arginine side chains (Arg22 and Arg24). Two of the three lysine residues are in very close proximity to arginine residues (Lys21 and Lys23) and are highly unlikely to carry protons due to electrostatic repulsion exacerbated by the absence of dielectric screening. In fact, these two residues are very likely to be uncharged even in the presence of solvent, as has been demonstrated by titration studies *(21)*. Of the two remaining sites, Lys7 appears to be the more favorable candidate, as it is very remote from Arg22 and Arg24, and its side chain flexibility would allow for better intramolecular charge stabilization in both extended and helical conformations. Based on these considerations, we assume that the protonation sites in the triply charged melittin are

Lys7, Arg22, and Arg24. In this case, the fully extended conformation would result in rather modest electrostatic repulsion between the y_n^{2+} fragment ion and the complimentary $b_{26-n}{}^+$ ion (0.52 eV). The KER distribution curve (**Fig. 4B**) indicates that the population of fragment ions with such low KER values is virtually nonexistent. Dynamic simulation studies on melittin indicated that the helical conformation of this peptide is rather stable in the absence of solvent *(19)*, yielding a relatively narrow range of calculated electrostatic repulsion (1.05–1.40 eV). Most of the fragment ions have KER values that fall within this region, suggesting that the native-like helical conformation of melittin is retained in the gas phase. It appears that the highly oriented network of hydrogen bonds maintains the α-helical structure despite the destabilizing influence exerted by unshielded electrostatic repulsion and a rather high degree of vibrational excitation of metastable ions.

3.4.2. Stability of Multi-Strand β-Pleated Sheets in the Gas Phase

One of the major factors contributing to the stability of β-sheets in proteins is interstrand hydrogen bonding. We have attempted to evaluate the stability of such hydrogen bonding networks using relatively short peptides that form stable multistrand β-sheets in solution. One of the examples of a designed polypeptide that forms stable β-pleated sheets in solution has been named βpep-4 *(16)*. This 33-residue polypeptide (*see* **Subheading 2.**) has been shown to form a self-association-induced three-strand β-sheet structure in solution under physiological conditions *(16)*. Although NMR data suggest that βpep-4 aggregates in solution to form dimers and tetramers *(16)*, interstrand interaction is strong enough to promote folding of monomeric peptides into three-strand β-sheet structures. We have used MIKE spectrometry to examine the stability of such "zigzag" structures in the solvent-free environment. The most abundant fragment ion peak in the MIKE spectrum of fragment ions formed by unimolecular dissociation of $[MH_2]^{2+}$ ions of βpep-4 is $b_{26}{}^+$ *(22)*, originating from the cleavage of the amide bond between residues Asp26 and Gly27 located in the turn region of the reported β-sheet structure *(16)*. A comparison of recent MIKE data *(22)* with the results of molecular modeling studies suggests that this peptide is very likely to retain its multistrand β-pleated sheet structure in the absence of solvent. Analogous results have been obtained for a β-hairpin peptide *(15)*, another model β-sheet former *(22)*.

3.4.3. Collapse of Incipient β-Strands in the Gas Phase

3.4.3.1. Unstructured Single Strands with β-Sheet Propensity

The gas-phase behavior of single-stranded peptides with high β-sheet propensities was examined using several models. First, we evaluated solvent-free structures of two peptides (dynorphin fragments [1–17] and [7–15]) that have

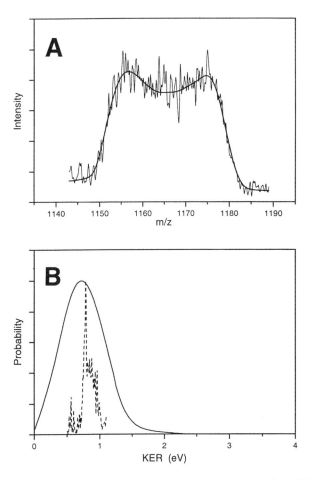

Fig. 4. (A) MIKE spectrum of unimolecular dissociation of an $[MH_2]^{2+}$ ion of bombesin, b_{10}^+ fragment ion region, and a KER distribution curve deconvoluted from this spectrum (B). Dashed line represents the distribution of electrostatic repulsion energy generated during the dynamics simulation cycle.

been found to be mostly unstructured in solution (random coil), despite their very high β-sheet propensities (23). The KER distributions were calculated based on the unimolecular dissociation of the metastable triply charged ion of dynorphin A fragment [1–17] and the doubly charged metastable ion of dynorphin A fragment [7–15]. Comparison of these distributions with the results of molecular modeling studies (data not shown) suggest that only a small fraction of these ions have KER values consistent with an extended conformation. Interestingly, analysis of the dynamics simulation studies indicates that such conformations are highly unstable in the absence of solvent due to the lack of internal stabilization of incipient β-strands. This leads to a collapse of extended chains to more compact structures stabilized by internal hydrogen bonding.

3.4.3.2. SINGLE-STRANDED SUBUNIT OF β-SHEET AGGREGATES

It has been shown that even rather short peptides can assume a β-strand conformation in solution due to *inter*molecular strand-strand stabilization, thus forming peptide aggregates *(14,24)*. It is believed that in most cases hydrophobic interaction is playing a major role in stabilizing such conformations *(14)*. Since the hydrophobic interaction is all but eliminated in the gas phase, such multipeptide aggregates are usually not observed in the electrospray spectra (apparent dissociation of multimers to monomers upon desorption). Nevertheless, the gas-phase conformation of resulting subunits is not clear. To address that question, we have evaluated the three-dimensional structure of monomers of bombesin, a peptide that aggregates in solution, exhibiting high β-sheet content *(24)*. The most abundant fragment ion in the MIKE spectrum of unimolecular dissociation of a doubly charged ion of bombesin is b_{10}^+ (**Fig. 4A**). The KER distribution curve calculated based on the MIKE data (**Fig. 4B**) indicates that the most probable KER value is 0.73 eV, corresponding to an intercharge distance of 19.3 Å. This same distance calculated for the extended conformation model of bombesin (assuming the charges are located on the side chains of Arg^3 and His^{12}, the only two basic residues of this peptide) is much higher (30.3 Å). Again, the molecular dynamics simulation yields the collapsed structure stabilized by internal hydrogen bonding. A dashed curve overlaid with the KER distribution (**Fig. 4B**) represents the distribution of electrostatic repulsion energy recalculated from the trajectory file acquired during the dynamics simulation cycle. The qualitative agreement between the two distributions is very notable; therefore it is evident that the experimentally measured KER distribution curve represents a collapsed structure of bombesin.

It must be noted that the KER distribution curve will always be broader, compared with the calculated electrostatic repulsion distributions, even for rigid model systems *(8)*. Similar results (collapsed gas-phase structures) have been recently reported *(21)* for another β-aggregate former, a polyalanine peptide *(14)*.

3.5. Concluding Remarks

The instability of single β-strands in the gas phase, regardless of their origin from random coil or β-aggregate, indicates that hydrogen bonds compete successfully with electrostatic repulsion even in the absence of the dielectric shielding provided by the solvent molecules. Such self-solvation through formation of random intramolecular hydrogen bonds leads to formation of collapsed structures in the gas phase. In contrast to that, secondary structural elements that are stabilized by *intra*molecular hydrogen bonding networks (α-helices and multi-strand β-pleated sheets) appear to be rather stable in the gas phase (at least on the microsecond time scale). Such cooperative interaction appears to compensate for the destabilizing effects of both the elevated internal energy of metastable ions *(25)* and electrostatic repulsion. Therefore, the relevance of gas-phase structures to the native conformations of polypeptides and proteins depends largely on the nature of the interactions that maintain a particular conformation or a part of it. As has been already noted, internal networks of hydrogen bonding are usually the major determinants of the secondary structural elements, ensuring survival of these elements on transition from the native environment (solution) to the gas phase. However, since hydrophobic interactions are often implicated in maintaining tertiary contacts within proteins, it is expected that the stabilities of the tertiary structures of proteins will be less predictable and reliable in the solvent-free environment. Therefore, it appears that questions involving the relation of gas-phase and solution phase higher order structures should be approached with extreme caution and examined on a case-by-case basis.

References

1. Przybylski, M. and Glocker, M. O. (1996) Electrospray mass spectrometry of biomacromolecular complexes with noncovalent interactions—new analytical perspectives for supramolecular chemistry and molecular recognition processes. *Angew. Chem. Int. Ed. Engl.* **35,** 806–826.
2. Wood, T. D., Chorush, R. A., Wampler, F. M., Little, D. P., O'Connor, P. B., and McLafferty, F. W. (1995) Gas phase folding and unfolding of cytochrome *c* cations. *Proc. Natl. Acad. Sci. USA* **92,** 2451–2454.
3. Cassady, C. J. and Carr, S. R. (1996) Elucidation of isomeric structures for ubiquitin [M+12H]$^{12+}$ ions produced by electrospray ionization mass spectrometry. *J. Mass Spectrom.* **31,** 247–254.
4. Clemmer, D. E. and Jarrold, M. F. (1997) Ion mobility measurements and their applications to clusters and biomolecules. *J. Mass Spectrom.* **32,** 577–592.
5. Cooks, R. G., Beynon, J. H., Caprioli, R. M., and Lester, G. R. (1973) *Metastable Ions.* Elsevier, New York.
6. Kaltashov, I. A. and Fenselau, C. (1995) A direct comparison of the "first" and "second" gas phase basicities of an octapeptide RPPGFSPF. *J. Am. Chem. Soc.* **117,** 9906–9910.
7. Adams, J., Strobel, F., Reiter, A., and Sullards, M. C. (1996) The importance of charge-separation reactions in tandem mass spectrometry of doubly protonated angiotensin II formed by electrospray ionization: experimental considerations and structural implications. *J. Am. Soc. Mass Spectrom.* **7,** 30–41.
8. Szilágyi, Z., Drahos, L., and Vékey, K. (1997) Conformation of doubly protonated peptides studied by charge-separation reactions in mass spectrometry. *J. Mass Spectrom.* **32,** 689–696.

9. Rumpf, B. A. and Derrick, P. J. (1988) Determination of translational energy release distributions through analysis of metastable peaks. *Int. J. Mass Spectrom. Ion Proc.* **82,** 239–257.
10. Koyanagi, K. G., Wang, J., and March, R. E. (1990) Deconvolution of kinetic energy release signals from singly, doubly and triply charged metastable ions. *Rapid Commun. Mass Spectrom.* **4,** 373–375.
11. Kim, M. S. and Yeh, I. C., (1992) Analysis of mass-analyzed kinetic energy profile. II. Systematic determination of the kinetic energy release distribution. *Rapid Commun. Mass Spectrom.* **6,** 293–297.
12. Rockwood, A. L., Busman, M., and Smith, R. D. (1991) Coulombic effects in the dissociation of large highly charged ions. *Int. J. Mass Spectrom. Ion Processes* **111,** 103–129.
13. Avbelj, F. and Moult, J. (1995) Role of screening in determining protein main chain conformational preferences. *Biochemistry* **34,** 755–764.
14. Forood, B., Perez-Paya, E., Houghten, R. A., and Blondelle, S. E. (1995) Formation of an extremely stable polyalanine β-sheet macromolecule. *Biochem. Biophys. Res. Commun.* **211,** 7–13.
15. Ramirez-Alvarado, M., Blanco, F. J., and Serrano, L. (1996) De novo design and structural analysis of a model β-hairpin peptide system. *Nature Struct. Biol.* **3,** 604–612.
16. Ilyina, E., Roongta, V., and Mayo, K. H. (1997) NMR structure of a de novo designed peptide 33mer with two distinct compact β-sheet folds. *Biochemistry* **36,** 5245–5250.
17. Szulejko, J. E., Amaya, A. M., Morgan, R. P., Brenton, A. G., and Beynon, J. H. (1980) A method for calculating the shapes of peaks resulting from fragmentations of metastable ions in a mass spectrometer. I. Peak shapes arising from single valued kinetic energy releases. *Proc. R. Soc. Lond.* A **373,** 1–11.
18. Szilágyi, Z. and Vékey, K. (1995) A simple algorithm for the calculation of kinetic energy release distribution. *Eur. Mass Spectrom.* **1,** 507–518.
19. Kaltashov, I. A. and Fenselau, C. (1997) Stability of secondary structural elements in a solvent-free environment: the α-helix. *Proteins* **27,** 166–171.
20. Wu, Z. and Fenselau, C. (1992) Proton affinity of arginine measured by the kinetic approach. *Rapid Comm. Mass Spectrom.* **6,** 403–405.
21. Quay, S. and Tronson, L. P. (1983) Conformational studies of aqueous melittin: determination of ionization constants of lysine-21 and lysine-23 by reactivity toward 2,4,6–trinitrobenzenesulfonate. *Biochemistry* **22,** 700–707.
22. Li, A., Fenselau, C., and Kaltashov, I. A. (1998)) Stability of secondary structural elements in a solvent-free environment. II. The β-pleated sheets. *Proteins* **Suppl. 2,** 22–27.
23. Taylor, J. W. (1990) Peptide models of dynorphin A (1–17) incorporating minimally homologous substitutes for the potential amphiphilic β-strand in residues 7–15. *Biochemistry* **29,** 5364–5373.
24. Carmona, P., Lasagabaster, A., and Molina, M. (1995) Conformational structure of bombesin studied by vibrational and circular dichroism spectroscopy. *Biochim. Biophys. Acta* **1246,** 128–134.
25. Drahos, L. and Vékey, K. (1998) Determination of the thermal energy and its distribution in peptides. *J. Am. Soc. Mass Spectrom.* **10,** 323–328.

9

Preparation and Mass Spectrometric Analysis of S-Nitrosohemoglobin

Pasquale Ferranti, Gianfranco Mamone, and Antonio Malorni

1. Introduction

The interaction of nitric oxide (NO) with proteins plays a critical role in several physiological systems extending from blood pressure regulation to neurotransmission *(1–3)*. In addition, NO produced in infected and inflamed tissue can contribute to the process of carcinogenesis *(4,5)*. The targets for NO on proteins are Cys and Tyr residues and bound metals such as heme-Fe^{2+} (for a recent review see ref. *6*), through which NO exerts its effects by covalently modifying or oxidizing critical thiols or transition metals in proteins. Although much work has been performed in characterizing the NO-Fe^{2+} interaction *(7,8)*, there is considerably less data on NO-Cys interactions. One reason is the labile nature of nitrosothiols (RSNOs), which are unstable in aqueous solution. It has been assumed that the lability of RSNOs is due to their propensity to undergo homolytic cleavage of the S-N bond with release of NO *(9)*.

Recent studies have demonstrated the biological importance of the interaction of RSNOs, which can be considered as NO donors, with proteins at the level of Cys residues *(10,11)*. Low-mass RSNOs are one class of endogenous compounds capable of protein S-nitrosylation, as demonstrated in the case of hemoglobin *(12)*. The β-chain of hemoglobin possesses a highly reactive thiol group at position 93 *(13)*, which is conserved among mammalian species, although its function remains unknown. It has been recently proposed that its interaction with endogenous RSNOs to form S-nitrosohemoglobin (HbSNO) *(12)* may serve to regulate blood pressure and to facilitate efficient delivery of oxygen to tissues *(14)*. Recently, structural data have been produced to support the preferential in vitro formation of HbSNO at βCys93 *(15)* in the presence of

RSNO concentrations close to that in vivo, indicating that physiological HbSNO is S-nitrosylated at that site. Other forms of HbSNO, in which βCys112 and αCys104 are also nitrosylated, are produced in vitro at higher RSNO concentration *(15)*.

Current methods of RSNO colorimetric detection require a large amount of sample. Other spectrophotometric methods, including ultraviolet (UV), infrared (IR), and nuclear magnetic resonance (NMR) techniques, can also be used. Again, sample amount, purity, and quantitation are serious difficulties associated with these methods. Electrospray ionization mass spectrometry (ESI-MS) can detect RSNOs in very small amounts and can also determine the stoichiometry of substitution *(15)*. Furthermore, this method allows coupling to high-performance liquid chromatography (HPLC), which enables on-line quantitative analysis and peptide mapping of the sites of RSNO formation.

In this chapter, procedures are described for the chemical or enzymatic synthesis and stabilization of HbSNO and for the analysis of nitrosylated globin chains using liquid chromatography coupled to ESI-MS (LC/ESI-MS). This methodology can be used to monitor the formation of HbSNO and can therefore be applied (1) to investigate the basic features of NO-hemoglobin interaction, and (2) to identify and quantify HbSNO as well as other S-nitrosoproteins in blood.

Conditions for the enzymatic hydrolysis of HbSNO to peptides are also described. The subsequent peptide mapping by LC/ESI-MS allows the determination of the cysteine residue(s) involved in nitrosothiol formation. The application of mass spectrometric methodologies to locate the position and to evaluate the strength of the S-NO bond constitute the basis for establishing the site of nitrosylation in native proteins. This approach can be extended to the characterization of S-nitrosoproteins from different sources. This makes it possible to undertake structural studies on extremely labile species such as protein nitrosothiols.

2. Materials

The following list of materials and regents is only representative. Several NO donors, e.g., S-nitrosocysteine (Cys-NO), S-nitrosoglutathione, or S-nitrosohomocysteine, synthesized as described below, can be used. Other columns and mobile phase modifiers, chromatographic and mass spectrometric instruments, are used routinely. The use of Analar-grade, or even Aristar-grade, reagents together with water produced by a Milli-Q (Millipore, Bedford, MA) system is recommended.

2.1. Blood and Hemolysate Samples

According to the aim of the researcher, HbSNO can be produced in intact red cells (blood erythrocytes) for functional studies, activity assays, and so forth, or in hemolysates (corresponding to a hemoglobin solution). In the

Preparation and MS Analysis of S-Nitrosohemoglobin

former case, blood samples are used directly after washing with physiological saline solution; in the latter case, the hemolysate is prepared as described in **Subheading 3.1.**

1. Blood samples should be collected from nonpathologic subjects immediately prior to their use for in vitro incubation in anticoagulation tubes containing 20 mM EDTA. These samples are stable for several hours stored at 4°C.
2. 0.45-µm membrane filters (Millipore).
3. Buffers for blood sample preparation: (1) physiological isotonic solution: NaCl (Fluka, Switzerland) 0.9%, pH 6.8, prepared daily; and (2) 20 mM sodium dihydrogen phosphate, NaH_2PO_4 (Fluka), pH 7.4, containing 0.5 mM sodium ethylenediamine tetraacetate (Na_2EDTA) (Fluka, Buchs, Switzerland).

2.2. Chemical Synthesis of S-Nitrosohemoglobin

2.2.1. Chemicals

1. Sodium nitrite, $NaNO_2$, (Sigma, St. Louis, MO) 100 mM in water, pH 7.4 (*see* **Note 1**).
2. Hydrochloric acid (HCl) (Carlo Erba, Milan, Italy) 250 mM in water.
3. Stock solution of L-cysteine (Sigma) 100 mM in 250 mM HCl containing 0.1 mM Na_2EDTA.
4. NaOH (Fluka) 1 M in water.
5. Physiological 10 mM sodium dihydrogen phosphate (Fluka), pH 7.4.
6. Tris buffer (Sigma) 10 mM containing NaCl (Fluka) 0.15 M in water, pH 7.4.
7. Trifluoroacetic acid (TFA; Carlo Erba) 1% (v/v) in water.

2.3. Enzymatic Synthesis of S-Nitrosohemoglobin

2.3.1. Chemicals

1. Recombinant inducible mouse macrophage NO-synthase (iNOS) (Cayman, Ann Arbor, MI, # 60802) stored in 50 mM HEPES (BioRad, Richmond, CA), pH 7.4, containing 10% glycerol (*see* **Note 2**).
2. Cell type
   a. For synthesis in intact cells: freshly prepared erythrocyte sample (2 mM hemoglobin; *see* **Subheading 3.1.1.**)
   b. For synthesis in hemolysate: freshly prepared hemolysate sample (1 mM hemoglobin; *see* **Subheading 3.1.2.**).
3. 10 mL of incubation medium (assay mixture), pH 7.4, containing: HEPES 50 mM (120 mg), $MgSO_4$ (Carlo Erba) 0.1 mM (1.2 mg), L-arginine (Sigma) 1 mM (0.3 mg), flavin adenine dinucleotide (FAD) (Sigma) 12 µM (0.1 mg), flavin mononucleotide (FMN) (Sigma) 12 µM (60 µg), NADPH (Sigma) 0.1 mM (0.8 mg), tetrahydrobiopterin (Sigma) 12 µM (0.4 mg), dithiothreitol (DTT; Sigma) 170 µM (25 µg).
4. Stock solution of L-cysteine (Sigma) 1 mM in water (*see* **Note 1**).
5. TFA (Carlo Erba) 1% (v/v) in water.

2.4. LC/MS of S-Nitrosohemoglobin and Peptide Mapping

1. 0.45-µm filter (Millipore) for sample preparation.
2. Columns (25 cm × 2.1 mm ID) for both analytical and micropreparative scale procedures. The columns are Vydac (Hesperia, CA) C4 (#214TP52) for globin analysis, and Vydac C18 (#218TP52) for peptide mapping.
3. Hewlett-Packard series 1100 modular HPLC system with an integrated diode array detector and Hewlett-Packard Chemstation Rev. A. 02.00 software for data analysis. Samples (10–50 µg) are loaded onto the HPLC column with a Hamilton glass syringe through a Rheodyne injector The loop size is 20 µL. The column effluent is split 1:25 using a Valco tee. The bulk column effluent flows through the detector for peak collection.
4. Platform (Micromass, Manchester, UK) mass spectrometer equipped with a standard electrospray source for on-line HPLC analysis via a 75-µm ID fused-silica capillary. Masslynx (Micromass) software is used for data processing.
5. HPLC-grade solvents and reagents are from Carlo Erba. These are made up as follows: (1) for globin chain analysis, Buffer A is 80:20:0.02 (v/v/v) water + acetonitrile + TFA, Buffer B is 40:60:0.02 (v/v/v) water + acetonitrile + TFA; (2) for peptide mapping, Buffer A is 0.03% (v/v) TFA in water, and Buffer B is 0.02% (v/v) TFA in acetonitrile.

2.5. Enzyme Digestions

2.5.1. Pepsin

1. Porcine stomach mucosa pepsin (Sigma) 1 µg/µL in water, freshly prepared.
2. Formic acid (Carlo Erba) 5% (v/v) in water as buffer solution.

2.5.2. Endoproteinase Glu-C

1. Endoproteinase Glu-C (Sequencing grade, Boehringer, Mannheim, Germany) from *Staphylococcus aureus* (strain V-8), 1 µg/µL in water, aliquoted and stored at –20°C.
2. Ammonium acetate 0.4% (50 mM) in water, pH 4.0, as buffer solution.

3. Methods
3.1. Preparation of Blood and Hemolysate Samples
3.1.1. Preparation of Erythrocyte Solution

1. Transfer aliquots of 1 mL of blood to centrifuge tubes. Centrifuge at 1100g. Decant away plasma (supernatant), taking care to avoid disturbing the pellet.
2. Wash erythrocytes 3 times with 2 mL isotonic 0.9% NaCl solution by centrifugation at 1100g for 15 min at 4°C.
3. Suspend 2 mL of packed erythrocytes in 2 mL of 20 mM sodium dihydrogen phosphate, pH 7.4.
4. Aerate samples with a gentle stream of air for 5 min to convert hemoglobin fully to the oxy-form. This procedure results in a hemoglobin concentration of about 2 mM.

3.1.2. Preparation of the Hemolysate

1. Prepare the hemolysate by suspending 1 mL of washed erythrocytes in 2 mL of deionized water. Mix well. After hemolysis has occurred, pellet the debris by centrifugation (12,000g, 2 × 20 min at 4°C) and use the supernatant fraction at once or store frozen (–20°C for up to a week or –70°C for up to 3 mo) until required. This hemolysate has a hemoglobin concentration of about 1 mM.
2. Aerate the sample with a gentle stream of air for 5 min. This oxy-Hb (about 1 mM) is estimated by sodium dodecyl sulfate-gel electrophoresis and by isoelectric focusing (IEF) to be about 95% pure and is prepared immediately prior to use *(16)* *(see* **Note 3**).
3. Filter through a 0.45-µm membrane filter (Millipore).

3.2. Chemical Synthesis of S-Nitrosohemoglobin

3.2.1. Preparation of S-Nitrosothiol (S-nitrosocysteine) Solutions

Due to the instability of *S*-nitrosocysteine in water, with a half-life of the order of minutes *(9)*, solutions must be prepared, using the following procedure, immediately before use *(see* **Note 1**).

1. Prepare a stock solution of 100 mM L-cysteine by dissolving the compound in 250 mM HCl.
2. Prepare a solution (50 mM) of Cys-NO by mixing equal volumes of 100 mM L-cysteine in 250 mM HCl containing 0.1 mM Na$_2$EDTA with 100 mM NaNO$_2$ in water at 25°C. Check that the solution turns from clear to various shades of red instantaneously, indicating completion of the reaction. Previous studies have established that essentially no nitrite remains in the nitrosothiol solutions using this synthetic method *(9)*.
3. Titrate the solution to pH 7.4 by adding 1 M NaOH and then 10 mM Tris 0.15 M buffer containing NaCl (pH 7.4). Measure the RSNO concentration by UV-visible adsorption spectroscopy *(9)* *(see* **Note 1**).
4. Make dilutions as necessary in the same buffer.

3.2.2. Preparation of S-Nitrosohemoglobin

3.2.2.1. In Hemolysate

1. Add 200 µL of fresh hemolysate to 800 µL of aerated phosphate buffer (pH 7.4) containing 0.5 mM Na$_2$EDTA.
2. Incubate at 25°C.
3. Add the Cys-NO solution (prepared immediately before use), using 40 µL (if a 1:10 hemoglobin/Cys-NO ratio is desired) or 400 µL (if a 1:100 hemoglobin/Cys-NO ratio is desired) Cys-NO solution. Mix continuously.
4. Incubate at 25°C for 30 min.
5. Stop the reactions by diluting samples with 400 µL 1% TFA. The pH decrease due to the acid has been found by mass spectrometry to slow the degradation of HbSNO by extending the half-life of HbSNO to about 48 h *(see* **Note 4**). This

step is also denaturing and has to be avoided if HbSNO is required in the native form for structural or functional studies.
6. Analyze HbSNO immediately by LC/ESI-MS.
7. Provide, as a suitable reference, a blank containing all the components listed in the previous steps, but to which the sodium nitrite solution has not been added.

3.2.2.2. IN INTACT RED CELLS

1. Use blood samples directly after washing with physiological saline as described in **Subheading 3.1.1., step 2**.
2. Aerate the sample with a gentle stream of air for 5 min to convert hemoglobin fully to the oxy-form.
3. Add 100 µL of this solution to 800 µL aerated phosphate buffer (pH 7.4) containing 0.5 mM EDTA.
4. Incubate with the necessary amount of Cys-NO at 25°C for 30 min, using 40 µL or 400 µL Cys-NO solution (corresponding to a 1:10 or 1:100 hemoglobin/Cys-NO ratio, respectively) as required.
5. To avoid HbSNO degradation, after the reaction has occurred, lyse red cells by directly adding 300 µL 1% aqueous TFA. This step is denaturing and has to be avoided if HbSNO is required in the native form.
6. Remove stroma by centrifugation (twice) at 12,000g for 30 min at 4°C.
7. Analyze the HbSNO produced immediately by LC/ESI-MS.
8. Provide, as a suitable reference, a blank containing all the components listed in the previous steps, but to which the sodium nitrite solution has not been added.

3.3. Enzymatic Synthesis of S-Nitrosohemoglobin

Before its use, the activity of iNOS is quantitated spectrophotometrically using the oxyhemoglobin assay *(17)*. This assay measures the reaction of nitric oxide with oxyhemoglobin to yield metahemoglobin ($\varepsilon = 60000$ M^{-1} cm^{-1} at 401 nm). One unit of enzyme produces 1 nmol of nitric oxide per minute at 37°C in 50 mM HEPES (pH 7.4) containing 1 mM L-arginine, 1 mM magnesium acetate, 5 µM oxyhemoglobin, 0.1 mM NADPH, 12 µM tetrahydrobiopterin, and 170 µM DTT.

3.3.1. Production of HbSNO (see **Note 5**)

1. Add 30 µL 1 mM L-cysteine in water to 250 µL incubation buffer.
2. Add 9.8 µL of enzyme (iNOS) solution (corresponding to 1 enzyme unit).
3. Allow the formation of Cys-NO to take place for 30 min at 37°C.
4. After **step 3** is complete, synthesize HbSNO by adding 3 µL of hemolysate or 1.5 µL red cell sample (both previously flushed with a gentle air stream) and incubate at 37°C for 30 min.
5. Stop the reaction with 200 µL 1% TFA.
6. Immediately analyze the HbSNO product by LC/ESI-MS.
7. Prepare a blank in which all the above components are present except the enzyme solution.

3.4. LC/ESI-MS Analysis of S-Nitrosohemoglobin (see Note 6)

3.4.1. HPLC Separation of Globin Chains

Globin chains from HbSNO synthesized either chemically or enzymatically are purified by reversed-phase (RP)-HPLC using the procedure developed by Shelton et al. *(18)* with the following modifications:

1. Before the analysis, filter the globin sample through a 0.45-µm filter (Millipore).
2. Carry out liquid chromatography, under thermostatically controlled conditions at 45°C, with a 5-µm C4 RP, 250 × 2.1 mm ID, column (Vydac) and using the following buffer system: Buffer A 80:20:0.02 (v/v/v) water + acetonitrile + TFA; Buffer B 40:60:0.02 (v/v/v) water + acetonitrile + TFA.
3. Set the flow rate at 0.2 mL/min and a split of 1:25 of the column effluent by using a Valco tee, to give a flow rate of about 8 µL/min into the electrospray nebulizer. The bulk of the flow is run through the detector for peak collection as determined by measuring the absorbance at 220 nm.
4. Equilibrate the column with 52% Buffer B
5. Apply the sample (50 µg of freshly synthesized HbSNO dissolved in 20 µL volume) onto the column with a glass syringe by means of a 20 µL loop injector (*see* **Note 7**).
6. After a 2-min hold, raise Buffer B to 66% over 50 min.
7. For further structural studies, collect HPLC fractions containing nitrosylated globins manually or by use of an automated sample collector. Concentrate using a Speed-Vac vacuum evaporating system, reducing the volume to 30–50 µL, but taking care not to dry completely to avoid nitrosothiol decomposition. Store at –20°C.

3.4.2. Electrospray Mass Spectrometry

1. Set the electrospray needle at a voltage of 3.6 kV, the cone voltage at 20V, the sheath gas pressure (nitrogen 99.997%) at 75 psi, and the source temperature at 100°C.
2. Just before starting the analysis, calibrate the mass scale using 5 µL myoglobin solution (from a solution of 10 pmol/µL) as reference, introduced by means of a separate flow injection at a flow rate of 5 µL/min.
3. Set the effluent split from the HPLC separation to a flow rate of 8 µL/min. Connect the HPLC column to the mass spectrometer via a 75-µm ID fused-silica capillary. Inject the sample as previously described.
4. Start data acquisition. Scan the mass spectrometer from mass-to-change ratio (m/z) 1600 to 600 with a scan cycle of 5 s/scan.
5. Stop the acquisition after 50 min.
6. Identify the components by deconvolution of the multiply charged ion data corresponding to each species. Nitrosylated α- and β-globins are identified by a +29 mass unit shift with respect to the native chains.
7. Perform quantitative analysis of components by integration of the multiply charged ions for a single species. For both the α- and β-chains use the ions carrying 13–22 charges, which in the analytical conditions used are the most intense and can be easily identified in the spectra (*see* **Note 8**).

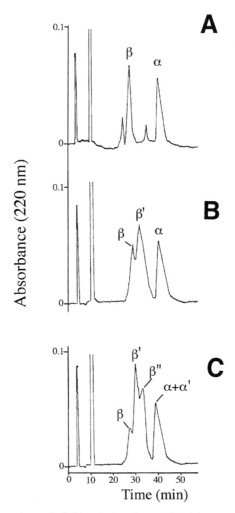

Fig. 1. HPLC separation of globin chains from HbSNO prepared by the chemical procedure. Chromatography was performed at 45°C using a Vydac C4 column at a total flow rate of 0.2 mL/min. Elution of the globin chains used a linear gradient from 52 to 66% Solvent B *(see text)* over 50 min and UV absorbance was monitored at 220 nm. The nitrosylated globins were eluted at a slightly more hydrophobic position, which resulted in broadening of the peaks. (**A**) Standard hemolysate. (**B**) Sample incubated with a 10-fold molar (10:1) Cys-NO/hemoglobin ratio. (**C**) Sample incubated with a (100:1) Cys-NO/hemoglobin ratio; β, β', β" represent β-globin and nitrosylated derivatives; α and α' represent α-globin and nitrosylated derivative.

Figure 1 shows the HPLC separation of the constituent globins of the hemolysate incubated in the presence of Cys-NO in a 1:10 (**Fig. 1B**) and 1:100 (**Fig. 1C**) molar ratio. Compared with a standard hemolysate (**Fig. 1A**), the

HPLC profile shows a double peak at the β-position in the case of the 10-fold Cys-NO/hemoglobin ratio, whereas a greater shift to a more hydrophobic position for both α- and β-globin in the case of incubation with a 100:1 ratio was observed. This indicates that with a low concentration of the reagent, it is possible to obtain a more selective modification.

Figure 2 shows the electrospray spectrum of globin chains from HbSNO obtained by the chemical method with a 10-fold excess of NO-Cys. The mass measured for the α-chain (15,126.5 ± 1.1; **Fig. 2B**) perfectly matches with that of the native globin (15,126.4). Of the two species present at the β-position (**Fig. 2A**), the first one is the native β-globin (15,868.5 ± 1.4, expected mass 15,867.2), whereas the second has a mass of 15,897.8 ± 1.1 with a shift of +29 mass units compared with that measured for the β-globin, indicating binding to the protein of a single NO moiety *(see* **Note 9***)*.

In the case of HbSNO obtained by treatment of hemoglobin with a 100-fold excess of Cys-NO, a different pattern of modification is observed (spectrum not shown). The α-chain is present both as the unmodified and the mononitrosylated species (measured masses 15,125.5 ± 1.1 and 15,164.1 ± 2.2, respectively), whereas in the ES spectra of the β-chain, components are observed containing zero, one, or two nitrosyl-groups (15,867.2 ± 0.7, 15,896.0 ± 0.5 and 15,924.0 ± 1.6, respectively). Therefore, using this level of nitrosating agent, a nonselective modification, which involves the two cysteines of the β-chain and the single cysteine of the α-chain, is obtained.

In the case of HbSNO prepared by the enzymatic method using iNOS *(see* **Subheading 3.3.***)*, only the mononitrosyl-protein at the level of βCys93 is obtained. The amounts of the native and nitrosylated β-globins are in the ratio 1:0.3. This quantitative analysis must be carried out at a low cone voltage (20 V) and temperature (100°C) *(see* **Subheading 3.7.***)*, conditions that minimize nitrosothiol decomposition in the ion source and furnish the best quantitative data.

3.5. Enzymatic Digests

3.5.1. Sample Preparation

Concentrate HPLC-purified globin samples from HbSNO by using a Speed-Vac evaporating system, to a volume of 30–50 µl. These samples will contain about 15 µg of globin and can be directly digested with pepsin or endoproteinase Glu-C for structural characterization *(see* **Note 10***)*.

3.5.2. Peptic Digest

1. Dissolve sample in 100 µL of 5% formic acid.
2. Add the enzyme solution (1 µL) corresponding to a substrate:enzyme ratio of 15:1 (w/w) and incubate at 37°C for 1 h. Stop the reaction by freezing at –20°C.

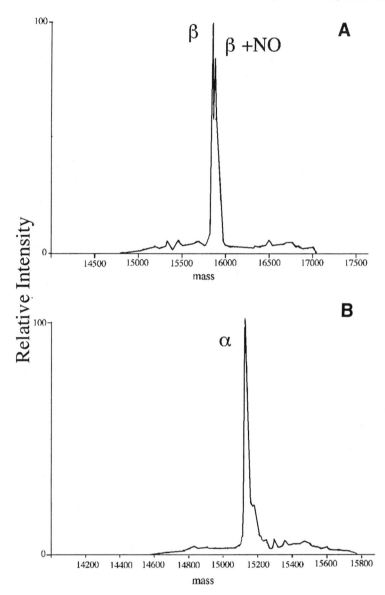

Fig. 2. ESI spectra of globin chains from HbSNO on a transformed mass scale. Globin chains from hemoglobin, prepared by the chemical method using a 10-fold molar Cys-NO/hemoglobin ratio, were separated by HPLC as shown in **Fig. 1B**. The column effluent was split on-line to give a flow rate of 8 μL/min into the electrospray ion source, which was held at 100°C. ESI spectra were scanned from m/z 1600 to 600 at a scan rate of 5 s/scan. Average masses are reported. The mass spectra of the peaks eluting at 30 min (β-globin and nitrosylated derivative) and at 40 min (α-globin) are shown in **A** and **B**, respectively.

Preparation and MS Analysis of S-Nitrosohemoglobin

3.5.3. Endoproteinase Glu-C Digest

1. Dissolve samples in 100 µL of 0.4% ammonium acetate, pH 4.0.
2. Add the enzyme solution (1 µL) corresponding to a substrate:enzyme ratio of 15:1 (w/w) and incubate at 40°C for 2 h. Stop the reaction by freezing at –20°C.

3.6. LC/ESI-MS Peptide Mapping

3.6.1. HPLC Separation of Peptides

1. Carry out liquid chromatography, under thermostatically controlled conditions at 45°C, with a 5-µm C4 RP, 250 × 2.1 mm ID, column (Vydac) and a flow rate of 0.2 mL/min using the following solvent system: Solvent A 0.03% TFA (v/v) in water; Solvent B 0.02% TFA in acetonitrile.
2. Equilibrate the column with 5% Solvent B.
3. Apply the sample (20 µL of peptic or endoproteinase Glu-C digest) onto the column with a glass syringe via a 20-µL loop injector.
4. Split the column effluent by 1:25 with a Valco tee, to give a flow rate of about 8 µL/min into the electrospray nebulizer. Run the bulk of the flow through the detector for peak collection as measured by following the absorbance at 220 nm.
5. Separate the peptides with a gradient of 5–40% Solvent B over 50 min.
6. If required, collect HPLC fractions containing nitrosylated globins manually or by use of an automated sample collector. Concentrate using a Speed-Vac vacuum evaporating system, reducing the volume to 30–50 µL, but taking care not to dry samples completely to avoid nitrosothiol degradation. Store at –20°C.

3.6.2. Electrospray Mass Spectrometry of Nitrosylated Peptides

1. Repeat **steps 1–3** of **Subheading 3.4.2.**
2. Start data acquisition. Scan the mass spectrometer from m/z 1800 to 400 with a scan cycle time of 5 s (for analysis in the selected-ion monitoring mode, set a cycle time of 2 s and an acquisition time window of ±25 s around the expected elution time of the peptide). Stop data acquisition after 60 min.
3. Assign peptide masses by deconvolution of the multiply charged ion data corresponding to each species. Perform quantitative analysis of components by integration of the multiply charged ions for a single species.
4. Identify the components by the use of a dedicated software program (Peptide tools, Hewlett-Packard). Just as for globins, nitrosylated peptides are identified by a +29 mass unit shift with respect to the native chains (*see* **Note 11**).

The LC/ESI-MS analysis of a peptic digest of the nitrosylated β-globin after incubation with 1:10 RSNO is shown in **Fig. 3**. HPLC peaks can be assigned to peptides within the globin sequence on the basis of the measured molecular weight *(15)*. It is possible to confirm the entire β-globin sequence by this procedure.

The peak at 30.9 min in the chromatogram of **Fig. 3** contains a peptide with a molecular weight of 2052.6 ± 0.2 (identified by the doubly and triply charged

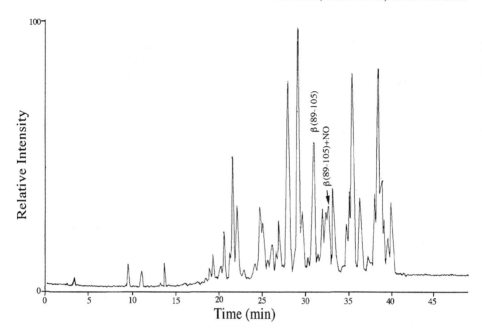

Fig. 3. LC/ESI-MS analysis of the peptic digest of mononitrosyl β-globin. Fractionation of peptides used a Vydac C18 column (250 × 2.1 mm × 5 μm). The column effluent was split 1:25 to give an approximate 8 μL/min flow rate into the electrospray ion source. The bulk of the column effluent flows through the detector for peak collection. ESI spectra were scanned from m/z 1800 to 400 at a scan rate of 5 s/scan. The source temperature was kept at 100°C and the cone voltage at 20 V.

ions at m/z 1027.2 and m/z 685.1, respectively), which is due to peptide 89–105 (molecular weight 2052.3) containing βCys93. A novel peak is also detected at higher retention time (31.9 min) and contains a peptide of mass 2081.3 ± 0.7 (again identified by the doubly and triply charged ions at m/z 1041.7 and m/z 694.8 respectively). This value corresponds to the mass of peptide 89–105 increased by 29 mass units, due to nitrosylation at the level of βCys93. The procedure used to identify the site of nitrosylation is described in **Subheading 3.7.**

Globin chains modified following incubation with a 100:1 Cys-NO/hemoglobin ratio are analyzed using the same procedure. At this higher ratio, all the cysteine residues in the α- and β-chains are at least partially nitrosylated. Peptides containing βCys112 are all found in the native form.

Nitrosylated peptides from HbSNO can also be identified from the analysis of an endoproteinase Glu-C digest by the characteristic +29 mass unit shift. Endoproteinase Glu-C hydrolysis of HbSNO (chemically or enzymatically prepared) produces, as the only modified peptide, the sequence 91–101 having βCys93 nitrosylated and a resultant mass of 1335.1.

3.7. Identification of Modified Cysteine Residues

Strategies for the identification of nitrosyl-peptides all rely on the lability of the S-NO bond, as a result of which they can be easily identified in a mixture of ordinary peptides. Mass spectrometric analysis can be instrumental in the accurate location of nitrosylated Cys residues on proteins.

3.7.1. Identification by Varying the Cone Voltage or the Source Temperature

In ESI-MS experiments, the degree of energy input into protein or peptide ions can be controlled by two independent parameters. One variable is the ion source capillary temperature, and the other is the potential difference between the capillary and the tube lens (cone) of the ion source. In a peptide containing a single cysteine residue, a nitrosothiol can be identified and its thermodynamic properties examined by altering this voltage or the temperature of the electrospray capillary transfer tube.

For example, in the case of the peptic digest of nitrosyl β-globin, the peak found at 31.9 min in the peptic digest is collected and analyzed by ESI-MS at different cone voltage values. In **Fig. 4** the ESI mass spectra obtained at 20 and 80 V are shown. At low voltage, essentially only the nitrosylated peptide is present (2081.6 ± 0.7). Increasing the cone voltage leads to fragmentation until at 80 V only the fragment peak at 2052.5 ± 0.3 is observed.

The same results can be obtained by comparing two experiments at different temperatures, as suggested by previous studies *(19)*. The first analysis is carried out at a source temperature of 100°C, and the second at 180°C. In the first case, the nitrosylated peptides remain intact, whereas at higher temperature the S-NO bond is broken and only fragmented peptides are observed. In addition, selected-ion monitoring, using ions corresponding to nitrosylated peptides, can be used to identify and quantify the level of HbSNO present.

3.7.2. By Alkaline Degradation

The HPLC fractions containing the nitrosylated peptides are collected, lyophilized, and then redissolved in 0.4% ammonium bicarbonate (pH 9.0) for 2 h. Samples are finally analyzed by ESI-MS. Thus, in the case of the peptic digest of the HbSNO obtained by incubation with a 10-fold Cys-NO excess, ESI-MS analysis shows that the peptide 89–105 shifts from mass 2081.6 to 2052.3, because of hydrolysis of the S-NO bond in the alkaline buffer.

3.8. Applications

3.8.1. Identification of the Unique Reactive Cysteine of Hemoglobin

The present procedure for the synthesis of HbSNO, involving RSNO intermediates, is able to nitrosylate selectively the thiol group of cysteine residues in hemoglobin by varying the ratio of reagent to protein. In practice, the use of

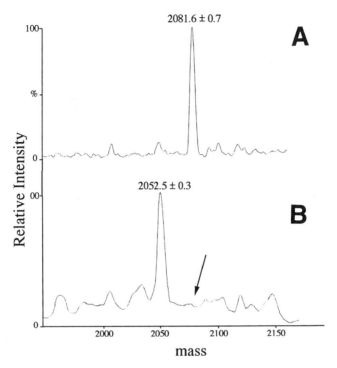

Fig. 4. Transformed ESI spectra of nitrosylated peptide (89–105) obtained at different values of cone voltage: **(A)** 20 V and **(B)** 80 V. The source temperature was kept at 100°C. At 20 V only the nitrosylated peptide is present (2081.6 ± 0.7 mass units). At 80 V only the fragment peak at 2052.5 ± 0.3 mass units is observed.

RSNOs at different concentrations allows the modification of hemoglobin to be controlled. Thus, at a low RSNO concentration, a single cysteine residue, namely βCys93, is almost completely modified in vitro after 30 min. At longer times (2 h), the reaction is complete and no further modification is observed. It is worth mentioning that, by use of a higher concentration (100:1) of nitrosothiol, modification at βCys112 and αCys104 is also observed. However, this concentration appears to be far from a concentration of real physiological significance *(12)*.

In the case of the enzymatic synthesis of HbSNO, only the mononitrosylated species is produced, even after a long (5-h) incubation time. By the chemical method, selective modification is strictly dependent on the RSNO concentration, whereas a product closer to the physiological one is obtained by the enzymatic procedure.

3.8.2. Analytical Model for Monitoring Cysteine Nitrosylation in Proteins

Ideally, an analytical method will easily and rapidly detect any modification in intact proteins and identify their specific nature and site(s). Mass spectro-

Preparation and MS Analysis of S-Nitrosohemoglobin 161

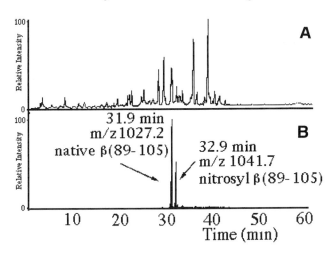

Fig. 5. LC/ESI-MS analysis of the peptic digest of globin chains from a blood sample treated with iNOS. (A) UV trace. (B) Chromatogram, acquired in SIM mode, of the ions at m/z 1027.2 and m/z 1041.7 (doubly charged ions of native and nitrosylated peptide 89–105 containing βCys 93, respectively). An acquisition window of ±25 s around the expected elution time for each of the two peptides was used.

metric methods are virtually independent of the type of modification to be identified and therefore ideally suited, even to the study of labile modifications of proteins. Modified species can be identified, even in the case of NO, in which the mass difference due to the reaction is only 29 mass units. Molecular weight measurement by ESI-MS is instrumental in the immediate evaluation of the degree of nitrosylation of the globin chains. In the analysis described for HbSNO, LC/ESI-MS is used to monitor the reactivity of cysteine residues within the hemoglobin sequence. This approach can be applied to the analysis of purified globins as well as to direct examination of hemolysate and blood samples without the need for adduct purification.

Once HbSNO has been enzymatically digested to peptides, the reactivity of specific cysteine residues can be measured by selected-ion monitoring (SIM) of nitrosylated peptides. In **Fig. 5** the LC/ESI-MS analysis of the peptic digest of globin chains from a blood sample treated with iNOS in vitro is shown. **Figure 5A** shows the UV trace and **Fig. 5B** the SIM chromatogram from the ions at m/z 1027.2 (doubly charged ion of the native peptide 89–105 containing βCys93) and m/z 1041.7 (doubly charged ion of the nitrosylated peptide 89–105). Data for each ion are acquired in the LC/ESI-MS experiment at the elution time expected for the corresponding peptide, previously determined by inspection of **Fig. 3**. By this measurement, the ratio of modified and native species can be derived. This method can be applied to the quantitative measurement of nitrosylated proteins in the blood of subjects with pathologies involving an altered NO level (traced by the increase of HbSNO). It also constitutes an example of application of the methodology for characterizing, under physiological conditions, proteins whose function is modulated by nitrosylation, as recently discovered for tissue plasminogen activator *(20)*, calcium-dependent potassium channel *(21)*, pertussis toxin-sensitive G proteins *(22)*, and cardiac calcium release channel *(23)*.

4. Notes

1. Several NO donors can be used other than Cys-NO, for example *S*-nitrosoglutathione or *S*-nitrosohomocysteine, each having a different life time *(24–26)*. Owing to the propensity of the thiol solutions to autooxidize, even at acidic pH, make fresh preparations daily, protected from light, and store them on ice at 4°C until just before their use *(24)*. The synthesis of RSNOs is monitored, and the concentration of the RSNO produced is measured, by UV-visible absorption spectroscopy, these compounds having characteristic absorption maxima at 320 and 550 nm *(25,26)*.

    Another potential problem that arises in the analysis of biological thiols and their *S*-nitrosothiol derivatives is the potential formation of RSNOs as artifacts at acidic pH and high temperature. However, it has been demonstrated *(25)* that by working at temperatures lower than 37°C and at pH ≥ 3, the formation of RSNOs as artifacts is <5%.

2. The enzyme frozen at –80°C is stable for several months. The specific activity is typically 5–7 U/mg. One unit corresponds to about 9.8 µL of enzyme preparation and must be checked before its use. A lower value indicates that the enzyme is degraded and must be discarded.

3. Isoelectric focusing is performed with the Resolve-Hb IEF system gel, pH 6–8 (Isolab., Akron, OH) at 15°C for 40 min at constant power (10 W with a 7-well gel or 15 W with a 13-well gel) in a Pharmacia (Uppsala, Sweden) Multiphor II electrophoresis apparatus. The gel is then soaked in 10% trichloroacetic acid for 20 min and washed with five changes of distilled water over a period of 2–3 h. After drying in a Beckman (Berkeley, CA) Paragon dryer for 60 min, the gel is stained for 10 min in a solution containing 0.2% bromophenol blue + 30% ethanol + 5% glacial acetic acid. The destaining solution consists of 30% ethanol + 5% glacial acetic acid solution.

4. The formation and stability of the HbSNO can confirmed by mass spectrometry. In physiological phosphate buffer (pH 7.4) at 25°C, mass spectrometric analysis shows that the half-life of HbSNO is about 24 h. The inherent stability of HbSNO, and in general of protein RSNOs, is quite remarkable and contrasts strikingly with that of low molecular weight RSNOs, which are exceedingly unstable under physiological conditions *(9,10)*.

5. iNOS is responsible for the biosynthesis of nitric oxide from L-arginine. iNOS is a soluble enzyme found in a variety of tissues/cell types including macrophages, hepatocytes, and vascular smooth muscle *(27)*. iNOS activity is essentially inde-

pendent of calcium and has a K_m of 16 µM for L-arginine. Recombinant mouse macrophage iNOS, stored frozen at –80°C, in 50 mM HEPES (pH 7.4) containing 10% glycerol, is stable for at least six months, but loses approximately 10–15% of its activity during a single freeze/thaw cycle. In the event that only a portion of the enzyme is to be used in a single experiment, it is recommended that the enzyme be aliquoted into smaller sizes and frozen at –80°C. iNOS is a relatively unstable enzyme. Hence, to prevent loss of activity, keep the stock vial of the enzyme on ice (0–4°C) at all times. The enzyme should be added to the incubation medium (assay mixture) just before the start of the experiment.

6. The performance of the protein/peptide identification system is assessed in advance by the use of protein/peptide calibration samples. Different amounts of purified human hemoglobin sample, ranging from 100 to 300 ng as determined by amino acid composition analysis, are digested with pepsin or endoproteinase Glu-C; the peptide mixtures are then separated by RP-HPLC and detected on-line by UV absorbance and ESI-MS. The reliability of the results obtained is discussed in **Notes 7** and **8**.

7. UV absorbance detection. In general agreement with our previous experience *(15)*, use of neither the chemical nor the enzymatic synthesis protocol is found to interfere significantly with the UV absorbance detection of proteins and peptides. A typical result is shown in **Fig. 4**. After elution of the large injection peak, the peptides derived from the digestion of 300 ng of HbSNO are detected as a smooth baseline with a sensitivity limit in the picomole range and a typical peak width of 30–40 s. Although not essential for protein identification, the UV chromatogram provides valuable information on the quality of the sample analyzed and the level of contaminants and is useful for trouble-shooting the HPLC system. The use of a microbore on-column detection system, low ID fused-silica capillaries, and PEEK tubing minimizes the loss of chromatographic resolution between the flow cell and the ion source of the mass spectrometer.

8. Mass spectrometry detection. The column effluent monitored by UV absorbance detection is also analyzed by on-line ESI-MS. The minimum amount of protein required for unambiguous protein/peptide identification is approx 2 pmol (120 ng) of hemoglobin. The high accuracy of molecular mass determination using ESI-MS (standard deviation 0.02% or ±0.2 Daltons for peptides and ±1.5 Daltons for globins) allows determinations of modified proteins/peptides that differ from the native species by 29.0 ± 0.2 Daltons. Therefore, RSNO formation in proteins and peptides can be unambiguously identified.

9. The different behavior exhibited by βCys93, compared with the other cysteines, is in agreement with the already reported in vivo reactivity of this residue in oxy-hemoglobin, which readily undergoes oxidation to form mixed disulphides and other thioethers due to its position within the protein quaternary structure *(12)*. A different reactivity of cysteine residues in hemoglobin has been recently observed for another highly reactive electrophilic agent, methyl bromide *(28)*.

10. Once formed, Hb-NO adducts are readily degraded at neutral or alkaline pH *(9)*. For this reason, nitrosylated peptides are not observed after hydrolysis of HbSNO

carried out with enzymes working at neutral or alkaline pH, such as trypsin *(15)*. Therefore, a procedure based on the use of pepsin or endoproteinase Glu-C as the proteolytic agent is chosen, which allows the nitrosylated peptides to be preserved during hydrolysis at low pH values. In principle, any other enzyme working at pH < 7 can be chosen for the production of nitrosylated peptides.

11. Peptic or endoproteinase Glu-C digestion of HbSNO results in a mixture of native and nitrosylated peptides, the relative abundance depending on the extent of overall nitrosylation. The presence of overlapping peptides due to incomplete digestion is useful in obtaining a complete peptide map and in the accurate location of the modified cysteines *(15)*. Nitrosylated peptides are readily identified from their molecular weight compared with that of the native peptides, and their identity can be confirmed by selective fragmentation at the S-NO bond by increasing the cone voltage or source temperature (*see* **Subheading 3.7.**).

References

1. Umans, G. and Levi R. (1995) Nitric oxide in the regulation of the blood flow and arterial pressure. *Annu. Rev. Physiol.* **57,** 771–790.
2. Wink, D. A., Grisham, M. B., Mitchell, J. B., and Ford, P. C. (1996) Direct and indirect effects of nitric oxide in chemical reactions relevant to biology. *Methods Enzymol.* **268,** 12–31.
3. Rand, M. J. and Li, C. J. (1995) Nitric oxide as a neurotransmitter in peripheral nerve: nature of transmitter and mechanism of transmission. *Annu. Rev. Physiol.* **57,** 659–682.
4. Nussler, A. K., Liu, Z.-Z., Hatakeyama, K., Geller, D., Billiar, T. R., and Morris, S. M., Jr. (1996) A cohort of supporting metabolic enzymes is coinduced with nitric oxide synthase in human tumor cell lines. *Cancer Lett.* **103,** 79–84.
5. Ohshima, H., Bandaletova, T. Y., Brouet, I., Bartsch, H., Kirby, G., Ogunbiyi, F., et al. (1994) Increased nitrosamine and nitrate biosynthesis mediated by nitric oxide syinthase induced in hamsters infected with liver fluke *(Opisthorchis viverrini). Carcinogenesis* **15,** 271–275.
6. Stamler, J. S. (1994) Redox signaling: nitrosylation and related target interactions of nitric oxide. *Cell* **78,** 931–936.
7. Doyle, M. P., Pickering, R. A., and da Conceicao, J. (1984) Structural effects in alkyl nitrite oxidation of human haemoglobin. *J. Biol. Chem.* **259,** 80–87.
8. Gorbunov, N. V., Osipov, A. N., Day, B. W., Zayas-Rivera, B., Kagan, V. E., and Elsayed, N. M. (1995) Reduction of ferrylmyoglobin and ferrylehaemoglobin by nitric oxide: a protection mechanism against ferryl hemoprotein-induced oxidation. *Biochemistry* **34,** 6689–6699.
9. Arnelle, D. R. and Stamler, J. S. (1995) NO^+, $NO^\bullet$, and NO^- donation by *S*-nitrosothiols: implications for regulation of physiological functions by *S*-nitrosylation and acceleration of disulphide formation. *Arch. Biochem. Biophys.* **318,** 279–285.
10. Stamler, J. S., Simon, D. I., Osborne, J. A., Mullins, M. E., Jaraki, O., Michel, T., et al. (1992) *S*-nitrosylation of proteins with nitric oxide: synthesis and characterization of biologically active compounds. *Proc. Natl. Accad. Sci. USA* **89,** 444–448.

11. Gaston, B., Reilley, J., Drazen, J. M., Fackler, J., Ramdev, P., Arnelle, D., et al. (1993) Endogenous bronchodilator S-nitrosothiol in human airways. *Proc. Natl. Acad. Sci. USA* **90**, 10,957–10,961.
12. Jia, L., Bonaventura, C., Bonaventura, J., and Stamler, J. S., (1996) S-Nitrosohaemoglobin: a dynamic activity of blood involved in vascular control. *Nature* **380**, 221–226.
13. Bunn, H. F. and Forget, B. G., eds. (1986) *Hemoglobin: Molecular, Genetic and Clinical Aspects* W. B. Saunders, Philadelphia.
14. Stamler, J. S., Jia, L., Eu, J. P., McMahon, T. J., Demchenko, I. T., Bonaventura, J., et al. (1997) Blood flow regulation by S-nitrosohaemoglobin in the physiologic oxygen gradient. *Science*, **276**, 2034–2037.
15. Ferranti, P., Malorni, A., Mamone, G., Sannolo, N. and Marino, G. (1997) Characterisation of S-nitrosohaemoglobin by mass spectrometry. *FEBS Lett.* **400**, 17–24.
16. Hevel, J. M. and Marletta, M. A. (1994) Nitric-oxide synthase assays. *Methods Enzymol.* **233**, 250–258.
17. Moriguchi M., Manning L. R., and Manning M. (1992) Nitric oxide can modify amino acid residues in proteins. *Biochem. Biophys. Res. Commun.* **183**, 598–604.
18. Shelton, J. B., Shelton, J. R., and Schroeder, W. A. (1994) High performance liquid chromatography separation of globin chains on a large-pore column. *J. Liquid Chromatogr.* **7**, 1969–1977.
19. Mirza, U. A., Chait, B. T., and Lander, H, M. (1995) Monitoring reactions of nitric oxide with peptides and proteins by electrospray ionization-mass spectrometry. *J. Biol. Chem.* **270**, 17,185–17,188.
20. Stamler, J. S., Simon, D. I., Jaraki, O., Osborne, J. A., Francis, S., Mullins, M., et al. (1992) S-nitrosylation of tissue-type plasminogen activator confers vasodilatory and antiplatelet properties on the enzyme. *Proc. Natl. Acad. Sci. USA* **89**, 8087–8091.
21. Bolotina, V. M., Najibi, S., Palacino, J. J., Pagano, P. J., and Cohen, R. A. (1994) Nitric oxide directly activates calcium-dependent potassium channels in vascular smooth muscle. *Nature* **368**, 850–853.
22. Lander, H. M. Sehajpal, P. K., and Novogrodsyk, A. (1993) Nitric oxide signaling: a possible role for G proteins. *Immunology* **151**, 7182–7187.
23. Xu, L., Eu, J. P., Meissner, G., and Stamler, J. S. (1998) Activation of the cardiac calcium release channel (ryanodine receptor) by poly-S-nitrosylation. *Science* **279**, 234–237.
24. *Handbook of Biochemistry and Molecular Biology* (1982) CRC, Boca Raton, FL.
25. Stamler, J. S. and Loscalzo, J. (1992) Capillary zone electrophoretic detection of biological thiols and their S-nitrosated derivatives. *Anal. Chem.* **64**, 779–785.
26. Feelish, M. and Stamler, J. S., eds. (1996) *Methods in Nitric Oxide Research*. Wiley, London, U. K.
27. Knowles, R. G. and Moncada, S. (1994) Nitric oxide synthases in mammals. *Biochem. J.* **298**, 249–258.
28. Ferranti, P., Sannolo, N., Mamone, G., Fiume, I, Carbone, V., Tornqvist, M., et al. (1996) Structural characterization by mass spectrometry of haemoglobin adducts formed after *in vitro* exposure to methyl bromide. *Carcinogenesis* **17**, 2661–2671.

10

Multiple and Subsequent MALDI-MS On-Target Chemical Reactions for the Characterization of Disulfide Bonds and Primary Structures of Proteins

H. Peter Happersberger, Marcus Bantscheff, Stefanie Barbirz, and Michael O. Glocker

1. Introduction

Matrix-assisted laser desorption/ionization mass spectrometry (MALDI-MS) analyses have been successfully applied in a vast number of examples for precise molecular mass determinations of biomacromolecules such as proteins. In combination with chemical and/or proteolytic derivatization and degradation reactions, MALDI-MS enables the analysis of the primary structural details of proteins such as posttranslational modifications. A well-established methodology for the investigation of partial peptides of proteins is the combination of proteolytic degradation with mass spectrometry *(1–3)*. Hydrolytic cleavage is carried out under conditions such that particular peptide bonds are cleaved with a highly specific protease, e.g., trypsin, hence creating a specific "peptide map."

On-target chemical reactions, such as proteolytic digestions and reduction of disulfide bond-linked peptides (*see* **Subheading 2.3.** and following), combined with MALDI-MS analysis of the resulting peptides have been reported as a useful method for minimization of the amount of peptide or protein sample required *(4–11)*. Digestions are performed by spotting samples and protease solutions directly on the MALDI target and then carrying out data acquisition after incubation for a short period. Most suitable solvent systems for direct MALDI-MS characterization that have been used hitherto for *on-target* reactions are volatile buffer systems, such as NH_4HCO_3 or NH_4OAc, and organic solvents, e.g., those from high-performance liquid chromatography (HPLC) analyses. In addition, scattered reports have described the application of reversed-phase chromatography materials and GELoader tips for the microscale purification of peptide or protein samples prior to mass spectrometry *(7,12–14)*.

One of the most common primary structure modifications in proteins is the formation of disulfide bonds. Proper formation of disulfide bonds is crucial in attaining the correct three-dimensional structure of proteins *(15,16)*. Mass spectrometric methods have been employed successfully for the study of disulfide bridges of peptides and proteins *(3,8,17–20)*, and several comprehensive reviews are available on this subject *(21–24)*. However, to determine whether or not an observed peak in a recorded mass spectrum of a (complex) peptide mixture is due to material that contains any disulfide linkages, the corresponding ion signal should be completely, or at least mostly, eliminated by reduction. This is important as the recently discovered MALDI-induced cleavage of disulfide bonds *(25–28)* may otherwise obscure the results.

Up to now, a major drawback of the MALDI-MS characterization of protein and peptide structures was the fact that this method had been considered unidirectional, i.e., once the sample had been mixed with matrix, the protein of interest was lost for subsequent biological experiments such as in vitro bioassays. This is particularly unfortunate as only a minor portion of the analyte is consumed during mass spectrometric data acquisition.

In this chapter, we describe the *on-target* performance of multiple and different subsequent chemical experiments (schematically depicted in **Fig. 1**), which enable the determination of different and complementary molecular protein structure details from one and the same sample spot on the MALDI-MS target. In addition, we describe an efficient method of sample recovery, purification, and fractionation based on easy-to-use microscale reversed-phase columns.

2. Materials
2.1 MALDI-MS Instrumentation and Acquisition Conditions

1. MALDI-MS analyses were carried out with a Bruker Biflex linear time-of-flight spectrometer (Bruker Franzen, Bremen, Germany), equipped with a multiprobe SCOUT source, video system, ultraviolet (UV)-nitrogen laser (337 nm), and dual microchannel-plate detector. The acceleration voltage was set to 20 kV, and spectra were calibrated using the singly and doubly charged ions of insulin. Spectra were recorded after evaporation of the solvent and processed using the X-MASS data system (Bruker Franzen).
2. 4-Hydroxy-α-cyanocinnamic acid (HCCA, Aldrich Chemie, Steinheim, Germany) and 6-aza-2-thiothymine (ATT, Aldrich Chemie) were used as sample matrix materials for MALDI-MS. Matrix materials were dissolved in CH_3CN + 0.1% trifluoroacetic acid (TFA) (2:1 v/v) resulting in concentrations of 10 µg/µL each.

MALDI-MS On-Target Reactions

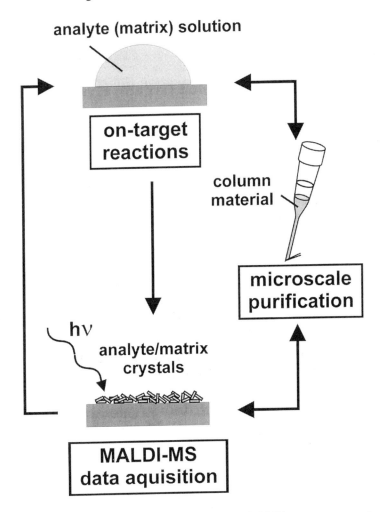

Fig. 1. Scheme of multiple and subsequent MALDI-MS *on-target* reactions. The microscale reversed-phase columns may be applied for purification, fractionation, and sample recovery after *on-target* reactions.

2.2. On-Target Proteolytic Degradation Reactions

1. Tightly sealing glass vial coated with wetted tissue that is capable of holding the MALDI-MS multiprobe target.
2. Buffer A: 50 mM NH$_4$HCO$_3$, 1 mM EDTA (pH 7.8).
3. Buffer B: 200 mM NH$_4$HCO$_3$, 1 mM EDTA (pH 7.8).
4. Trypsin (EC 3.4.21.4; Sigma-Aldrich Chemie, Deisenhofen, Germany). Stock solution (1 µg/µL) in 1 mM HCl. This solution could be stored for at least 1 mo at −20°C.

5. Asp-N (EC 3.4.24.33; Boehringer Mannheim, Germany) at a concentration of 0.04 μg/μL prepared by dissolving the commercially available lyophilized powder in water. This solution could be stored for at least 1 mo at −20°C.

2.3. On-Target Reduction and Alkylation Reactions

1. Glass vial (see **Subheading 2.2.**).
2. Buffer A: 50 mM NH$_4$HCO$_3$, 1 mM EDTA (pH 7.8).
3. Buffer B: 200 mM NH$_4$HCO$_3$, 1 mM EDTA (pH 7.8).
4. Stock solution containing 25 mM Tris(2-carboxyethyl)phosphine hydrochloride (TCEP; Pierce, Rockford, IL) in buffer A (see **Note 1**).
5. Stock solution containing 100 mM 2-mercaptoethanol (ME; Merck, Darmstadt, Germany) in water (see **Note 1**).
6. 25 mM iodoacetamide (IAA; Fluka, Buchs, Switzerland) stock solution in water (see **Note 1**).
7. Solution D: 5 μL TCEP stock solution and 5 μL CH$_3$CN mixed with 10 μL buffer B (see **Note 2**).
8. Solution E: 25 μL ME stock solution and 5 μL CH$_3$CN mixed with 10 μL buffer B (see **Note 2**).
9. Solution F: 5 μL TCEP and 5 μL IAA stock solutions and 5 μL CH$_3$CN mixed with 10 μL buffer B (see **Notes 2** and **3**).

2.4. Sample Purification, Fractionation, and Recovery Using Microscale Reversed-Phase Columns

1. Commercially available GELoader tips (Eppendorf, Hamburg, Germany).
2. Prepare suspensions of C18- and C4-stationary phase (YMC-gel, 120 Å pore size, 50-μm; YMC Europe, Schermbeck, Germany) in methanol by addition of 1 g of either reversed-phase material, to 2 mL methanol. These suspensions are stored at 4°C.
3. A flow of the mobile phase during microscale reversed-phase purification was achieved using a conventional single-use 10-mL syringe (Braun Melsungen, Melsungen, Germany), a 100-μL microliter syringe (Unimetrics, Shorewood, IL) or a 1–10-μL Eppendorf pipette (Hamburg, Germany).
4. 0.1% TFA was used for equilibration of the stationary phase. Samples were eluted stepwise using solvent mixtures that consisted of 0.1% TFA and increasing concentrations (20, 40, 60, and 80%) of CH$_3$CN.

3. Methods
3.1. On-Target Proteolytic Degradation Reactions

In this section we describe *on-target* proteolytic digestion methods as they are applied in our laboratory. We distinguish two procedures. In the first, digestions are carried out in the absence of the MALDI matrix, whereas in the second procedure proteolytic degradation reactions are performed in the presence of the matrix.

3.1.1. In the Absence of Matrix

The methods described here enable the digestion of proteins dissolved not only in volatile buffers but also in buffers containing large amounts of nonevaporable substances, such as detergents, salts, glycerol, or chaotropes.

1. Spot 1–2 µL of the peptide/protein solution (10–100 pmol) on the MALDI target and allow complete evaporation of the solvent (*see* **Note 4**).
2. Dilute the protease stock solution with buffer A (*see* **Note 5**) immediately before use to adjust to an enzyme-to-substrate ratio between 1:10 and 1:100. A final volume of approx 2 µL should be spotted on the target and thoroughly mixed in order to redissolve all precipitated material.
3. Put the MALDI target into a tightly sealing glass vial that contains a layer of wetted tissue to maintain a humid atmosphere. Incubate the glass vial at 37°C (water bath) for 30 min to 4 h (*see* **Note 6**). In the case of evaporation of the solvent add approx 1 µL of water per hour.
4. Purification of sample.
   a. If both the peptide/protein and the protease are dissolved in a volatile buffer system, such as 50 mM NH_4HCO_3 (pH 7.8) or 50 mM NH_4OAc (pH 4), with or without organic solvents, the remaining solvent can be evaporated to dryness. After addition of 1 µL of matrix solution the sample is then ready for MALDI-MS analysis (*see* **Note 9**).
   b. If the protein sample contains nonvolatile compounds, purification of the sample can be performed using the microcolumn system prior to MALDI-MS analysis as described in **Subheading 3.3.**

The methodology described above was developed with melanine-concentrating hormone (MCH; M_r 2386.9), consisting of 19 amino acids in which one disulfide bond connects residues Cys7 and Cys15 (sequence is given in **Fig. 2**). Thus, MCH (41 pmol) was dissolved in a volatile buffer (buffer A) and digested *on-target* with trypsin (E/S = 1:25) in the absence of matrix for 1 h at 37°C. MALDI-MS analysis was performed after addition of HCCA matrix. The MALDI mass spectrum (**Fig. 2**) shows the complete digestion of MCH resulting in signals for the protonated peptides [1–6] (mass-to-charge ratio *[m/z]* 797) and the disulfide bond-linked peptide [7–11]-S–S-[12–19] (*m/z* 1628). Small signals at *m/z* 580 and *m/z* 1051 are observed for the cysteine-containing peptides [7–11] and [12–19] due to MALDI-induced cleavage of disulfide bonds (*see* **Subheading 1.**). Peptide signals due to oxidation at methionine residues Met4 and Met8 ($\Delta m = 16$) and tryptophan residue W17 ($\Delta m = 32$) are indicated by asterisks *(29,30)*.

An application example of *on-target* proteolytic degradation in the presence of nonvolatile buffer components is given by the Asp-N digestion of NtrC. NtrC is a nitrogen assimilation regulatory protein from *Escherichia coli* that contains 469 amino acids ($M_r = 52254.8$). NtrC (30 pmol) is dissolved in 50 mM Tris buffer (pH 7) containing 50 mM NaCl, 0.1 mM EDTA, and 25% glycerol.

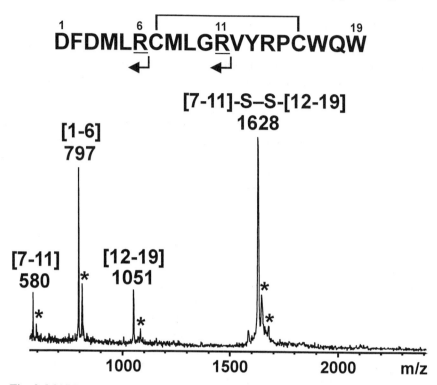

Fig. 2. MALDI mass spectrum of MCH peptides obtained by *on-target* tryptic digestion. The disulfide bond and the tryptic cleavage sites (arginines are underlined) of MCH are indicated on the amino acid sequence. Derivatives resulting from methionine or tryptophan oxidation are indicated by *.

Enzyme-to-substrate ratio is adjusted to 1:50, and digestion is performed for 2 h. The applied buffer system makes direct MALDI-MS analysis of the resulting peptide mixture impossible. Therefore the digestion mixture is transferred onto a microscale reversed-phase (C18) column and washed with 50 µL 0.1% TFA (*see* **Subheading 3.3.1.**). The retained peptides are eluted in two fractions (≈ 2–3 µL each) by using first 40% and then 60% CH_3CN mixed with water that contains 0.1% TFA in each case. Spectra from these fractions are then recorded after addition of HCCA matrix solution (1 µL). The peptide mixtures obtained in this manner have different compositions (**Fig. 3**). For example, peptide [124–138] (*m/z* 1710) elutes in the 40% CH_3CN fraction (**Fig. 3A**) whereas peptide [109–123] (*m/z* 1700) elutes in the 60% CH_3CN fraction (**Fig. 3B**).

3.1.2. In the Presence of ATT Matrix

In the presence of the water-soluble ATT matrix, it is possible to perform proteolytic digestions of the protein of interest *(4,5)*, after acquiring

MALDI-MS On-Target Reactions

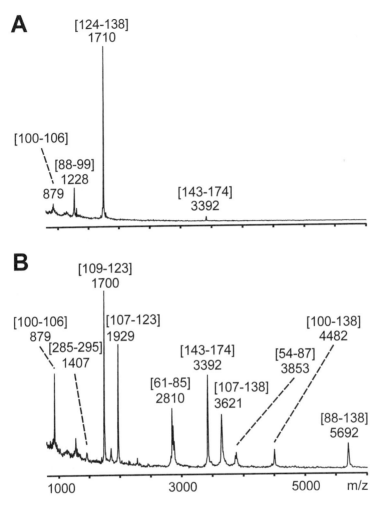

Fig. 3. MALDI mass spectra after *on-target* Asp-N digestion of NtrC and fractionation. Fractions were eluted with (A) 40% CH_3CN and (B) 60% CH_3CN in 0.1% TFA.

spectra of the intact protein, from the same sample spot. The *on-target* tryptic digestion, following molecular weight determination, presents a convenient tool for the sequence characterization of synthetic polypeptides and proteins, particularly when the N-terminus is blocked and Edman degradation prevented.

1. Mix 1 µL of the peptide/protein solution (10–100 pmol sample) with 1 µL of the ATT matrix solution directly on the MALDI-MS target (*see* **Note 4**). After solvent evaporation, the residue is available for MALDI-MS analysis and then for further proteolytic digestion.

Table 1
Peptide Mapping Analysis of an *On-Target* Tryptic Digest of Cytochrome c in the Presence of ATT Matrix *(4)*

Peptide	Partial sequence	[M+H]+ Calc.	[M+H]+ Found
T1[a]	1–5	589	589
T4	9–13	635	635
T4/5[b]	9–22	2249	2249
T5[b]	14–22	1634	1634
T7	28–38	1169	1169
T9	40–53	1472	1472
T11/12/13	56–73	2210	2210
T12	61–72	1497	1497
T12/13/14	61–79	2284	2284
T14	74–79	679	679
T15	80–86	780	780
T17/18/19[c]	88–97	1238	1238
T23	101–104	433	433

[a]N-terminus acetylated.
[b]Contains covalently linked heme group.
[c]Cleavage after Tyr97, due to intrinsic chymotryptic activity.

2. Perform proteolytic digestion (*see* **Subheading 3.1.1., steps 2** and **3**) and MALDI-MS analysis (*see* **Subheading 3.1.1., step 4**).

As an example, the peptides obtained from a cytochrome c digestion with trypsin in the presence of ATT matrix are listed in **Table 1**. The catalytic activity of proteolytic enzymes under the conditions employed here seems comparable to that in solution, and the proteolytic specificity appears to be unchanged *(4)*.

3.2. On-Target Proteolytic Reduction and Alkylation Reactions

3.2.1. On-Target Reduction in the Presence of Matrix

In this section we describe MALDI-MS *on-target* reduction and alkylation reactions for the rapid and unequivocal identification of disulfide bridges in peptides and proteins.

1. Dissolve the peptide or protein sample (10–40 pmol) in an evaporable buffer system such as 50 mM NH_4HCO_3 (pH 7.8) or 50 mM NH_4OAc (pH 4) perhaps with organic co-solvents.
2. Mix 1 µL of the peptide/protein sample solution with 1 µL of the matrix solution directly on the MALDI-MS target (*see* **Note 4**). After evaporation of the solvent, the sample is ready for MALDI-MS analysis.

3. Completely dissolve the peptide/protein + matrix mixture on the target by adding 2 µL of solution D or E, and mixing gently. The solution adopts a pH value of 7–8. The HCCA matrix solution turns yellow (*see* **Note 7**). The amount of reducing agent needed depends on the nature of the reducing agent (TCEP or ME). Typically, a 10–50-*M* excess of reducing agent over the number of disulfide bonds is applied.
4. Put the MALDI target that holds the reaction mixture into a sealed glass vial and make an air-tight closure (*see* **Subheading 2.2., item 1**). Incubate the glass vial at 37°C (water bath) for 30 min to 2 h. Add approx 1 µL of water per hour if the solvent evaporates.
5. After evaporation of the solvent the precipitate is recrystallized by adding 2 µL of CH_3CN + 0.1% aqueous TFA (2:1, v/v) (pH 2). M ix gently to avoid sample spillage due to formation of large bubbles. After complete evaporation of the solvent, the sample is ready for MALDI-MS analysis (*see* **Notes 8** and **9**).

For developing *on-target* reduction in the presence of HCCA matrix, the tryptic digest of MCH (cf. **Fig. 2** and **Subheading 3.1.1.**) was used as a model system. Thus, after *on-target* digestion and MALDI-MS data acquisition, the mixture that contains the tryptic MCH peptides and HCCA was resolubilized on the target. The MALDI mass spectrum (**Fig. 4**) after reduction with TCEP (*see* **Note 10**) shows the complete disappearance of the disulfide bond-linked peptide [7–11]-S-S-[12–19] indicating complete reduction. A strong signal at *m/z* 1051 is present for peptide [12–19]. The signal at *m/z* 1240 corresponds to the HCCA matrix-adduct of peptide [12–19]. The formation of significant amounts of HCCA adducts is frequently observed after *on-target* reduction in the presence of the HCCA matrix (*see* **Note 11**). In addition, the ion signal for peptide [1–6] (*m/z* 797) is present in the mixture.

As an application example we show the characterization of a disulfide bond of the protein α-amylase (MalS) from *Escherichia coli*. MalS has a molecular mass ≈ 74 kDa *(8)*. After tryptic digestion and HPLC fractionation of the peptides, the disulfide-linked peptide [103–106]-S-S-[517–534] is observed at *m/z* 2544 together with further peptides (**Fig. 5A**). *On-target* reduction is performed with TCEP in the presence of the HCCA matrix. The ion signal at *m/z* 2544 disappears to a great extent, and two new ion signals for the peptide [517–534] (*m/z* 2070) and its HCCA matrix adduct at *m/z* 2259 are observed (**Fig. 5B**). Thus, this method proved successful for identification of disulfide bonds, even with very little material, e.g., after several subsequent chromatographic separation steps *(8)*.

3.2.2. Combined On-Target Reduction and Alkylation

The formation of peptide-matrix adducts and of mixed disulfides (*see* **Note 11**) during *on-target* reduction may make the assignment of disulfide-linked peptides difficult. Thus, we also carry out *on-target* reduction with TCEP combined with alkylation of the thiol groups using iodoacetamide (IAA) in the presence of the HCCA matrix.

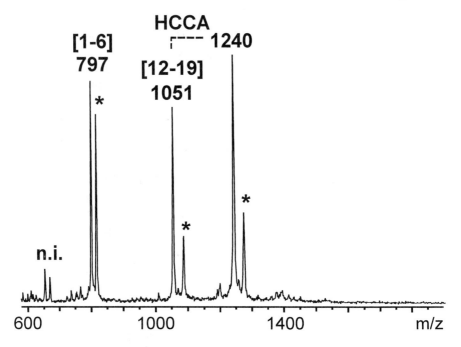

Fig. 4. MALDI mass spectrum of the tryptic peptide mixture from MCH after *on-target* reduction. *On-target* reduction was performed with TCEP in the presence of HCCA matrix. Signals corresponding to matrix adducts are indicated by dashed lines. Derivatives resulting from methionine or tryptophan oxidation are indicated by *.

1. Completely dissolve the peptide/protein mixture and the matrix on the target by adding 2 μL of solution F and mixing gently. The solution adopts a pH value of 7–8.
2. Perform **steps 4** and **5** of the protocol given in **Subheading 3.2.1.**

The MALDI mass spectrum (**Fig. 6**) of an *on-target* tryptic digest of MCH (cf. **Fig. 2B** and **Subheading 3.3.1.**) shows, after subsequent *on-target* reduction and alkylation, the nearly complete disappearance of the disulfide bond-linked peptide [7–11]-S-S-[12–19] (*m/z* 1628) but a strong signal at *m/z* 1108 for the reduced and carboxamidomethylated (CAM-modified) peptide [12–19] ([12–19]-CAM). In contrast to the reduction experiment (cf. **Subheading 3.2.1.**), in which the signal for the reduced and protonated peptide [7–11] (*m/z* 580) is not detectable, a strong peptide ion signal for the reduced and CAM-modified peptide [7–11] ([7–11]-CAM) is observed at *m/z* 637. Only small signals for HCCA matrix-adducts of peptide [12–19] are observed at *m/z* 1240.

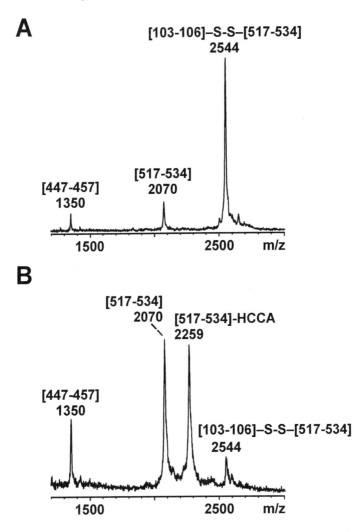

Fig. 5. MALDI mass spectra of an HPLC-fractionated tryptic digest of a disulfide bond-containing peptide from MalS. (**A**) before and (**B**) after *on-target* reduction with TCEP in the presence of HCCA matrix *(8)*.

3.3. Sample Recovery and Purification/Fractionation Using Microscale Reversed-Phase Columns

3.3.1. Purification and Fractionation of Samples Containing High Amounts of Salts, Glycerol, Detergents

Sample purification is critical for successful mass spectrometric analyses in order to achieve high sensitivity, high resolution, and mass accuracy. Here we

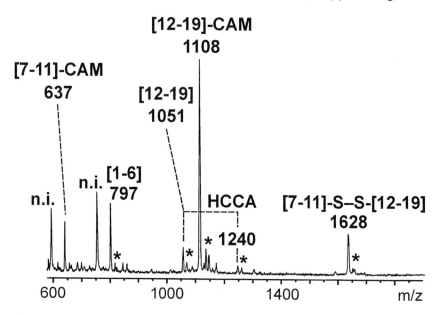

Fig. 6. MALDI mass spectrum after combined *on-target* reduction/alkylation of the tryptic peptide mixture from MCH. *On-target* reduction was performed with TCEP in the presence of IAA and HCCA matrix. Signals corresponding to matrix adducts are indicated by dashed lines. Derivatives resulting from methionine and tryptophan oxidation are indicated by *. Signals marked as "n.i." were not identified.

describe methods used in our laboratory to reduce the amount of low molecular weight impurities such as salts, glycerol, detergents, chaotropes, or mixtures of such materials.

1. Cut off half of the extended capillary-like outlet of a GELoader tip and make a bend (approx 90°) about 3 mm above the cut end.
2. Fill a microliter syringe with 50 µL of methanol and add 2–4 µL of a suspension of reversed-phase material to the syringe (C18 material for peptides, C4 material for proteins).
3. Transfer the suspension completely into the tip and quickly push through the excess of methanol so that the column material stays wet (*see* **Note 12**).
4. To equilibrate the column, add 50 µL of 0.1% aqueous TFA using the microliter syringe. The TFA solution should then be pushed slowly through the microcolumn by applying pressure with a 1–10-µL Eppendorf pipette.
5. Load 1–2 µL of sample solution onto the column from a second GELoader tip. Push the solution gently onto the reversed-phase material by applying pressure with a 1–10-µL pipette.
6. Wash the sample by pushing through 50 µL of aqueous 0.1% TFA (*see* **Note 12**).
7. Elution.

a. Elute the peptide/protein sample with 4–6 µL of elution solvent. In most cases a mixture of CH_3CN + aqueous 0.1% TFA (80:20) will be suitable (see **Note 13**). The eluate should be split into two spots on the MALDI target (approx 2–3 µL for each spot).
   b. Alternatively, a stepwise elution of the peptides can be achieved by using solvents based on aqueous 0.1% TFA with increasing concentrations of CH_3CN (e.g., 20, 40, 60, and 80%, approx 3–4 µL each). This procedure is particularly advantageous when complex peptide mixtures are to be analyzed.
8. Add 1 µL of matrix solution to each sample spot on the MALDI target and perform MALDI-MS analyses after evaporation of the solvent.

The applicability of this procedure to fractionating peptides from an *on-target* tryptic digest of cytochrome c (80 pmol in buffer A; E/S = 1:20; 1 h at 37°C), in the absence of matrix, is shown. After addition of HCCA matrix, a MALDI-MS analysis was carried out. The identified peptides are summarized in **Table 2** (see column entitled "mixture"). The same sample spot on the target was resolubilized as described above, transferred onto a microscale reversed-phase (RP)-C18 column, and eluted in three fractions with 0.1% aqueous TFA containing 20, 40, and 60% CH_3CN, respectively. Subsequent MALDI-MS analysis of the resulting fractions shows different compositions of peptides depending on the amount of CH_3CN in the elution solvent (see **Table 2**; columns entitled "microscale RP separation"). The quite complex peptide mixture is, by applying this methodology, conveniently fractionated and the components unambiguously assigned. It is noteworthy that cytochrome c that is still undigested is not detectable in the MALDI-MS mixture analysis prior to the microseparation procedure but is the major component of the fraction that is eluted using 60% CH_3CN.

3.3.2. Sample Recovery from Matrix Preparations

MALDI matrix materials can also be removed from peptide/protein samples, after mass spectrometric analyses, by use of the microscale reversed-phase columns.

1. Solubilization.
   a. Redissolve water-soluble matrix + sample mixtures in 2–3 µL aqueous 0.1% TFA.
   b. For complete solubilization of HCCA, add 2–3 µL of buffer B to the matrix + sample mixture on the target and mix thoroughly (see **Note 14**).
2. Perform **steps 5–8** of the protocol in **Subheading 3.3.1.** (see **Note 15**).

This procedure enables fast and simple recovery of samples after MALDI-MS analyses. The recovered sample is then ready for further *on-target* reactions and/or biological assays.

Table 2
Peptide Mapping Analyses of *On-Target* Tryptic Digestion of Cytochrome c With and Without Subsequent Fractionation by Microscale Reversed-Phase Chromatographic Separation

		[M+H]$^+$ (found)			
			Microscale RP separation (% CH$_3$CN)		
Peptide	[M+H]$^+$ (calc.)	Mixture	20	40^a	60^b
1–5^c	589.6	—	590	—	—
8–22^d	2378.6	2379	—	2379	—
8–25	2620.9	2621	—	—	—
9–22^d	2250.5	2250	—	2250	2250
9–25^d	2492.7	2492	—	2492	2492
14–22^d	1634.7	—	—	1635	—
23–27	526.6	—	526	—	—
26–38	1434.6	1434	1435	—	—
26–39	1562.8	1562	1563	—	—
28–29	1297.5	1297	1297	1297	1297
28–38	1169.3	1169	1169	1169	1169
39–53	1599.8	—	1600	—	—
39–55	1842.0	1840	1842	—	—
39–72	4033.5	4036	—	—	—
40–53	1471.6	—	1471	—	—
40–55	1713.8	—	1714	—	1714
56–72^e	2082.4	2083	—	2082	2083
56–73^f	2210.6	2211	—	2210	2211
61–72	1496.7	—	—	1497	—
61–73	1624.8	—	—	1625	—
73–79	807.0	—	807	—	—
73–91^f	2210.7	2211	—	2210	2211
74–91^e	2082.6	2083	—	2082	2083
80–87	908.2	—	908	—	—
80–104	2912.5	2911	—	2911	2912
87–100	1736.1	—	1736	—	—
88–99/89–100^g	1479.7	1478	—	1478	1478
88–100/89–101^h	1607.9	1607	—	1607	1608
89–99	1351.5	1352	—	1351	—

^{a}Molecular ions of undigested cytochrome c were observed at *m/z* 12361, 6182, and 4124.
^{b}The undigested cytochrome c was the major compound of this fraction.
^{c}N-terminus acetylated.
^{d}Contains covalently linked heme group.
^{e,f}Peptides have nearly the same mass and could not be distinguished in the MALDI spectrum.
^{g,h}Peptides have the same mass and could not be distinguished in the MALDI spectrum.

4. Notes

1. These solutions can be stored for at least 1 mo at –20°C.
2. These solutions should be freshly prepared prior to use. In the case of water-soluble matrices, e.g., ATT, the addition of CH_3CN is not necessary for the complete solubilization of the matrix and it can be replaced by water.
3. In contrast to TCEP, the reducing power of ME is deactivated due to alkylation with IAA. Hence premixing of the latter two compounds is not recommended.
4. The stainless steel target should be cleaned thoroughly and finally wiped with pentane in order to maintain droplets with a small diameter ($\approx$ 2 mm) on the target surface.
5. The buffer system used depends on the protease, but most of the commercially available proteases, e.g., trypsin, Lys-C, Asp-N, and Glu-C, have sufficient activity in 50 mM NH_4HCO_3 (pH 7.8).
6. The required digestion time depends on protease activity and enzyme-to-substrate ratio. In general, low amounts of protease should be used to minimize autoproteolytic cleavages and thereby avoid peptide ions from the protease, which might overlap with peptide ions from the sample.
7. This color of the solution disappears at pH values below 7 and may be used as an indication of pH value during *on-target* reactions.
8. Because of significant amounts of NH_4HCO_3 and of nonevaporable contents (e.g., TCEP, IAA) it is often necessary to perform washing and/or recrystallizing steps as described in **Note 9**.
9. When HCCA is used as the matrix, the signal-to-noise ratio of the MALDI mass spectra can often be improved by a washing step and/or subsequent recrystallization of the sample/matrix mixture on the target. Washing is performed by the addition of 2 µL of 0.1% aqueous TFA to the peptide/protein + matrix mixture (HCCA is barely soluble in 0.1% TFA). After 1–2 min the droplet is removed by absorbing it with a small strip of tissue dipped into the droplet. This procedure decreases the amount of water-soluble material (buffer salts, detergents, chaotropes, and so forth). If necessary, 0.5 µL of HCCA solution may be added prior to recrystallization or data acquisition. Recrystallization is effected by adding 1–2 µL CH_3CN + 0.1% TFA (2:1, v/v) (pH 2) to the sample of peptide/protein + matrix and mixing gently. After final evaporation of the solvent, the sample is ready for MALDI-MS data acquisition.
10. In the case of the *on-target* reduction with TCEP in the presence of HCCA, a strong signal for the HCCA trimer is seen at m/z 565.
11. When ME is used for reduction of disulfide bonds, formation of mixed disulfides between the cysteinyl thiol-containing peptide and ME is observed.
12. In this case pressure can be applied by mounting the microcolumn onto the Luer tip of a 10-mL single-use syringe to obtain a slightly higher flow rate.
13. The solvent volume that is necessary for complete elution of peptides increases with the use of larger amounts of reversed-phase material as stationary phase.
14. HCCA is nearly insoluble in 0.1% aqueous TFA but is soluble in basic buffer systems (pH > 7.5).

15. During the washing step of the microscale reversed-phase sample purification protocol (*see* **Subheading 3.3.1.**), matrices elute first. In the case of HCCA, this is indicated by the yellow color of the solvent.

Acknowledgments

We are grateful to Dr. Michael Przybylski in whose laboratory parts of this work were carried out. We extend our thanks to Dr. Verena Weiss in whose laboratory NtrC was purified. We thank Priv. Doz. Dr. Michael Ehrmann for providing MalS. This work was supported by the Deutsche Forschungsgemeinschaft (Bonn, Germany).

References

1. Glocker, M. O., Kalkum, M., Yamamoto, R., and Schreurs, J. (1996) Selective biochemical modification of functional residues in recombinant human macrophage colony-stimulating factor β (rhM-CSFβ): identification by mass apectrometry. *Biochemistry* **35**, 14,625–14,633.
2. Denzinger, T., Przybylski, M., Savoca, R., and Sonderegger, P. (1998) Mass spectrometric characterization of primary structure, sequence heterogeity, and intramolecular disulfide loops of the cell adhesion protein axonin-1 from chicken. *Eur. Mass Spectrom.* **3**, 379–389.
3. Bures, E. J., Hui, J. O., Young, Y., Chow, D. T., Katta, V., Rohde, M. F., et al. (1998) Determination of disulfide structure in Agouti-Related (AGRP) by stepwise reduction and alkylation. *Biochemistry* **37**, 12,172–12,177.
4. Glocker, M. O., Bauer, S. H. J., Kast, J., Volz, J., and Przybylski, M. (1996) Characterization of specific noncovalent protein complexes by UV matrix-assisted laser desorption ionization mass spectrometry. *J. Mass Spectrom.* **31**, 1221–1227.
5. Glocker, M. O., Jetschke, M. R., Bauer, S. H. J., and Przybylski, M. (1998) Characterization of tertiary structures and specific noncovalent complexes of proteins by UV-matrix assisted laser-desorption/ionization mass spectrometry, in *New Methods for the Study of Biomolecular Complexes* (Ens, W., Standing, K. G., and Chernushevich I. V., eds.), Kluwer Academic, Dordrecht, pp. 193–208.
6. Wu, J., Gage, D. and Watson, J. T. (1996) A strategy to locate cysteine residues in proteins by specific chemical cleavage followed by matrix-assisted laser desorption/ionization time-of-flight mass spectrometry. *Anal. Biochem.* **235**, 161–174.
7. Kussmann, M., Nordhoff, E., Rahbek-Nielsen, H., Haebel, S., Rossel-Larsen, M., Jakobsen, L., et al. (1997) Matrix-assisted laser desorption/ionization mass spectrometry sample preparation techniques designed for various peptide and protein analytes. *J. Mass Spectrom.* **32**, 593–601.
8. Spiess, C., Happersberger, H. P., Glocker, M. O., Spiess, E., Rippe, K., and Ehrmann, M. (1997) Biochemical characterization and mass spectrometric disulfide bond mapping of periplasmic α-amylase MalS of *Escherichia coli. J. Biol. Chem.* **272**, 22,125–22,133.
9. Happersberger, H. P., Przybylski, M., and Glocker, M. O. (1998) Selective chemical bridging of bis-cysteinyl residues by arsonous acid derivatives as an approach

to the characterization of protein tertiary structures and folding pathways by mass spectrometry. *Anal. Biochem.* **264,** 237–250.
10. Happersberger, H. P., Stapleton, J., Cowgill, C., and Glocker, M. O. (1998) Characterization of the in vitro folding pathway of Recombinant Human macrophage-colony stimulating factor (rhM-CSF) by bis-cysteinyl modification and mass spectrometry. *Proteins Struct. Funct. Genet.,* **Suppl.2,** 50–62.
11. Bantscheff, M., Weiss, V., and Glocker, M. (1998) Identification of linker regions and domain borders of the response regulator protein NtrC from *E. coli* by limited proteolysis, in-gel digestion, and mass spectrometry. *Biochemistry* **38,** 11,012–11,020.
12. Wilm, M. and Mann, M. (1996) Analytical properties of the nanoelectrospray ion source. *Anal. Chem.* **68,** 1–8.
13. Winston, R. L. and Fitzgerald, M. C. (1998) Concentration and desalting of protein samples for mass spectrometry analysis. *Anal. Biochem.* **262,** 83–85.
14. Gevaert, K., Demol, H., Sklyarova, T., Vandekerckhove, J., and Houthaeve, T. (1998) A peptide concentrating and purification method for protein characterization in the sub-picomole range using matrix assisted laser desorption/ionization-postsource decay (MALDI-PSD) sequencing. *Electrophoresis,* **19,** 909–917.
15. Glocker, M. O., Arbogast, B., and Deinzer, M. L. (1995) Characterization of disulfide linkages and disulfide bond scrambling in recombinant human macrophage colony stimulating factor by fast-atom bombardment mass spectrometry of enzymatic digests. *J. Am. Soc. Mass Spectrom.* **6,** 638–643.
16. Creighton, T. (1995) Disulfide-coupled protein-folding pathways. *Phil. Trans. R. Soc. Lond. B,* **348,** 5–10.
17. Schütte, C. G., Lemm, T., Glombitza, G. J., and Sandhoff, K. (1998) Complete localization of disulfide bonds in GM2 activator protein. *Protein Sci.* **7,** 1039–1045.
18. Glocker, M. O., Arbogast, B., Schreurs, J., and Deinzer, M. L. (1993) Assignment of the inter- and intramolecular disulfide linkages in recombinant human macrophage colony stimulating factor using fast atom bombardment mass spectrometry. *Biochemistry,* **32,** 482–488.
19. Sorensen, H. H., Thomsen, J., Bayne, S., Hojrup, P., and Roepstorff, P. (1990) Strategies for determination of disulfide bridges in proteins using plasma desorption mass-spectrometry. *Biomed. Environ. Mass Spectrom.* **19,** 713–720.
20. Yazdanparast, R., Andrews, P. C., Smith, D. L., and Dixon, J. E. (1987) Assignment of disulfide bonds in proteins by fast atom bombardment mass spectrometry. *J. Biol. Chem.* **262,** 2507–2513.
21. Sun, Y., Bauer, M. D., Keough, T. W., and Lacey, M. P. (1996) Disulfide bond location in proteins, in *Protein and Peptide Analysis by Mass Spectrometry* (Chapman, J. R., ed.), Humana, Totowa, NJ, pp. 185–210.
22. Carr, S. A., Bean, M. F., Hemling, M. E., and Roberts, G. D. (1990) Integration of mass spectrometry in biopharmaceutical research, in *Biology and Mass Spectrometry* (Burlingame, A. L. and McCloskey, J. A. eds.), Elsevier Science, Amsterdam, pp. 621–652.
23. Smith, D. L. and Zhou, Z. (1990) Strategies for locating disulfide bonds in proteins. *Methods Enzymol.* **193,** 374–389.

24. Morris, H. R. and Greer, F. M. (1988) Mass spectrometry of matural and recombinant proteins and glycoproteins. *Trends Biotechnol.* **6,** 140–147.
25. Gehrig, P. M. and Biemann, K. (1996) Assignment of the disulfide bonds in napain, a seed storage protein from *Brassica napus,* using matrix-assisted laser desorption ionization mass spectrometry. *Peptide Res.* **9,** 308–314.
26. Zhou, J., Ens, W., Poppe-Schriemer, N., Standing, K. G., and Westmore, J. B. (1993) Cleavage of interchain disulfide bonds following matrix-assisted laser desorption. *Int. J. Mass Spectrom. Ion Processes* **126,** 115–122.
27. Crimmins, D. L., Saylor, M., Rush, J., and Thoma, R. S. (1995) Facile *in situ* matrix-assisted laser desorption ionization-mass spectrometry analysis and assignment of disulfide pairing in heteropeptide molecules. *Anal. Biochem.* **226,** 355–361.
28. Patterson, S. D. and Katta, V. (1994) Prompt fragmentation of disulfide-linked peptides during matrix-assisted laser desorption ionization mass spectrometry. *Anal. Chem.* **66,** 3727–3732.
29. Volkin, D. B., Mach, H., and Middaugh, C. R. (1995) Degradative covalent reactions important to protein stability, in *Protein Stability and Folding* (Shirley B. A., ed.), Humana, Totowa, NJ, pp. 35–63.
30. Creighton, T. E., ed. (1993) *Proteins, Structures and Molecular Properties.* W. H. Freeman, New York.

11

Epitope Mapping by a Combination of Epitope Excision and MALDI-MS

Carol E. Parker and Kenneth B. Tomer

1. Introduction

Two approaches have been used for the mass spectrometric identification of functional epitopes on antigens bound to antibodies: epitope extraction *(1,2)* and epitope excision *(3,4)*. In epitope extraction, a protein is subjected to enzymatic digestion, and the fragments are then presented to either an immobilized antibody or an antibody in solution. In epitope excision, the enzymatic digestion is done while the protein is bound to the antibody. In theory, the antibody prevents either proteolysis *(5)*, or chemical modification *(6)* of sites on the antigen that are situated in the antibody binding pocket. An important feature of this approach is that, because nondenaturing conditions are used, the antigen retains its native conformation so that conformational epitopes can be determined; it also offers the possibility of determining discontinuous epitopes, or epitopes that contain an enzymatic cleavage site (**Fig. 1**).

If an immobilized antibody is used in either technique, unbound fragments from enzymatic cleavage of the protein are washed away, and the bound fragments are identified. If the antibody is in solution, the enzymatic digest fragment containing the antigenic region is immunoprecipitated along with the antibody. The antibody with attached antigen is separated from unbound fragments based on size. The antibody-antigen complex is then dissociated, and the previously bound antigen-containing fragments are separated from the antibody and identified by mass spectrometry *(7)*.

Epitope excision analyses have been reported using both immobilized and non-immobilized antibodies *(7,8)*. Using immobilized antibodies and matrix-assisted laser desorption/ionization (MALDI) can simplify the procedure in several ways. The MALDI matrix solution desorbs the affinity-bound frag-

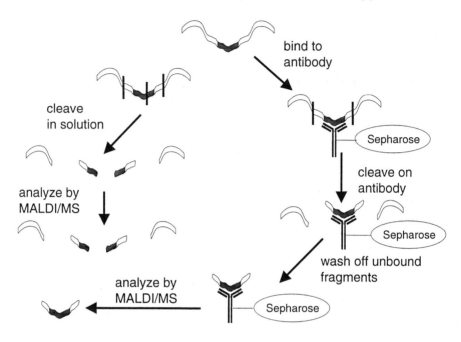
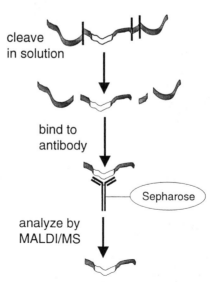

Fig. 1. Epitope excision and epitope extraction. (Adapted with permission from **ref. 4.**)

Epitope Mapping

ments, which can then be mass analyzed without additional sample workup. The sensitivity of MALDI-mass spectrometry (MS) is such that this analysis can be implemented using only a small portion of the sample. The remainder of the sample, with the peptide still bound to the antibody, can be subjected to additional enzymatic reactions and washes in order to map the epitope more finely. The particular enzymes and sequence of enzymatic digestions chosen will, of course, depend on the particular protein being digested.

Both of these approaches rely on the fact that the antibody is fairly resistant to enzymatic digestion, possibly because it is highly glycosylated. Also, for either of these approaches to work, it is best if the sequence of the antigen is known and the enzymes used have specific cleavage sites. In this case, the peptides generated by the enzyme can be predicted and the molecular weights of these peptides can be calculated. This list of predicted peptide masses can then be compared with the masses observed in the MALDI spectra, and the peptides containing the epitope can be identified. Obviously, if there is sufficient material for tandem sequence determination by mass spectrometry (MS/MS), the sequence of the antigen does not have to be known in advance.

In this chapter, we describe experimental details of the epitope excision analysis of an acquired immunodeficiency syndrome (AIDS)-related protein, intact HIV-1$_{IIIB}$ p26, affinity-bound to an immobilized antibody, using MALDI/MS for the determination of the affinity-bound fragments *(9)*.

2. Materials

2.1. MALDI Matrix and Mass Calibration Standards

1. α-Cyano-4-hydroxycinnamic acid (Aldrich, Milwaukee, WI), cat. no. 14-550-5. The α-cyano-4-hydroxycinnamic acid was recrystallized from hot methanol and stored in the dark at room temperature.
2. Ethanol (ethyl alcohol) (Pharmco Products, Inc., Brookfield, CT), cat. no. 111000200CSGL.
3. Bovine serum albumin (BSA), (Sigma, St. Louis, MO), cat. no. A-0281.
4. Carbonic anhydrase I, (Sigma), cat. no. C-4396.
5. Insulin B oxidized (Sigma), cat. no. I-6383.
6. Angiotensin I (human) (Sigma), cat. no. A-9650.

2.2. HIV Protein

HIV-1$_{IIIB}$ p26 protein was expressed in our laboratory as a glutathione-*S*-transferase fusion protein and was cleaved with factor Xa to give free p26 *(10)*. The procedures are given in **Subheading 3.1.** below.

2.3. Antibody

Mouse monoclonal anti-p24 antibody, IgG, purified from ascites fluid (Advanced Biotechnologies, Columbia, MD), cat. no. 13-102-100. This clone is identical *(11)* to Gallo/Veronese antibody M26 *(12,13)*. The antibody was kept frozen at –30°C and was thawed immediately before use.

2.4. Enzymes

1. Restriction factor Xa (Boehringer Mannheim), cat. no. 1 585 924.
2. Carboxypeptidase Y (CPY) (Boehringer Mannheim), cat. no. 1 111 914.
3. Aminopeptidase M (Boehringer Mannheim), cat. no. 102 768 .
4. Trypsin-TPCK (Worthington Biochemical, Freehold, NJ), cat. no. 3740.
5. Lysyl endopeptidase (Lys-C) (Wako, Dallas, TX), cat. no. 129-02541.

2.5. Reagents and Solvents

1. Deionized water was prepared on a Hydro Service and Supplies (Research Triangle Park, NC) model RO 40 water system.
2. Methanol (Fisher Scientific, Fairlawn, NJ), High-performance liquid chromatography (HPLC) grade, cat. no. 9093-03.
3. Acetonitrile (J.T. Baker, Phillipsburg, NJ), HPLC grade, cat. no. A998^{SK}-4.
4. 88% Formic acid (Baker), cat. no. 0129-01.
5. Ammonium acetate (Mallinckrodt, Paris, KY), cat. no. 3272.
6. Ammonium carbonate (Aldrich), ACS grade, cat. no. 20,786–1.
7. *N*-methyl morpholine (Sigma), cat. no. M-7889.
8. 4-Ethyl morpholine (Aldrich), cat. no. 23,952-6.
9. Hydrochloric acid (HCl; Mallinckrodt Specialty Chemicals), cat. no. H613.
10. Tris (Trizma Base), (Tris[hydroxymethyl]aminomethane) (Sigma), cat. no. T-8529
11. Tris-HCl (Trizma Hydrochloride; Sigma), cat. no. T-6791.
12. CNBr-activated Sepharose 4B beads (Pharmacia Biotech, Piscataway, NJ), cat. no. 17-0430-01.
13. Ethylenediamine tetraacetic acid (EDTA; Sigma), cat. no. E5513.
14. Pefablok SC (Boehringer Mannheim), cat. no. 1 429 868.
15. Phenylmethylsulfonyl fluoride (PMSF), (Boehringer Mannheim), cat. no. 236 608.
16. Leupeptin (Boehringer Mannheim), cat. no. 1 017 128.
17. Pepstatin (Boehringer Mannheim), cat. no. 1 359 053.
18. Trifluoroacetic acid (TFA; Pierce) cat. no. 28903.
19. Acetic acid (Fisher), cat. no. A-38.
20. Compact reaction columns (CRCs; USB, Cleveland, OH); columns: cat. no. 13928; 35 µ*M* compact column filters, cat. no. 13912.
21. Glutathione sepharose (GSH-sepharose) beads (Pharmacia Biotech), cat. no. 17-0765-01.

2.6. MALDI Instrumentation

The MALDI mass spectrometer used to acquire the mass spectra was a Voyager-RP (PerSeptive Biosystems, Framingham, MA). The instrument was equipped with a nitrogen laser (λ = 337 nm) to desorb and ionize the samples. The accelerating voltage used was 30 kV. Mass calibration was effected external to the sample, using two points that bracketed the mass range of interest. The MALDI instrument uses a stainless-steel target, on which the samples are deposited and dried. The Voyager-RP instrument is equipped with a video camera, which displays a real-time image on a monitor, and the laser can be aimed at specific features within the area of the target. For the experiments described here, the laser was aimed at or near to the affinity beads on the target. Relative signal abundances in the MALDI spectra are not directly related to the relative abundances of the species, due to potential differences in the sensitivities of different peptides.

3. Methods
3.1 Preparation of Recombinant p26
3.1.1. Preparation of p26 for Binding (see **Note 1**)

For these studies, a recombinant protein was expressed as a glutathione-S-transferase (GST) fusion protein in *E. coli* (*14*). The bacteria were grown in ampicillin-containing LB media and were induced with isopropyl-*p*-D-thiogalactopyranoside (IPTG) (*10*).

1. Pellet the cells by centrifugation in a Sorvall model RC-3B Refrigerated Centrifuge (DuPont Instruments Co., Newtown, CT) for 10 min at 5000*g* at 4°C. Remove the supernatant, and resuspend the cells in approximately 15 mL ice-cold pH 7.2 phosphate-buffered saline (PBS). Add protease inhibitors (10 μL of a solution of 100 μ*M* EDTA, 1 μ*M* leupeptin, and 1 μ*M* pepstatin). At this stage, the cells can be stored frozen at –30°C until needed.
2. Lyse the cells by sonication. Sonicate the cells in a cut-off 50-mL Falcon tube (Becton Dickinson Labware, Franklin Lakes, NJ) using a Branson Sonifier Cell Disrupter (Branson, Melville, NY) at 50% duty cycle, for 3 min at 60% power.
3. Transfer the lysate to a 15-mL Corex (Corning, Corning, NY) centrifuge tube, and centrifuge in a Sorvall Model RC-5 Superspeed Centrifuge (DuPont) at 12000*g* for 10 min at 4°C.
4. Transfer the supernatant (approx 15 mL) to a 15 mL Falcon tube, and store on ice.
5. Wash the glutathione (GSH)-sepharose beads 3 times with pH 7.2 PBS to remove the ethanol solution in which the beads are provided.
6. Add the beads (approx 0.6-mL bead slurry) to the cell lysate in a 15-mL Falcon tube, and rotate the mixture slowly for 2 h at room temperature.
7. Transfer (*see* **Note 2**) the beads to compact reaction columns (CRCs), which have been fitted with 35-pM bottom filters. Drain (*see* **Note 3**), and rinse 3 times with 0.4 mL pH 7.2 PBS. Add enough buffer solution to the CRCs to keep the beads moist.
8. Analyze a 1-μL aliquot of the settled beads by MALDI to confirm the presence of the fusion protein (*see* **Note 4**).

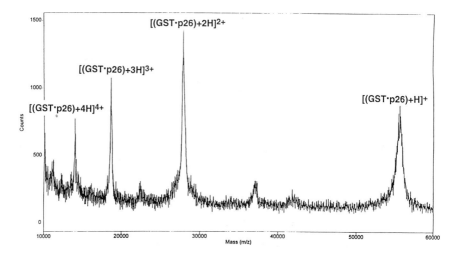

Fig. 2. MALDI mass spectrum of the GST-p26 affinity bound to glutathione-sepharose beads.

A MALDI spectrum of the affinity-bound GST-p26 is shown in **Fig. 2**. It should be noted that only the affinity-bound analytes appear in the spectra. As observed in an earlier study *(8)*, the covalently bound antibody is not volatilized by the laser and does not appear in the spectrum.

3.1.2. Cleavage of p26 from the Fusion Protein

1. Rinse the GSH-sepharose beads containing the adsorbed GST-p26 fusion protein with 3 × 0.5-mL factor Xa cleavage buffer (50 mM Tris-HCl, 150 mM NaCl, and 1 mM CaCl$_2$, pH 8.0).
2. Dissolve approximately 100 µg factor Xa in 100 µL deionized H$_2$O.
3. Add 0.5-mL of cleavage buffer to each CRC, and add 15 µL of factor Xa solution.
4. Incubate the beads overnight at 37°C to cleave the p26 from the affinity-bound GST.
5. Add Pefablok SC, a protease inhibitor, and drain the recombinant p26-containing supernatant directly into one of the reaction columns containing the immobilized antibody beads.

3.2. Preparation of CNBr-Activated Sepharose-Immobilized Antibody Columns

CNBr-activated Sepharose columns were prepared according to the procedures provided with the packing material, but on a smaller scale.

1. Put approx 0.2 g of dry CNBr-sepharose beads into a Falcon tube.
2. Add 10 mL of 1 mM HCl. Swirl the mixture and leave to equilibrate for 15 min.
3. Add approx 150 µL of the wet beads from the bottom of the tube to each of two CRCs, and drain the columns.

4. Wash the CNBr-activated sepharose column with 6 × 0.8 mL 1 mM HCl, followed by 6 × 0.4-mL washes with coupling buffer (0.1 M NaHCO$_3$, pH 8.3).
5. Thaw the antibody solution (which was shipped and stored on dry ice) immediately before use. Check the buffer that the antibody is shipped in. If it is an amine buffer, it will interfere with the coupling reaction, and the buffer solution will have to be exchanged before using. Incubate both columns with the mouse monoclonal antibody solution (≈ 20–50 µg antibody) in the coupling buffer for 2 h at room temperature, with slow rotation.
6. Block any unreacted sites by rinsing with 0.4 mL of 0.1 M Tris-HCl, pH 8.0, quenching buffer, drain and add 0.4 mL quenching buffer and rotate at room temperature for 2 h.
7. Remove unbound antibody with a series of 0.4-mL washes at alternating pH (0.1 M sodium acetate/0.5 M NaCl at pH 4.0 and 0.1 M Tris HCl at pH 8.0) 3 times. Add enough buffer solution to keep the beads moist.
8. Spot a 1-µL aliquot of the settled beads on the MALDI target to see if there are any interference peaks originating from the antibody.

3.3. Affinity-Binding of p26 to the Immobilized Antibody
3.3.1. Binding of the Antigenic Protein to the Antibody

1. Incubate one of the columns containing antibody beads with the recombinant p26 for 2 h at room temperature, with slow rotation. Incubate the second column with PBS to serve as a control. Carry it through the same series of enzymatic digestions as the protein-containing column.
2. Wash the columns 3 times with 0.4 mL PBS, pH 7.2, to remove any unbound products, and remove an aliquot for MALDI analysis (*see* **Note 5**). If you are using a valuable protein, save the eluant! If, for some reason, the antibody doesn't bind the protein, or if the amount of protein used exceeds the binding capacity of the column, unbound protein will pass through the antibody column.

The MALDI spectrum of p26 affinity-bound to the immobilized antibody is shown in **Fig. 3**. Four proteins are present—the intact p26 (at mass-to-change ratio *[m/z]* 29251) and three truncated p26 proteins (mol wt$_{(obs)}$ 23,380, 25,368 and 27,295), resulting from proteolysis by bacterial enzymes which occurred even in the presence of protease inhibitors. Since these truncated proteins are still bound to the antibody, they must still contain the epitope, so this bacterial proteolysis is not a problem in this study.

3.4. Proteolytic Footprinting of Affinity-Bound p26
3.4.1. Initial Enzymatic Digestion of the Bound Protein with Achromobacter Protease (Endoproteinase Lys-C) (see **Notes 6–10**)

1. Dissolve ≈ 50 µg of endoproteinase Lys-C in 50 µL deionized water, to make a solution of about 1 µg/µL.
2. Wash the beads containing p26 affinity-bound to the monoclonal antibody with 0.4 mL × 3 Lys-C buffer (50 mM Tris HCl, 0.25 mM EDTA, pH 8.0)

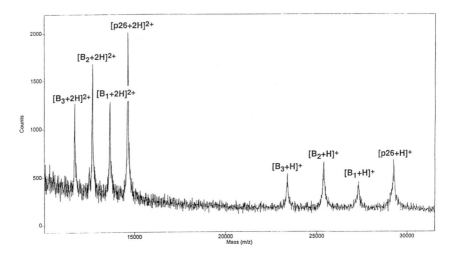

Fig. 3. MALDI mass spectrum of HIV-1$_{IIIB}$ p26 and three bacterial proteolytic cleavage products, affinity-bound to immobilized anti-p24 (13-102-100). The spectrum shows singly and doubly charged molecular ions for the four proteins (spectrum from a 1–μL aliquot of the column bed). (Reproduced with permission from ref. **9**.)

3. Add 10 μL enzyme solution (*see* **Note 9**) to each CRC and incubate overnight at 37°C, with slow rotation.
4. Wash the column bed with 0.4 mL × 3 of pH 7.0 PBS. Add enough buffer solution to keep the beads moist, and remove aliquots of the washed column bed for MALDI analysis, to determine which Lys-C fragments are still bound to the immobilized antibody.

The sequence of p26 and the predicted enzymatic cleavage sites are shown in **Fig. 4**. Lys-C was chosen as the first enzyme used because it should give a "reasonable" number of moderately sized cleavage products from the p26. It is preferable to choose an initial enzyme that does not cut the protein into very small pieces. It is sometimes difficult to assign mass spectral peaks to specific enzymatically-generated peptides if very small peptides are generated at this stage in the procedure (*see* **Note 8**).

The MALDI spectrum showing the Lys-C fragments still affinity-bound to the immobilized antibody beads is shown in **Fig. 5**. Of the 12 possible Lys-C fragments, only three Lys-C fragments remain attached to the antibody beads after the nonbound fragments have been removed by washing. These peptides can be identified by comparison of the predicted Lys-C fragment molecular weights with those observed in the mass spectrum. In this case, they correspond to L4, L5, and L8, with mol wt 6691, 1113, and 1361, respectively.

Epitope Mapping

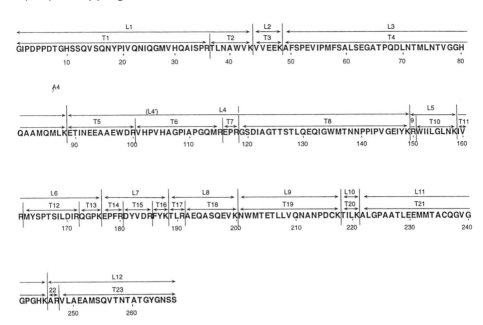

Fig. 4. Amino acid sequence of HIV-1$_{IIIB}$ p26, showing Lys-C (L1-L12), and tryptic (T1-T23) cleavage products. (Reproduced with permission from **ref. 9**.)

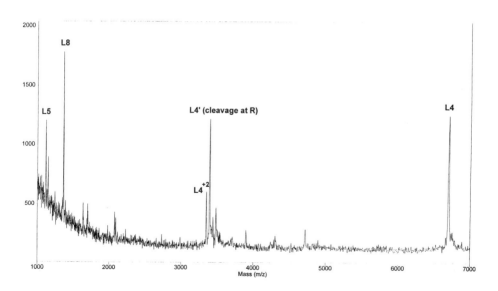

Fig. 5. MALDI mass spectrum obtained from a 1-µL aliquot of the rinsed column bed after Lys-C digest of the affinity-bound p26. (Reproduced with permission from **ref. 9**.)

L4: ETINEEAAEWDRVHPVHAGPIAPGQMREPRGSDIAGTTSTL
QEQIGWMTNNPPIPVGEIYK
L5: RWIILGNLK
L8: TLRAEQASQEVK

An ion of m/z 3395 was also observed still bound to the antibody. This mass corresponds to an anomalous cleavage at residue 118 (**R**), giving an additional cleavage within L4. This fragment (designated L4′) was also observed in the solution digestions of p26.

L4′: ETINEEAEWDRVHPVHAGPIAPGQMREPR

3.4.2. Second Enzymatic Digestion of the Affinity-Bound Peptides

The Lys-C fragments still bound to the anti-p24 monoclonal antibodies are then subjected to digestion over a 2-h period with trypsin-TPCK.

1. Rinse the beads with 0.4 mL × 3 aliquots of 50 mM NH$_4$HCO$_3$ buffer, pH 7.8.
2. Incubate at 37°C for 2 h with slow rotation.
3. Rinse the beads x 3 with buffer solution, and analyze a 1-µL aliquot of the beads by MALDI to determine which tryptic fragments are still bound to the immobilized antibody.

The MALDI spectrum of the beads after rinsing is shown in **Fig. 6**. Now we are looking for tryptic fragments of the Lys-C fragments found in the first digestion. After a 2-h trypsin-TPCK digestion of the Lys-C-treated beads, L5(=T9+T10) was no longer bound, but some L8(=T17+T18) was still present, although at lower relative abundance (**Fig. 6**). Most of L4(=T5+T6+T7+T8) and L4′(=T5+T6+T7) had been digested, leaving only the tryptic fragment T5+T6: **ETINEEAAEWDRVHPVHAGPIAPGQMR** (residues 89-115) and some T6: **VHPVHAGPIAPGQMR** (residues 101-115). The cleavage site between T5 and T6 thus seems to be partially blocked, indicating that the epitope may be near the N-terminal end of T6.

3.4.3. Third Enzymatic Digestion of the Affinity-Bound Peptides

The beads from the tryptic digest were divided into two portions with the first portion being subjected to carboxypeptidase Y digestion for fine-scale mapping of the epitope from the C terminus.

This is the third enzymatic digestion on the affinity-bound peptide—first, Lys-C, then trypsin, now carboxypeptidase Y. It had been noted previously that repeated enzymatic digestions on the affinity-bound peptides led to a reduction in MALDI signal (*see* **Note 10**). For this reason, the third (this **Subheading**) and fourth (**Subheading 3.4.4.**) enzymatic digestions were done "in parallel" on two different sets of beads, rather than "in series."

Epitope Mapping

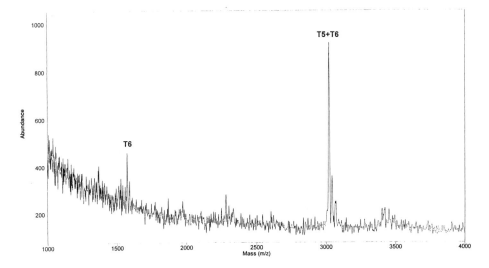

Fig. 6. MALDI mass spectrum obtained from a 1-µL aliquot of the rinsed column bed after consecutive Lys-C and tryptic digests of the affinity-bound p26. (Reproduced with permission from **ref. 9**.)

1. Rinse the columns with 0.4 mL × 3 of 50 mM N-ethylmorpholine, pH 6.3.
2. Prepare carboxypeptidase Y solution by adding 20 µg enzyme to 20 µL deionized water.
3. Add 100 µL 50 mM N-ethylmorpholine, pH 6.3, to each column.
4. Add a 2-µL aliquot of the enzyme solution to the beads.
5. Incubate at 37°C overnight, with slow rotation.
6. Record a MALDI spectrum of a 1-µL aliquot of the rinsed beads.
7. Repeat **steps 3–5** until no further digestion is observed (*see* **Note 11**). (In these experiments, two overnight digestions were used.)

Carboxypeptidase Y cleavage of the Lys-C/trypsin-cleaved residues removed three residues (R, M, and Q) from the C-terminal end of T6, leaving **VHPVHAGPIAPG (Fig. 7)**. The remaining peptide has G at the C terminus: The cleavage of G by carboxypeptidase Y is known to be slower than that of other residues *(15)*, and cleavage of affinity-bound peptides is slower than that of proteins in solution. To check the rate of proteolysis of this residue when it is "unprotected" (i.e., free in solution), the peptide VHPVHAGPIAPGQ was synthesized and subjected to

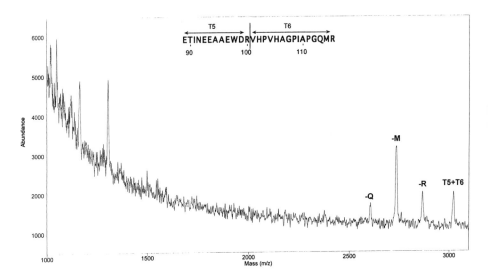

Fig. 7. MALDI mass spectrum obtained from a 1-μL aliquot of the rinsed column bed after consecutive Lys-C, tryptic, and carboxypeptidase Y digests of the affinity-bound p26. (Reproduced with permission from **ref. 9**.)

carboxypeptidase Y digestion. Cleavage of five residues occurred within 2 h. The fact that no cleavage past G was observed in the affinity-bound T5+T6 during two overnight digestions was therefore taken to mean that the G was protected from enzymatic cleavage because it was bound to the antibody as part of the epitope.

3.4.4. Fourth Enzymatic Digestion of the Affinity-Bound Peptides

The second portion of the beads from the tryptic digestion was subjected to aminopeptidase M digestion for fine-scale mapping of the epitope from the N terminus.

1. Rinse the columns with 0.4 mL × 3 of 5 m*M* *N*-methylmorpholine, pH 7.2.
2. Dissolve a 10-μL aliquot of the enzyme solution (in ammonium sulfate buffer) in 100 μL buffer.
3. Add 100 μL 5 m*M* *N*-methylmorpholine, pH 7.2, to each column. On-column digestions of the tryptic fragments with aminopeptidase *M* were done in 5 m*M* *N*-methylmorpholine, pH 7.2.
4. Add a 25-μL aliquot of enzyme solution to each compact reaction column.
5. Incubate overnight at 37°C, with slow rotation.
6. Analyze a 1-μL aliquot of the rinsed beads by MALDI.

The overnight digestion with aminopeptidase *M* of the affinity-bound T5+T6 peptide cleaved all of the residues of T5. There was also a peak of low relative abundance corresponding to cleavage of one residue (V) from the N-terminal end of T6, leaving the fragment **HPVHAGPIAPGQMR** (**Fig. 8**).

Epitope Mapping

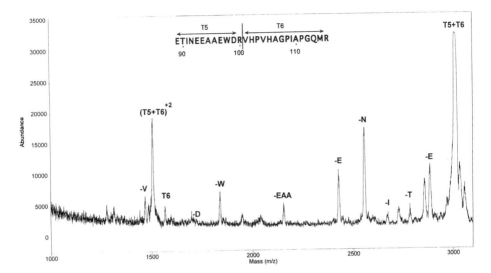

Fig. 8. MALDI mass spectrum obtained from a 1-µL aliquot of the rinsed column bed after consecutive tryptic and aminopeptidase M digests of the affinity-bound p26. (Reproduced with permission from **ref. 9**.)

3.4.5. Results

The carboxypeptidase Y and aminopeptidase *M* data taken together show that the fragment **HPVHAGPIAPG** is protected from proteolytic digestion. This leaves the fragment **HPVHAGPIAPG** as the epitope (**Fig. 9**).

3.4.6. Structural Relevance

In an earlier study by Marcus-Sekura et al. *(16,17)* on M26, the cleavage points for the restriction enzymes used to create some of the DNA fragments used in the ELISA studies happen to lie within the region that codes for residues 102-112 in p26. Thus, the DNA fragments generated coded for only a portion of the region we have found to be the epitope, and binding of these expressed protein fragments to the monoclonal antibody did not occur. Only clone 5HB, expressing residues 7–193, contained an intact amino acid 102-112 sequence, and this was the only clone showing binding in their study. This is consistent with our conclusion that the epitope is **HPVHAGPIAPG**.

These results are also consistent with more recent affinity capillary electrophoresis (ACE) data *(18)* from our laboratory, which shows the strongest binding for the synthetic peptide **VHPVHAGPIAP**. In the ACE experiments, this peptide bound a little more tightly than did **VHPVHAGPIAPG** or **HPVHAGPIAP**, which showed binding as well. Although the peptide that *exactly* corresponded to the epitope determined by epitope excision was not

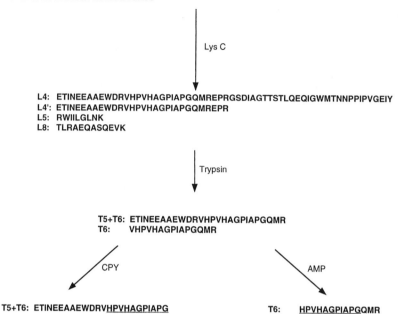

Fig. 9. Summary of epitope excision results for p26 bound to immobilized anti-p24 (13-102-100).

analyzed by ACE, the different techniques seem to agree within one residue on each end of the peptide. Because the ACE experiments more closely approximate epitope extraction than epitope excision, it is possible that some of the differences may be conformational—in epitope excision, the peptide is held in position by the rest of the protein when binding takes place. The end residues may help shape a peptide floating free in solution into a more favorable conformation for binding. Alternatively, it is also possible that the peptide comprising the epitope is not *completely* protected, even when bound, so that the proteases are still able to cleave off the end residues.

A nuclear magnetic resonance (NMR) study of HIV p24 *(19)* and an X-ray crystallography study of the HIV capsid dimer *(20)* were published after the mass spectrometric epitope mapping studies on M26 (13-102-100) antibody were completed. Both of these studies show that the epitope found in our studies is located on a long exposed loop of the p24 protein. Interestingly, this epitope also includes the binding site for cyclophilin (AGPI, Ala-Gly-Pro-Ile), which had been determined in earlier studies *(21)*.

4. Notes

1. If a commercially available protein is to be bound to antibody, this part of the procedure would not be necessary. Approximately 10–50 µg purified protein can simply be dissolved in 0.4 mL pH 7.2 PBS and added directly to the prepared antibody column.
2. If the end of the pipette tip is very small, it may be difficult for the beads to get into the pipette tip. If this happens, cut off the tip to enlarge the opening.
3. The compact reaction columns were drained by removing the column bottom plugs, placing the tubes into 1.5-mL Eppendorf centrifuge tubes to catch the eluate, and centrifuging (2–3 min) at the lowest speed setting (<80g) in an Eppendorf Microfuge (Westbury, NY). This procedure is gentle enough to avoid crushing the beads and allows several columns to be drained and/or rinsed in the centrifuge at the same time.
4. Check under a microscope to see whether a good sample of beads has been caught—about 10–20 beads is sufficient. Wet agarose beads look like transparent spheres under the microscope; they become small and opaque when dry.
5. Rinse the unbound fragments off the beads with the next digestion buffer, if you know what your next enzyme will be.
6. With affinity-bound peptides, there is the option of washing away a buffer that is not compatible with MALDI, but it is always safer to use MALDI-compatible buffers. When using an different enzyme for the first time, start with the product literature or a literature procedure and modify the buffers as needed.
    a. If the coupling reaction involves amino groups, do not use amine buffers.
    b. The standard recipes for buffer solutions designed to digest proteins often contain dithioerythritol (DTT) or other reducing/denaturing agents. These are left out of the buffers used here because the native conformation of the protein must be maintained.
    c. Glycerol is also not included because it can interfere with crystallization in MALDI experiments.
    d. A recommended salt can often be substituted by one that is known not to interfere with MALDI-MS analysis so long as the same pH is maintained.
    e. High concentrations of Ca^{2+} can interfere with MALDI. If the enzyme requires calcium (or another cation), and it interferes, try a lower concentration. For example, 2 M $CaCl_2$ interferes with MALDI, but 1 mM $CaCl_2$ is acceptable). The ion concentrations in the "recipes" that are provided with the enzymes are often much higher than the enzyme actually needs. Test the conditions on a model protein.
7. It is a good idea to practice *all* digestions on a readily available protein to get a feel for how fast the enzyme works in solution. Reaction rates will be slower with affinity-bound substrates.
8. The preferred first enzyme is a slow enzyme, which generates larger fragments. Once the epitope has been partly characterized and reaction rates are known, a faster enzyme may be used to confirm the initial results. In this study, Lys-C was used as the first enzyme, with trypsin as the second enzyme. These results were subsequently confirmed using trypsin as the first enzyme.

9. Enzyme-to-substrate ratios used for proteolytic footprinting are often higher than those used in free solution, because reaction rates of bound analytes are often slower than those in free solution, especially when the cleavage site is close to the epitope.
10. Antibody beads with peptides or proteins attached can be stored for a few days in the refrigerator. The MALDI spectra suggest degradation with longer sample storage.
11. The advantage of the epitope excision method, using immobilized antibodies and MALDI, is that small aliquots of beads can be analyzed periodically throughout the digestion procedure, to follow the course of the reaction, to see how quickly the enzyme is working, and to determine when the reaction is complete.

Acknowledgments

Support for the purchase of the Voyager-RP MALDI mass spectrometer by the NIH Office of AIDS Research is gratefully acknowledged. We would like to thank Dr. Ian M. Jones (NERC Institute of Virology, Oxford, UK) for the gift of the *E. coli* that express HIV-p26.

References

1. Zhao, Y. and Chait, B. T. (1994) Protein epitope mapping by mass spectrometry. *Anal. Chem.* **66,** 3723–3726.
2. Zhao, Y., Muir, T. W., Kent, S. B. H., Tisher, E., Scardina, J. M., and Chait, B. (1996) Mapping protein-protein interactions by affinity-directed mass spectrometry. *Proc. Natl. Acad. Sci. USA* **93,** 4020–4024.
3. Suckau, D., Kohl, J., Karwath, G., Schneider, K., Casaretto, M., Bitter-Suermann, D., and Przybylski, M. (1990) Molecular epitope identification by limited proteolysis of an immobilized antigen-antibody complex and mass spectrometric peptide mapping. *Proc. Natl. Acad. Sci. USA* **87,** 9848–9852.
4. Przybylski, M. (1994) Mass spectrometric approaches to the characterization of tertiary and supramolecular structures of biomolecules. *Adv. Mass Spectrom.* **13,** 257–283.
5. Jemmersen, R. and Paterson, Y. (1986) Mapping epitopes on a protein antigen by the proteolysis of antigen-antibody complexes. *Science* **232,** 1001–1004.
6. Burnens, A., Demotz, S., Corradin, G., Binz, H., and Bosshard, H. R. (1987) Epitope mapping by chemical modification of free and antibody-bound protein antigen. *Science* **235,** 780–783.
7. Macht, M., Feidler, W., Kurzinger, K., and Przybylski, M. (1996) Mass spectrometric mapping of protein epitope structures of myocardial infarct markers myoglobin and troponin T. *Biochemistry* **35,** 15,633–15,699.
8. Papac, D. I., Hoyes, J., and Tomer, K. B. (1994) Epitope-mapping of the gastrin-releasing peptide/anti-bombesin monoclonal antibody complex by proteolysis followed by matrix-assisted laser desorption mass spectrometry. *Protein Sci.* **3,** 1485–1492.
9. Parker, C. E., Papac, D. I., Trojak, S. K. and Tomer, K. B. (1996) Epitope mapping by mass spectrometry. Determination of an epitope on HIV-1,,,,p26 recognized by a monoclonal antibody. *J. Immunol.* **157,** 198–206.

10. Parker, C. E., Papac, D. I., and Tomer, K. B. (1996) Monitoring cleavage of fusion proteins by matrix-assisted laser desorption Ionization/mass spectrometry: recombinant HIV-1IIIB p26. *Anal. Biochem.* **239**, 25–34.
11. Hoekzema, D. T., Director, Laboratory Operations, Advanced Biotechnologies, Inc., Personal communication.
12. Veronese, F. D., Sarngadharan, M. G., Rahman, R., Markham P. D., Popovic, M., Bodner, A. J., and Gallo, R. C. (1985) Monoclonal antibodies specific to p24, the major core protein of human T-cell leukemia virus type III. *Proc. Natl. Acad. Sci. USA* **82**, 5199–5202.
13. Veronese, F. D., Copeland T. D., Oroszlan, S., Gallo, R. C., and Sarngadharan, M. G. (1988) Biochemical and immunological analysis of human immunodeficiency virus *gag* gene products p17 and p24. *J. Virol.* **62**, 795–801.
14. Smith, D. B. and Corcoran, L. M. (1991) Expression and purification of glutathione-*S*-transferase fusion proteins, in *Current Protocols in Molecular Biology*, vol. 2 (Ausubel, F. M., Brent, R., Kingston, R. E., Moore, D. D., Seidman, J. G., Smith, J. A., and Struhl, K., eds.), Green and Wiley-Interscience, New York, pp. 16.7.1–16.7.8.
15. Carboxypeptidase Y product literature, Boehringer Mannheim Biochemica, Indianapolis, IN.
16. Marcus-Sekura, C. J., Woerner, A. M., Klutch, M., and Quinnan, G. V., Jr. (1988) Reactivity of an HIV *gag* gene polypeptide expressed in *E. coli* with sera from AIDS patients and monoclonal antibodies to *gag*. *Biochem. Biophys. Acta* **949**, 213–223.
17. Marcus-Sekura, C. J., Woerner, A. M., Zhang, P.-F., and Klutch, M. (1990) Epitope mapping of the HIV-1 *gag* region by analysis of *gag* gene deletion fragments expressed in *Escherichia coli* defines eight antigenic determinants. *AIDS Res. Hum. Retroviruses* **6**, 317–327.
18. Qian, X.-H. and Tomer, K. B. (1998) Affinity capillary electrophoresis investigation of an epitope on human immunodeficiency virus recognized by a monoclonal antibody. *Electrophoresis* **19**, 415–419.
19. Luban, J., Bossolt, K. L., Franke, E. K., Kalpana, G. V., and Goff, S. P. (1993) Human immunodeficiency virus type 1 *gag* protein binds to cyclophilins A and B. *Cell* **73**, 1067–1078.
20. Momany, C., Kovari, L. C., Prongay, A. J., Keller, W, Gitti, R. K., Lee, B. M., et al. (1996) Crystal structure of dimeric HIV-1 capsid protein. *Nature Structural Biology* **3**, 763–770.
21. Gitti, R. K., Lee, B. M. Walker, J., Summers, M. K., Yoo, S., and Sundquist. W. I. (1996) Structure of the amino-terminal core domain of the HIV-1 capsid protein. *Science* **273**, 231–234.

12

Identification of Active Site Residues in Glycosidases by Use of Tandem Mass Spectrometry

David J. Vocadlo and Stephen G. Withers

1. Introduction

The glycosidases are a class of enzymes that are responsible for the hydrolysis of glycosidic bonds. Such glycosidic linkages occur in a wide range of contexts, including polysaccharides, oligosaccharides, glycolipids, glycoproteins, lipopolysaccharides, proteoglycans, saponins, and a range of other glycoconjugates. Corresponding to this diverse collection of substrates there is a very large assortment of glycosidases responsible for their selective hydrolysis. Amino acid sequences are now available for well over 2000 of these enzymes, and these have been arranged into families on the basis of sequence similarities *(1–4)*. At the last count (November 1999) there were 76 such families, and a regularly updated list of these is readily available at the URL http://afmb.cnrs-mrs.fr/~pedro/CAZY/db.html. A large amount of effort has been expended on structural studies of these enzymes in the past 10 yr with the result that three-dimensional X-ray crystal structures are now available for representatives of at least 27 of these families *(4–6)*. These structures are remarkably diverse, with monomer sizes ranging from approx 14,000 to 170,000 Daltons, and compositions ranging from essentially completely α-helical to almost exclusively β-sheet. The reasons for this diversity are probably twofold; the diverse nature of the substrates themselves, and different evolutionary pathways to the construction of an active site.

Mechanistically, however, these enzymes are all closely related, there being two principal ways in which the glycosidic bond is cleaved, viz. either with net retention or with net inversion of the anomeric configuration of the sugar. Likely mechanisms to explain these two different stereochemical outcomes

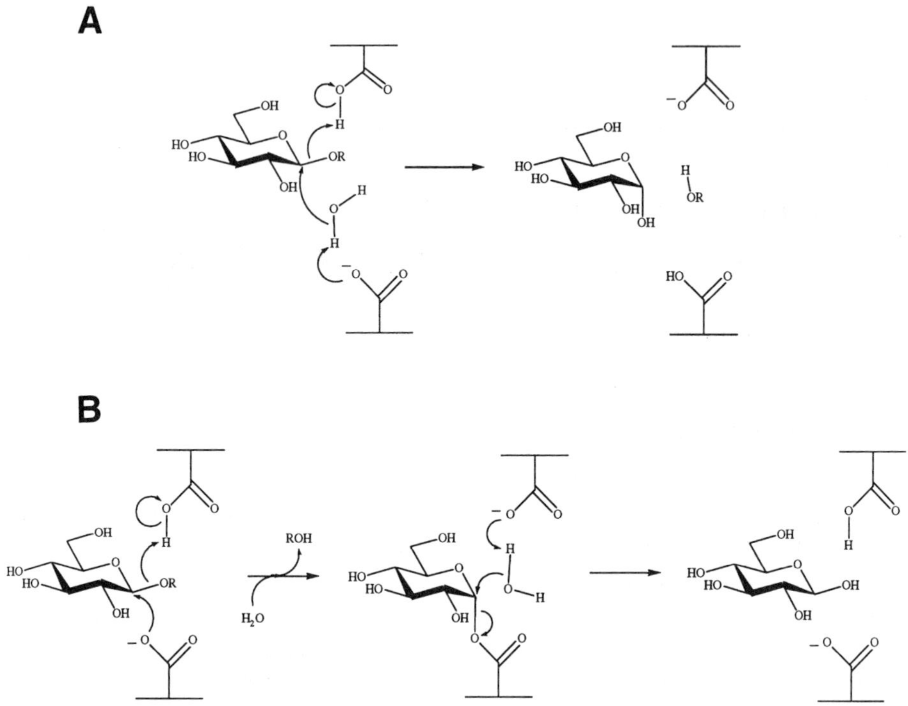

Fig. 1. Catalytic mechanisms of (**A**) inverting and (**B**) retaining glycosidases.

were proposed some 45 yr ago by Koshland, and these mechanisms have largely stood the test of time, although the details have been considerably expanded on in the interim *(6–10)*. As shown schematically in **Fig. 1**, the active sites of these enzymes contain a pair of carboxylic acids that are intimately involved in the catalytic mechanism, although the separation between these two acid groups is quite different in the two cases. In inverting glycosidases the two carboxyl groups are approximately 10 Å apart (O to O), whereas in the retaining glycosidases a 5 Å separation is found. These different geometries are consistent with the two mechanisms.

Inverting glycosidases employ a single displacement mechanism (**Fig. 1A**) in which a water molecule attacks at the bound substrate anomeric center with general base catalytic assistance from one of the active site carboxylates. Simultaneously, the second carboxyl group provides general acid catalytic assistance to the departure of the leaving group. The reaction follows a dissociative mechanism, proceeding via an oxocarbenium ion-like transition state (**Fig. 2**). Retaining glycosidases employ a double-displacement mechanism in which a covalent glycosyl-enzyme intermediate is formed and is then hydrolyzed with general acid/base catalytic assistance (**Fig. 1B**). In the first step, one of

A OR = Aglycone, OR' = H₂O
B OR = Aglycone, OR' = Enzyme-COO⁻
C OR = H₂O, OR' = Enzyme-COO⁻

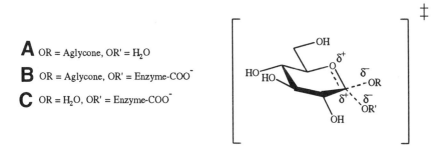

Fig. 2. Generalized transition state for glycosidase-catalyzed hydrolysis of a β glycoside. **(A)** inverting glycosidase; **(B)** glycosylation step of a retaining glycosidase; **(C)** deglycosylation step of a retaining glycosidase.

the carboxyl groups protonates the glycosidic oxygen while the other attacks directly at the anomeric center, forming a covalent glycosyl-enzyme intermediate. Once this intermediate has formed, removing the charge on the nucleophilic carboxylate, the pK_a of the acid/base carboxyl group drops and leaves it in a deprotonated state, ready to act as a general base catalyst for the second step in catalysis. The second step therefore involves the general base-catalyzed attack of water at the anomeric center of the glycosyl-enzyme intermediate, which releases the product sugar with the same anomeric configuration as that of the original substrate. Both steps of this mechanism proceed through oxocarbenium ion-like transition states (**Fig. 2**). The different active site geometries are therefore simply a natural consequence of the different mechanistic requirements.

The two active site carboxylic acids are the key players in catalysis and, being so important, are generally completely conserved within families of these enzymes. Knowledge of the identities of these residues for each enzyme family is therefore of considerable importance. This knowledge provides information not only on the location of the active site, but also, through mutational studies, on the importance of these residues to catalysis. It also allows the generation of glycosynthases, recently described mutant glycosidases that are useful in oligosaccharide synthesis since they can form glycosidic bonds but do not degrade them *(11)*. Although the X-ray crystal structure, when available, will most likely pinpoint the two residues, it is often not able to distinguish their specific roles. In the absence of crystallographic data, the only clue as to the identities of these residues is the presence of highly conserved amino acid residues within the sequence, provided that sufficient structural information is available to make this possible. There has, therefore, been a need for a reliable method for labeling these active residues, particularly the nucleophile, and then identifying and sequencing the labeled peptide in proteolytic digests. This chapter describes a well-evolved method for reliably identifying the active site nucleophile in retaining glycosidases.

1.1. Labeling the Active Site Nucleophile

A number of reagents have been developed over the years in the hope of reliably tagging the active site nucleophile; these include glycosyl epoxides, conduritol epoxides, and glycosylmethyl triazenes *(7,9)*. However, despite the clever design of these reagents, which it was hoped would render them highly selective for the active site nucleophile, these are essentially affinity labels that place a reactive group in the active site, which can then react with whatever is closest. A better method for tagging this residue involves taking more direct advantage of the actual enzyme mechanism and simply stabilizing the covalent species that is inherently formed during catalysis by somehow slowing its decomposition. This has now been achieved in several different ways. One general approach is to synthesize modified substrates that will relatively rapidly form an intermediate, which then only slowly hydrolyzes. This approach will be the major focus of this chapter. The other approaches include selectively slowing the second step by mutation of the enzyme *(12–19)*, or, in the case of glycosyl transferases that transfer to acceptors other than water, using an activated glycosyl donor in conjunction with an incompetent glycosyl acceptor that has no nucleophilic hydroxyl group *(16,20)*. These latter approaches will not be discussed, but the above references will provide the interested reader with the necessary information.

Both formation and hydrolysis of the glycosyl-enzyme intermediate proceed through transition states with substantial oxocarbenium ion character. Indeed, secondary deuterium kinetic isotope effects measured for the hydrolysis of aryl glycosides by β-glycosidases suggest that the second step has more oxocarbenium ion character than the first *(5,21–23)*. Inductive destabilization of this positively charged transition state by the introduction of a very electronegative substituent close to the reaction center should therefore slow both steps, but particularly the second step for β-glycosidases. The first such approach involved replacement of the sugar 2-hydroxyl with a sterically conservative fluorine atom *(24,25)*. This places the highly electronegative fluorine adjacent to the developing positive charge on the anomeric carbon. Such a substitution has a second, equally important, destabilizing effect on the transition state, which arises as a consequence of the removal of the very important hydrogen bonds (worth up to 40 kJ/mol) that have been shown to form with the 2-hydroxyl group in the transition state *(26–31)*. These interactions appear to be principally between the 2-hydroxyl and the carbonyl oxygen of the nucleophile itself as well as with another highly conserved active site residue, which in many cases is an asparagine side chain *(17,32)*. The net consequence of the substitution of the

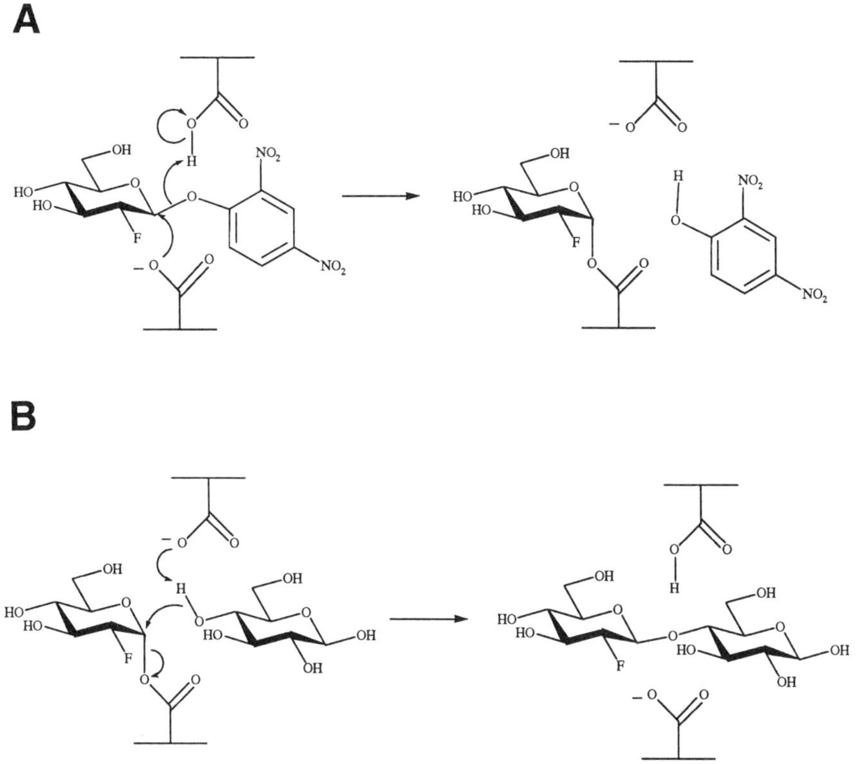

Active Site Residues in Glycosidases 207

Fig. 3. Inactivation and reactivation by 2-deoxy-2-fluoro glycosides. (**A**) Inactivation by formation of the fluoroglycosyl-enzyme intermediate. (**B**) Reactivation by transglycosylation of the fluoroglycosyl-enzyme intermediate.

2-hydroxyl is therefore the destabilization of the transition state both inductively and also through the removal of key hydrogen-bonding interactions. Since the *formation* of the intermediate is also slowed, it was necessary to incorporate a good leaving group, typically dinitrophenolate or fluoride, at the anomeric center of the sugar to speed up the first step again, thereby rendering the intermediate kinetically accessible. Fortunately, most glycosidases have a relatively loose specificity for the aglycone, and even the most specific enzymes will generally accommodate the very small, but reactive, fluoride leaving group. When incubated with such a substrate analog, the enzyme relatively rapidly forms a fluoroglycosyl-enzyme intermediate, as shown in **Fig. 3A**.

The intermediate so trapped is surprisingly stable, turning over by hydrolysis only extremely slowly. Typical half-lives are measured in days. However, if inactive enzyme that has been freed of excess inactivator is incubated in the presence of a sugar that can be accommodated in the aglycone (+1) site of the

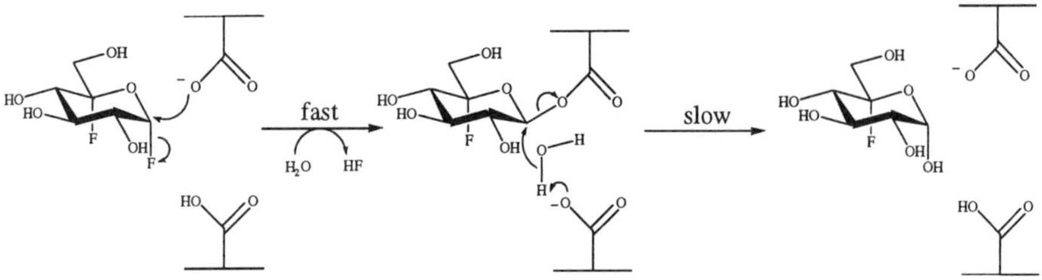

Fig. 4. Inactivation of a retaining α-glucosidase by 5-fluoro-α-D-glucopyranosyl fluoride and subsequent hydrolysis of the intermediate.

enzyme, then reactivation occurs rapidly through a transglycosylation process, as shown in **Fig. 3B** *(33)*. Such transglycosylations are typical reactions with normal substrates and thereby demonstrate the kinetic competence of the trapped intermediate. Other evidence for this mechanism of inactivation is the stoichiometric (with enzyme) burst of released dinitrophenolate upon inactivation, the observation by ^{19}F-nuclear magnetic resonance (NMR) spectroscopy of a signal corresponding to an α-linked 2-fluoroglycosyl-enzyme derivative, the increase in mass of the enzyme by exactly that of the fluorosugar, and most recently, the solution of the three-dimensional crystal structures of several such intermediate complexes *(32–34)*.

This "2-fluorosugar strategy" has proved to be extremely useful for trapping intermediates of retaining β-glycosidases. However, it does not appear to work with retaining α-glycosidases, the compounds functioning rather as slow substrates, their rate-determining step being glycosylation rather than deglycosylation *(35)*. Furthermore, the approach cannot be expected to work for *N*-acetylhexosaminidases since the fluorine is an inadequate replacement for the *N*-acetylamino group. A second, similar, strategy, involving placement of a fluorine substituent at C-5 in place of the hydrogen was therefore devised *(36–38)*. This approach means that the 2-substituent is kept intact, and the electronegative substituent is placed close to the ring oxygen, arguably the principal center of (relative) positive charge. These compounds are more challenging synthetically, resulting in less flexibility in the choice of leaving group. However, synthetic routes have now been developed to a number of different 5-fluoroglycosyl fluorides of both anomeric configurations and also of both configurations at C-5. These compounds have been proved to function much as anticipated, forming a relatively stable 5-fluoroglycosyl-enzyme intermediate and subsequently turning over via hydrolysis or transglycosylation as shown in **Fig. 4**.

The lifetimes of these 5-fluoroglycosyl intermediates tend to be much shorter than those of their 2-fluoro counterparts, consistent with the fact that

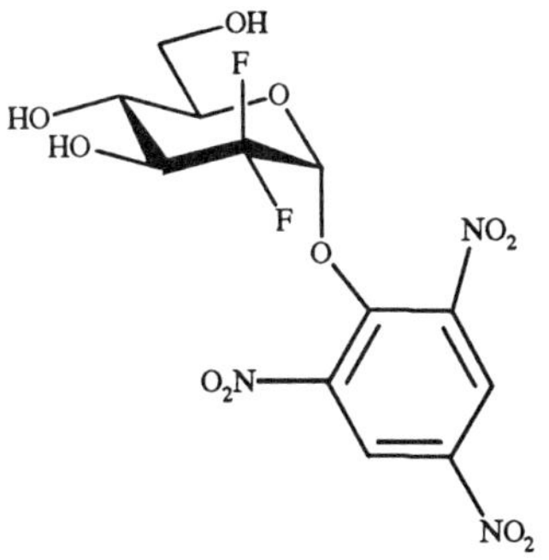

Fig. 5. 2,4,6-Trinitrophenyl 2-deoxy-2,2-difluoro-α-D-glucopyranoside.

only inductive destabilization of the transition state is being employed, all interactions at the important 2-position remaining intact. Indeed, in many cases these compounds act as slow substrates, but ones for which the deglycosylation step is rate limiting; thus the intermediate accumulates. This is reflected in very low K_m values measured for these compounds as substrates. Direct evidence for the accumulation of the intermediate can be obtained by directly loading samples of enzyme, reacted with these reagents, onto a reversed-phase high-performance liquid chromatography (RP-HPLC) column from which material is eluted directly into the ion source of an electrospray ionization mass spectrometer (ESI-MS). The mass measured is that of the covalent glycosyl-enzyme adduct or of some mixture of free enzyme and glycosyl-enzyme intermediate, the relative proportions reflecting the relative rates of intermediate formation and breakdown along with breakdown on the column. This approach has proved to be successful with a range of α- and β-glycosidases. Surprisingly the C-5 epimers of these reagents are also highly effective inactivators, trapping the intermediate in the same manner as above.

Other approaches have proved useful in trapping intermediates. One approach that was developed for α-glycosidases involved the introduction of *two* fluorine substituents at C-2 and an excellent (trinitrophenolate) leaving group at the anomeric center, as shown in **Fig. 5**. The idea here was that, since the 2-fluoroglycosides functioned as slow substrates for which the glycosylation step was rate limiting, it was necessary to slow the deglycosylation step further to allow the intermediate to accumulate. The "brute force" method of doing this was to substitute a second fluorine at C-2 in place of the hydrogen. In addition it was now necessary to install a much better leaving group at C-1 to allow formation of the intermediate, hence the trinitrophenolate. The compound shown in **Fig. 5** proved to be an effective time-dependent inactivator of yeast α-glucosidase, whereas the *malto*-analog is an effective inactivator of human pancreatic α-amylase *(39)*.

1.2. Identification of the Labeling Site

Once a method and reagents for accumulating the intermediate have been established, the stage is set for identification of the site of labeling through digestion of the labeled enzyme followed by HPLC separation of the peptide digest, localization of the labeled peptide, and sequencing thereof. The earlier approaches involved the use of radiolabeled sugars to locate the peptide and will not be discussed here *(40–42)*. The more recently developed approach, which is the subject of this chapter, involves the use of tandem mass spectrometric (MS/MS) methods to localize and sequence the labeled peptide *(9,43)*.

Proteolytic digestion, in our hands, is best achieved using a protease that functions under acidic conditions, pepsin having been our choice in all cases to date. The reasons for this are twofold. The first is that the ester linkage between the sugar and the peptide is much more stable under acidic than under neutral conditions, where it has a half-life of only a few hours, as shown much earlier by Legler *(44,45)*. The second reason is that the glycosyl-enzyme is exceedingly stable toward proteolytic degradation compared with the free enzyme. This stability is a consequence of the fact that the species trapped is a true intermediate and the enzyme presumably stabilizes such species. Therefore, conversely, the presence of the intermediate must stabilize the protein structure *(33)*. By performing the proteolysis at low pH (pH 1–2), conditions under which pepsin is fully active, the protein is sufficiently unfolded that it can be degraded. In some cases we have found it is necessary to boil the labeled protein in the low-pH buffer for a short time (typically 2 min), then cool the sample, and immediately add pepsin *(46)*. This procedure typically denatures the labeled protein sufficiently, without significant loss of label, to allow the pepsin to function.

Once good digestion conditions have been found, the digest is separated on an RP-HPLC column (typically C18) and eluted with a gradient of trifluoroacetic acid (TFA) in acetonitrile (typically starting at 0.05% TFA, 2% acetonitrile in water, and finishing at 0.045% TFA, 80% acetonitrile in water). The effluent is interfaced with the ionspray (electrospray) ion source of the mass spectrometer, generally via a splitter which introduces approximately 15% of the eluate into the mass spectrometer, the other 85% being retained for further analysis. When scanned in the normal LC/MS mode, the total-ion chromatogram will show a large number of peaks, each of which corresponds to one or more of the peptides within the proteolytic digest. The task then is to determine which of the peaks contains the labeled peptide. This can be done in either (or both) of two ways.

One approach is to run a parallel control experiment in which a nonlabeled sample of the enzyme is subjected to identical digestion conditions and chromatographic separation. The individual peptide masses within the two chromatograms are then compared in a search for a peptide present in the labeled digest (but not in the unlabeled) that is heavier by the expected mass than one that is present in the digest of the unlabeled enzyme (but absent from the labeled digest).

The other approach is to take advantage of the fact that the bond between the sugar and the peptide is an acylal ester that undergoes relatively facile cleavage within the collision cell of a quadrupole mass spectrometer, releasing a neutral sugar species and leaving the peptide with the same charge. This neutral loss is readily detected by scanning the two analytical quadrupoles in a linked mode to permit passage to the detector of only those peptides that lose a mass, equal to the mass of the sugar moiety, in passing between the first and third (i.e., analytical) quadrupoles. In this way, most of the other signals are removed and only those from peptides of interest are detected. Since the peptide may be doubly or triply charged, it may be necessary to scan for the loss of one-half or one-third of the sugar mass if the first scan is unsuccessful. Typically, neutral-loss scans will reveal one, or possibly two, significant peaks with some background of lower intensity signals. This background arises from real fragmentations in the unlabeled protein, as is shown by running the same neutral-loss scan on the digest of the unlabeled enzyme. These background signals occur because the mass of a 2-deoxy-2-fluorohexosyl group (165 Daltons) is identical to that of a phenylalanyl moiety, thus fragmentations of terminal phenylalanyl residues will be detected. Fortunately, the control scan reproducibly accounts for this. Furthermore, neutral-loss scans in which larger sugars, e.g., a disaccharide, are lost, typically have much smaller background signals since the probability of the neutral loss of a peptidyl moiety of that mass is greatly reduced.

Confirmation that the peptide identified is indeed the peptide of interest wherein the sugar is linked via an ester is obtained by treating the purified, or partially purified, labeled peptide with aqueous ammonia. This cleaves the ester via aminolysis, converting the amino acid residue to the corresponding amide with a corresponding decrease in mass. Amines other than ammonia can be substituted if desired.

The mass of the unlabeled peptide is readily calculated by subtraction of the mass of the label and allowing for the degree of protonation of the peptide. If the sequence of the protein is known, then candidate peptides can be identified by searching the sequence for all possible peptides of that mass. Obviously, if a more selective protease could be used, then this mass alone would probably identify the peptide of interest. However, this is not feasible with pepsin. The identity of the peptide, and the location of the label therein, must then be made either by purifying the peptide and subjecting it to Edman degradation or by direct sequencing through MS/MS experiments on the labeled peptide. This latter route is the method of choice, not only because it is much more rapid and does not require purification of the peptide, but also because the fragments observed can, in some cases, provide direct proof of the site of attachment of the label. This was not the case for the example given below, but provided excellent confirmation in several other cases *(36,47)*. This methodology has now been applied to the identification of the catalytic nucleophiles of a number of glycosidases *(36,37,43,46–55)*.

2. Materials
2.1. Enzymes and Reagents

1. *Thermoanacrobacterium saccharolyticum* β-xylosidase was purified as a recombinant enzyme from *Escherichia coli* R1360 as previously described *(56)*.
2. 2,4-Dinitrophenyl 2-deoxy-2-fluoro-β-D-xylopyranoside (2F-DNPX) and benzyl 1-thio-β-D-xylopyranoside (BTX) were synthesized as previously described *(57)*.

2.2. Mass Spectrometry Instrumentation

1. ESI spectra were recorded on a PE-Sciex API 300 triple quadrupole mass spectrometer (Sciex, Thornhill, Ontario, Canada) equipped with an Ionspray ion source.
2. HPLC separations were performed on an Ultrafast Microprotein Analyzer (Michrom BioResources, Pleasanton, CA) directly interfaced with the mass spectrometer.

3. Methods
3.1 Procedures
3.1.1. Enzyme Kinetics

1. Prepare 50 mM sodium phosphate (pH 6.5) containing 0.1% bovine serum albumin (buffer A) as the medium for kinetic studies. All kinetic studies were performed at 37°C.
2. Monitor the inactivation of β-xylosidase by 2F-DNPX (*see* **Note 1**) by incubation of the enzyme (0.68 mg/mL) in buffer A at 37°C in the presence of various concentrations of the inactivator (35.5–532.3 μM).
3. Assay residual enzyme activity at timed intervals by addition of an aliquot (5–10 μL) of the inactivation mixture to a solution (267 μM, 750 μL) of *p*-nitrophenyl-β-D-xylopyranoside (pNPX) in buffer A. Follow the reaction by using pNPX (K_m = 26 μM, k_{cat} = 242 min^{-1}) as the assay substrate and monitoring its absorbance change at 400 nm ($\Delta\varepsilon_{400}$ 4610 M^{-1} cm^{-1}).
4. Fit the inactivation curve to a first-order rate equation to determine pseudo-first order rate constants at each inactivator concentration (k_{obs}) *(58)*. Determine the inactivation rate constant (k_i) and the dissociation constant for the inactivator (K_i) by fitting to the equation $k_{obs} = k_i [I]/(K_i + [I])$.

Active Site Residues in Glycosidases

3.1.2. Labeling, Proteolysis, and Aminolysis

1. Incubate β-xylosidase (3.4 mg/mL solution in 15 µL 50 mM sodium phosphate [pH 6.5]) with 2F-DNPX (2.13 mM solution in 3 µL 50 mM sodium phosphate [pH 6.5]) at 37°C for 1 h, complete inactivation (> 99%) being confirmed by enzyme assay.
2. Immediately carry out a peptic digestion on the inactivated β-xylosidase by mixing the product from **step 1** (2-fluoro-xylosyl-enzyme, 12 µL, 3.4 mg/mL) with 18 µL of 150 mM sodium phosphate buffer, pH 2.0, and pepsin (10 µL at 0.4 mg/mL in 150 mM sodium phosphate buffer, pH 2.0) (*see* **Note 2**).
3. Incubate the peptide digestion mixture at room temperature for 20 min, if necessary freeze, and analyze immediately by LC/ESI-MS (*see* **Subheading 3.1.3.**) on thawing.
4. Perform aminolysis of the labeled peptide by adding concentrated ammonium hydroxide (5 µL) to a sample of the inactivated 2F-xylosyl enzyme (20 µL, 0.85 mg/mL). Incubate the mixture for 15 min at 50°C, acidify with 50% TFA, and analyze by ESI-MS as described in **Subheading 3.1.3**.

3.1.3. Electrospray Mass Spectrometry

1. Load the proteolytic digest onto a C18 column (Reliasil, 1 × 150 mm) equilibrated with solvent A (0.05% TFA and 2% acetonitrile in water).
2. Elute the digest components from the column into the mass spectrometer ion source using a gradient (0–60%) of solvent B (solvent B is 0.045% TFA and 80% acetonitrile in water) in solvent A over 60 min followed by 100% solvent B over 2 min at a constant flow rate of 50 µL/min.
3. Record electrospray spectra in either the single-quadrupole scan mode (**step 4**), the MS/MS neutral-loss scan mode (**step 5**), or the MS/MS product-ion scan mode (**step 6**).
4. In the single-quadrupole mode (conventional scanning LC/MS), the quadrupole mass analyzer was scanned over a mass-to-charge ratio (m/z) range of 400–1800 with a step size of 0.5 mass units and a dwell time of 1.5 ms/step. The ion source voltage (ISV) was set at 5.5 kV, and the orifice energy (OR) was 45 V.
5. In the neutral-loss scanning mode, MS/MS spectra were obtained by searching for a mass loss of m/z = 67.5 mass units, corresponding to the loss of the 2F-xylose label from a peptide ion in the doubly charged state, between the first quadrupole (Q1) and the third quadrupole (Q3) analyzers.
6. In the product-ion scan mode, MS/MS spectra were obtained by selectively introducing the parent ion (m/z = 1103; *see* **Note 3**) from the first quadrupole (Q1) into the collision cell (Q2), for collision-induced decomposition, and analyzing the fragment ions produced by this means in the third quadrupole (Q3). Thus, Q1 was locked on to m/z = 1103 (*see* **Note 3**), Q3 scan range was m/z 50–1120, scan step size was 0.5 mass units, dwell time was 1 ms, the ion source voltage was 5 kV, OR was 45 V, and Q0, and IQ2 was set at –10V and –48V, respectively.

3.2 Results and Discussion

3.2.1. Identification of the Catalytic Nucleophile of Thermoanaerobacterium Saccharolyticum β-*Xylosidase*

T. saccharolyticum β-xylosidase belongs to a sequence-related family of glycosidases (family 39) composed of both β-D-xylosidases and α-L-iduronidases. Members of this family share a section of high sequence similarity in a region encompassing the proposed acid/base catalyst and lesser similarities elsewhere. This family, by extension from stereochemical outcome studies with *T. saccharolyticum* β-xylosidase *(59)*, operates by a retaining mechanism. Only five members of family 39 are currently known, none of which have had their structure elucidated or their nucleophile identified. However, the catalytic nucleophile of *T. saccharolyticum* β-xylosidase has been predicted to be E277 on the basis of hydrophobic cluster analysis *(60)*. The results described below on the direct identification of the catalytic nucleophile have been reported in full previously *(54)*.

Reaction of this enzyme with 2,4-dinitrophenyl 2-deoxy-2-fluoro-β-D-xylopyranoside (2F-DNPX) resulted in time-dependent inactivation of the enzyme according to pseudo-first-order kinetics. Analysis of these data as described in **Subheading 3.1.1.** allowed calculation of the inactivation rate constant ($k_i = 0.089 \pm 0.001$ min^{-1}) and of the dissociation constant ($K_i = 65 \pm 4$ μM) for the inactivator. Evidence for the notion that inactivation was occurring via trapping of an intermediate at the active site came from the fact that a competitive inhibitor, benzyl 1-thio-β-D-xylopyranoside (BTX) protected the enzyme against this inactivation. Incubation of the enzyme with 2F-DNPX (213 μM) in the presence of BTX (11.33 mM, $K_i = 5.3$ mM) resulted in a decrease in the apparent inactivation rate constant from 0.060 min^{-1} in the absence of BTX to 0.047 min^{-1} in its presence.

The conclusion that inactivation occurs via accumulation of a stable, covalent 2-deoxy-2-fluoro-α-D-xylosyl-enzyme intermediate is supported by mass spectral analysis of the inactivated enzyme. The mass of the native xylosidase was found, by ESI-MS, to be 58,666 ± 6 Daltons whereas, after inactivation with 2F-DNPX, only one species with a mass of 58,800 ± 6 Daltons was observed. The mass difference between the native and inactivated enzymes is 134 Daltons, a value consistent, within error, with the addition of a single 2F-xylosyl label (135 Daltons).

3.2.2. Identification of the Labeled Active Site Peptide by ESI-MS

The total-ion chromatogram (TIC) derived from reversed-phase conventional scanning LC/MS analysis (*see* **Subheading 3.1.3., step 4**) of the peptic digest of the labeled enzyme showed a large number of peaks, each corre-

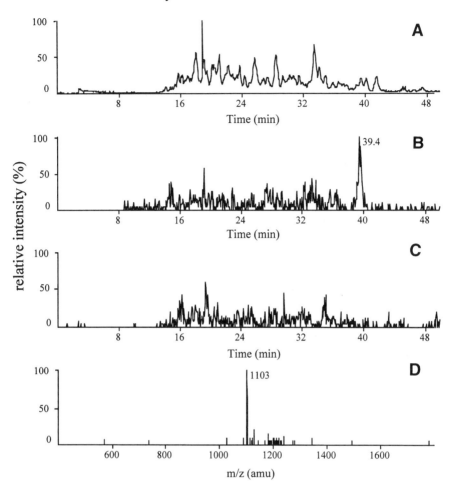

Fig. 6. ESI-MS experiments on a peptic digest of *Thermoanaerobacterium saccharolyticum* β-xylosidase. (**A**) Enzyme labeled with 2F-DNPX, TIC from mass spectrometer in normal scanning mode. (**B**) Enzyme labeled with 2F-DNPX, TIC in the neutral-loss mode; (**C**) Unlabeled enzyme. TIC in the neutral-loss mode. (**D**) Mass spectrum of peptide eluting at 39.4 min.

sponding to one or more peptides in the digest mixture (**Fig. 6A**). Localization of the 2F-xylosyl-peptide within this chromatogram was achieved through a second experiment, operating the mass spectrometer in the neutral-loss MS/MS mode (*see* **Subheading 3.1.3., step 5**). The instrument was first set up to scan for the loss of 135 mass units in the collision cell, but no significant peaks were observed when scanned in this manner. However, when scanned in the neutral-loss MS/MS mode searching for a mass loss of *m/z* 67.5, a single major peak was observed in the total-ion chromatogram (**Fig. 6B**), with no such peak being observed in the neutral-loss scan of the unlabeled xylosidase digest (**Fig. 6C**). These results indicate that a doubly charged peptide ion (m/z 1103; **Fig. 6D**) bears the 2-fluoroxylose; thus the mass of the labeled peptide is 2204 Daltons (2[1103] − 2 H). Since the mass of the label is 135 Daltons the unlabeled peptide must have a mass of 2070 Daltons (2204−135 + 1 H). Aminolysis of the isolated peptide bearing the 2-F-xylose resulted in a single new peak in the conventional scanning ion-chromatogram with an ion at m/z 1035.5 (data not shown). No peak corresponding to the labeled peptide was observed. The mass loss of 135 Daltons (2[1103−1035.5]) is consistent with the expected mass loss from cleavage of the ester-linked 2F-xylose label. When the enzyme amino acid sequence *(56)* was searched for all possible peptides of mass 2070 ± 2.0 Daltons, 18 possible candidates were identified.

3.2.3. Peptide Sequencing *(see* **Note 4***)*

With the mass spectrometer set to transmit the doubly charged parent ion of the peptide of interest (m/z 1103) to the collision cell, operation in the production scan mode (**Fig. 7**) results in that both the labeled parent ion (m/z 1103) and the unlabeled intact peptide arising from the loss of the 2F-xylose label (m/z 1036) appearing as doubly charged species. In addition, singly charged peaks corresponding to b-type cleavage of the unlabeled peptide are seen and correspond to fragments II (m/z 227), IIK (m/z 355), IIKN (m/z 469), IIKNSH (m/z 693), IIKNSHF (m/z 840), IIKNSHFPN (m/z 1051), IIKNSHFPNL (m/z 1165), IIKNSHFPNLPF (m/z 1409), IIKNSHFPNLPFH (m/z 1546), and IIKNSHFPNLPFHI (m/z 1659). Both labeled and unlabeled y"-type ions are observed in the spectrum. Peaks arising from y" ions bearing the label include KNSHFPNLPFHITEY (m/z 1979), NSHFPNLPFHITEY (m/z 1850), SHFPNLPFHITEY (m/z 1736), HFPNLPFHITEY (m/z 1650), FPNLPFHITEY (m/z 1512), and PNLPFHITEY (m/z 1365). Unlabeled y" ions include only SHFPNLPFHITEY (m/z 1601) and PNLPFHITEY (m/z 1230). The doubly charged fragments, apart from the labeled and unlabeled parent ions, remain unassigned. The peptide containing the active site nucleophile is therefore $_{261}$IIKNSHFPNLPFHITEY$_{278}$, the number being derived from the amino acid sequence of the protein derived from the gene sequence.

The fact that the sugar label can be cleaved by treatment with ammonia indicates that the linkage between the two is an ester bond, as has been found for all other glycosyl hydrolases to date *(6)*. We can therefore assign Glu277 as the nucleophile residue since it is the only carboxylic acid present in the peptide. Furthermore, this residue is highly conserved within this family and resides within a sequence (ITE) found in several other related families. Moreover, it was indeed the residue predicted.

Active Site Residues in Glycosidases

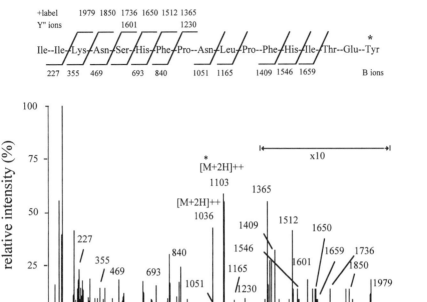

Fig. 7. ESI-MS/MS daughter-ion spectrum of the 2F-xylosyl peptide (m/z 1103, in the doubly charged state). Observed b- and y"-fragments are shown below and above the peptide sequence, respectively.

4. Notes

1. Binding of 2F-DNPX at the active site was proved by demonstrating protection against inactivation by a competitive inhibitor. Inactivation mixtures (60 µL) containing enzyme (0.68 mg/mL) and 2F-DNPX (213 µM) were incubated in the presence and absence of BTX (11.33 mM, K_i = 5.3 mM). Aliquots (5 µL) of the mixture were assayed for residual activity at timed intervals as described in **Subheading 3.1.1.**
2. Underivatized, native β-xylosidase is subjected to peptic digestion in parallel with the derivatized material, to provide a comparison specimen (**Fig. 6C**). The amounts to be used are native β-xylosidase (10 µL, 3.4 mg/mL) with 150 mM sodium phosphate buffer, pH 2.0, (20 µL) and pepsin (10 µL at 0.4 mg/mL in 150 mM sodium phosphate buffer, pH 2.0).
3. In this particular example, the parent ion is found at m/z 1103 (*see* **Subheading 3.2.3.** for a more detailed explanation).
4. Typically, peptide sequencing works best using a partially purified fraction (as in the case described) although it could be carried out using a complete digest so long as no other peptides of the same mass are present.

Acknowledgments

We are indebted to numerous colleagues for their hard work in the development of these methodologies. Their names appear in the references quoted. Particular thanks go to Ruedi Aebersold (University of Washington) for a valuable collaboration that started us into the world of biological mass spectrometry. We thank the Natural Sciences and Engineering Research Council of Canada and the Protein Engineering Network of Centres of Excellence for financial support.

References

1. Henrissat, B. (1991) A classification of glycosyl hydrolases based on amino acid sequence similarities. *Biochem. J.* **280,** 309–316.
2. Henrissat, B. and Bairoch, A. (1993) New families in the classification of glycosyl hydrolases based on amino acid sequence similarities. *Biochem. J.* **293,** 781–788.
3. Henrissat, B. and Bairoch, A. (1996) Updating the sequence-based classification of glycosyl hydrolases. *Biochem. J.* **316,** 695–696.
4. Henrissat, B. and Davies, G. (1997) Structural and sequence-based classification of glycoside hydrolases. *Curr. Opin. Struct. Biol.* **7,** 637–644.
5. Davies, G. and Henrissat, B. (1995) Structures and mechanisms of glycosyl hydrolases. *Structure* **3,** 853–859.
6. Davies, G., Sinnott, M. L., and Withers, S. G. (1998) Glycosyl transfer, in *Comprehensive Biological Catalysis,* vol. 1 (Sinnott, M. L., ed.), Academic, New York, pp. 119–208.
7. Legler, G. (1990) Glycoside hydrolases: mechanistic information from studies with reversible and irreversible inhibitors. *Adv. Carb. Chem. Biochem.* **48,** 319–385.
8. McCarter, J. D. and Withers, S. G. (1994) Mechanisms of enzymatic glycoside hydrolysis. *Curr. Opin. Struct. Biol.* **4,** 885–892.
9. Withers, S. G. and Aebersold, R. (1995) Approaches to labeling and identification of active site residues in glycosidases. *Protein Sci.* **4,** 361–372.
10. Sinnott, M. L. (1990) Catalytic mechanisms of enzymic glycosyl transfer. *Chem. Rev.* **90,** 1171–1202.
11. Mackenzie, L. F., Wang, Q., Warren, R. A. J., and Withers, S. G. (1998) Glycosynthases: mutant glycosidases for oligosaccharide synthesis. *J. Am. Chem. Soc.* **120,** 5583–5584.
12. Gebler, J. C., Trimbur, D. E., Warren, R. A. J., Aebersold, R., Namchuk, M., and Withers, S. G. (1995) Substrate-induced inactivation of a crippled β-glucosidase mutant: identification of the labeled amino acid and mutagenic analysis of its role. *Biochemistry* **34,** 14,547–14,533.
13. Lawson, S. L., Wakarchuk, W. W., and Withers, S. G. (1997) Positioning the acid/base catalyst in a glycosidase: studies with *Bacillus circulans* xylanase. *Biochemistry,* **36,** 2257–2265.
14. MacLeod, A. M., Lindhorst, T., Withers, S. G., and Warren, R. A. J. (1994) The acid/base catalyst in the exoglucanase/xylanase from *Cellulomonas fimi* is Glu127: evidence from detailed kinetic studies of mutants. *Biochemistry* **33,** 6371–6376.

15. MacLeod, A. M., Tull, D., Rupitz, K., Warren, R. A. J., and Withers, S. G. (1996) Mechanistic consequences of mutation of active site carboxylates in a retaining β-1,4-glycanase from *Cellulomonas fimi*. *Biochemistry* **35**, 13,165–13,172.
16. Mosi, R., He, S., Uitdehaag, J., Dijkstra, B. W., and Withers, S. G. (1997) Trapping and characterization of the reaction intermediate in cyclodextrin glycosyltransferase by use of activated substrates and a mutant enzyme. *Biochemistry* **36**, 9927–9934.
17. Notenboom, V., Birsan, C., Nitz, M., Rose, D. R., Warren, R. A. J., and Withers, S. G. (1998) Active site mutations lead to covalent intermediate accumulation in *Cellulomonas fimi* β-1,4-glycosidase Cex: insights into transition state stabilization. *Nature Struct. Biol.* **5**, 812–818.
18. Wang, Q., Trimbur, D., Graham, R., Warren, R. A. J., and Withers, S. G. (1995) Identification of the acid/base catalyst in *Agrobacterium faecalis* β-glucosidase by kinetic analysis of mutants. *Biochemistry* **34**, 14,554–14,562.
19. Withers, S. G. (1995) Probing of glycosidase active sites through labeling, mutagenesis and kinetic studies, in *Carbohydrate Bioengineering* (Petersen S. B., Svensson, B., and Pedersen, S., eds.), Elsinore, Denmark, pp. 97–111.
20. Braun, C., Lindhorst, T., Madsen, N. B., and Withers, S. G. (1996) Identification of Asp 549 as the catalytic nucleophile of glycogen-debranching enzyme via trapping of the glycosyl-enzyme intermediate. *Biochemistry* **35**, 5458–5463.
21. Sinnott, M. L. and Souchard, I. J. L. (1973) The mechanism of action of β-galactosidase: effect of aglycone nature and α-deuterium substitution on the hydrolysis of aryl galactosides. *Biochem. J.* **133**, 89–98.
22. Kempton, J. B. and Withers, S. G. (1992) Mechanism of *Agrobacterium faecalis* β-glucosidase: kinetic studies. *Biochemistry* **31**, 9961–9969.
23. Tull, D. and Withers, S. G. (1994) Mechanisms of cellulases and xylanases: a detailed kinetic study of the exo-β-1,4–glycanase from *Cellulomonas fimi*. *Biochemistry* **33**, 6363–6370.
24. Withers, S. G., Street, I. P., Bird, P., and Dolphin, D. H. (1987) 2-Deoxy-2-fluoroglucosides: a novel class of mechanism-based glucosidase inhibitors. *J. Am. Chem. Soc.* **109**, 7530–7531.
25. Withers, S. G., Rupitz, K., and Street, I. P. (1988) 2-Deoxy-2-fluoro-D-glycosyl fluorides. A new class of specific mechanism-based glycosidase inhibitors. *J. Biol. Chem.* **263**, 7929–7932.
26. Wolfenden, R. and Kati, W. M. (1991) Testing the limits of protein-ligand binding discrimination with transition-state analogue inhibitors. *Acc. Chem. Res.* **24**, 209–215.
27. Wentworth, D. F. and Wolfenden, R. (1974) Slow binding of D-galactal, a "reversible" inhibitor of bacterial β-galactosidase. *Biochemistry* **13**, 4715–4720.
28. Roeser, K. R. and Legler, G. (1981) Role of sugar hydroxyl groups in glycoside hydrolysis. Cleavage mechanism of deoxyglucosidases and related substrates by β-glucosidase A_3 from *Aspergillus wentii*. *Biochim. Biophys. Acta* **657**, 321–333.
29. Namchuk, M. N. and Withers, S. G. (1995) Mechanism of *Agrobacterium* β-glucosidase: kinetic analysis of the role of noncovalent enzyme/substrate interactions. *Biochemistry* **34**, 16,194–16,202.

30. McCarter, J. D., Adam, M. J., and Withers, S. G. (1992) Binding energy and catalysis. Fluorinated and deoxygenated glycosides as mechanistic probes of *Escherichia coli* (*lacZ*) β-galactosidase. *Biochem. J.* **286**, 721–727.
31. Street, I. P., Rupitz, K., and Withers, S. G. (1989) Fluorinated and deoxygenated substrates as probes of transition-state structure in glycogen phosphorylase. *Biochemistry* **28**, 1581–1587.
32. White, A., Tull, D., Johns, K., Withers, S. G., and Rose, D. R. (1996) Crystallographic observation of a covalent catalytic intermediate in a β-glycosidase. *Nature Struct. Biol* **3**, 149–154.
33. Street, I. P., Kempton, J. B., and Withers, S. G. (1992) Inactivation of a β-glucosidase through the accumulation of a stable 2-deoxy-2-fluoro-α-D-glucopyranosyl-enzyme intermediate: a detailed investigation. *Biochemistry* **31**, 9970–9978.
34. Withers, S. G. and Street, I. P. (1988) Identification of a covalent α-D-glucopyranosyl enzyme intermediate formed on a β-glucosidase. *J. Am. Chem. Soc.* **110**, 8551–8553.
35. McCarter, J. D., Adam, M. J., Braun, C., Namchuk, M., Tull, D., and Withers, S. G. (1993) Syntheses of 2-deoxy-2-fluoro mono- and oligo-saccharide glycosides from glycals and evaluation as glycosidase inhibitors. *Carbohydr. Res.* **249**, 77–90.
36. Howard, S., He, S., and Withers, S. G. (1998) Identification of the active site nucleophile in jack bean α-mannosidase using 5-fluoro-β-l-gulosyl fluoride. *J. Biol. Chem.* **273**, 2067–2072.
37. McCarter, J. D. and Withers, S. G. (1996) Unequivocal identification of Asp-214 as the catalytic nucleophile of *Saccharomyces cerevisiae* β-glucosidase using 5-fluoro glycosyl fluorides. *J. Biol. Chem.* **271**, 6889–6894.
38. McCarter, J. D. and Withers, S. G. (1996) 5-Fluoro glycosides: a new class of mechanism-based inhibitors of both α- and β-glucosidases. *J. Am. Chem. Soc.* **118**, 241–242.
39. Braun, C., Brayer, G. D., and Withers, S. G. (1995) Mechanism-based inhibition of yeast α-glucosidase and human pancreatic α-amylase by a new class of inhibitors. 2-Deoxy-2,2-difluoro-(-glycosides. *J. Biol. Chem.* **270**, 26,778–26,781.
40. Withers, S. G., Warren, R. A. J., Street, I. P., Rupitz, K., Kempton, J. B., and Aebersold, R. (1990) Unequivocal demonstration of the involvement of a glutamate residue as a nucleophile in the mechanism of a 'retaining' glycosidase. *J. Am. Chem. Soc.* **112**, 5887–5889.
41. Gebler, J. C., Aebersold, R., and Withers, S. G. (1992) Glu-537, not Glu-461, is the nucleophile in the active site of (*lac Z*) β-galactosidase from *Escherichia coli. J. Biol. Chem.* **267**, 11,126–11,130.
42. Tull, D., Withers, S. G., Gilkes, N. R., Kilburn, D. G., Warren, R. A., and Aebersold, R. (1991) Glutamic acid 274 is the nucleophile in the active site of a "retaining" exoglucanase from *Cellulomonas fimi. J. Biol. Chem.* **266**, 15,621–15,625.
43. Tull, D., Miao, S., Withers, S. G., and Aebersold, R. (1994) Identification of derivatised peptides without radiolabels: tandem mass spectrometric localisation of the tagged active site nucleophiles of two cellulases and a β-glucosidase. *Anal. Biochem.* **224**, 509–514.

44. Legler, G. and Hasnain, S. N. (1970) Markierung des aktiven Zentrums der β-Glucosidasen A und B aus dem Sußmandel-emulsin mit [³H]6-Brom-6-desoxy-condurit-B-epoxid. *Hoppe-Seylers Z. Physiol. Chem.* **351**, 25–31.
45. Legler, G. (1968) Untersuchungen zum Wirkungsmechanismus glykosidspaltender Enzyme, III. Markierung des aktiven Aentrums einer β-Glucosidase aus *Aspergillus wentii* mit [¹⁴C]Condrurit-B-epoxid. *Hoppe-Seyler's Z. Physiol. Chem.* **349**, 767–774.
46. Miao, S., Ziser, L., Aebersold, R., and Withers, S. G. (1994) Identification of glutamic acid 78 as the active site nucleophile in *Bacillus subtilis* xylanase using electrospray tandem mass spectrometry. *Biochemistry* **33**, 7027–7032.
47. Wong, A. W., He, S., Grubb, J. H., Sly, W. S., and Withers, S. G. (1998) Identification of Glu540 as the catalytic nucleophile of human β-glucuronidase using electrospray mass spectrometry. *J. Biol. Chem.* **273**, 34,057–34,062.
48. He, S. and Withers, S. G. (1997) Assignment of sweet almond β-glucosidase as a family 1 glycosidase and identification of its active site nucleophile. *J. Biol. Chem.* **272**, 24,864–24,867.
49. Mackenzie, L. F., Brooke, G. S., Cutfield, J. F., Sullivan, P. A., and Withers, S. G. (1997) Identification of glu-330 as the catalytic nucleophile of *Candida albicans* exo-β-(1,3)-glucanase. *J. Biol. Chem.* **272**, 3161–3167.
50. Mackenzie, L. F., Davies, G. J., Schulein, M., and Withers, S. G. (1997) Identification of the catalytic nucleophile of endoglucanase I from *Fusarium oxysporum* by mass spectrometry. *Biochemistry* **36**, 5893–5901.
51. Mackenzie, L. F., Sulzenbacher, G., Divne, C., Jones, T. A., Woldike, H. F., Schulein, M., et al. J. (1998) Crystal structure of the family 7 endoglucanase I (CelB) from *Humicola insolens* at 2.2 A resolution and identification of the catalytic nucleophile by trapping of the covalent glycosyl-enzyme intermediate. *Biochem. J.* **335**, 409–416.
52. McCarter, J. D., Burgoyne, D. L., Miao, S. C., Zhang, S. Q., Callahan, J. W., and Withers, S. G. (1997) Identification of glu-268 as the catalytic nucleophile of human lysosomal β-galactosidase precursor by mass spectrometry. *J. Biol. Chem.* **272**, 396–400.
53. Miao, S., McCarter, J. D., Grace, M. E., Grabowski, G. A., Aebersold, R., and Withers, S. G. (1994) Identification of Glu340 as the active-site nucleophile in human glucocerebrosidase by use of electrospray tandem mass spectrometry. *J. Biol. Chem.* **269**, 10,975–10,978.
54. Vocadlo, D. J., Mackenzie, L. F., He, S., Zeikus, G., and Withers, S. G. (1998) Identification of Glu277 as the catalytic nucleophile of *Thermoanaerobacterium saccharolyticum* β-xylosidase using electrospray MS. *Biochem. J.* **335**, 449–455.
55. Zechel, D. L., He, S., Dupont, C., and Withers, S. G. (1998) Identification of Glu 120 as the catalytic nucleophile in *Streptomyces lividans* endoglucanase CelB. *Biochem. J.* **336**, 139–145.
56. Lee, Y.-E. and Zeikus, J. G. (1993) Genetic organization, sequence and biochemical characterization of recombinant beta-xylosidase from *Thermoanerobacterium saccharolyticum* strain B6A-R1. *J. Gen. Microbiol.* **139**, 1235–1243.

57. Ziser, L., Setyawati, I., and Withers, S. G. (1995) Syntheses and testing of substrates and mechanism-based inactivators for xylanases. *Carbohydr. Res.* **274,** 137–153.
58. Leatherbarrow, R. J. (1992) Gra-Fit, Version **2.0,** Erithacus Software, Staines, UK.
59. Armand, S., Vielle, C., Gey, C., Heyraud, A., Zeikus, J. G., and Henrissat, B. (1996) Stereochemical course and reaction products of the action of β-xylosidase from *Thermoanaerobacterium saccharolyticum* strain B6A-RI. *Eur. J. Biochem.* **263,** 706–713.
60. Henrissat, B., Callebaut, I., Fabrega, S., Lehn, P., Mornon, J. P., and Davies, G. (1995) Conserved catalytic machinery and the prediction of a common fold for several families of glycosyl hydrolases. *Proc. Natl. Acad. Sci. USA* **92,** 7090–7094.

13

Probing Protein–Protein Interactions with Mass Spectrometry

Richard W. Kriwacki and Gary Siuzdak

1. Introduction

Protease mapping is an established method for probing the primary structure of proteins *(1,2)* and has traditionally been performed through the use of chromatography and/or gel electrophoresis techniques in combination with Edman degradation NH_2-terminal sequencing *(3)*. More recently, mass spectrometry has been combined with protease mapping to perform "protein mass mapping." Definitively, protein mass mapping combines enzymatic digestion, mass spectrometry, and computer-facilitated data analysis to examine proteolytic fragments for protein structure determination. Protein mass mapping permits the identification of protein primary structure by applying sequence-specific proteases and performing mass analysis on the resulting proteolytic fragments, thus yielding information on fragment masses with accuracy approaching ±5 ppm, or ±0.005 Daltons for a 1000 Daltons peptide. The protease fragmentation pattern is then compared with the patterns predicted for all proteins within a database, and matches are statistically evaluated. Since the occurrence of Arg and Lys residues in proteins is statistically high, trypsin cleavage (specific for Arg and Lys) generally produces a large number of fragments, which, in turn, offer a reasonable probability for unambiguously identifying the target protein. The success of this strategy relies on the existence of the protein sequence within the database, but with the sequences of whole genomes for several organisms now complete (*Escherichia coli, Bacillus subtilis,* and *Archaeoglobus fulgidus*) and others approaching completion (*Saccharomyces cerevisiae, Saccaramyces pombe, Homo sapiens, Drosophila melanogaster,* and so forth), the likelihood for matches is reasonably high.

Fig. 1. Illustration of the use of proteolytic cleavage as a probe of protein structure. The arrows mark surface-exposed and flexible sites that would be susceptible to proteolytic cleavage. If a sequence-specific protease were used, the marked sites would also have to contain the protease recognition sequence to sustain cleavage. Mass analysis of all fragments together yields the cleavage "map" that provides information on secondary and tertiary structure.

Although exact matches are readily identified, homologous proteins are also identified (albeit with lower statistical significance), whereby a target protein is placed within a particular family in the absence of an exact match.

The basis for studying tertiary structure using protein mass mapping is the application of limited proteolytic digestion with mass spectrometry *(4,5)*. In the analysis of protein structure, a factor that governs the selectivity of cleavage is the sequence specificity of the enzyme. A sequence-specific protease reduces the number of fragments that are produced and, concomitantly, improves the likelihood of statistically significant matches between observed and predicted fragment masses and reduces the opportunities for spurious matches. Another factor, the accessibility/flexibility of the site to the protease *(6,7)* also plays an important role in the analysis of structure where, ideally, only a subset of all possible cleavages are observed due to the inaccessibility of some sites due to higher order protein structure. An example of this can be seen in **Fig. 1**, where arrows mark potential cleavage sites within a hypothetical protein; these sites are surface exposed and located in flexible loop regions. The distribution of amino acids in a protein guides the choice of protease to be used as a structural probe. Since amino acids with hydrophilic side chains are found in greater abundance on the surface of proteins (at the solvent interface), proteases that cleave at hydrophilic sites are preferred in structural analysis. Trypsin and V8 protease, which cleave basic (K, R) and acidic (D, E) sites, respectively, are good choices. In addition, non-sequence-specific proteases such as subtilisin Carlsberg are often used as structural probes.

Protein mass mapping can also be used to probe the quaternary structure of multicomponent assemblies, including protein-protein complexes *(4,5)* and protein-DNA complexes *(8)*. A common feature of these applications is that the protease is used to provide contrast between the associated and unassociated states of the system. The formation of an interface between a protein and another macromolecule will exclude both solvent molecules and macromolecules such as proteases and will also protect otherwise accessible sites from protease cleavage. Methods developed for primary sequence elucidation using mass spectrometry are particularly well suited to the analysis of higher order, native protein structure since they are directly transferable to the analysis of native structure. Analysis methods, however, must be modified to take into account the added spectral complexity due to incomplete proteolysis under limiting conditions.

1.1. Analysis of Protein–DNA Interactions

The first application of limited proteolysis and matrix-assisted laser desorption/ionization (MALDI) mass spectrometry to the study of a multicomponent biomolecular assembly was published by Chait and co-workers in 1995 *(8)*. This combined approach was used for the structural analysis of the protein transcription factor Max, both free in solution and bound to an oligonucleotide containing its specific DNA binding site. Max is a member of the basic helix-loop-helix (bHLH) family of DNA binding proteins and has also been the target of crystallographic studies. An extensive series of limited proteolysis experiments by Chait was conducted using free Max. The products of digestion reactions were analyzed using MALDI time-of-flight mass spectrometry (TOF-MS), demonstrating the suitability of this mass spectrometric technique in the analysis of multicomponent biomolecular samples, both for identification of fragments and for their relative quantitation. The results showed that Max is generally very susceptible to proteolytic cleavage. However, Max is less susceptible to digestion by a variety of proteases at high ionic strength, indicating that salt stabilizes Max structure. Since cleavage requires both accessibility and flexibility, this result suggested that Max structure is more highly ordered in the presence of higher salt concentrations, with loop regions in less flexible states. These results, indicating that Max may be relatively flexible in the absence of DNA, are consistent with the inability to crystallize Max in the free state. Much more dramatic stabilization of Max was observed in the presence of specific DNA. In this case, cleavage rates were reduced 100-fold, indicating that the Max protein is significantly stabilized in the presence of DNA. This stabilization stems partly from the protection of potential cleavage sites on formation of the Max-DNA interface and partly from the added thermodynamic stability imparted to Max by association with DNA. The cleavage pattern within the Max-DNA complex revealed that the bHLH domain is the minimal requirement for DNA binding and, importantly, that the leucine-zipper domain is dispensable for this activity. Furthermore, the locations of the Max-DNA interaction sites were identified. These results provided valuable insights into Max-DNA binding and the structure-function relationships that guided the successful crystallization and structure determination of the Max-DNA complex.

The following describes the application of the approach outlined above to protein-protein interactions, specifically its use to probe the solution structure of a protein-protein complex between cell cycle regulatory proteins, p21 and Cdk2. Analysis of proteolytic digests of the p21-Cdk2 complex revealed a segment of between 22 and 36 amino acids of p21 that is protected from trypsin cleavage, identifying this as the Cdk2 binding site on p21. This approach was further utilized for the protein-protein complexes on viral capsids, including those of the common cold virus where, in addition to structural information, protein mass mapping revealed mobile features of the viral proteins.

2. Materials

2.1. Protein Preparation

1. Expression vector, pET-24a (Novagen, Madison, WI).
2. *Drosophila* cells (Schneider's line 1).
3. Schneider's insect medium.
4. Fetal bovine serum (Atlanta Biologicals, Norcross, GA).

2.2. Labeled Reagents

1. ^{15}N-labeled ammonium chloride (Martek, Columbia, MD).

2.3. Protein Chromatography

1. Ni^{2+}-affinity chromatography (Chelating sepharose, Amersham Pharmacia Biotech, Piscataway, NJ).
2. Anion-exchange chromatography (Q-sepharose, Amersham Pharmacia).
3. Reversed-phase high-performance liquid chromatography (HPLC; C4, Vydac, Hesperia, CA).

2.4. Proteases

1. Trypsin, clostripain (Arg-C), V8 protease (Glu-C), and Lys-C (Promega, Madison, WI).
2. Asp-N (Calbiochem, La Jolla, CA).
3. Carboxypeptidase Y (CPY; Sigma, St. Louis, MO).

2.5. Buffers

1. Tris-HCl, Tris (pH 7.5).
2. 0.1 M HEPES (pH 7.0).

2.6. General Reagents and Solvents

Phenylmethylsulfonyl fluoride (PMSF), trichloroacetic acid (TCA), 2-mercaptoethanol, sucrose, trifluoroacetic acid (TFA), ethylenediamine tetraacetic acid (EDTA), bovine serum albumin (BSA), CaCl$_2$, tosyllysine chloromethyl ketone (TLCK), NaCl, dithiothreitol (DTT), Nonidet P-40, and CH$_3$CN.

2.7. MALDI Matrices

2,5-Dihydroxybenzoic acid, 3,5-dimethoxy-4-hydroxycinnamic acid (Aldrich, Milwaukee, MI).

2.8. Instrumentation

1. PerSeptive Voyager Elite MALDI-MS with delayed extraction and nitrogen laser.
2. Kratos MALDI-IV MALDI-MS with delayed extraction and nitrogen laser.
3. Mass spectrometer sample plate derivatized with trypsin (Intrinsic Bioprobes [Tempe, AZ]).

2.9. Data Processing

1. Protein Analysis Worksheet (PAWS, Macintosh version 6.0b2, copyright © 1995, Dr. Ronald Beavis) available on the Internet.

3. Methods

3.1. p21-B, ^{15}N-p21-B, and p21-B/Cdk2 Complex Experiments

p21-B was overexpressed in *E. coli* after the insertion of the gene segment for amino acids 9–84 of p21 plus an N-terminal (His)$_6$ purification tag into the expression vector, pET-24a *(9)*. ^{15}N-labeled p21-B was prepared using a "minimal media" recipe based on that originally developed by Neidhardt et al. *(10)* using ^{15}N-labeled ammonium chloride. Cdk2 was obtained after overexpression in Sf9 insect cells using a bacculovirus provided by Dr. David Morgan *(11)* and was kindly provided by Drs. Mark Watson and Steve Reed.

3.1.1. Purification and Preparation of p21-B and ^{15}N-p21-B

1. Purify p21-B or ^{15}N-p21-B in three steps:
   a. Ni^{2+}-affinity chromatography using chelating sepharose (Amersham Pharmacia Biotech).
   b. Anion-exchange chromatography using Q-sepharose (Amersham Pharmacia).
   c. Reversed-phase HPLC using a C4 column (Vydac).
2. Prepare p21-B solutions for proteolysis/MALDI-MS experiments by dissolving lyophilized material in distilled water at 1 mg/mL. Dilute this solution 1:10 (v/v) into 50 mM Tris (pH 7.5), 500 mM NaCl, 10–10 mM DTT, and 1 mM EDTA.

3.1.2. Preparation of p21-B/Cdk2 and ^{15}N-p21-B/Cdk2 Complexes

1. Prepare the p21-B/Cdk2 or ^{15}N-p21-B/Cdk2 complex by adding a highly concentrated Cdk2 solution (≈ 300 µM) to the appropriate p21-B solution to achieve a final concentration of 20 µM for each component.

3.1.3. Proteolysis of Free p21-B and p21-B/Cdk2 and ^{15}N-p21-B/Cdk2 Complexes

1. Mix a 20–60-μM solution of free p21-B or of the complex, prepared as in **Subheading 3.1.2.** above, together with trypsin to give a protein-to-trypsin ratio of 1000:1–50:1 (w/w). Also, 50 mM Tris (pH 7.5), 500 mM NaCl, 10 mM DTT, and 1 mM EDTA are present.
2. Carry out proteolysis at 23°C for 30 min and then follow with the addition of PMSF to 1 mM and TCA to 10% (wt/vol).
3. Recover protein fragments by centrifugation and then dissolve in 50:50 CH_3CN+H_2O with 0.1% TFA for MALDI-MS analysis.

3.1.4. MALDI Mass Measurements of Proteolysis Products

1. Combine digest samples (40 μM) with 2,5-dihydroxybenzoic acid matrix ($\approx$ 0.2 M in 50:50:0.5 $H_2O+CH_3CN+TFA$) at a 1:1 volume ratio.
2. Deposit 2 μL of the sample-matrix mixture onto the MALDI instrument (PerSeptive Voyager Elite MALDI-MS) solution sample plate and insert into the ionization source.
3. Irradiate samples with a nitrogen laser (Laser Science, Franklin, MA) operated at 337 nm (*see* **Note 1**).

3.2. Flock House and Human Rhinovirus Experiments

3.2.1. Preparation of FHV and HRV14

Flock House virus (FHV) was prepared in *Drosophila* cells (Schneider's line 1) suspended to 4×10^7 cells/mL in a complete growth medium containing Schneider's insect medium with 15% fetal bovine serum (CGM). FHV was added at a multiplicity of 120 plaque-forming units/cell and allowed to attach for 1 h at 26°C. HRV14 was prepared as previously described (*see* **Note 2**) to a final concentration of 1 mg/mL in 10 mM Tris buffer at pH 7.6.

1. Sediment cells and then resuspend them to 5×10^6 cells/mL in CGM.
2. Distribute aliquots onto 100-mm tissue culture plates and incubate at 26°C.
3. Remove the medium at 15 h post infection and rinse monolayers with 10 mL of ice-cold HE buffer (0.1 M HEPES [pH 7.0] 10 mM EDTA, 0.1% 2-mercaptoethanol, 0.1% BSA).
4. Lyse cells in 2 mL of ice-cold HE buffer containing 1% (v/v) Nonidet P-40.
5. Remove nuclei and cell debris from the lysate by centrifugation for 5 min at 4°C in a table-top centrifuge. Pellet the supernatant containing the virus through 2-mL sucrose gradients (10–30% [w/w] in HE buffer without BSA) at 100,000 rpm for 13 min.
6. Resuspend the pellets in 400 μL 5 mM $CaCl_2$ containing 0.1% 2-mercaptoethanol and buffered with HEPES (pH 7.0).

3.2.2. Proteolytic Digests of FHV

1. Set up proteolytic digests at 25°C and 1 mg/mL virus using clostripain (Arg-C), V8 protease (Glu-C), Lys-C, or Asp-N in the manufacturer's recommended reaction buffer or trypsin in 25 mM Tris-HCl (pH 7.7) containing 1 mM EDTA (*see* **Note 3**).
2. Adjust enzyme-to-virus ratio (w/w) to 1:3000 in a total volume of 10–20 µL to achieve time-resolved cleavage.
3. Withdraw 0.5 µL from the digest at separate time points over a period of 1 min to 24 h beginning with a sampling rate of one per minute.
4. Place each 0.5 µL reaction volume directly on the MALDI analysis plate and allow it to dry before the addition of matrix (*see* **Subheading 3.2.4.**).

To confirm the identity of trypsin-released fragments, the digest was further exposed to the exoprotease carboxypeptidase Y (CPY) to obtain C-terminal sequence information on each of the trypsin fragments as follows.

1. Inhibit the trypsin digest sample (**step 4** above) by the addition of TLCK 50 µg/mL and allow the mixture to dry on the MALDI sample plate.
2. Add CPY (1–3 µL of enzyme diluted to 1 mg/mL in water) to the dried trypsin digest and allow digestion to continue at room temperature until stopped by evaporation.

3.2.3. Proteolytic Digests of HRV14

1. Set up trypsin digest at 25°C with 1 mg/mL virus in 25 mM Tris HCl (pH 7.7).
2. Adjust the enzyme-to-virus ratio to 1:100 (w/w) in a total reaction volume of 10 µL.
3. Remove 0.5 µL samples from the reaction at separate time points (5, 10, and 60 min), place directly on the MALDI analysis plate, and allow to dry before the addition of matrix (*see* **Subheading 3.2.4.**).
4. Alternatively, carry out on-plate digestions (data not shown) at room temperature using a mass spectrometer sample plate previously derivatized with trypsin.

3.2.4. Mass Spectrometry of Proteolytic Digests

1. Add MALDI matrix (0.5 µL of 3,5-dimethoxy-4-hydroxycinnamic acid in a saturated solution of acetonitrile/water (50:50) containing 0.25% TFA) to the dried sample on the instrument (PerSeptive Biosystems Voyager Elite or Kratos MALDI-IV) (*see* **Note 4**).
2. Determine the identity of trypsin-released fragments using the Protein Analysis Worksheet (PAWS, Macintosh version 6.0b2, copyright © 1995, Dr. Ronald Beavis—available on the Internet; *see* **Note 5**).

3.3. Results and Discussion

3.3.1. Protein Mass Mapping of a Protein–Protein Complex

The approach outlined in the Introduction has also been applied to protein–protein interactions, specifically to probe quaternary structure of cell cycle

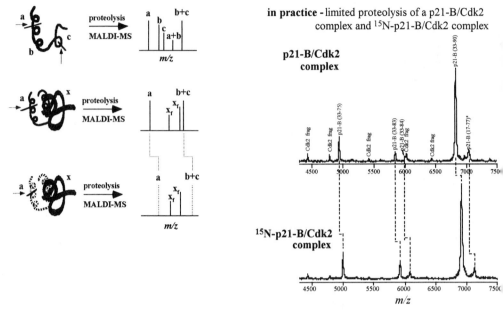

Fig. 2. Probing protein-protein interactions using proteolysis and MALDI-MS. Schematic view (left) of key concepts. Two cleavage sites are accessible for the protein of interest alone (top), yielding five fragments after limited digestion. In the complex with protein X, one site is protected (middle), yielding fewer fragments. However, fragments from protein X are also produced. Isotope labeling of the protein of interest causes fragments **a** and **b+c** to shift to higher m/z values, whereas fragments from protein X do not shift (bottom). Actual results are shown on the right.

regulatory proteins, Cdk2 and p21[Waf1/Cip1/Sdi1] (**Figs. 2** and **3**) *(4,5,9,12–21)*. A distinct and key advantage of this approach is that the masses of the fragments are obtained together in a *single* mass spectrum without the need for individual purification using, for example, HPLC or sodium dodecyl sulfate polyacrylamide gel electrophoresis (SDS-PAGE). Given a protein of known sequence and an enzyme with known sequence specificity, the mass usually identifies the exact fragment within the protein's sequence. The mapping of protein-protein complexes *in situ*, however, is complicated because peptide fragments are produced for all subunits within a complex.

We have recently demonstrated an alternative approach to mapping protein-protein interfaces that overcomes these complications *(4,5)*. The experimental scheme, illustrated in **Fig. 2**, exploits the high mass accuracy, resolution, and sensitivity of MALDI-MS in combination with the power of stable isotope

Protein–Protein Interactions 231

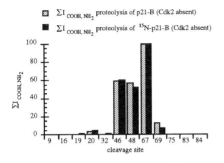
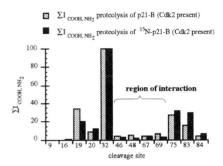

Fig. 3. Protected regions of p21-B within the p2l-B/Cdk2 complex. Trypsin accessibility at a particular cleavage site expressed as the sum of MALDI mass spectral peak intensities for all fragments with COOH- and NH_2-termini corresponding to scission at that particular site ($\Sigma I_{COOH,NH2}$), vs. position within the primary amino acid sequence (numbered with respect to the full-length p21 sequence). The left bar graph shows digestion results in the absence of Cdk2, and the right bar graph shows those in the presence of Cdk2. Results are given for natural isotopic abundance p2l-B (**hatched bars**) and ^{15}N-p21-B (**solid bars**).

labeling. Proteolysis reactions are performed for one component before and after formation of a multiprotein assembly (**Fig. 2** left, top, and middle). Proteolysis reactions for the complex are performed in duplicate, with one subunit prepared at natural isotopic abundance in one experiment and in isotope labeled form in a second (**Fig. 2** left, middle, and bottom, respectively). Other proteins within the assembly are used at natural isotopic abundance in both experiments. Reaction products for both experiments are analyzed using MALDI-MS. **Figure 2** (right) clearly illustrates the power of isotope labeling in identifying peaks in spectra from digests of multiprotein assemblies; one subset of peaks occurs at shifted positions compared with the upper spectrum, whereas another subset does not. The former group of peaks arises from the isotope-labeled component of the assembly.

The mass accuracy of current-day MALDI-MS instruments (0.0005%) allows reliable identification of most fragments from both p21 and Cdk2 without resorting to isotope labeling. In these experiments, the kinase inhibitory domain of p21, called p21-B, was used. However, even this level of accuracy will not allow identification in all cases due to the finite probability that fragments from the different subunits will have similar masses. Therefore, separate p21-B samples were prepared with natural isotopic abundance and ^{15}N-labeled to allow for unambiguous differentiation of p21-B and Cdk2 fragments.

The results obtained from protein mass mapping experiments on the p21-B/Cdk2 complex are summarized in the two bar graphs shown in **Fig. 3**. The left-hand bar graph shows a MALDI analysis of the tryptic fragments of p21-B, generated in the absence of Cdk2. The right-hand bar graph shows data after proteolysis of the p21/Cdk2 complex, with p21 at natural abundance (black) and ^{15}N-labeled (hatched). Cdk2 is unlabeled in both cases. These data revealed a segment of 24 amino acids in p21-B that is protected from trypsin cleavage, thus identifying the segment as the Cdk2 binding site on p21-B.

For proteins of known and potentially of unknown sequence, the protein mass mapping technique offers access to detailed information with only a modest investment in time and material. The accuracy and speed of this approach surpasses traditional methods based on HPLC/SDS-PAGE analysis and Edman degradation. Furthermore, due to the high resolution and sensitivity of MALDI mass analysis, a much greater number of protein fragments can be identified than was previously possible, offering more detailed "maps" of protein–protein structure. Based on the success of protein mass mapping with the DNA–protein and protein complexes, the MALDI method has recently been applied to the more highly structured and complex protein–protein viral assemblies *(22,23)*.

3.3.2. Time-Resolved Protein Mass Mapping of Viruses

Mass spectrometry has recently been recognized as a valuable source of information on both local and global viral structure *(22–27)*. For instance, the mass measurement of viral capsid proteins is now straightforward and has even allowed the identification of posttranslational modifications *(23,25,27)*. Since the viral capsids represent an interesting and important noncovalent quaternary association of protein subunits, viral analysis has been a logical step in the development of protein mass mapping. For instance, cleavage sites that reside on the exterior of the virus will be most accessible to the enzyme and therefore be among the first digestion fragments observed. Since proteolysis is performed in solution and can detect different conformers, this method can contribute to an understanding of the dynamic domains within the virus structure.

Virus particles are stable, yet exhibit highly dynamic character given the events that shape their life cycle. Isolated from their hosts, the nucleoprotein particles are macromolecules that can be crystallized and studied by X-ray diffraction. During assembly, maturation, and entry, however, they are highly dynamic and display remarkable plasticity. These dynamic properties can only be inferred from the X-ray structure and must be studied by methods that are sensitive to mobility.

Limited proteolysis/MALDI-MS experiments have been performed on human rhinovirus 14 (HRV14) and flock house virus (FHV). Virus HRV14, a causative agent of the common cold, is a member of a family of animal viruses called the picornaviruses, whose other members include the polio, hepatitis A, and foot-and-mouth disease viruses. The HRV14 virion consists of an icosahedral protein shell, or viral capsid, surrounding an RNA core. The capsid is

Protein–Protein Interactions

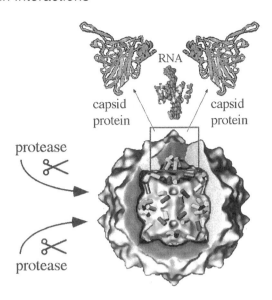

Fig 4. A nonenveloped icosahedral virus with a portion of the capsid proteins and RNA magnified above the virus. The experiments performed involved exposing viruses to limited proteolysis followed by mass analysis of the proteolytic fragments. Time-resolved proteolysis allowed for the study of protein capsid mobility. This encapsulated RNA virus (FHV) belongs to a structural class that includes thousands of viruses responsible for plant and animal diseases such as polio and the common cold.

composed of 60 copies of each of four structural proteins, VP1-VP4. Based on crystal structure data *(28)*, VP1, VP2, and VP3 compose the viral surface, whereas VP4 lies in the interior at the capsid-RNA interface. Virus FHV, like HRV14, is also a nonenveloped, icosahedral, RNA animal virus. The mature protein coat or capsid is composed of 180 copies of β-protein and γ-peptide (**Fig. 4**).

Using time-resolved proteolysis followed by MALDI-MS analysis, it was expected that the reactivities of virus particles to different proteases would reveal the surface-accessible regions of the viral capsid and offer a new way of mapping the viral surface. In these studies, identification of the viral capsid protein fragments was facilitated by sequential digestion, in which proteins were first digested by an endoprotease, such as trypsin, and then exposed to an exoprotease such as carboxypeptidase *(22,23)*. When these experiments were performed on both HRV *(23)* and FHV *(22)*, cleavages on the surface-accessible regions were generated; however, cleavages internal to the viral capsids (based on the crystal structures) were also generated. Observation of digestion fragments resulting from internal protein regions was initially perplexing. After further examination, these results, along with the X-ray data, indicated that

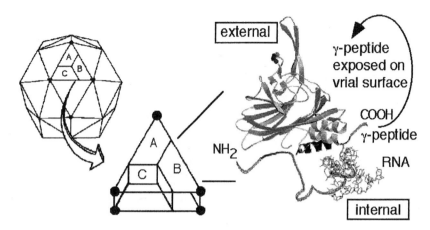

Fig 5. Crystal structure of Flock House virus shows that the γ-peptide and the N- and C termini of β-protein are localized internal to the virus. Yet, proteolytic time-course experiments demonstrated that these domains are transiently exposed on the viral surface.

portions of the internal proteins are transiently exposed on the surface of the virus (**Fig. 5**).

The effects of drug binding on the viral capsid dynamics was also examined by protein mass mapping. Enzymatic digestions were performed in the presence of the antiviral agent WIN 52084 *(23)*. The X-ray crystal structure of HRV14 reveals a 25-Å deep canyon on the surface of the virion at each five-fold axis of symmetry that has been identified as the site of cell surface receptor attachment. WIN compounds bind to these hydrophobic pockets, which lie beneath the canyon floor. Previous studies have shown that the binding of WIN compounds blocks cell attachment of some rhinovirus serotypes, inhibits the uncoating process, and stabilizes the viral capsid to thermal and acid inactivation.

MALDI-MS analysis of the virus following digestion in the presence of the WIN 52084 drug was significantly retarded after exposure to the enzyme for 3 h (**Fig. 6**) and, even after 18 h, only two digestion fragments were observed. In control studies, mass mapping of other proteins and viruses in the presence of WIN 52084 revealed no inhibition of viral capsid degradation. Based on these results, it was concluded that WIN inhibition of protease activity was due to specific effects on the availability of HRV14 cleavage sites and not on the protease itself. The WIN drug is thought to inhibit the capsid mobility, thereby effecting the digestion of the virus.

The protein mass mapping experiments offer a complementary approach to the inherently static methods of crystallography and electron microscopy and reveal dynamic structural changes in solution that may fundamentally alter the way we look at viruses. However, these observations are consistent with the

Protein–Protein Interactions 235

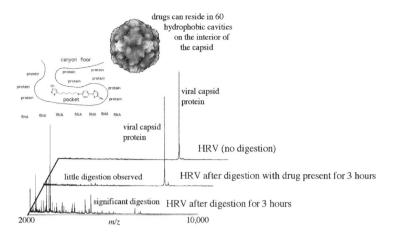

Fig 6. Inhibition of dynamics with drug present. MALDI-MS analyses were performed on HRV, HRV following proteolysis with drug present, and HRV following proteolysis. Inset is an electron micrograph of HRV. (Inset adapted from **ref. 29**.)

events that shape the viral life cycle (cell attachment, cell entry, and nucleic acid release), a life cycle that demands a highly mobile viral surface.

3.3.3. Conclusions and Future Directions

In this era of the bioinformation revolution, biological scientists are in a position to obtain information on intermolecular interactions without the need for site-directed mutagenesis and free of the caveats associated with that technique. Clearly, the examples presented in this chapter illustrate the utility of combining proteolysis and mass spectrometric analysis in structural studies of proteins and multicomponent protein-based assemblies. Because the mass spectrometry component of the experiments still requires sophisticated instrumentation and the computer analysis of mapping data requires chemical knowledge of underlying cleavage and modification reactions, these approaches currently remain in the realm of scientists experienced with these methods, such as those working in support laboratories. Importantly, however, the key concepts of the methods, as illustrated here, are simple, and the probing reactions are also simple to perform. In essence, these methods are within reach of the entire biological community.

Beyond the methods discussed herein, what are the future directions and potential for mass spectrometry-based proteolytic mapping methods? Although current mass spectrometric instrumentation is limited by low resolution and low accuracy at the upper mass limits, indirect mass analysis (in which only the products of probing reactions rather than whole molecular assemblies need be transmitted through the mass spectrometer) makes the window of opportunity quite broad indeed. We have discussed applications to biomolecular assemblies that have been conducted both with and without high-resolution structural data. In the early stages of structural studies, mass spectrometry-based probing methods are particularly well suited to provide rapid access to low-resolution maps that can then be used to guide subsequent high-resolution studies. However, this early stage may itself be an end-point in some investigations in which the simple identification of interacting residues is the desired information. As a complement to high-resolution structural information (either from X-ray crystallography or NMR spectroscopy), probing studies have already been shown to provide valuable and startling insights into protein structural dynamics and rearrangements. These investigations mark the take-off point for follow-up studies that seek to quantitate molecular parameters related to the kinetics and thermodynamics of the underlying dynamic phenomena. We are all familiar with the prominence of gel electrophoresis methods in molecular biology studies; we posit that mass spectrometry methods will capture this position as a primary research tool in coming years due to their superior sensitivity, precision, accuracy, and throughput. The applications discussed here are just a few examples of current approaches that will see expanded use in the future. Only time will tell what others will be invented.

4. Notes

1. Spectra shown are typically an average of data from 128 laser pulses.
2. See **ref.** *21* for preparation of HRV14.
3. The activity of all the enzymes was verified with control peptides.
4. MALDI-MS mass analysis of on-plate trypsin digestions was conducted using the PerSeptive Biosystems instrument.
5. External calibration was typically accurate to 0.05% and allowed unequivocal assignment of most proteolytic fragments.

Acknowledgments

The authors thank Jennifer Boydston for helpful editorial comments and Drs. Tom Steitz of (Yale University), Peter Wright and Steve Reed (The Scripps Research Institute), Ludger Hengst (Max-Planck Institute, Martinsried, Germany), and Mark Watson (Scripps) for their insightful comments and helpful suggestions. This work was supported by grants from the NIH 1 R01 GM55775–01A1 and 1 S10 RR07273–01 (G.S.), the American Lebanese Syrian Associated Charities (RWK), and the American Cancer Society (RWK).

References

1. Fontana, A., Fassina, G., Vita, C., Dalzoppo, D., Zamai, M., and Zambonin, M. (1986) Correlation between sites of limited proteolysis and segmental mobility in thermolysin. *Biochemistry* **25,** 1847–1851.

2. Hubbard, S. J., Eisenmenger, F., and Thornton, J. M. (1994) Modeling studies of the change in conformation required for cleavage of limited proteolytic sites. *Protein Sci.* **3,** 757–768.
3. Edman, P. and Begg, G. (1967) A protein sequenator. *Eur. J. Biochem.* **1,** 80–91.
4. Kriwacki, R. W., Wu, J., Siuzdak, G., and Wright, P. E. (1996) Probing protein-protein interactions by mass spectrometry: analysis of the p21/Cdk2 complex. *J. Am. Chem. Soc.* **118,** 5320.
5. Kriwacki, R. W., Wu, J., Tennant, T., Wright, P. E., and Siuzdak, G. (1997) Probing protein structure using biochemical and biophysical methods: proteolysis, MALDI mass analysis, HPLC, and gel-filtration chromatography of p21Waf1/Cip1/Sdil. *J. Chromatogr.* **777,** 23–30.
6. Fontana, A., Zambonin, M., Polverino de Laureto, P., De Filippis, V., Clementi, A., and Scaramella, E. (1997) Probing the conformational state of apomyoglobin by limited proteolysis. *J. Mol. Biol.* **266,** 223–230.
7. Fontana, A., Polverino de Laureto, P., De Filippis, V., Scaramella, E., and Zambonin, M. (1997) Probing the partly folded states of proteins by limited proteolysis. *Fold Des.* **2,** R17–26.
8. Cohen, S. L., Ferre-D'Amare, A. R., Burley, S. K., and Chait, B. T. (1995) Probing the solution structure of the DNA-binding protein Max by a combination of proteolysis and mass spectrometry. *Protein Sci.* **4,** 1088–1099.
9. Kriwacki, R. W., Hengst, L., Tennant, L., Reed, S. I., and Wright, P. E. (1996) Structural studies of p21Waf1/Cip1/Sdil in the free and Cdk2-bound state: conformational disorder mediates binding diversity. *Proc. Natl. Acad. Sci. USA* **93,** 11,504–11,509.
10. Neidhardt, F. C., Bloch, P. F., and Smith, D. F. (1974) Culture medium for enterobacteria. *J. Bacteriol.* **119,** 736–747.
11. De Bondt, H. L., Rosenblatt, J., Jones, H. D., Morgan, D. O., and Kim, S. (1993) Crystal structure of cyclin-dependent kinase 2. *Nature* **363,** 595–602.
12. Harper, J. W., Adami, G. R., Wei, N., Keyomarsi, K., and Elledge, S. J. (1993) The p21 Cdk-interacting protein Cip1 is a potent inhibitor of G1 cyclin-dependent kinases. *Cell* **75,** 805–816.
13. El-Deiry, W. S., Tokino, T., Velculescu, V. E., Levy, D. B., Lin, D., Mercer, W. E., et al. (1993) WAF1, a potential mediator of p53 tumor suppression. *Cell* **75,** 817–825.
14. Toyoshima, H. and Hunter, T. (1994) p27, a novel inhibitor of G1 cyclin-Cdk protein kinase activity, is related to p21. *Cell* **78,** 67–74.
15. Polyak, K., Lee, M., Erdjument-Bromage, H., Koff, A., Roberts, J. M., Tempst, P., et al. (1994) Cloning of p27Kip1, a cyclin-dependent kinase inhibitor and a potential mediator of extracellular antimitogenic signals. *Cell* **78,** 59–66.
16. Lee, M.-H., Reynisdottir, I., and Massagué, J. (1995) Cloning of p57KIP2, a cyclin-dependent kinase inhibitor with unique domain structure and tissue distribution. *Genes and Dev.* **9,** 639–649.
17. Matsuoka, S., Edwards, M. C., Bai, C., Parker, S., Zhang, P. B., A., Harper, W. J., and Elledge, S. (1995) p57KIP2, a structurally distinct member of the p21CIP1 Cdk inhibitor family, is a candidate tumor suppressor gene. *Genes Dev.* **9,** 650–662.

18. Nakanishi, M., Robetorye, R. S., Adami, G. R., Pereira-Smith, O. M., and Smith, J. R. (1995) Identification of the active region of the DNA synthesis inhibitory gene p21Sdil/CIP1/WAF1. *EMBO J.* **14,** 555–563.
19. Chen, J., Jackson, P. K., Kirschner, M. W., and Dutta, A. (1995) Separate domains of p21 involved in the inhibition of Cdk kinase and PCNA. *Nature* **374,** 386–388.
20. Luo, Y., Hurwitz, J., and Massagué, J. (1995) Cell-cycle inhibition by independent CDK and PCNA binding domains in p21Cip1. *Nature* **375,** 159–161.
21. Russo, A. A., Jeffrey, P. D., Patten, A. K., Massague, J., and Pavletich, N. P. (1996) Crystal structure of the p27Kip1 cyclin-dependent-kinase inhibitor bound to the cyclin A-Cdk2 complex. *Nature* **382,** 325–331.
22. Bothner, B., Dong, X.-F., Bibbs, L., Johnson, J. E., and Siuzdak, G. (1998) Evidence of Viral Capsid Dynamics Using Limited Proteolysis and Mass Spectrometry. *J. Biol. Chem.* **273,** 673–676.
23. Lewis, J. K., Bothner, B., Smith, T. J., and Siuzdak, G. (1998) Antiviral Agent Blocks Breathing of Common Cold Virus. *Proc. Natl. Acad. Sci. USA* **95,** 6774–6778.
24. Siuzdak, G., Bothner, B., Yeager, M., Brugidou, C., Fauquet, C. M., Hoey, K., et al. (1996) Mass spectrometry and viral analysis. *Chem. Biol.* **3,** 45–48.
25. Siuzdak, G. (1998) Probing viruses with mass spectrometry. *J. Mass Spectrom.* **33,** 203–211.
26. Gorman, J. J., Ferguson, B. L., Speelman, D., and Mills, J. (1997) Determination of the disulfide bond arrangement of human respiratory syncytial virus attachment (G) protein by matrix-assisted laser desorption/ionization time-of-flight mass spectrometry. *Protein Sci.* **6,** 1308–1315.
27. Gorman, J. J. (1992) Mapping of post-translational modifications of viral proteins by mass spectrometry. *Trac-Trends Anal. Chem.* **11,** 96–105.
28. Rossmann, M. G., Smith, T. J., and Rueckert, R. R. (1993) The structure of human rhinovirus 14. *Structure* Intro Issue xxiv-xxv.
29. Spencer, S. M., Sgro, J. Y., Dryden, K. A., and Baker, M. L. *J. Struct. Biol.* **120,** 11–21.

14

Studies of Noncovalent Complexes in an Electrospray Ionization/ Time-of-Flight Mass Spectrometer

Andrew N. Krutchinsky, Ayeda Ayed, Lynda J. Donald, Werner Ens, Harry W. Duckworth, and Kenneth G. Standing

1. Introduction

The formation of noncovalent complexes by macromolecules is one of the most important molecular processes in biology. It is intimately involved in such recognition phenomena as enzyme-substrate interaction, receptor-ligand binding, formation of oligomeric proteins, assembly of transcription complexes, and the formation of cellular structures themselves. Established methods for studying these complexes suffer from various difficulties such as large sample requirements (X-ray crystallography), limited mass range (nuclear magnetic resonance spectroscopy), lack of specificity (ultracentrifuge), or poor mass resolution (gel electrophoresis).

Mass spectrometry offers very high sensitivity and good mass resolution. Because of its unparalleled ability to distinguish species of different molecular masses, it has the potential to analyze macromolecular complexes with multiple components in equilibrium. Thus, mass spectrometry can give unique insights into the formation and properties of these complexes, yielding information that is complementary to that obtained by established methods and can be particularly valuable when the amount of sample is limited. However, mass spectrometry has not been applicable to the study of noncovalent complexes until fairly recently, because these entities are often weakly bound, and thus may be destroyed during ionization or transfer into the gas phase. This problem has now been greatly alleviated by the development of electrospray ionization (ESI), a very gentle ionization method *(1)*, and, as a result, an increasing number of measurements on noncovalent complexes are being carried out using this technique. The subject has been discussed recently in a previous volume in this series *(2)*, in other reviews *(3,4)*, and in conference proceedings *(5)*. In a number of cases, ESI has been found to preserve native structure and higher-order noncovalent interactions.

A serious practical problem for mass spectrometric measurements of noncovalent complexes is the limited mass-to-charge (*m/z*) range of most commercial mass analyzers (usually < 4000 for quadrupole mass filters), which is inadequate for the observation of many interesting but large complexes. In the previous volume the use of an extended *m/z* range quadrupole mass filter for such measurements was described, but this instrument suffered from a number of problems, especially low mass resolution (< 100) *(2)*. A time-of-flight (TOF) mass analyzer provides an attractive alternative, since TOF instruments have, in principle, an unlimited *m/z* range without sacrificing resolution (*see* **Note 1**). They also demonstrate high sensitivity as a result of parallel detection of all masses (i.e., without scanning). The need to convert the continuous beam from an electrospray ion source to the pulsed beam required for TOF analysis was a difficulty that impeded construction of an efficient ESI-TOF spectrometer for some time. However, a solution to this problem was provided by the technique of orthogonal injection *(6)*, in which the electrosprayed ions move perpendicular to the spectrometer axis at low energy to enter a storage region at the entrance to the TOF analyzer (**Fig. 1**). The stored ions are then injected into the flight path as a series of pulses.

Several years ago we constructed such an ESI-TOF mass spectrometer *(7,8)*, and instruments using similar techniques are now available commercially *(9)*. As an illustration of the use of an ESI-TOF instrument for examining noncovalent interactions, we describe below the analysis of a system of complexes with *m/z* values close to 10,000 *(10)*. The system studied involves *Escherichia coli* citrate synthase (CS). This is the enzyme that initiates the citric acid cycle; as such, it plays a key role in cellular metabolism, catalyzing the reaction between oxaloacetate and acetyl coenzyme A to produce citrate and coenzyme A. The interaction with its allosteric inhibitor, the reduced form of nicotinamide adenine dinucleotide (NADH), involves simultaneous protein-protein and ligand-protein equilibria.

2. Materials

1. Water was prepared freshly using a NANOpure II system (Sybron/Barnstead, Dubuque, Iowa).
2. Plastic labware was new (Nalgene, Rochester, NY, *see* **Note 2**).
3. β-NADH (Sigma, St. Louis, MO).
4. Substance P (Sigma).

Studies of Noncovalent Complexes

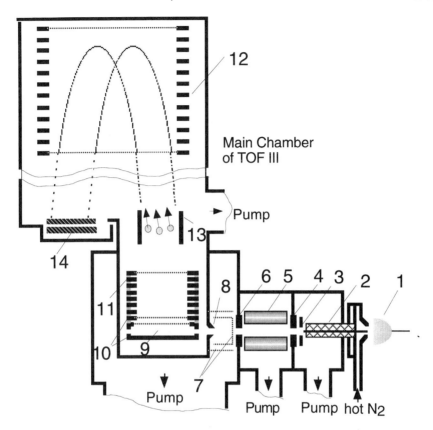

Fig. 1. Schematic diagram of the time-of-flight instrument. 1, ESI ion source; 2, heated capillary; 3, focusing electrode; 4, first aperture plate; 5, RF quadrupole; 6, second aperture plate; 7, grids; 8, slit; 9, the storage region; 10, extraction electrodes; 11, acceleration column; 12, electrostatic mirror; 13, deflection plates; 14, detector.

5. Dialysis tubing 500 MWCO cellulose ester membrane (Spectrum, Laguna Hills, CA).
6. Centricon 30 (Amicon/Millipore, Bedford, MA).
7. *E. coli* citrate synthase was prepared on site (*see* **Note 3**).
8. Nanospray capillaries (Protana, Odense, Denmark).
9. GELoader tips (Eppendorf/Brinkmann, Hamburg, Germany).
10. Ammonium bicarbonate, certified grade (Fisher, Fair Lawn, NJ).

3. Methods and Discussion
3.1. Sample Preparation and Measurement
3.1.1. Citrate Synthase Preparation

1. Pass freshly prepared 20 mM NH_4HCO_3 through a 20-µm filter to remove particulates.

2. Transfer an aliquot of pure CS protein (100–200 μL at 1 mg/mL in 20 mM Tris-Cl (pH 7.8), 1 mM EDTA, and 50 mM KCl) to a Centricon 30 containing 2 mL of 20 mM NH$_4$HCO$_3$ and centrifuge at 4°C, following the manufacturer's instructions.
3. Repeat the washing and centrifuging 6–8 times to ensure removal of unwanted buffer components.
4. Determine the concentration spectrophotometrically using the known extinction coefficient (47,000 M^{-1} cm^{-1} at 278 nm).
5. Prepare solutions of the protein so as to dilute the buffer to 5 mM at the same time (*see* **Note 4**).

3.1.2. β-NADH Preparation

1. Take up the disodium salt of β-NADH into water and determine the concentration spectrophotometrically from the known extinction coefficient (6220 M^{-1} cm^{-1} at 340 nm).
2. Assemble aliquots of 200 μL each in "waterbugs" *(11)*, using 500 MWCO dialysis membrane. Use at least three changes of buffer, each of 1 L, at 4°C.

3.1.3. CS-NADH Complex Preparation

1. Prepare a 100-μL sample of the desired combination of CS and NADH 1 d prior to mass spectrometric measurements of the CS-NADH complex.
2. Dilute the samples to give a final buffer concentration of 5 mM and a CS concentration of about 10 μM monomer. The chosen concentrations of NADH were 0, 4.5, 9.0, 18, and 108 μM.
3. Allow samples to equilibrate overnight at 4°C.

3.1.4. Sample Preparation for Measurement Using Nanospray

1. Prepare CS samples for nanospray measurements as in **Subheading 3.1.1.**, but using a 0.2 M NH$_4$HCO$_3$ buffer (as an alternative to the use of a conventional electrospray ion source, *see* **Subheading 3.2.**).
2. Dilute both protein and buffer were just before mass spectrometric measurements so that the protein concentration is a 10 μM monomer in the chosen concentration of buffer (*see* **Note 4**).
3. Cut the nanospray capillary to the required length and wash with 2.5 μL of the sample using a ultramicropipet tip. Introduce a second 2.5-μL aliquot of the sample to be used for experiments. Machine conditions are similar to those used for ordinary ESI.

3.1.5. Mass Calibration

Calibration used the singly and doubly charged ions of substance P.

3.2. Mass Spectrometry

The original ESI atmosphere/vacuum interface used for orthogonal ion injection into our TOF mass spectrometer *(7)* has been modified more recently by the addition of a heated metal capillary to provide another stage of pumping and desolvation, a new section containing a small radio frequency (RF) quadrupole has been installed to provide collisional cooling of the ions *(12)*. A schematic diagram of the main elements of the instrument and the new interface is shown in **Fig. 1**.

Most of the results reported here were obtained with a conventional electrospray ion source. In this device a solution of an analyte is delivered to the sharpened tip of a stainless steel needle ([**1**]; bold numbers refer to **Fig. 1**) (0.45 mm OD, 0.11 mm ID) by a syringe pump with a typical flow rate of 0.17–0.25 µL/min. Alternatively, we have employed a nanospray source (using Protana capillaries), with a flow rate approx 20 nL/min produced by a backing pressure of air, at approx 10^3–10^4 Pa, instead of the syringe pump. Electrospraying of the analyte against a counterflow of heated (50–70°C) nitrogen (*see* **Note 5**) was performed at 3–3.5 kV potential difference between the tip of the needle and the inlet of a heated metal capillary (**2**) (0.5 mm ID, 12 cm long). The ions and droplets produced in this manner are swept into the capillary (**2**) along with the nitrogen, where they undergo some additional desolvation.

Expansion of this mixture into the next region produces a supersonic jet. This region is pumped to a pressure of approx -2.5 Torr by a 5.7 L/s mechanical pump. A declustering voltage is applied to a focusing electrode (**3**) installed approx -10 mm downstream from the end of the heated capillary to which it has an electrical connection. Up to 300 V potential difference (for nitrogen) can be maintained between the focusing electrode and the flat aperture plate (**4**) located 3 mm further downstream. The 0.35-mm diameter aperture in the plate connects this region to the second pumping stage containing the RF quadrupole (**5**); which is evacuated to a pressure of approx 0.1 Torr by a 12.6 L/s mechanical pump. In this region the ions oscillate in the two-dimensional potential well produced by the RF quadrupole and are cooled to near-thermal energies by collisions with the ambient gas *(12)*.

After passing through the quadrupole the ions enter the third region through a 0.75-mm-diameter aperture in the flat plate (**6**). This region is pumped to a pressure of approx 10^{-5} Torr by a 450 L/s turbomolecular pump. The electric field between this aperture plate and the grids (**7**) forms an ion lens used to shape the ion beam prior to its entry, through the horizontal slit (**8**) (6 mm width, 1.5 mm height), into the main chamber of the mass spectrometer.

At the beginning of the injection cycle, the electric field in the gap of the storage region (**9**) is zero. After a group of ions has filled the storage region, the process of ion acceleration starts by applying pulses to the extraction electrodes (**10**) of the accelerator column (**11**). Ions then leave the acceleration column in the direction of the electrostatic mirror (**12**) with approx 4 keV kinetic energy per charge. The direction of ion motion can be corrected, if necessary, by the deflection plates (**13**), so as not to miss the detector (**14**). The pressure in the main chamber is kept at approx 3×10^{-7} Torr by a cryopump (approx 1000 L/s), recently replaced by an turbomolecular pump (approx 450 L/s). The voltages applied to the aperture plates and slits, as well as to the RF quadrupole, are adjusted to obtain optimum ion transmission through the interface.

The quadrupole rods (0.8 cm diameter and 3.5 cm long) are supported by an insulator whose surfaces are screened from the ion beam. The quadrupole is driven by a small sine-wave signal generator coupled through a broadband RF power amplifier. An RF coupling transformer constructed in our laboratory gives an output voltage range from 0 to 1000 volts peak-to-peak. The transformer also provides the required 180° phase difference between the pole pairs. The quadrupole rods are normally "offset" to some DC potential intermediate between the potentials on the first and second aperture plates.

In its original configuration, the whole ESI interface assembly was elevated to 4 kV DC acceleration potential so that the flight path of the mass spectrometer was at ground *(7)*. In that case, the relatively high pressure in the first stage inhibited discharges to the grounded mechanical pump; a 1-m length of plastic hose was sufficient. The pressure in the third and fourth regions was low enough to prevent discharges. The section containing the quadrupole was more of a problem. Here it was necessary to run both internal and external voltage dividers (each consisting of 27 1MΩ resistors) along the approx 1.5-m polyethylene hose; these were sufficient to prevent electrical arcing. However, the spectrometer has been modified recently by the insertion of an internal shield enclosing the flight path *(13)*. The shield is now electrically floated at the acceleration potential so that the ESI source operates at near ground potential. This modification makes it much easier to operate, but it does not change the essential features of the ion guide operation.

3.3. Electrospray Analysis of CS in the Absence of NADH

When an electrospray spectrum of CS is obtained from acidic solution, it yields a broad distribution of charge states, carrying approx 30–40 positive charges centered around m/z approx 1100, and corresponding to the monomeric form (47,887 Daltons). In contrast to this, the spectra shown in **Fig. 2**, obtained under nondenaturing conditions, contain only a few charge states, with m/z values up to approx 10,000. These correspond to dimers and hexamers, as expected for the native forms *(10)*. Deconvoluted spectra for the dimer and hexamer, illustrated in **Fig. 3**, yield masses in good agreement with the theoretical values of 95,770 Daltons (dimer) and 287,310 Daltons (hexamer). Some tetramers are also observed, but with considerably lower abundance. Their intensity appears to follow the dimer abundance, suggesting that they arise from a nonspecific interaction between two dimers. The dependence of the dimer/hexamer molar ratio on the concentration of CS subunits in the

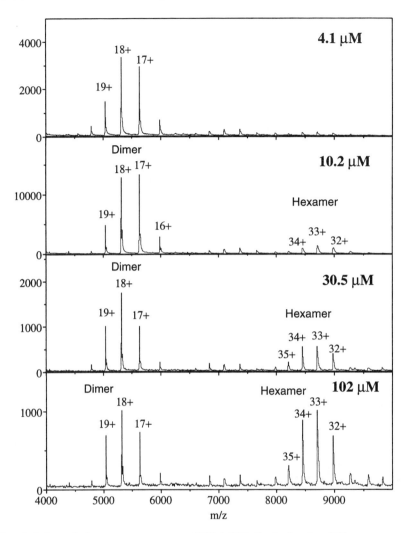

Fig. 2. Selected electrospray spectra of CS (pH 7.5) obtained at different concentrations of CS subunits in the solution. The concentration is indicated in each spectrum.

solution can be used to determine an association constant (K_A), assuming a simple equilibrium between dimers and hexamers *(10)*.

3.4. Electrospray Analysis of CS in the Presence of NADH Inhibitor

Figure 4 shows ESI spectra of CS in the presence of increasing concentrations of NADH. Although an allosteric mechanism of NADH inhibition involving the hexamer has been proposed *(14)*, how this occurs has been

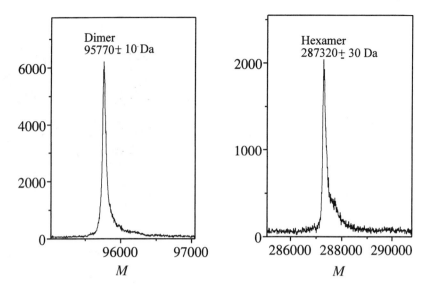

Fig. 3. Deconvoluted spectra of the CS dimer and hexamer (approx 30 μM subunit concentration) from 5 mM ammonium bicarbonate buffer (pH approx 7.5).

unknown. The mass spectra now provide an explanation. From **Fig. 4** it is clear that increasing concentrations of NADH both shift the oligomeric ratio toward the hexamer and bind to the hexamer to form a set of (inactive) NADH-hexamer complexes, almost fully resolved from one another, and containing from 0 to as many as 18 NADH molecules. A set of NADH-dimer complexes also appears, but only at higher NADH concentrations. The pattern suggests that nonspecific binding occurs to both dimers and hexamers, but that specific binding is exclusive to hexameric CS. The dependence of bound NADH on the free NADH concentration has been used to determine dissociation constants K_D, assuming 6 "tight" binding sites and 12 "loose" binding sites per hexamer *(10)*.

Our experience with this rather complicated system suggests that ESI-TOFMS will be widely useful in obtaining fundamental physicochemical information about macromolecular interactions. In particular, it appears to be a powerful method for identifying the noncovalent complexes formed in biological systems and determining their stoichiometry *(15)*, especially when the amount of sample available is limited.

4. Notes

1. A comparison of mass spectra obtained in our ESI-TOF spectrometer with those from several other types of instrument for the well-known test compound streptavidin and for the streptavidin-biotin complex is given in **ref.** *16*.

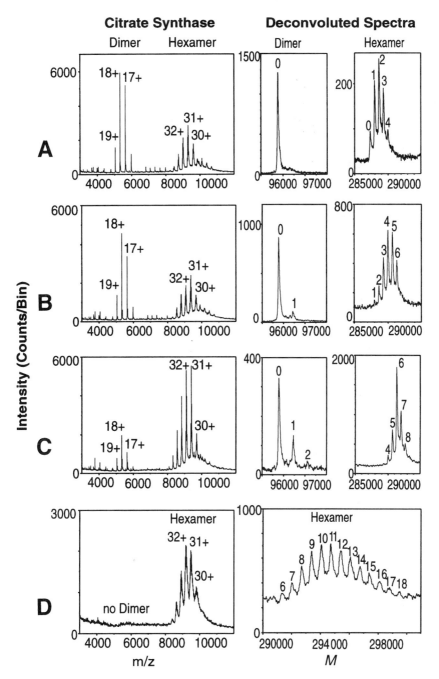

Fig. 4. Selected raw and deconvoluted electrospray mass spectra of CS (9-μM subunit concentration in 5 mM ammonium bicarbonate at pH 7.5) in the presence of increasing concentrations of NADH. NADH concentrations were (**A**) 4.5 mM, (**B**) 9 mM, (**C**) 18 mM, and (**D**) 108 mM. The digits labeling the deconvoluted spectra correspond to the number of NADH molecules bound.

2. It is important to avoid glass for preparation and storage of buffers used in the final workup of samples before ESI. We have found that a dedicated set of plastic beakers is ideal for this task.
3. The expression plasmid pES*glt*A in host cell MOB154 was used to produce *E. coli* CS *(17,18)*. The enzyme was purified from cell extracts by ion-exchange chromatography DE52 (Whatman, Kent, England) followed by size-exclusion chromatography on a Sepharose 6B column (Pharmacia, Uppsala, Sweden). Pure enzyme was concentrated to about 100 mg/mL using vacuum dialysis, and stored at 4°C in 20 mM Tris-Cl (pH 7.8), 1 mM EDTA, 50 mM KCl.
4. For efficient spraying in conventional electrospray, it was necessary to keep the buffer concentration less than about 20 mM. However, nanospray is much more tolerant of high buffer concentrations. We have obtained nanospray spectra of CS comparable to those shown in **Fig. 2** with a 50-mM ammonium bicarbonate solution. For other complexes we have observed good nanospray spectra at buffer concentrations up to approx 500 mM, as also reported by Rostom and Robinson *(3)*.
5. More recently we have used SF_6 instead of nitrogen as the curtain gas. SF_6 has two advantages for declustering. First, it is considerably more efficient in removing adducts and dissociating complexes because of its larger mass; a typical dissociation pattern at 250 V declustering voltage (the voltage between capillary and aperture plate) resembles the pattern at 300 V with nitrogen. Second, it permits the use of larger declustering voltages than nitrogen (< 400 V compared with < 300 V), because of its superior insulating properties.

Acknowledgments

We thank Igor Chernushevich and Noelle Potier for helpful discussions and Victor Spicer for technical assistance. This work was supported by a Collaborative Project Grant to H.W.D., W.E., and K.G.S., and also by individual research grants to H.W.D. and to K.G.S., all from the Natural Sciences and Engineering Research Council of Canada.

References

1. Cole, R. B., ed. (1997) *Electrospray Ionization Mass Spectrometry*, John Wiley & Sons, New York.
2. Schwartz, B. L., Gale, D. C., and Smith, R. C. (1996) Noncovalent interactions observed using electrospray ionization, in *Methods in Molecular Biology*, vol. 61: *Peptide and Protein Analysis by Mass Spectrometry* (Chapman, J. R., ed.), Humana, Totawa, NJ, pp. 115–127.
3. Rostom, A. A. and Robinson, C. V. (1999) Disassembly of intact multiprotein complexes in the gas phase, *Curr.Opin. Struct. Biol.* **9,** 135–141.
4. Loo, J. A. (1997) Studying noncovalent protein complexes by electrospray ionization mass spectrometry. *Mass Spectrom. Rev.* **16,** 1–23, and references therein.
5. Ens, W., Standing, K. G., and Chernushevich, I. V., eds. (1998) *New Methods for the Study of Biomolecular Complexes,* Kluwer, Dordrecht, NL.

6. Chernushevich, I. V., Ens, W., and Standing, K. G. (1999) Orthogonal-injection time-of-flight mass spectrometry for analysis of biomolecules. *Anal. Chem.* **71,** 452A–461A.
7. Verentchikov, A. N., Ens, W., and Standing, K. G. (1994) Reflecting time-of-flight mass spectrometer with an electrospray ion source and orthogonal extraction. *Anal. Chem.* **66,** 126–133.
8. Chernushevich, I. V., Ens, W., and Standing, K. G. (1997) Electrospray ionization time-of-flight mass spectrometry, in *Electrospray Ionization Mass Spectrometry* (Cole, R. B., ed.), John Wiley & Sons, New York, pp. 203–234.
9. Henry, C. M. (1999) Electrospray in flight. *Anal. Chem.* **71,** 197A-201A.
10. Ayed, A., Krutchinsky, A. N., Ens, W., Standing K. G., and Duckworth, H. W. (1998) Quantitative evaluation of protein-protein and ligand-protein equilibria of a large allosteric enzyme by electrospray ionization time-of-flight mass spectrometry. *Rapid Commun. Mass Spectrom.* **12,** 339–344.
11. Orr, A., Ivanova, V. S., and Bonner, W. M. (1995) Waterbug dialysis. *Biotechniques* **19,** 204–206.
12. Krutchinsky, A. N., Chemushevich, I. V., Spicer, V. L., Ens, W., and Standing, K. G. (1998) A collisional damping interface for an electrospray ionization time-of-flight mass spectrometer. *J. Am. Soc. Mass Spectrom.* **9,** 469–579.
13. Shevchenko, A., Chernushevich, I. V., Ens, W., Standing, K. G., Thomson, B., Wilm, M., et al. (1997) Rapid *de novo* peptide sequencing by a combination of nanoelectrospray, isotopic labeling, and a quadrupole/time-of-flight mass spectrometer. *Rapid Commun. Mass Spectrom.* **11,** 1015–1024.
14. Tong, E. K. and Duckworth, H. W. (1975) The quaternary structure of citrate synthase from *Escherichia coli* K12. *Biochemistry* **14,** 235–241.
15. Zhang, Z., Krutchinsky, A., Endicott, S., Realini, C., Rechsteiner, M., and Standing, K. G. (1999) The proteasome activator 11S REG or PA28; recombinant REGc/REGp hetero-oligomers are heptamers *Biochemistry* **38,** 5651–5658.
16. Chernushevich, I. V., Ens, W., and Standing, K. G. (1998) in *New Methods for the Study of Biomolecular Complexes,* Kluwer, Dordrecht, pp. 101–116.
17. Duckworth, H. W. and Bell, A. W. (1982) Large-scale production of citrate synthase from a cloned gene. *Can. J. Biochem.* **60,** 1143–1147.
18. Anderson, D. H. and Duckworth, H. W. (1988) *In vitro* mutagenesis of Escherichia coli citrate synthase to clarify the locations of ligand binding sites. *J. Biol. Chem.* **263,** 2163–2169.

15

Kinetic Analysis of Enzymatic and Nonenzymatic Degradation of Peptides by MALDI-TOFMS

Fred Rosche, Jörn Schmidt, Torsten Hoffmann, Robert P. Pauly, Christopher H.S. McIntosh, Raymond A. Pederson, and Hans-Ulrich Demuth

1. Introduction

Currently used methods for investigating kinetics of peptide degradation such as refractive index monitoring, radioimmunoassay (RIA), high-performance liquid chromatography (HPLC), or capillary electrophoresis (CE) are time consuming, need large amounts of substrate, and are often too insensitive. Moreover, as in the case of RIA, HPLC, and CE, it is often impossible to interpret the observed results with confidence in the integrity of the analyte. To circumvent such obstacles, we found matrix-assisted laser desorption/ionization (MALDI) used with time-of-flight mass spectrometry (TOFMS) not only useful for qualitative analysis of reaction pathways but also for quantification. In the following chapter, we give two examples of kinetic reaction course evaluation, one non-enzymatic and one enzymatic.

To be able to study the kinetics of the nonenzymatic degradation of a protease inhibitor Z-Phe-Pro-(isonicotinic acid methyl ester) *(1)* as well as the hydrolysis of protease substrates, in this case two peptide hormones, glucose-dependent insulinotropic polypeptide (GIP_{1-42}) and glucagon-like peptide-1-(7–36)-amide ($GLP-1_{7-36}$) *(2)*, we have developed protocols for the quantification of MALDI-TOF-MS signals.

A small number of applications of MALDI-TOFMS for quantitative analysis have recently been published. In these cases, it was difficult to obtain good quantitative results by simply determining an unknown concentration of a substance from the signal height. Thus, quantification using mass spectrometer signals typically involves the incorporation of a specific internal standard into the sample mixture. In our kinetic studies of peptide degradation by quantifying substrate and product peaks relative to each other, these peaks served as their own internal standard *(3–5)*.

The close correlation between MALDI-TOFMS-derived kinetic constants and those previously evaluated by other means, or those determined in our study using a comparative spectrophotometric assay, validate MALDI-TOFMS as a reliable method for kinetic analysis.

2. Materials
2.1. Instruments

1. MALDI-MS was carried out using a Hewlett-Packard (Palo Alto, CA) G2025 LD-TOF system with a linear time-of-flight analyzer. The instrument was equipped with a 337-nm nitrogen laser, an ion source operating at 5-kV acceleration potential and a 1.0-m flight tube. The detector was operated in the positive-ion mode, and signals were recorded and filtered using a LeCroy (Chestnut Ridge, NY) 9350M digital storage oscilloscope linked to a personal computer. The spectrometer was calibrated externally using molecular weight standards (Hewlett-Packard).
2. A Hewlett-Packard G2024A Sample Prep Accessory was used to ensure rapid and homogenous sample crystallization.

2.2. Enzymes and Serum

1. Prolyl endopeptidase from *Flavobacterium meningosepticum* (*see* **Note 1**).
2. Dipeptidyl peptidase IV from porcine kidney (*see* **Note 2**).
3. Human serum (*see* **Note 3**).

2.3. Peptides

1. Pyridinium methyl ketone derivatives were synthesized in our laboratory (*see* **Note 4**).
2. Synthetic porcine glucose-dependent insulinotropic polypeptide/gastric inhibitory polypeptide (GIP_{1-42}) (Peninsula, San Carlos, CA).
3. Synthetic human glucagon-like peptide-1-(7–36) amide ($GLP-1_{7-36}$) (Bachem, Bubendorf, Switzerland).

2.4. Chemical Reagents

1. 2',6'-dihydroxyacetophenone (DHAP) (Aldrich, Deisenhofen, Germany).
2. Diammonium hydrogen citrate (Fluka, Deisenhofen, Germany).
3. H-Gly-Pro-Pro-4-nitroanilide was synthesized in our laboratory.
4. H-Gly-Pro-4-nitroanilide (Bachem).

2.5. Solvents and Buffers

1. 0.1 mM Tricine-HCl (pH 7.6).
2. Acetonitrile, high-performance liquid chromatography (HPLC) grade.

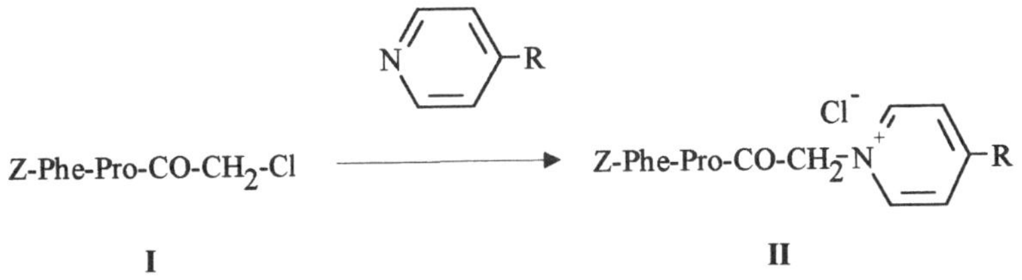

Degradation of Peptides 253

$$\text{Z-Phe-Pro-CO-CH}_2\text{-Cl} \longrightarrow \text{Z-Phe-Pro-CO-CH}_2\overset{+}{\text{N}}\underset{}{\diagdown}\text{R} \;\; \text{Cl}^-$$

I II

R = H, COOH, COOCH$_3$ (II-1, II-2, II-3)

Scheme 1. Synthesis of peptidyl pyridinium methyl ketone derivatives.

3. Trifluoroacetic acid (TFA), HPLC grade.
4. 0.04 M HEPES (pH 7.6).

3. Methods

3.1. Investigation of the Hydrolytic Stability of the Prolyl Endopeptidase Inhibitor Z-Phe-Pro-(Isonicotinic Acid Methyl Ester) Methyl Ketone

Prolyl endopeptidase (PEP) is a serine enzyme cleaving highly specific post-prolyl and post-alanyl peptide bonds. The protease is involved in peptide hormone processing and regulation *(6)*. We introduced peptidyl ammonium methyl ketones as powerful slow-binding inhibitors of proline-specific peptidases capable of being structurally modified by C-terminal elongation *(7)*. These inhibitors should be applicable as potential diagnostic materials in bioassays and animal experiments.

Z-Phe-Pro-(isonicotinic acid methyl ester) methyl ketone is a highly potent inhibitor ($K_i^* = 18$ nM) *(1)* of prolyl endopeptidase from *Flavobacterium meningosepticum*. It acts as a slow-binding inhibitor and decomposes to compounds still possessing inhibitory activity. The pathway of degradation was analyzed qualitatively and quantitatively by using MALDI-TOF mass spectrometry.

During an earlier analysis of the influence of the compound Z-Phe-Pro-COCH$_2$N$^+$C$_5$H$_4$COOCH$_3$ Cl$^-$ (**II-3**, *see* **Note 4** and **Scheme 1**), on PEP-catalyzed hydrolysis at fairly low substrate and inhibitor concentrations (*see* **Note 5**), the curve displayed partial restoration of enzyme activity after about 100 min of incubation time (**Fig. 1**, curve 2). Numerical analysis of the part of the curve, designated C in **Fig. 1**, yields a corresponding pseudo-first-order rate constant, independent of the inhibitor concentration between 1.0 µM and 10 nM and independent of the substrate concentration between 0.4 mM and 80 µM. This value of $k_{obs} = 0.018$ min^{-1} (±0.0019) corresponds to an apparent half-life of 38.8 min for this unexpected process. To investigate the cause

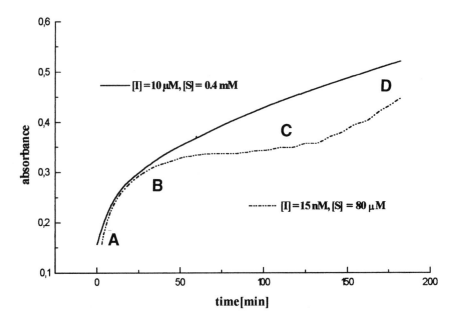

Fig. 1. Curves of PEP-catalyzed hydrolysis of H-Gly-Pro-Pro-4-nitroanilide in the presence of different concentrations of Z-Phe-Pro-(isonicotinic acid methyl ester) methyl ketone (II-3) monitored at 390 nm ([PEP] = 15 nM). **A**, initial competitive inhibition of the enzyme-catalyzed substrate hydrolysis before formation of the tighter enzyme-inhibitor complex; **B**, pseudo-first-order rate constant of complex formation; **C**, steady state between formation of the tight enzyme-inhibitor complex and inhibitor degradation; **D**, degradation of the inhibitor.

of this kinetic behavior, MALDI-TOFMS was applied as a means of observing the reaction.

In **Fig. 2**, the qualitative assignment clearly demonstrates the stepwise decomposition of compound **II-3** into compound **II-1** via the Z-Phe-Pro-isonicotinic acid methyl ketone (compound **II-2**) in a typical sequential reaction (A → B → C), which was confirmed by an analysis of the separately synthesized compounds **II-1** and **II-2** (*see* **Note 4** and **Scheme 1**).

To study their stability, mass spectra of the synthetic inhibitors were obtained as follows:

1. Incubate the chosen inhibitor compound (**II-1–II-3**) at 30°C, at a concentration ≈ 0.6 nmol/mL, in 0.04 M HEPES buffer (pH 7.6) with an ionic strength of 0.125 (KCl).
2. Remove samples of the incubation mixture (4 µL containing 2.5 pmol) at timed intervals up to 200 min and mix with an equal volume of 0.2 M DHAP in acetonitrile + water (1:1, v/v) containing 44 g/L diammonium hydrogen citrate (to suppress alkali ion adducts) as matrix solution (*see* **Note 6**).

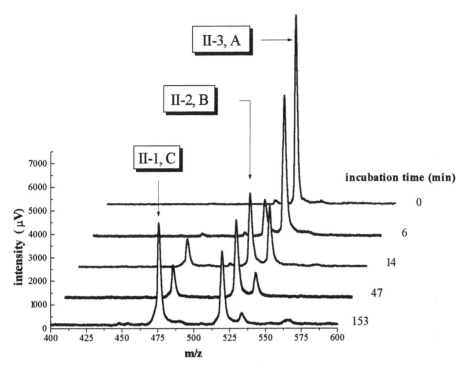

Fig. 2. MALDI-TOF spectra of Z-Phe-Pro-(isonicotinic acid methyl ester) methyl ketone taken at different time intervals after incubation at pH 7.6 and 30°C.

3. Transfer a small volume (< 1 µL) of this mixture to the probe tip and immediately evaporate it in a vacuum chamber.
4. Accumulate spectra by averaging 250 single shots with a laser power between 1.5 and 4.5 µJ. The signal-to-noise ratio was chosen to be between 1.5 and 35. The laser power was increased from 1.5 to 2.5 µJ automatically with a statistical algorithm up to a signal-to-noise ratio in the desired range.
5. Determine relative peak heights by measuring the distance between the baseline and the peak top in the region of the protonated molecule ion and dividing by the intensity sum over all the peaks. Adduct peaks such as [M+Na]$^+$ were not detected.

In a separate experiment to determine the relationship between inhibitor concentration and mass spectral signal intensity, various concentrations of all three compounds were mixed with buffer and water, as detailed above, and 1-µL samples, ranging from 0.5 to 10 pmol/sample, were analyzed by mass spectrometry. Spectra for each peptide concentration were generated in triplicate. Quantification of the signal was accomplished by dividing the peak intensity by the baseline intensity resulting in a signal intensity normalized to the spec-

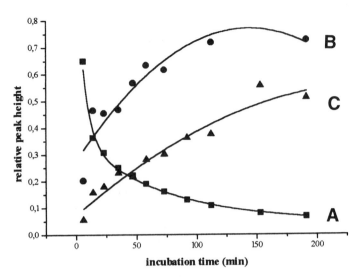

Fig. 3. Curves of the time-dependent decomposition of compound **II-3** monitored by a MALDI-TOF kinetic analysis protocol (see text). **A**, exponential decay of Z-Phe-Pro-(isonicotinic acid methyl ester) methyl ketone; **B**, formation and decay of intermediate Z-Phe-Pro-(isonicotinic acid) methyl ketone; **C**, formation of the final stable Z-Phe-Pro-(pyridinium) methyl ketone.

trum baseline. Although the range of linearity of the relationship between concentration and mass spectral signal intensity depends on the chemical properties of a given compound, we found a linear relationship up to 14 µM concentration for the compounds **II-1–II-3** (concentration window; *see* **Subheading 3.2.2.**).

Fig. 3 displays the change of intensity of the mass signals of compounds **II-1–II-3** depending on the incubation time of compound **II-3** in reaction buffer. Each individual point in the graphs of **Fig. 3** represents the mean relative intensity of three separately recorded mass spectra, which show a standard error ranging from 3.3 to 10.6% of the mean. As expected, the ester derivative (compound **II-3**, curve A in **Fig. 3**) is unstable under weakly basic conditions. It hydrolyzes to an intermediate compound **II-2**, the corresponding carboxyl derivative, which itself decarboxylates in a subsequent spontaneous reaction to form the stable compound **II-1**, Z-Phe-Pro-(pyridinium) methyl ketone. From the data of the curve A (**Fig. 3**), we calculated the pseudo-first-order rate constant of the hydrolysis of Z-Phe-Pro-(isonicotinic acid methyl ester) methyl ketone as k_{obs} = 0.0185 min^{-1} (±0.0009), which is in excellent agreement with the value obtained in the enzyme kinetic experiments (part C of curve 2, **Fig. 1**).

3.2. Monitoring and Kinetic Analysis of Enzymatic Hydrolysis of the Incretins GIP_{1-42} and $GLP-1_{7-36}$ by Dipeptidyl Peptidase IV

3.2.1. The Hydrolysis Process and MALDI-TOF Protocol

Dipeptidyl peptidase IV (DP IV) is a highly specific exopeptidase that selectively cleaves peptides at penultimate N-terminal proline and alanine residues *(8)*. It is believed to play a pivotal role in the activation and inactivation of polypeptides in vivo *(9,10)*. GIP_{1-42} and $GLP-1_{7-36}$, hormones that potentiate glucose-induced insulin secretion from the endocrine pancreas *(11)*, are substrates of DP IV and are rendered biologically inactive on removal of their N-terminal dipeptides *(12–14))*. Evidence has accumulated that serum inactivation is the initial step in the metabolism of these hormones *(15)*. We used MALDI-TOFMS for the study of DP IV-catalyzed hydrolysis of GIP_{1-42} and $GLP-1_{7-36}$, including enzyme kinetic analysis. Mass spectra indicated that serum-incubated GIP_{1-42} and $GLP-1_{7-36}$ were primarily cleaved by DP IV reaction with only minor subsequent degradation due to other serum protease activity. A protocol for the quantification of the MALDI-TOFMS signal was developed allowing kinetic constants for both porcine kidney and human serum DP IV-catalyzed GIP_{1-42} and $GLP-1_{7-36}$ hydrolysis to be calculated from mass spectrometric data. The general protocol to obtain mass spectra of GIP_{1-42} and $GLP-1_{7-36}$, in the presence or absence of DP IV is as follows:

1. Incubate substrates at 30°C with 0.1 mM Tricine buffer (pH 7.6) and either enzyme or water in a 2:2:1 ratio.
2. Remove samples (4 µL) of the incubation mixture at timed intervals and mix with an equal volume of matrix solution (DHAP/DHC as described in **step 2** of **Subheading 3.1.**).
3. Transfer a small volume (< 1 µL) of this mixture to a probe tip and immediately evaporate in a vacuum chamber.
4. Acquire mass spectra by accumulating data generated from 250 single shots with a laser power between 2.5 and 3.5 µJ.

3.2.2. Dependence of MALDI-TOFMS Signal on the Concentration of GIP_{1-42} or $GLP-1_{7-36}$

Various concentrations of synthetic porcine GIP_{1-42} or synthetic human $GLP-1_{7-36}$ were mixed with buffer and water as described above, and 1-µL samples ranging from 0.5 to 6 pmol GIP_{1-42} or from 3.75 to 10 pmol $GLP-1_{7-36}$ were analyzed by mass spectrometry to determine the relationship between concentration of hormone and mass spectral signal intensity. Spectra were generated in triplicate at each peptide concentration. Quantification of GIP_{1-42} and $GLP-1_{7-36}$ signals was accomplished by dividing the peak intensity by

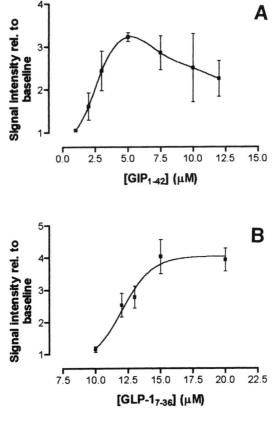

Fig. 4. (A, B) Concentration dependence of signal intensity relative to base-line intensity.

the baseline intensity resulting in a signal intensity normalized to the spectrum baseline.

Polypeptide concentration was plotted versus the GIP_{1-42} or $GLP-1_{7-36}$ MALDI-TOFMS signal intensity normalized to the spectrum baseline (**Fig. 4**). This simple approach resulted in graphs that indicated a polypeptide concentration range over which signal intensity increased with increasing concentration of substance so that concentration could be measured without any need for internal standards. By knowing this unique concentration window, bound by the limit of detection and by the highest normalized signal intensity, the optimum analyte-to-matrix ratio for subsequent sample dilution, was chosen. The molar GIP_{1-42}-to-matrix ratio was optimum at $2.5:10^5$, and the $GLP-1_{7-36}$-to-matrix optimum was $7.5:10^5$.

Degradation of Peptides

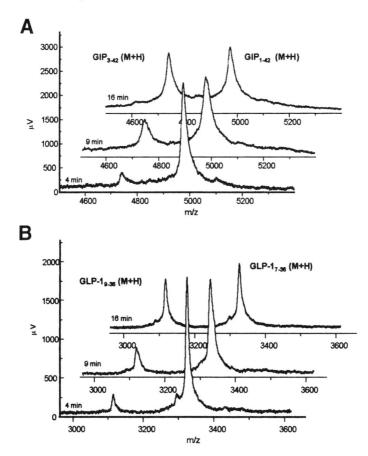

Fig. 5. MALDI-TOFMS analysis of DP IV-catalyzed hydrolysis of GIP_{1-42} (**A**) and $GLP-1_{7-36}$ (**B**).

3.2.3. Monitoring In-Vitro Degradation of GIP_{1-42} and $GLP-1_{7-36}$ by DP IV Using MALDI-TOFMS

3.2.3.1. INCUBATION OF PEPTIDE WITH PURIFIED KIDNEY DP IV

To study the hydrolysis of GIP_{1-42} (5 µM) or $GLP-1_{7-36}$ (15 µM) by DP IV, these peptides were incubated in buffer and enzyme (0.58 nM for GIP_{1-42}, 2.9 nM for $GLP-1_{7-36}$ incubations, respectively) under standard conditions (**step 1, Subheading 3.2.1.**). Samples of GIP_{1-42} (2.5 pmol) or $GLP-1_{7-36}$ (7.5 pmol) were removed from the incubation mixture at 4, 9, and 16 min and prepared for mass spectral analysis as described above (**steps 2–4, Subheading 3.2.1.**).

Figure 5 shows the mass spectra of GIP_{1-42} and $GLP-1_{7-36}$ and their DP IV reaction products at various time intervals during incubation with purified DP

IV. The relative heights of the substrate signal (GIP_{1-42} or $GLP-1_{7-36}$) decreased as the relative heights of the peaks corresponding to the DP IV hydrolysis products (GIP_{3-42} or $GLP-1_{9-36}$) increased. The average m/z values of GIP_{1-42} and GIP_{3-42} were measured as 4980.1 and 4745.2, respectively, representing errors of 0.09 and 0.10% relative to $[M+H]^+_{calc.}$. The errors between $[M+H]^+_{calc.}$ and $[M+H]^+_{exp.}$ for $GLP-1_{7-36}$ and $GLP-1_{9-36}$ were 0.05 and 0.06%, respectively.

3.2.3.2. INCUBATION IN HUMAN SERUM

To gain insight into the identity of the major metabolites found in the circulation, GIP_{1-42} and $GLP-1_{7-36}$ were incubated, as above, in human serum. The mass spectra generated at stated time intervals are shown in **Fig. 6**. Metabolites were identified on the basis of their m/z value. **Table 1** summarizes the $[M+H]^+_{exp.}$ versus the $[M+H]^+_{calc.}$ of possible metabolite sequences. Indistinct minor peaks were not considered for analysis nor were sequences in which the error between $[M+H]^+_{exp.}$ and $[M+H]^+_{calc.}$ was > 0.20%.

Data on the proteolytic degradation of GIP_{1-42} and $GLP-1_{7-36}$ in serum were obtained by incubating each peptide (30 µ*M*) in buffer containing 40% human serum (*see* **Note 4**) under standard conditions (**step 1, Subheading 3.2.1.**). Samples of the respective peptides (15 pmol) were removed from the incubation mixtures at hourly intervals for 15 h and analyzed by mass spectrometry (**steps 2–4, Subheading 3.2.1.**).

Over a 15-h period, serum-incubated GIP_{1-42} showed a consistent and gradual decrease in the relative height of the peak corresponding to the intact peptide with a complementary increase in the relative peak height of a degradation product having an m/z value corresponding to GIP_{3-42}. After only approximately 3 h, by which time more than half of the GIP_{1-42} was already converted to GIP_{3-42}, minor peaks due to secondary stepwise degradation by other serum proteases were observed *(2)*. These results support the hypothesis that DP IV is the primary serum protease acting on GIP. Similarly, serum-incubated $GLP-1_{7-36}$ was degraded by serum DP IV activity to $GLP-1_{9-36}$ (*see* **Note 7**).

3.2.4. Kinetic Analysis Using MALDI-TOFMS

3.2.4.1. HYDROLYSIS WITH VARYING CONCENTRATIONS OF DP IV

The following procedure was adopted to determine the feasibility of studying the time dependence of an enzymatic reaction using MALDI-TOFMS and to establish a convenient DP IV concentration for subsequent kinetic analysis.

1. Incubate GIP_{1-42} (5 µ*M*) or $GLP-1_{7-36}$ (15 µ*M*) under standard conditions (**step 1, Subheading 3.2.1.**) with varying concentrations of purified DP IV. Concentrations of DP IV ranged from 0.29 to 5.8 n*M* for GIP_{1-42} incubations and from 1.5 to 12 n*M* for $GLP-1_{7-36}$ incubations.

Degradation of Peptides

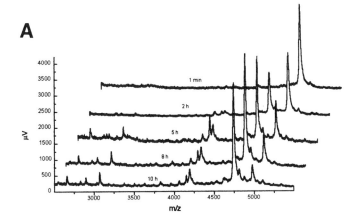

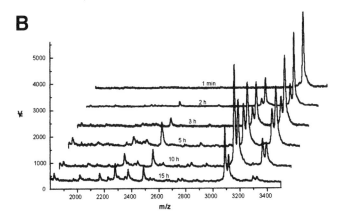

Fig. 6. MALDI-TOFMS analysis of GIP_{1-42} (**A**) and $GLP-1_{7-36}$ (**B**) degradation in serum.

2. Remove samples at timed intervals after the start of the reaction and dilute these so that the final amount of peptide on the sample probe tip is 2.5 pmol for GIP_{1-42} metabolites and 7.5 pmol for $GLP-1_{7-36}$ metabolites.
3. Transfer sample to the probe and record the mass spectrum. Calculate the relative amounts of GIP_{1-42} and $GLP-1_{7-36}$ from net substrate peak intensity divided by the sum of net substrate and net product peak intensities, and plot versus time (net peak intensity is defined as peak intensity minus baseline intensity).
4. Establish the linearity between the rate of hydrolysis and enzyme concentration from a plot of the initial slope of substrate turnover (μmol/L/min) vs. enzyme concentration.

Table 1
GIP and GLP-1 Degradation Products of Serum Protease Activity[a]

GIP degradation (10 h)				GLP-1 degradation (15 h)			
$[M+H]^+$ exp. (m/z)	Sequence	$[M+H]^+$ calc. (m/z)	Difference (%)	$[M+H]^+$ exp. (m/z)	Sequence	$[M+H]^+$ calc. (m/z)	Difference (%)
4975.3	1–42	4975.5	0.00	3323.7	(7–36) + ethyl ester	3325.7	−0.06
4872.4	(1–41) + Na	4869.4	0.06	3296.1	7–36	3297.7	−0.05
4809.8	2–42	4811.4	−0.03	3115.2	(9–36) + ethyl ester	3117.6	−0.06
4740.9	1–40	4745.4	−0.09	3087.9	9–36	3089.6	−0.06
	3–42	4740.4	0.01	2884.3	9–34	2879.4	0.17
4527.2	1–38	4518.3	0.20		(7–32) + ethyl ester	2885.4	−0.04
4462.8	2–39	4469.2	−0.14	2855.7	7–32	2857.4	−0.06
4192.4	8–42	4193.1	−0.02	2771.0	(7–31) + ethyl ester	2772.3	−0.04
4149.7	3–37	4147.1	0.06		(10–34) + ethyl ester	2775.4	−0.16
	4–38	4155.1	−0.13		(11–35) + ethyl ester	2775.4	−0.16
4062.0	1–35	4067.0	−0.12	2743.8	7–31	2744.3	−0.02
	8–41	4065.1	−0.08		9–33	2748.3	−0.16
3955.5	1–34	3952.0	0.09		10–34	2747.4	−0.13
	5–37	3961.0	−0.14		11–35	2747.4	−0.16
3824.6	11–42	3828.0	−0.09	2562.3	7.30	2558.2	0.16
3740.6	1–32	3736.9	0.10		11–33	2562.3	0.00
	8–38	3736.9	0.10		(9–31) + ethyl ester	2564.2	−0.07
	12–42	3741.0	−0.01	2534.9	9–31	2536.2	−0.05
3629.6	3–33	3630.9	−0.04	2487.2	7–29	2487.2	0.00
	13–42	3627.9	0.05	2375.0	7.28	2374.1	0.04
3560.2	14–42	3556.9	0.09		(8–29) + ethyl ester	2378.1	−0.13
3502.9	3–32	3502.8	0.00		(9–30) + ethyl ester	2378.1	−0.13
	4–33	3501.8	0.03		(11–31) + ethyl ester	2378.1	−0.13
	13–41	3499.8	0.09	2353.4	8–29	2350.1	0.14

3421.2	1–29	3423.7	-0.07		9–30	2350.1	0.14
	15–42	3425.8	-0.13		11–31	2350.1	0.14
3373.2	3–31	3374.7	-0.04		16–36	2352.3	0.05
	4–32	3373.7	-0.01	2305.8	(9–29) + ethyl ester	2307.6	-0.08
	11–38	3371.8	0.04	2279.5	9–29	2279.6	0.00
3069.0	7–32	3067.6	0.05	2226.1	7–27	2227.0	-0.04
	18–42	3068.6	0.01	2166.3	9.28	2166.0	0.01
2896.1	6–29	2901.5	-0.19		11–30	2164.0	0.11
2827.5	8–31	2826.4	0.04		18–36	2166.2	0.00
	12–35	2832.5	-0.18	2019.2	9–27	2018.9	0.01
	15–37	2831.5	-0.14	1971.0	7–25	1969.9	0.06
2657.2	20–41	2656.4	0.03	1828.5	7–23	1826.8	0.09
	21–42	2656.4	0.03		8–25	1831.8	-0.18
2544.4	22–42	2441.1	0.12		11–27	1831.9	-0.19
					18–33	1825.0	0.19

[a]GIP_{1-42} and $GLP-1_{7-36}$ were incubated with 40% human serum in 0.1 mM tricine buffer, pH 7.0, at 30°C for 10 and 15 h, respectively. MALDI-TOFMS analysis after this incubation period showed serum degradation products identified on the basis of their m/z ratios. Where more than one possible sequence of similar m/z is possible, all alternatives are given.

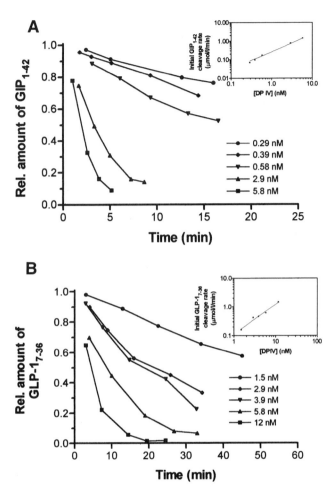

Fig. 7. Quantification of DP-IV catalyzed hydrolysis of GIP_{1-42} (**A**) and $GLP1_{7-36}$ (**B**) using MALDI-TOF-MS.

Under normal circumstances increasing the concentration of an enzyme while maintaining a constant substrate concentration results in an increased rate of product formation. **Figure 7** illustrates that MALDI-TOFMS analysis of DP IV-catalyzed hydrolysis of GIP_{1-42} or $GLP-1_{7-36}$ can be used to demonstrate this relationship. Peptide turnover varies linearly with increasing concentrations of DP IV (**Fig. 7**, insets; $R^2 = 0.9986$ and 0.9849 for GIP_{1-42} and $GLP-1_{7-36}$ hydrolyses respectively).

3.2.4.2. Determination of Kinetic Constants

The kinetic constants of DP IV-catalyzed GIP_{1-42} and $GLP-1_{7-36}$ hydrolysis were determined by introducing a specific and kinetically characterized DP IV inhibitor into the incubation mixture and observing the relative reaction rates of inhibited and uninhibited substrate hydrolysis as described by Crawford et al. *(16)*. Thus, GIP_{1-42} (20 µM) or $GLP-1_{7-36}$ (30 µM) were incubated with DP IV (0.59 and 2.9 nM, respectively) under standard conditions (0.1 mM Tricine buffer at pH 7.6 and 30°C, see **Subheading 3.2.1.**), in the presence or absence of either Ala-thiazolidide (20 µM—K_i [inhibition binding constant] of 3.4 µM) or Ile-thiazolidide (20 µM—K_i of 0.126 µM). Both are specific, competitive inhibitors of DP IV that have been synthesized in our laboratory *(17,18)*. Similarly, GIP_{1-42} (30 µM) and $GLP-1_{7-36}$ (30 µM) were incubated with 40% human serum (*see* **Subheading 3.2.3.2.**) in the presence or absence of the same inhibitors. Samples were appropriately diluted and then assayed by mass spectrometry. Quantification of the relative amounts of substrate after timed intervals was performed as described in **Subheading 3.2.4.1.**

The initial slopes of plots showing peptide turnover with purified DP IV or human serum DP IV in the presence and absence of inhibitors were used to calculate reaction velocities. To validate this approach, it was applied using different competitive inhibitors exhibiting fairly different K_i values (more than one order of magnitude different). The K_m value of DP IV-catalyzed peptide hydrolysis was calculated according to the equation:

$$K_m = [(v_o/v_i - 1)S]/[1 + (I/K_i) + (v_o/v_i)]$$

where v_o and v_i are the uninhibited and inhibited relative reaction rates, respectively, S is the substrate concentration, I is the inhibitor concentration, and K_i is the inhibition binding constant. V_{max} was then calculated according to the equation:

$$V_{max} = (v/S)(K_m + S)$$

Values for k_{cat} were calculated using M = 110 kDa per catalytically active subunit as the molar mass of DP IV. To estimate these kinetic constants for serum DP IV activity, it was necessary to determine the concentration of purified DP IV equivalent to human serum DP IV activity (*see* **Note 8**).

MALDI-TOFMS was used to demonstrate that GIP_{1-42} and $GLP-1_{7-36}$ turnover was attenuated by Ala-thiazolidide and Ile-thiazolidide inhibition of purified DP IV or of serum DP IV as predicted by the inhibitor binding constants (K_i) (**Fig. 8**). These results lend more credibility to MALDI-TOFMS as a feasible method for quantitative kinetic analysis, as well as allowing the K_m values for purified porcine kidney-catalyzed GIP_{1-42} and $GLP-1_{7-36}$ hydrolysis to be calculated. These results are summarized in **Table 2** and, where appropriate, expressed as a range derived from the two inhibitors.

Serum DP IV was determined to have an equivalent activity of $1.3 \cdot 10^{-5}$ mg/mL of purified porcine kidney (*see* **Note 8**), as measured by the rate of H-Gly-Pro-

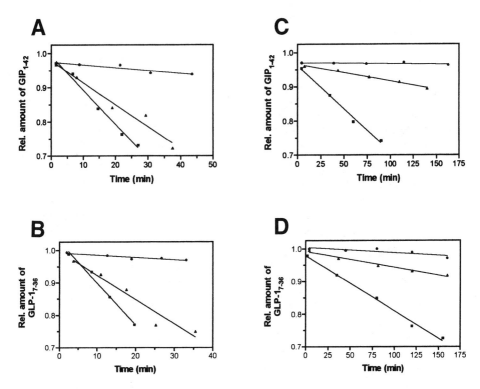

Fig. 8. Quantification of DP IV-catalyzed GIP_{1-42} and $GLP-1_{7-36}$ hydrolysis in the presence of specific DP IV inhibitors using MALDI-TOFMS. (**A, B**) GIP_{1-42} and $GLP-1_{7-36}$ were incubated with purified porcine kidney DP IV. (**C, D**) GIP_{1-42} and $GLP-1_{7-36}$ were also incubated in 40% human serum under the same experimental conditions. *Squares*, substrate turnover in the absence of inhibitor; *triangles* and *circles*, turnover in the presence of alanine-thiazolidide and isoleucine-thiazolidide, respectively.

4-nitroanilide hydrolysis using the standard curve in **Fig. 9**. The kinetic constants (k_{cat}) for GIP_{1-42} and $GLP-1_{7-36}$ hydrolysis for serum DP IV activity were calculated and are compared in Table 2. The binding constant of GIP_{1-42} derived from the competitive inhibition of porcine kidney DP IV-catalyzed hydrolysis of H-Gly-Pro-4-nitroanilide (*see* **Note 9**) was found to be 54 ± 8 μM (mean ± standard error).

4. Notes

1. Prolyl endopeptidase was purchased as lyophilized powder from Miles. Its specific activity was 14 U/mg using H-Gly-Pro-Pro-4-nitroanilide as chromogenic substrate *(19)*.

Table 2
Kinetic Constants for the DP IV-Catalyzed Hydrolysis of GIP_{1-42} and $GLP-1_{7-36}$ by DP IV as Determined by Quantitative MALDI-TOFMS

Peptide	DP IV source	K_m (μM)	V_{max} ($\mu mol \cdot min^{-1} \cdot mg^{-1}$)	k_{cat} (s^{-1})	k_{cat}/K_m ($M^{-1} \cdot s^{-1}$)	Reference
GIP_{1-42}	Porcine kidney	1.8 ± 0.3	13.6 ± 0.2	23	13 · 10⁶	This study
GIP_{1-42}	Human serum	39 ± 29	27 ± 12	22	0.56 · 10⁶	This study
GIP_{1-42}	Human placenta	34 ± 3	3.8 ± 0.2	7.6	0.22 · 10⁶	Mentlein et al., 1993 (9)
$GLP-1_{7-36}$	Porcine kidney	3.8 ± 0.3	5.45 ± 0.05	9	2.3 · 10⁶	This study
$GLP-1_{7-36}$	Human serum	13 ± 9	11 ± 2	14	1.1 · 10⁶	This study
$GLP-1_{7-36}$	Human placenta	4.5 ± 0.6	0.97 ± 0.05	1.9	0.43 · 10⁶	Mentlein et al., 1993 (9)

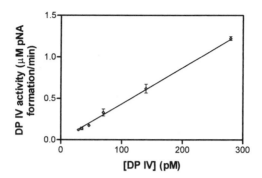

Fig. 9. Standard curve for adjusting human serum DP IV activity using purified porcine kidney DP IV.

2. The DP IV used in this study was purified from porcine kidney according to a previously described method *(19)*. The specific activity measured using H-Gly-Pro-4-nitroanilide as a chromogenic substrate, was 45 U/mg.
3. Serum was pooled from three individuals and was obtained from the Medical Science Division (courtesy of Dr. S. Heins, Department of Child Diseases), Martin-Luther University, Halle-Wittenberg, Germany.
4. The pyridinium methyl ketone derivatives **II** were synthesized by treatment of the corresponding Z-Phe-Pro-chloromethyl ketone **I** with the appropriate base in acetone or dimethylsulfoxide for several days (**Scheme 1**). The reaction was monitored by MALDI-TOFMS. The peptidyl chloromethyl ketone as starting material was prepared according to known standard procedures *(7,20)*. The purity and the structure of the compounds were verified by HPLC, mass spectrometry, and NMR (**Table 3**). The hydrolytic stability of all compounds dissolved in buffer-free, slightly acidic water was analyzed by HPLC and by MALDI-TOFMS. It was found that all compounds are stable under these conditions up to 24 h at 1°C.
5. The enzyme kinetic experiments were conducted at 30°C, in 0.04 M HEPES buffer, having a pH of 7.6 and an ionic strength of 0.125 (KCl). Enzyme-catalyzed hydrolysis of H-Gly-Pro-Pro-4-nitroanilide by bacterial PEP in the presence of inhibitor was analyzed on a Kontron (Zürich, Switzerland) Uvicon 930 UV-VIS spectrophotometer, equipped with a thermostated cell compartment. The slow-binding kinetics of the inhibition of PEP by the peptidyl pyridinium methyl ketone derivatives were analyzed as extensively described in refs. *7* and *21*.

Variation of the inhibitor concentration in the presence of a fixed substrate concentration gives a hyperbolic dependence of k on I according to the equation:

$$k = k_{off} \{1 + I/[K_i^* (1 + S/K_m)]/[1 + I/(K_i (1 + S/K_m)]\},$$

where k_{off} is the rate constant for complex dissociation and k_{on} may be calculated from $k_{on} = k_{off} ((K_i/K_i^*) - 1)$ (*see* footnotes to **Table 4** for additional information).

Table 3
Analytical Data for Various Z-Phe-Pro-pyridinium Methyl Ketone Derivatives

No.	Compound	Mw.[a]	MS [M+H]$^+$	HPLC-R_t (min)[b]
II-1	-H	507.5	472.9	3.4
II-2	-COOH	551.5	516.8	2.7
II-3	-COOCH$_3$	565.5	531.1	3.9

[a]Masses calculated including Cl⁻ as counter ion.
[b]Retention time on an HPLC column, RP 18, LiChrospher 100, 125 × 4 mm, 5-μm particle size (Merck). Solvent system: acetonitrile/water containing 0.04% TFA. Gradient: 30–80% acetonitrile in 25 min. Flow rate: 1.5 mL/min.

Table 4
Kinetic Parameters of the Slow-Binding Inhibition of PEP by Z-Phe-Pro-pyridinium Methyl Ketones (Analysis of Curves According to Morrisson and Walsh [21])

No.	$K_i{}^a$ (M)	K_i^{*b} (M)	$k_{on}{}^c$ (s^{-1})	k_{on}/K_i (M^{-1}s^{-1}) · 10^{-3}
II-1	8.4E-7 (±1.3E-7)	1.1E-7 (±1.8E-8)	1.42E-3 (±2.6E-4)	12.8 E+3 (±2.3)
II-2	2.0E-7 (±3.0E-9)	6.7E-8 (±4.1E-7)	1.68E-3 (±3.0E-4)	25.0 E+3 (±4.5)
II-3	4.8E-8 (±6.9E-9)	1.8E-8 (±2.9E-9)	3.00E-3 (±5.3E-4)	167.0 E+3 (±29.9)

[a]K_i, represents the competitive inhibition of the enzyme-catalyzed substrate hydrolysis before forming the tighter enzyme-inhibitor complex (compare **Fig. 1**, curve 2, designated part A).

[b]K_i^*, represents the complex competitive inhibition of the enzyme-catalyzed substrate hydrolysis after forming the tighter enzyme-inhibitor complex and equilibrium of all enzyme species.

[c]k_{on} (see **Note 5**), pseudo-first-order rate constant of complex formation (compare **Fig. 1**, curve 2, designated part B).

Incubation of prolyl endopeptidase with the three differently substituted Z-Phe-Pro-pyridinium methyl ketones in the presence of substrate resulted in time-dependent delay of the PEP-catalyzed hydrolysis of H-Gly-Pro-Pro-4-nitroanilide (see **Fig. 1**). The time course of curve 1 is characteristic of a slow-binding type of enzyme inhibition *(7,21)*. By variation of the inhibitor concentrations at different substrate concentrations, the kinetic parameters of slow-binding inhibition according to Morrison and Walsh *(21)* were estimated (**Table 4**).

6. We found that DHAP in a mixture with diammonium hydrogen citrate (DHC) showed a behavior superior to that of other matrices, because of its excellent crystallization, resulting in good mass resolution and highly reproducible mass spectra.

7. The serum degradation spectra for GLP-1$_{7-36}$ at different time periods are illustrated in **Fig. 5B** and show doublet peaks for both GLP-1$_{7-36}$ and GLP-1$_{9-36}$. The m/z (mass) difference between these doublets was consistently 28, corresponding

to an ethyl group most likely attached as a protecting group to a glutamate residue during peptide synthesis of the commercial product.

As the incubation time increased, the height of the mass peaks corresponding to [M+H]$^+$ for the ethyl ester of GLP-1$_{7-36}$ decreased relative to the height of the peaks corresponding to unesterified GLP-1$_{7-36}$. This suggests that nonspecific serum esterases remove the ethyl group over time. Parallel studies of GLP-1$_{7-36}$, using the same commercially available substance, but with purified DP IV did not result in doublet peaks, and only [M+H]$^+$ for the ethyl ester of GLP-1$_{7-36}$ was seen (**Fig. 5**). Presumably this occurs because the purified enzyme preparation is free of contaminating nonspecific esterases.

8. A standard curve of DP IV activity versus DP IV concentration was generated by incubating 50 µL of various DP IV concentrations (ranging from 29.3 to 293 pM) in 0.04 M HEPES buffer, pH 7.6, at 30°C and monitoring the rate of H-Gly-Pro-4-nitroanilide (4 µL) hydrolysis (**Fig. 9**; R^2 = 0.9979). Data acquisition was carried out using a Kontron 930 Uvicon UV-VIS spectrophotometer, equipped with thermostated cells, at 390 nm (ε = 11,500 $M^{-1} \cdot cm^{-1}$). An equivalent volume of serum was assayed under identical conditions, allowing the purified DP IV concentration equivalence of serum DP IV activity to be determined using the standard curve. It was assumed that the major serum DP IV isoenzyme has a molecular mass of M = 110 kDa per catalytically active subunit.

9. To confirm the kinetic constants determined using MALDI-TOFMS, the inhibition constant of GIP$_{1-42}$ as a competitive substrate of the DP IV-catalyzed hydrolysis of a chromogenic substrate was determined spectrophotometrically. Three concentrations of H-Gly-Pro-4-nitroanilide corresponding to $K_m/2, K_m$, and $2K_m$ (5.0 · 10^{-5}, 1.0 · 10^{-4} and 2.0 · 10^{-4} M) were incubated in 0.04 HEPES buffer (pH 7.6) at 30°C in the presence of a range of GIP$_{1-42}$ concentrations (1.0 · 10^{-7} to 1.0 · 10^{-5} M). Hydrolysis of the chromogenic substrate was monitored using the Kontron 930 Uvicon UV-VIS spectrophotometer as outlined in **Note 8**. Data were analyzed using nonlinear regression (Graphfit 3.01) yielding an inhibition binding constant (K_i) for GIP$_{1-42}$. Since GIP$_{1-42}$ is simultaneously an inhibitor to DP IV-catalyzed H-Gly-Pro-4-nitroanilide hydrolysis, as well as a substrate of DP IV, this inhibition binding constant should be an approximation to the K_m value for DP IV-catalyzed GIP$_{1-42}$ hydrolysis.

References

1. Schmidt, J., Wermann, M., Rosche, F., and Demuth, H.-U. (1996) The use of MALDI-TOF mass spectrometry in quantification of the stability of prolyl endopeptidase inhibitors. *Protein Pep. Lett.*, **3**, 385–392.
2. Pauly, R., Rosche, F., Wermann, M., McIntosh, C. H. S., Pederson, R. A., and Demuth, H. U. (1996) Investigation of glucose-dependent insulinotropic polypeptide (1–42) and glucagon-like peptide-1-(7–36) degradation *in vitro* by dipeptidyl peptidase IV using matrix-assisted laser desorption/ionization—time of flight mass spectrometry: a novel kinetic approach. *J. Biol. Chem.* **271**, 23,222–23,229.

3. Jespersen, S., Niessen, W. M. A., Tjaden, U. R., and van der Greef, J. (1995) Quantitative bioanalysis using matrix-assisted laser desorption/ionization mass spectrometry. *J. Mass Spectrom.* **30,** 357–364.
4. Wu, J. Y., Chatman, K., Harris, K., and Siudzak, G. (1997) An automated MALDI mass spectrometry approach for optimizing cyclosporin extraction and quantitation. *Anal. Chem.* **69,** 3767–3771.
5. Wilkinson, W. R., Gusev, A. I., Proctor, A., Houalla, M., and Hercules, D. M. (1997) Selection of internal standards for quantitative analysis by matrix-assisted laser desorption-ionization (MALDI) time-of-flight mass spectrometry. *Fresenius J. Anal. Chem.* **357,** 241–248.
6. Yaron, A. and Naider, F. (1993) Proline-dependent structural and biological properties of peptides and proteins. *Crit. Rev. Biochem. Mol. Biol.* **28,** 31–81.
7. Steinmetzer, T., Silberring, J., Mrestani-Klaus, C., Fittkau, S., Barth, A., and Demuth, H.-U. (1993) Ammonium methyl ketones as substrate analog inhibitors of proline-specific peptidases. *J. Enzym. Inhib.* **7,** 77–85.
8. Demuth, H.-U. and Heins, J. (1995) On the catalytic mechanism of dipeptidyl peptidase IV, in *Dipeptidyl Peptidase IV* (CD26) *in Metabolism and the Immune Response* (Fleischer, B., ed.), R. G. Landes, Georgetown, pp. 1–37.
9. Mentlein, R., Gallwitz, B., and Schmidt, W. E. (1993) Dipeptidyl-peptidase IV hydrolyzes gastric inhibitory polypeptide, glucagon-like peptide-1(7–36)amide, peptide histidine methionine and is responsible for their degradation in human serum. *Eur. J. Biochem.* **214,** 829–835.
10. Kieffer, T. J., McIntosh, C. H. S., and Pederson, R. A. (1995) Degradation of glucose-dependent insulinotropic polypeptide and truncated glucagon-like peptide 1 in vitro and in vivo by dipeptidyl peptidase IV. *Endocrin*ology **136,** 3585–3596.
11. Brown, J. C. (1994) Enteroinsular axis, in *Gut Peptides* (Walsh, J. H. and Dochray, G. J., eds.), Raven Press, New York, pp. 765–784.
12. Schmidt, W. E., Siegel, E. G., Ebert, R., and Creutzfeldt, W. (1986) N-terminal tyrosine-alanine is required for the insulin-releasing activity of glucose-dependent insulinotropic polypeptide (GIP). *Eur. J. Clin. Invest.* **16,** A9.
13. Brown, J. C., Dahl, M., McIntosh, C. H. S., Otte, S. C., and Pederson, R. A. (1981) Actions of GIP. *Peptides* **2** (Suppl. 2), 241–245.
14. Deacon, C. F., Johnsen, A. H., and Holst, J. J. (1995) Degradation of glucagon-like peptide-1 by human plasma *in vitro* yields an N-terminally truncated peptide that is a major endogenous metabolite *in vivo*. *J. Clin. Endocrinol. Metab.* **80,** 952–957.
15. Pederson, R. A., White, H. A., Schlenzig, D., Pauly, R. P., McIntosh, C. H. S., and Demuth, H. U. (1998) Improved glucose tolerance in Zucker fatty rats by oral administration of the dipeptidyl peptidase IV inhibitor isoleucine thiazolidide. *Diabetes* **47,** 1253–1258.
16. Crawford, C., Mason, R. W., Wikström, P., and Shaw, E. (1988) The design of peptidyldiazomethane inhibitors to distinguish between the cysteine proteinases calpain II, cathepsin L and cathepsin. B. *Biochem. J.* **253,** 751–758.
17. Demuth, H.-U. (1990) Recent developments in the irreversible inhibition of serine and cysteine proteases. *J. Enzym. Inhib.* **3,** 249–278.

18. Schön, E., Born, I., Demuth, H.-U., Faust, J., Neubert, K., Steinmetzer, T., et al. (1991) Dipeptidyl peptidase IV: Einfluß von DP IV Effektoren auf die Enzymaktivität und die Proliferation humaner Lymphozyten. *Biol. Chem. Hoppe-Seyler* **372,** 305–311.
19. Wolf, B., Fischer, G., and Barth, A. (1978) Kinetics of dipeptidyl peptidase IV. *Acta Biol. Med. Germ.* **37,** 409–420.
20. Wünsch, E. (1974) in *Synthesis of Peptides, Houben-Weyl, vol. 15/I, Methods in Organic Chemistry* (Müller, E., ed.), Georg-Thieme-Verlag, Stuttgart.
21. Morrison, J. and Walsh, C. T. (1988) The behavior and significance of slow-binding enzyme inhibitors. *Adv. Enzymol.* **61,** 201–301.

16

Characterization of Protein Glycosylation by MALDI-TOFMS

Ekaterina Mirgorodskaya, Thomas N. Krogh, and Peter Roepstorff

1. Introduction

Glycosylation is one of the most common modifications in proteins. Among all characterized proteins, more than 50% are glycoproteins. During the last two decades, extensive knowledge about functional aspects of the glycans attached to the glycoproteins has been obtained *(1,2)*. Structural characterization of glycoproteins, including glycan structure elucidation, is important for understanding their functions. Furthermore, characterization of glycosylation patterns of recombinant glycoproteins is of importance for both industry and pharmaceutical research.

The structural diversity of glycoproteins, with respect to the glycan structures and the attachment sites, is enormous *(1,2)*. Each attachment site may be occupied by different glycans, due to the ability of monosaccharides to combine with each other in a variety of ways differing in sequence, linkages and branching points, resulting in different glycoforms of the same protein. In addition, because glycosylation is a posttranslational modification, it is species and cell specific. According to the nature of the linkage between the oligosaccharide chains and the polypeptide, the glycans of glycoproteins can be divided into two major groups, the N-linked and the O-linked glycans.

N-linked glycans are generally large structures (7–25 monosaccharide units), linked to the amide group of asparagine residues (Asn) at the well-defined consensus sequence Asn-Xxx-Ser/Thr/Cys, where Xxx can be any amino acid except Pro. N-linked glycans share the common core structure, shown in bold type in **Fig. 1**. Based on the glycan structures attached to the common core, mammalian N-linked glycans can then be classified into three major subgroups:

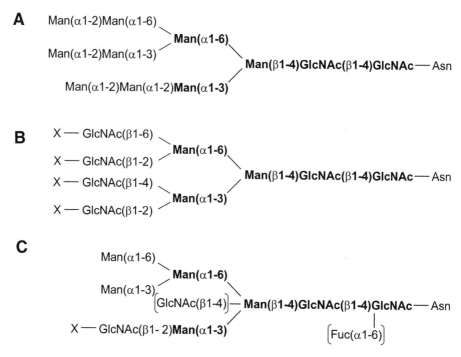

Fig. 1. The three major subgroups of N-linked glycans (**A**) High-mannose-type oligosaccharides. (**B**) Complex-type oligosaccharides. (**C**) Hybrid-type oligosaccharides. The structures shown in bold type indicate a trimannosyl core common to all N-linked glycans. X can denote hydrogen or any monosaccharide residue. The most frequent elongations are galactose Galβ1→, or disaccharides such as NeuAc-Galβ1→ or Fuc-Galβ1→. In addition, complex- and hybrid-type glycans may exist in fucosylated form [Fuc(α1–6) linked to the innermost GlcNAc residue of the core] or bisected form [GlcNAc(β1–4) linked to the β-mannose residue of the core] as shown in (**C**) in square brackets.

high-mannose-type, complex-type, and hybrid-type oligosaccharides (**Fig. 1**). High-mannose-type oligosaccharides contain only α-mannosyl residues attached to the core structure. The total number of mannose residues is often five to nine (including the three in the core) in most mammalian glycoproteins. Complex-type oligosaccharides have *N*-acetylglucosamine residues (GlcNAc) linked to the two α-mannosyl residues of the core. These GlcNAc residues are often elongated by galactose (Gal) or by a disaccharide such as sialic acid → galactose (NeuAc → Gal). Hybrid-type oligosaccharides have structures composed of elements characteristic of both high-mannose and complex-type glycans. One or two α-mannosyl residues are linked to the core Man(α1–6) arm and one or more GlcNAc-initiated antennae are linked to the core Man(α1–3)

Protein Glycosylation by MALDI-TOFMS

arm. In addition, complex and hybrid-type glycans may exist in a fucosylated form (i.e., Fuc(α1–6) linked to the innermost GlcNAc residue of the core) or a bisected form (i.e., GlcNAc(β1–4) linked to the β-mannose residue of the core).

O-linked glycans are primarily linked to the hydroxyl groups of serine (Ser) and threonine (Thr) residues, but linkages to hydroxyproline (Hyp), hydroxylysine (Hyl), or tyrosine (Tyr) residues may also occur. Unlike N-linked glycans, O-linked oligosaccharides do not share a common core structure and vary from a single monosaccharide to large structures. Several different monosaccharides can form the O-glycosidic linkage to the polypeptide chain (1). Glycans with N-acetylgalactosamine (GalNAc) linked to the hydroxyl group of Ser or Thr residues form one of the most common types of mammalian O-glycosylation, also known as mucin-type. No universal consensus sequence has been found for O-linked glycosylation, although several sequence motifs for certain types of O-glycosylation have been reported. The sequence motif Cys-Xxx-Xxx-Gly-Gly-Thr/Ser-Cys exists for O-fucosyl-type glycosylation of Ser and Thr (3). Glycosylation of hydroxylysine in collagen is associated with the Gly-Xxx-Hyl-Gly sequence (4). The importance of certain amino acids in the sequence around mucin-type (5) and GlcNAc-type (6) O-glycosylated sites has been reported. Although computer programs for prediction of potential O-glycosylation sites are available (7), such predictions are difficult and, at best, tentative.

Since the development of matrix-assisted laser desorption/ionization (MALDI) and electrospray ionization (ESI) techniques, the role of mass spectrometry in the analysis of glycoproteins has increased dramatically (8–10). Both techniques have been successfully applied to the analysis of glycoproteins, glycopeptides, and oligosaccharides. The strength of coupled liquid chromatography ESI-mass spectrometry (LC/ESI-MS) is the ability to perform sensitive and selective detection of posttranslational modification in proteins (11,12). MALDI time-of-flight (TOF) mass spectrometry is advantageous for the direct analysis of complex mixtures. Tolerance to sample contaminants, such as buffer salts and detergents, allows direct monitoring of enzymatic digests. Mass spectrometric determination of glycosylated sites by peptide mapping and elucidation of glycan structure using specific exo- and endoglycosidases has proved to be an efficient strategy for characterization of glycoproteins (13–17). In addition, direct mass spectrometric sequencing of glycans derived from glycoproteins using postsource decay (PSD) analysis (18–20) or tandem mass spectrometry (MS/MS) (12,21,22) has been applied to glycan structure elucidation.

This chapter provides protocols for characterization of protein glycosylation, developed in our laboratory during structural characterization of a number of glycoproteins (16,23–28). Procedures and examples are given for determination of glycosylation sites based on mass spectrometric peptide mapping, and for glycan structure elucidation using sequential exoglycosidase digestion followed by MALDI-TOFMS analysis. A chemical method for the determination of O-glycosylated sites in mucin-type glycopeptides is also described. In addition, general analytical considerations for the analysis of glycopeptides and glycoproteins by MALDI-TOFMS are discussed.

2. Materials

2.1. Glycoproteins and Glycopeptides

1. Recombinant glucoamylase *(29)*.
2. Murine fetal antigen 1 (mFA1) *(26)*.
3. In vitro GalNAc-T3 glycosylated mucin-derived peptide Muc1a' *(30)*.

2.2. Solvents and Chemicals

1. Ultra-high-quality (UHQ) water (Millipore, Molsheim, France).
2. High-performance liquid chromatography (HPLC)-grade trifluoroacetic acid (TFA) and acetonitrile (CH_3CN), both from Rathburn, Walkerburn, Scotland).
3. 25 mM sodium phosphate buffer pH 7.4 (Merck, Darmstadt, Germany).
4. 50 mM ammonium acetate pH 5.0 (Merck).
5. 50 mM ammonium bicarbonate pH 7.8 (Merck).
6. Pentafluoropropionic acid (PFPA; Sigma, St. Louis, MO).
7. 1,4-Dithiothreitol (DTT; Sigma).

2.3. Glycosidases

All glycosidases were purchased from Boehringer Mannheim (Mannheim, Germany).

1. Peptide-N-glycosidase (PNGaseF).
2. Neuraminidase, *Arthrobacter ureafaciens*.
3. β-Galactosidase, *Diplococcus pneumonia*.
4. *N*-acetyl-β-D-glucosaminidase, *D. pneumonia*.

2.4. MALDI Matrices

1. 4-Hydroxy-α-cyanocinnamic acid (HCCA; Sigma), 10 μg/μL dissolved in 70% aqueous acetonitrile containing 0.1% TFA.
2. 2,5-Dihydroxybenzoic acid (DHB; Hewlett-Packard, Palo Alto, CA). Lyophilize 100 μL of 100 mM DHB solution (supplied in methanol + water) and redissolve in an equal volume of 30% aqueous acetonitrile containing 0.1% TFA.

2.5. Equipment

1. Eppendorf vials (500-μL capacity).
2. Glass vial with mininert valve, 22 mL capacity (Pierce, Rockford, IL).
3. Vacuum centrifuge.

Protein Glycosylation by MALDI-TOFMS

2.6. Mass Spectrometry

Mass spectra were acquired on either a Voyager-Elite MALDI-TOF (PerSeptive Biosystems, Framingham, MA) or a Bruker Reflex MALDI-TOF (Bruker-Franzen, Bremen, Germany) mass spectrometer, equipped with delayed extraction technology.

HPLC-purified samples were prepared for MALDI analysis by mixing 1 µL of an HPLC fraction with 0.5 µL of matrix solution (DHB) directly on the target. Samples containing buffer salts were acidified to improve matrix crystallization. Thus, 0.5 µL of sample solution (2–5 pmol/µL) was mixed with 0.5 µL of 0.1% TFA and 0.5 µL of matrix solution (DHB or HCCA) directly on the target.

3. Methods
3.1. Analytical Considerations

Due to glycan microheterogeneity, MALDI-TOF mass spectra of glycopeptides and glycoproteins generally contain multiple signals corresponding to different glycoforms. It is known that molecular ions formed during the MALDI process, especially from sialylated glycopeptides and glycoproteins, may undergo prompt or metastable fragmentation, resulting in the loss of one or several monosaccharide units *(13,16,31)*. It is therefore important to ascertain that any observed heterogeneity is initially present in the sample and is not an artifact of the MALDI process. The occurrence and degree of fragmentation depend on the analyte, instrumental parameters, and MALDI matrix. In general, the following items (*see* **Subheadings 3.1.1.–3.1.4.**) should be considered when analyzing glycopeptides and glycoproteins.

3.1.1. Choice of Matrix

Among the commonly used MALDI matrixes, such as HCCA, sinapinic acid, DHB, 2,4,6-trihydoxyacetophenone (THAP), and 3-hydroxypicolinic acid (HPA), DHB is often the matrix of choice for the analysis of glycopeptides and glycoproteins. However, this is dependent on the nature of the analyte and on the possible presence of certain buffer salts or detergents (*see* **Notes 1–4**).

3.1.2. Prompt Fragmentation

Prompt fragmentation occurs in the ion source within a time frame comparable with the generation time of the precursor ions. The generated fragments will have flight times corresponding to their authentic *m/z* values, in both a linear TOF or a reflectron TOF analyzer. However rarely observed, this fragmentation may result in detection of glycoforms that are not initially present in the sample. The presence of prompt fragmentation can be confirmed by increasing the laser fluence. If prompt fragmentation is present, this increase results in a decreased signal intensity of the precursor ions and an increased signal intensity for the corresponding fragment ions. Prompt fragmentation can be avoided by decreasing the laser fluence, or using an active matrix compound that does not induce fragmentation of glycoconjugates, such as HPA or THAP *(31)*.

3.1.3. Metastable Fragmentation

Metastable fragmentation occurs within a time frame that is long compared with the ion acceleration time, resulting in decomposition of precursor ions that have left the ion source and are in the field-free drift region of the analyzer. The resulting fragment ions travel with the same velocity as the corresponding precursor ions and have flight times that correspond to the m/z value of the precursor ion in linear TOF. Having lower kinetic energy (the same speed but less mass) they will be time-dispersed in a reflectron TOF instrument. Metastable fragmentation may thus result in the detection of artifact glycan structures in reflectron TOF analysis, due to the loss of labile groups. Comparison of linear and reflectron TOF mass spectra reveals the occurrence of metastable fragmentation. Again, just as for prompt fragmentation, metastable fragmentation can be suppressed by using an active matrix compound that does not induce extensive fragmentation of glycoconjugates.

3.1.4. Signal Suppression of Glycopeptides in Mixtures

When analyzing complex peptide mixtures, signal suppression of some peptides is often observed. In addition, due to the microheterogeneity of glycans, the total signal intensity of a glycopeptide is distributed over all glycoforms present, resulting in an overall lower detection sensitivity. The combination of these two factors may impair the detection of glycopeptides in a mixture. Therefore a separation step is often necessary prior to mass spectrometric analysis. Strong signal suppression is often observed for sialylated glycopeptides in positive-ion analysis due to the lower efficiency of positive-ion formation for acidic oligosaccharides, thus preventing their detection in a mixture *(13)*. Negative-ion analysis, using DHB, often considerably improves the detection of sialylated glycopeptides.

3.2. MALDI-TOFMS of Intact Glycoproteins

Studies of intact glycoproteins above 30 kDa are limited by the mass resolution of current TOF mass analyzers, in that different glycoforms of the glycoprotein cannot be resolved. However, even in such cases, the observed mass and peak shape can reveal alterations in the carbohydrate components and may be used, for example, for quality control of recombinant glycoproteins.

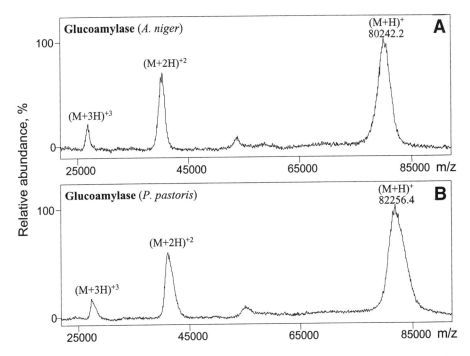

Fig. 2. MALDI-linear-TOF mass spectra of glucoamylase expressed in two different yeast hosts. (**A**) *Aspergillus niger* (commercial preparation). (**B**) *Pichia pastoris*. (Adapted from **ref. 29**.)

For example, MALDI-TOF mass spectra of a multi-domain N-and O-glycosylated glucoamylase, expressed in two different yeast hosts (*Pichia pastoris* and *Aspergillus niger*) are shown in **Fig. 2**. The average molecular mass of the recombinant glucoamylase expressed in *P. pastoris*, determined by MALDI-TOFMS analysis, was higher than the molecular mass of a commercial preparation of the same glucoamylase, expressed in *A. niger* (82,256 and 80,242 Daltons, respectively). The different average masses and peak widths indicate a difference in the carbohydrate components between these proteins. These differences may be due to variations in the efficiency of glycosylation during biosynthesis, or to the extent of deglycosylation by endogenous α-mannosidases. The difference in the observed M_r values corresponds to 14.4 hexose residues, in agreement with results obtained by neutral sugar analysis, which indicated a difference of 12.3 neutral hexose residues between the two glycoproteins *(29)*.

3.3. Site-Specific Characterization of Protein Glycosylation

Characterization of protein glycosylation includes localization of glycosylation sites, determination of the type of glycosylation, including microheterogeneity of the carbohydrate component (glycosylation profile), and detailed glycan structure elucidation.

3.3.1. Determination of the Glycosylation Profile

Glycoproteins often contain several glycosylation sites. For certain types of glycosylation, the glycosylation sites can be predicted by well-defined consensus sequences (*see* Introduction). Although the presence of a given motif is necessary, it does not guarantee the utilization of this site for glycosylation. Identification of glycosylated sites based on mass spectrometric peptide mapping using specific endoproteinases is a well-established procedure. The endoproteinase is chosen, based on the protein primary sequence, to map the potential glycosylation sites. The proteolytic fragments can be analyzed directly by MALDI-TOFMS, or alternatively, separated by reversed-phase (RP) HPLC prior to mass spectrometric analysis. HPLC separation of peptides in complex mixtures is often necessary to avoid signal suppression, as discussed in **Subheading 3.1.4**. Glycopeptides are easily recognized in the mass spectra due to the carbohydrate microheterogeneity, which gives rise to multiple peaks with mass differences corresponding mainly to Hex (162 Daltons), HexNAc (203 Daltons), or sialic acid (291 Daltons) residues. For peptides carrying neutral oligosaccharides, the signal intensities allow relative quantitation of the different glycoforms *(13,32)*. However, glycopeptides containing sialylated glycan structures only yield a qualitative pattern, due to different efficiency of positive- and negative-ion formation for neutral and acidic oligosaccharides, as well as due to the partial loss of sialic acid residues during MALDI-TOFMS analysis.

Differentiation between N- and O-linked glycans can be accomplished by incubation of the glycopeptides with PNGaseF, an enzyme that specifically liberates N-linked glycans. Furthermore, since the compositions of all mammalian N-linked glycans are restricted to certain types of monosaccharides, possible N-glycan structures can be suggested based on the mass differences obtained before and after enzymatic deglycosylation. Such structures can be retrieved from the Complex Carbohydrate Structure Database (CCSD; http://www.ccrc.uga.edu). A general strategy for glycosylation profile determination is outlined below.

1. Search the protein sequence for potential glycosylation motifs.
2. Choose specific protease(s) that will allow mapping of all potential glycosylation sites. Proteolysis is usually performed on reduced and alkylated proteins to improve the digestion. Dissolve the protein in a suitable digestion buffer (*see* **Note 5**). Add the protease (1–4% [w/w]) and incubate at 37°C for 18 h.
3. Analyze the digest by MALDI-TOFMS after HPLC separation. The presence of multiple peaks in the mass spectrum, with mass differences corresponding to

single monosaccharide residues (146, 162, 203, or 291 Daltons) is an indication that the given fraction contains a glycopeptide.
4. Lyophilize an aliquot of HPLC fractions containing glycopeptides (5–20 pmol). Redissolve in 5 µL of 25 mM Na$_2$HPO$_4$ (pH 7.4) and incubate with PNGaseF at 37°C for 18 h to differentiate between N- and O-linked glycans. Prepare the sample for MALDI-TOFMS analysis by mixing 0.5 µL of the sample solution, 0.5 µL of 0.1% TFA, and 0.5 µL of matrix (HCCA) directly on the target.
5. Identify the location of the glycosylated peptide in the protein primary sequence. N-glycosylated peptides can be directly identified by comparing the observed mass after PNGaseF digestion with the masses of the possible proteolytic cleavage products of the protein (*see* **Note 6**). O-glycosylated peptides are identified by comparing the observed mass of the intact glycopeptide with the masses of the possible proteolytic cleavage products of the protein, taking into account the mass increment corresponding to a possible glycan composition. If the identified peptide contains more than one potential glycosylation site, a subdigestion with a different protease may be required to identify the glycosylated site.
6. Calculate potential glycan compositions by correlating the molecular weight of the intact glycan with combinations of the masses of commonly occurring monosaccharide residues, viz. HexNAc (203 Daltons), Hex (162 Daltons), NeuAc (291 Daltons), and DeoxyHex (146 Daltons).

3.3.2. Glycan Structure Elucidation

Although potential glycan structures can be proposed based on the molecular mass of the carbohydrate component, confirmation of these structures is necessary for complete characterization of protein glycosylation. Such information can be obtained by monitoring the products of sequential exoglycosidase digestions, which selectively remove terminal monosaccharides depending on their type and linkage (**Table 1**). The relatively high tolerance of MALDI-MS to buffer salts and detergents allows direct monitoring of the glycan degradation without additional sample cleanup. Exoglycosidase digestions can be performed directly on glycosylated peptides *(14)* or on released glycans *(15)*. Analysis of glycopeptides has the advantage of higher ionization efficiency compared with underivatized oligosaccharides. Exoglycosidase degradation can be performed in two ways; sequential addition of exoglycosidase or parallel incubation with different exoglycosidase mixtures (the reagent-array analysis method [RAAM]). Parallel enzymatic analysis is faster, but a larger amount of exoglycosidases is required. Sequential exoglycosidase digestion is preferred in our laboratory. The procedure is outlined below.

1. Lyophilize an aliquot of an HPLC fraction containing an isolated glycopeptide (20–50 pmol) and reconstitute in 10 µL of digestion buffer (**Table 1**).
2. Add exoglycosidase (**Table 1**) to the sample and incubate at 37°C for 18 h. The choice of the exoglycosidase is based on the proposed glycan structures.

Table 1
Endo- and Exoglycosidases Commonly Used for Glycan Structure Analysis

Glycosidase	Origin	Cleavage specificity[a]	Supplier[b]	Amount[c] (units)	Digestion buffer	Digestion time (h)	Applications and comments
N-glycosidase F (PNGaseF)	*Flavobacterium menigosepticum*	GlcNAcβ → Asn (complex, hybrid, mannose)	BM	0.1–0.5 μ	25 mM Na$_2$HPO$_4$, pH 7.4	18	Differentiate between N- and O-linked glycans
Endoglycosidase F	*Flavobacterium menigosepticum*	GlcNAcβ1 → 4GlcNAc (complex, hybrid, mannose)	BM	5 μ	50 mM NH$_4$OAc, pH 5.0	18	Confirm the presence of N-linked glycans
Endoglycosidase H	*Streptomyces plicatus*	GlcNAcβ1 → 4GlcNAc (only hybrid and mannose type)	BM	5 μ	50 mM NH$_4$OAc, pH 5.0	18	Indicate the absence of complex-type structure
Neuraminidase	*Arthrobacter ureafaciens*	NeuAcα2 → X (α2–6 > α2–3, α2–8)	BM	10 μ	50 mM NH$_4$OAc, pH 5.0	2–18	Reduce degree of metastable fragmentation. Can be applied for both N- and O-linked glycans
	Newcastle disease virus	NeuAcα2 → 3X NeuAcα2 → 8NewAc	BM	5 μ	50 mM NH$_4$OAc, pH 5.0	18	Can be applied for both N- and O-linked glycans
β-Galactosidase	*Diplococcus pneumonia*	Galβ1 → 4GlcNAc/Glc	BM	1 μ	50 mM NH$_4$OAc, pH 5.0	18	
N-acetyl-β-D-glucosaminidase	*Diplococcus pneumonia*	GlcNAcβ1 → 2Man	BM	1 μ	50 mM NH$_4$OAc, pH 5.0	18	Differentiate between isomers of triantennary complex type structures
	Jack bean	GlcNAc/GalNAcβ1 → X	OG	10 μ	50 mM NH$_4$OAc, pH 5.0	18	
α-Fucosidase	Bovine epididymis	Fucα1 → X (α1–6 > α1–2, α1–3, α1–4)	OG	10 μ	50 mM NH$_4$OAc, pH 5.0	18	Works reliably only for smaller glycan structures
α-Mannosidase	Jack bean	Manα1 → X (α1–2, α1–6 > α1–3)	OG	50 μ	50 mM NH$_4$OAc, pH 5.0	18	The glycosidase is often contaminated. Include a control to differentiate between the sample and contamination
O-glycosidase	*Diplococcus pneumonia*	Galβ(1–3)GalNAcα1 → Ser/Thr	BM	0.5 μ	50 mM NH$_4$OAc, pH 5.0	18	

[a]Data were taken from Boehringer Mannheim and Oxford GlycoSystems product information.
[b]BM, Boehringer Mannheim (Germany); OG, Oxford GlycoSystems (UK).
[c]Glycosidase amounts listed are for deglycosylation of 20–50 pmol of a glycopeptide.

3. Prepare the sample for MALDI analysis by mixing 0.5 µL of the sample, 0.5 µL of 0.1% TFA, and 0.5 µL of matrix (DHB) directly on the target. Analyze the sample by MALDI-TOFMS.
4. Repeat **step 2** and so forth with the next exoglycosidase.

An example of glycosylation analysis with murine fetal antigen 1 (mFA1) is shown *(26)*. The protein sequence contains five potential N-glycosylation sites. Two of these are positioned in the typical N-glycosylation motif, Asn-Xxx-Thr/Ser (Asn77 and Asn111), and three are in the less common motif, Asn-Xxx-Cys (Asn142, Asn151, and Asn156). In addition, the sequence contains three potential *O*-glycosylation sites: Cys-Xxx-Ser-Xxx-Pro-Cys (Ser71 and Ser193), coding for the O-glycan (Xylα1–3)Xylα1–3Glcβ-*O*-Ser, and the *O*-fucosylation motif Cys-Xxx-Xxx-Gly-Gly-Thr/Ser-Cys (Thr201) *(3)*. The combination of two proteolytic digestions (Asp-N and Lys-C) was necessary to locate the glycosylation sites of mFA1. The glycan compositions proposed were based on the mass difference between the measured and the theoretically calculated masses of the peptide. For example, the Asp-N-generated peptide at position 65–81 in the sequence contains two potential glycosylation sites, Ser71 and Asn77. MALDI-TOFMS analysis of the peptide yielded a mass profile typical for glycopeptides (**Fig. 3A**), i.e., multiple signals with mass differences corresponding to monosaccharides. An aliquot of the sample was incubated with PNGaseF, which reduced the heterogeneity to a single product at *m/z* 2191.8 ([M+H]$^+$, data not shown) and verified the presence of a heterogeneous N-linked glycan with molecular masses of 2351.0, 2060.3, 1972.1, and 1768.8 Daltons. Using the masses of commonly occurring monosaccharide residues in mammalian N-linked glycans, the compositions of the observed glycans were proposed (**Table 2**). The difference of 162.5 Daltons between the calculated molecular mass of the peptide (2028.3 Daltons, Asn77 converted to Asp) and the observed molecular mass after PNGaseF treatment (2190.8 Daltons) indicated additional glycosylation of Ser71. The mass difference of 162.5 Daltons excluded the presence of the expected trisaccharide (Xylα1–3)Xylα1–3Glcβ-*O*-Ser, but indicated the presence of a hexose, most likely a glucose, residue.

Further N-glycan structure characterization was carried out using sequential exoglycosidase digestions. The proposed N-linked glycan composition indicated the presence of a fucosylated biantennary structure with differing degrees of sialylation (2351.0, 2060.3, and 1768.8 Daltons) as the major component and a fucosylated biantennary structure with bisected GlcNAc or fucosylated triantennary structure (1972.1 Daltons) as a minor component. Thus, first, to confirm the presence of the sialylated glycan structures, the sample was incubated with neuraminidase and analyzed by MALDI-TOFMS (**Fig. 3A** and **B**). This resulted in the disappearance of the signals at *m/z* 4251.1 and 4541.8, verifying the presence of one and two sialic acid residues, respectively, on the

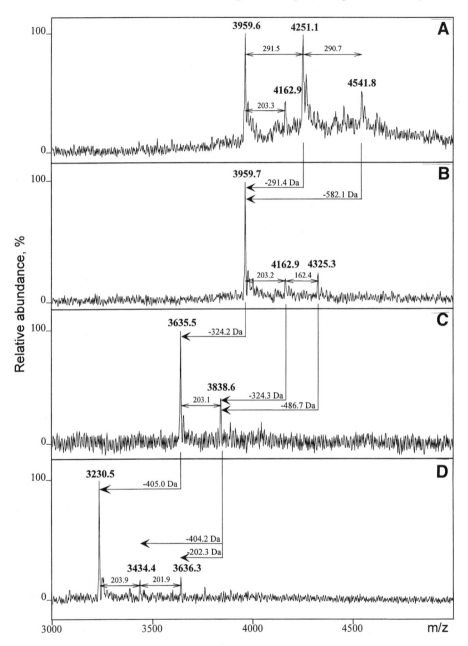

Fig. 3. MALDI-linear-TOF mass spectra of glycopeptide D8 from endoproteinase Asp-N digestion of mFA1 (**A**) intact glycopeptide, (**B**) after incubation with neuraminidase, (**C**) after sequential incubation with β-galactosidase, and (**D**) after sequential incubation with N-acetyl-β-D-glucosaminidase. (Adapted with permission from **ref. 26**.)

Table 2
Determination of Carbohydrate Composition of an Asp-N-Generated Glycopeptide from mFA1

Potential glycosylation sites	Proteolytic peptide fragment	Calculated M_r	Measured M_r		Proposed glycans composition
			Before PNGaseF	After PNGaseF	
Asn77	(65–81)	2027.3	4540.8	2190.8	Asn77
Ser71			4250.1		2351.0: $(HexNAc)_4(Hex)_5(DeoxyHex)_1(NeuAc)_2$
			4121.9		2060.3: $(HexNAc)_4(Hex)_5(DeoxyHex)_1(NeuAc)_1$
			3958.6		1972.1: $(HexNAc)_5(Hex)_5(DeoxyHex)_1$
					1768.8: $(HexNAc)_4(Hex)_5(DeoxyHex)_1$
					Ser71
					162.5: Hex

proposed biantennary structure. The analysis also revealed further glycan heterogeneity in that the ion corresponding to a molecular mass of 4324.3 Daltons indicated the presence of an additional antenna (HexNAc+Hex). Second, sequential digestion with β-galactosidase (**Fig. 3C**) reduced the molecular mass of the ion at m/z 3959.7 by 324.2 mass units (loss of two galactose residues from the biantennary structure) and also reduced the mass of the ion at m/z 4325.3 by 486.7 mass units (loss of three galactose residues from a triantennary structure). Finally, sequential incubation with N-acetyl-β-glucosaminidase, which only cleaves GlcNAc β1-2 linked to the mannose residue (**Fig. 3D**), further decreased the molecular mass of the ion at m/z 3635.5 by 405.0 mass units (loss of the two β1-2 linked GlcNAc residues from the biantennary structure). The molecular mass of the ion at m/z 3838.6 was decreased by 202.3 mass units (loss of one β1-2 linked GlcNAc residue) and by 404.2 mass units (loss of two β1-2 linked GlcNAc residues). N-acetyl-β-glucosaminidase (*D. pneumoniae*) does not hydrolyze GlcNAc β1-2 linked to the mannose having additional C-6 substitution or GlcNAc β1-2 linked to the α1-6 mannose arm of bisected structures. Therefore, the loss of one GlcNAc residue from the molecular ion at m/z 3838.6 can be ascribed to the presence of a 2,6-branched triantennary structure or to the bisected biantennary structure. The loss of two β1-2 linked GlcNAc residues from the molecular ion at m/z 3838.6 indicated the presence of a 2,4-branched triantennary structure. In total, sequential exoglycosidase digestions confirmed that the carbohydrate component at Asn77 was mainly the fucosylated biantennary structure with a minor content of 2,4-branched triantennary and 2,6-branched triantennary or bisected biantennary structures.

3.4. Localization of Glycosylated Sites in Mucin-Type Glycopeptides

Mucin-type glycosylation frequently occurs in regions of the protein with a high density of serine, threonine, and proline residues. Therefore, proteolytic cleavage between the potential glycosylation sites prior to mass spectrometric peptide mapping is often not possible. Mass spectrometric determination of O-glycosylation sites by PSD *(33,34)* and ESI-MS/MS analysis *(35–39)* has been reported. The success of these techniques is, however, highly dependent on the nature of the analyte and instrumentation.

Peptide hydrolysis using the vapor of PFPA followed by mass spectrometric analysis is an alternative for determining O-glycosylation sites in mucin-type glycopeptides *(40)*. This hydrolysis can be performed in a manner that generates extensive, nonspecific polypeptide backbone cleavage with minimal carbohydrate loss. Due to the partial cleavage of glycosidic bonds, glycosylated peptides generated by hydrolysis are detected as molecular ions corresponding to the fully glycosylated, partially deglycosylated, and completely deglycosylated peptides. Since the amino acid sequence of the peptides is normally known, comparison of the observed masses of the cleavage products with the calculated masses based on the peptide sequence allows identification of the glycosylated residue. Thus, the presence of an ion 203 mass units above the calculated $[M+H]^+$ value for a nonglycosylated peptide indicates the presence of a GalNAc residue on the peptide fragment. The same strategy can be applied to mannosyl-type glycosylation, which results in a difference of 162 mass units. The experimental setup is inexpensive and simple, and the method is compatible with most types of mass spectrometers. Performing the hydrolysis in the vapor phase eliminates the risk of contaminating the sample with impurities from reagents and solvents, thus allowing analysis of the reaction products, by both MALDI-TOFMS and ESI-MS, without the need for further purification.

1. Lyophilize purified peptide in a 500-µL Eppendorf vial. Remove the lid and place the vial in a 22 mL glass vial with a mininert valve (Pierce).
2. Add 100 µL of 20% aqueous PFPA, containing 100 µg DTT, to the glass vial and flush with argon.
3. Evacuate the glass vial to 1 mbar pressure and place it in an oven at 90°C for 60 min.
4. Vacuum centrifuge the hydrolysate for 15 min to remove remaining traces of acid.
5. Redissolve the sample in 0.1% TFA to a concentration of 1 pmol/µL, based on the initial amount of peptide subjected to hydrolysis, and prepare for MALDI-TOFMS analysis by mixing 1 µL of sample solution with 0.5 µL of matrix (DHB) solution directly on the target (*see* **Note 7**).

An illustrative example is based on the in vitro glycosylated mucin-derived peptide Muc1a' (AHGV<u>T</u>S<u>A</u>PD<u>T</u>R) with three potential glycosylation sites (underlined). Based on the difference between the observed (1313.3 Daltons) and calculated (1110.5 Daltons) monoisotopic molecular masses, the peptide was assumed to carry one GalNAc residue. Twenty picomoles of the peptide were hydrolyzed with 20% PFPA at 90°C for 1 h. The resulting MALDI-TOF mass spectrum, with assignment of the observed degradation products, is shown in **Fig. 4**. The observed N- and C-terminal sequence ladders unambiguously localize the glycosylation site to Thr_5. This is in agreement with the previously reported acceptor site for GalNAc-T1, -T2, and -T3 transferases *(30)*.

3.5. Concluding Remarks

MALDI-TOFMS used in combination with proteolytic cleavage and digestion with endo- and exoglycosidases is a method for site-specific assignment of N-glycosylated sites as well as for determination of glycan structures. Isomeric sugar residues as well as linkage types cannot be distinguished by mass spectrometry, and their determination relies on the use of specific glycosidases. A similar approach can be used for O-glycosylated sites, when the potential

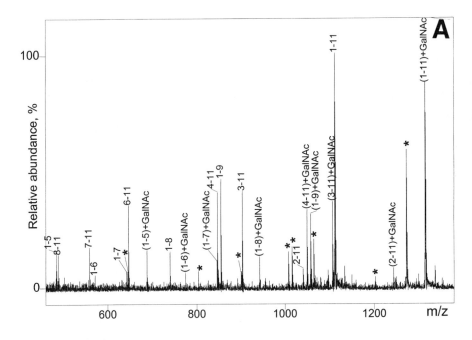

Fig. 4. **(A)** MALDI-reflectron-TOF mass spectrum of the glycosylated peptide Muc1a' (AHGVTSAPDTR) after hydrolysis with 20% PFPA at 90°C for 1 h. Loss of the acetyl group (-42 mass units) from the GalNAc residue is denoted by * (*see* **Note 8**). **(B)** The observed degradation products of the peptide Muc1a'. The peptide has three potential glycosylation sites, viz. Thr5, Ser6, and Thr10 (underlined in the peptide sequence). The glycosylated peptide fragments are detected as deglycosylated (NG) and glycosylated (+1GalNAc) product ions, accompanied by loss of the acetyl group from the GalNAc residue (masses given in parentheses). The presence of the GalNAc residue on the peptide (1–5), indicated in boldface, unambiguously identifies Thr5 as the glycosylated residue. (Adapted with permission from **ref. *40*.**)

glycosylation sites can be separated by specific proteolytic cleavages. Determination of O-glycosylated sites that are close or even adjacent in the protein primary sequence, as observed for mucin-type glycosylation, is possible by the generation of sequence ladders using partial acid hydrolysis followed by MALDI-TOFMS analysis of the resulting mixtures.

4. Notes

1. HCCA is the preferred peptide matrix, due to its high ionization efficiency and tolerance to buffer salts and detergents. However, HCCA often causes metastable fragmentation and, in some cases, prompt fragmentation, especially of O-glycosylated and sialylated glycopeptides.
2. DHB is a good glycopeptide and glycoprotein matrix. Depending on the analyte, a certain degree of metastable fragmentation may be observed. The main disadvantage of DHB is that the matrix is not as tolerant toward sample impurities (salts and detergents) as are other matrices. Therefore, when using DHB, a prior cleanup of the analyte may be necessary.
3. Sinapinic acid matrix can be used for glycopeptides and glycoproteins. This matrix may also cause fragmentation, but to a lesser extent then HCCA. The relatively high tolerance toward salts and detergents makes it a good compromise between HCCA and DHB.
4. HPA and THAP do not introduce fragmentation of glycopeptides and glycoproteins, but lower ionization efficiency and a tendency to form adducts are the major limitations associated with these matrices.
5. The use of volatile buffers is usually advantageous for mass spectrometric analysis. If detergents are necessary, n-octylglucoside (20 mM) is a good choice since it is compatible with MALDI-MS analysis.
6. Glycosylated asparagine residues undergo deamidation during enzymatic deglycosylation with PNGaseF, resulting in a mass increment of 1 Dalton.
7. Hydrolysis products may be accompanied by additional (M-18) peaks due to stable oxazolone intermediates formed during hydrolysis. Conversion of these intermediates to the final products can be accomplished by incubation with 10 µL of 25% aqueous ammonia for 10 min at room temperature *(40)*.
8. The loss of the acetyl group from GalNAc can be considerably minimized by performing hydrolysis in 20% HCl, at 50°C for 90–120 min, if necessary. On the other hand, the presence and intensity of peaks corresponding to acetyl losses have a considerable diagnostic value, e.g., the presence of an ion 42 mass units below the $[M+H]^+$ ion confirms that the molecular ion corresponds to the glycosylated form of the peptide. In addition, the relative abundance of a series of ions with mass differences of 42 mass units strongly supports the assignment of the number of GalNAc residues in a given hydrolytic product ion.

References

1. Lis, H. and Sharon, N. (1993) Protein glycosylation. Structural and functional aspects. *Eur. J. Biochem.* **218**, 1–27.

2. Varki, A. (1993) Biological roles of oligosaccharides: all of the theories are correct. *Glycobiology* **3**, 97–130.
3. Harris, R. J. and Spellman, M. W. (1993) O-Linked fucose and other post-translational modifications unique to EGF modules. *Glycobiology* **3**, 219–224.
4. Prockop, D. J., Kivirikko, K. I., Tuderman, L., and Guzman, N. A. (1979) Medical progress. The biosynthesis of collagen and its disorders. *N. Engl. J. Med.* **301**, 13–23.
5. Clausen, H. and Bennett, E. P. (1996) A family of UDP-GalNAc: polypeptide *N*-acetylgalactosaminyl-transferases control the initiation of mucin-type O-linked glycosylation. *Glycobiology* **6**, 635–646.
6. Haltiwanger, R. S., Blomberg, M. A., and Hart, G. W. (1992) Glycosylation of nuclear and cytoplasmic proteins. Purificaton and characterization of a uridine diphospho-N-acetylglucosamine: polypeptide beta-N-acetylglucosaminyltransferase. *J. Biol. Chem.* **267**, 9005–9013.
7. Hansen, J. E., Lund, O., Nielsen, J. O., Hansen, J.-E. S., and Brunak, S. (1996) O-GLYCBASE: a revised database of O-glycosylated proteins. *Nucleic Acids Res.* **24**, 248–252.
8. Harvey, D. J. (1996) Matrix-assisted laser desorption/ionization mass spectrometry of oligosaccharides and glycoconjugates. *J. Chromatogr.* **720**, 429–446.
9. James, D. C. (1996) Analysis of recombinant glycoproteins by mass spectrometry. *Cytotechnology* **22**, 17–24.
10. Burlingame, A. L. (1996) Characterization of protein glycosylation by mass spectrometry. *Curr. Opin. Biotechnol.* **7**, 4–10.
11. Carr, S. A., Huddleston, M. J., and Bean, M. F. (1993) Selective identification and differentiation of *N*-and *O*-linked oligosaccharides in glycoproteins by liquid chromatography-mass spectrometry. *Protein Sci.* **2**, 183–196.
12. Huddleston, M. J., Bean, M. F., and Carr, S. A. (1993) Collisional fragmentation of glycopeptides by electrospray ionization LC/MS and LC/MS/MS: methods for selective detection of glycopeptides in protein digests. *Anal. Chem.* **65**, 877–884.
13. Huberty, M. C., Vath, J. E., Yu, W., and Martin, S. A. (1993) Site-specific carbohydrate identification in recombinant proteins using MALD-TOF MS. *Anal. Chem.* **65**, 2791–2800.
14. Sutton, C. W., O'Neill, J. A., and Cottrell, J. S. (1994) Site-specific characterization of glycoprotein carbohydrates by exoglycosidase digestion and laser desorption mass spectrometry. *Anal. Biochem.* **218**, 34–46.
15. Küster, B., Naven, T. J. P., and Harvey, D. J. (1996) Rapid approach for sequencing neutral oligosaccharides by exoglycosidase digestion and matrix-assisted laser desorption/ionization time-of-flight mass spectrometry. *J. Mass Spectrom.* **31**, 1131–1140.
16. Mørtz, E., Sareneva, T., Julkunen, I., and Roepstorff, P. (1996) Does matrix-assisted laser desorption/ionization mass spectrometry allow analysis of carbohydrate heterogeneity in glycoproteins? A study of natural human interferon-γ. *J. Mass Spectrom.* **31**, 1109–1118.
17. Küster, B., Wheeler, S. F., Hunter, A. P., Dwek, R. A., and Harvey, D. J. (1997) Sequencing of N-linked oligosaccharides directly from protein gels: in-gel

deglycosylation followed by matrix-assisted laser desorption/ionization mass spectrometry and normal-phase high-performance liquid chromatography. *Anal. Biochem.* **250,** 82–101.
18. Spengler, B., Kirsch, D., Kaufmann, R., and Lemoine, J. (1995) Structure analysis of branched oligosaccharides using post-source decay in matrix-assisted laser desorption ionization mass spectrometry. *J. Mass Spectrom.* **30,** 782–787.
19. Talbo, G. and Mann, M. (1996) Aspects of the sequencing of carbohydrates and oligonucleotides by matrix-assisted laser desorption/ ionization post-source decay. *Rapid Commun. Mass Spectrom.* **10,** 100–103.
20. Rouse, J. C., Strang, A.-M., Yu, W., and Vath, J. E. (1998) Isomeric differentiation of asparagine-linked oligosaccharides by matrix-assisted laser desorption-ionization postsource decay time-of-flight mass spectrometry. *Anal. Biochem.* **256,** 33–46.
21. Reinhold, V. N., Reinhold, B. B., and Costello, E. (1995) Carbohydrate molecular weight profiling, sequence, linkage, and branching data: ES-MS and CID. *Anal. Chem.* **67,** 1772–1784.
22. Viseux, N., de Hoffmann, E., and Domon, B. (1997) Structural analysis of permethylated oligosaccharides by electrospray tandem mass spectrometry. *Anal. Chem.* **69,** 3193–3198.
23. Mørtz, E., Sareneva, T., Haebel, S., Julkunen, I., and Roepstorff, P. (1996) Mass spectrometric characterization of glycosylated interferon-γ variants separated by gel electrophoresis. *Electrophoresis* **17,** 925–931.
24. Kristensen, A. K., Schou, C., and Roepstorff, P. (1997) Determination of isoforms, N-linked glycan structure and disulfide bond linkages of the major cat allergen Fel d1 by a mass spectrometric approach. *Biol. Chem.* **378,** 899–908.
25. Rahbek-Nielsen, H., Roepstorff, P., Reischl, H., Wozny, M., Koll, H., and Haselbeck, A. (1997) Glycopeptide profiling of human urinary erythropoietin by matrix-assisted laser desorption/ionization mass spectrometry. *J. Mass Spectrom.* **32,** 948–958.
26. Krogh, T. N., Bachmann, E., Teisner, B., Skjødt, K., and Højrup, P. (1997) Glycosylation analysis and protein structure determination of murine fetal antigen 1 (mFA1). The circulating gene product of the delta-like protein (dlk), preadipocyte factor 1 (Pref-1) and stromal-cell-derived protein 1 (SCP-1) cDNAs. *Eur. J. Biochem.* **244,** 334–342.
27. Olsen, E. H. N., Rahbek-Nielsen, H., Thøgersen, I. B., Roepstorff, P., and Enghild, J. J. (1998) Posttranslational modifications of human inter-α-inhibitor: identification of glycans and disulfide bridges in heavy chains 1 and 2. *Biochemistry* **37,** 408–416.
28. Ploug, M., Rahbek-Nielsen, H., Nielsen, P. F., Roepstorff, P., and Danø, K. (1998) Glycosylation profile of a recombinant urokinase-type plasminogen activator receptor expressed in Chinese hamster ovary cells. *J. Biol. Chem.* **273,** 13,933–13,943.
29. Fierobe, H.-P., Mirgorodskaya, E., Frandsen, T. P., Roepstorff, P., and Svensson, B. (1997) Overexpression and characterization of *Aspergillus awamori* wild-type and mutant glucoamylase secreted by the methylotrophic yeast *Pichia pastoris*: comparison with wild-type recombinant glucoamylase produced using *Saccharomyces cerevisiae* and *Aspergillus niger* as hosts. *Protein Expression Purif.* **9,** 159–170.

30. Wandall, H. H., Hassan, H., Mirgorodskaya, E., Kristensen, A. K., Roepstorff, P., Bennett, E. P., et al. (1997) Substrate specificities of three members of the human UDP-N-acetyl-α-D-galactosamine:polypeptide N-acetylgalactosaminyltransferase family, GalNAc-T1, -T2, and -T3. *J. Biol. Chem.* **272,** 23,503–23,514.
31. Karas, M., Bahr, U., Strupat, K., Hillenkamp, F., Tsarbopoulos, A., and Pramanik, B. N. (1995) Matrix dependence of metastable fragmentation of glycoproteins in MALDI TOF mass spectrometry. *Anal. Chem.* **67,** 675–679.
32. Harvey, D. J. (1993) Quantitative aspects of the matrix-assisted laser desorption mass spectrometry of complex oligosaccharides. *Rapid Commun. Mass Spectrom.* **7,** 614–619.
33. Goletz, S., Thiede, B., Hanisch, F.-G., Schultz, M., Peter-Katalinic, J., Muller, S., et al. (1997) A sequencing strategy for the localization of O-glycosylation sites of MUC1 tandem repeats by PSD-MALDI mass spectrometry. *Glycobiology* **7,** 881–896.
34. Müller, S., Goletz, S., Packer, N., Gooley, A., Lawson, M., and Hanisch, F.-G. (1997) Localization of O-glycosylation sites on glycopeptide fragments from lactation-associated MUC1. All putative sites within the tandem repeat are glycosylation targets *in vivo*. *J. Biol. Chem.* **272,** 24,780–24,793.
35. Rademaker, G. J., Haverkamp, J., and Thomas-Oates, J. (1993) Determination of glycosylation sites in O-linked glycopeptides: a sensitive mass spectrometric protocol. *Org. Mass Spectrom* **28,** 1536–1541.
36. Medzihradszky, K. F., Gillece-Castro, B. L., Townsend, R. R., Burlingame, A. L., and Hardy, M. R. (1996) Structural elucidation of O-linked glycopeptides by high energy collision-induced dissociation. *J. Am. Soc. Mass Spectrom.* **7,** 391–328.
37. Greis, K. D., Hayes, B. K., Comer, F. I., Kirk, M., Barnes, S., Lowary, T. L., et al. (1996) Selective detection and site-analysis of O-GlcNAc-modified glycopeptides by β-elimination and tandem electrospray mass spectrometry. *Anal. Biochem.* **234,** 38–49.
38. Hanisch, F.-G., Green, B. N., Bateman, R., and Peter-Katalinic, J. (1998) Localization of O-glycosylation sites of MUC1 tandem repeats by QTof ESI mass spectrometry. *J. Mass Spectrom.* **33,** 358–362.
39. Rademaker, G. J., Pergantis, S. A., Bloktip, L., Langridge, J. I., Kleen, A., and Thomas-Oates, J. E. (1998) Mass spectrometric determination of the sites of O-glycan attachment with low picomolar sensitivity. *Anal. Biochem.* **257,** 149–160.
40. Mirgorodsakaya, E., Hassan, H., Wandall, H. H., Clausen, H., and Roepstorff, P. (1999) Partial vapor phase hydrolysis of peptide bonds: a method for mass spectrometric determination of O-glycosylated sites in glycopeptides. *Anal. Biochem.* **269,** 54–65.

17

Positive and Negative Labeling of Human Proinsulin, Insulin, and C-Peptide with Stable Isotopes

New Tools for In Vivo Pharmacokinetic and Metabolic Studies

Reto Stöcklin, Jean-François Arrighi, Khan Hoang-Van, Lan Vu, Fabrice Cerini, Nicolas Gilles, Roger Genet, Jan Markussen, Robin E. Offord, and Keith Rose

1. Introduction

Over 100 polypeptides and proteins are either already in use in clinics or are being intensively studied with a view to clinical application, and this number is likely to increase substantially over the next few years. Our central goal is to know more of the behavior and fate of therapeutic proteins and polypeptides in humans using the proinsulin, insulin, and C-peptide family as a model. Although diabetes is a major public health problem that has been widely investigated, much remains to be understood in this extremely complex field, especially as regards the in vivo metabolic pathways of the related proteins. Proinsulin is a hormonal precursor produced in the pancreatic β-cells of the islets of Langerhans, where its maturation is followed by a crucial enzymatic transformation into the active hormones insulin and C-peptide (1).

1.1. Isotope Dilution Assay

In previous work (2), we demonstrated that semisynthetically engineered deuterated analogs of insulin can be used in purified form as internal standards for precise insulin quantitation by isotope dilution mass spectrometry. More recently, we have been able to develop such a method for human insulin, and have applied it to clinical samples. This isotope dilution assay (IDA) method allowed us to determine human insulin levels accurately in several different samples of 1 mL of serum, down to the 20-pM basal level *(3)*. We further improved this methodology for simultaneous measurements of insulin, C-peptide, and proinsulin levels *(4)*. In a typical experiment, a known amount of an analog of the target compound is added to the biological sample to be analyzed (the internal standard having a different molecular mass; *see* **Notes 1–3**). After appropriate extraction and purification processes (*see* **Notes 4–6**), the final mass spectrometric analysis not only allows direct identification of the target compound and its analog by virtue of their respective molecular masses, but also enables a precise quantitation on the basis of the relative intensities of the observed signals (*see* **Notes 7–11**). To avoid losses by adsorption in the case of trace analysis, as is the case for insulin, another analog with yet another molecular mass (which does not interfere in the mass range of interest) may be added in large excess to serve as a carrier (*see* **Note 1**) and as a tracer on the ultraviolet (UV) detection record from the chromatographic steps. Such a strategy can in theory be applied to any biopolymer, and it should have considerable potential, especially in the field of in vivo pharmacokinetic and metabolic studies.

1.2. Stable-Isotope-Labeled Analogs

Production of pure and homogenous native proinsulin, insulin, and C-peptide has been achieved by several approaches including site-directed semisynthesis and biosynthesis *(5–9)*. Uniform labeling with ^{13}C, ^{15}N, or deuterium would give access to valuable analogs, and this has been achieved for several proteins for nuclear magnetic resonance (NMR) purposes *(10,11)*. However, the physicochemical properties of a fully deuterated protein might differ sufficiently to cause problems (for example, a later elution time on reversed-phase high-performance liquid chromatography [HPLC]). Less common or somewhat less exposed atoms, such as nitrogen, oxygen, or sulphur, thus seem to be better candidates for such experiments, but there is an alternative—negative labeling. This elegant and economical route involves the biosynthetic production of insulin from carbon in which the natural ^{13}C has not been enriched, but instead depleted to nearly zero.

1.3. Negative Labeling

A natural macromolecule consists of a mixture of atoms containing, in addition to their major isotopic form, different numbers of ^{13}C, ^{15}N, ^{18}O, and ^{34}S atoms (1.1%, 0.37%, 0.2% and 4.4% natural abundance, respectively). The mass-spectrometric peak envelopes at low resolution are correspondingly flattened and broadened and lead to the measurement of an average molecular weight (**Fig. 1A**). A negatively labeled analog (depleted in isotopes more massive than the major and lightest isotope) will, by contrast, have a slightly lower average molecular weight; the envelope will be narrower, since it contains

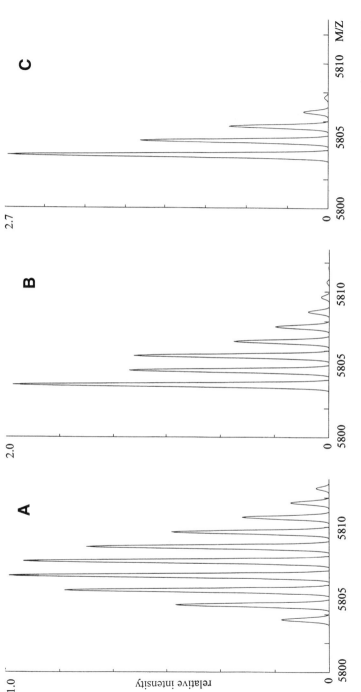

Fig. 1. Negative labeling of human insulin. Theoretical mass spectra of (**A**) native insulin; (**B**) enrichment in ^{12}C at 99.9% (depletion of the 1.1% natural abundance of ^{13}C down to 0.1%); and (**C**) enrichment in ^{12}C and ^{32}S at 99.9% each. Note the different intensity scales. The two negatively labeled forms clearly give a highly specific isotopic envelope of nonnatural aspect. They have the further advantage of increasing the height of the signal envelope by factors of 2.0 and 2.7, respectively. The peak at m/z 5802.65 corresponds to the calculated monoisotopic molecular mass of human insulin. At the relatively low resolution conditions typically used with ES-MS, the individual isotopic signals coalesce (*see* **Fig. 3**). In each spectrum, the monoisotopic peak is at the left. Spectra were calculated using a published program (**27**).

fewer isotopic signals, and these signals will have a nonnatural relative abundance (**Fig. 1B, C**). This results in an analog that has much greater similarity to its native unlabeled form than is the case for an analog possessing many heavy-atom substitutions, but that remains, nonetheless, easily recognizable. Isotopic depletion is well known as an NMR tool in small-molecule organic chemistry and has recently been used for accurate molecular weight determination of proteins *(12,13)*. We now propose the use of negative labeling in the field of macromolecular metabolic tracers. By virtue of their characteristic isotopic pattern, such tracers can easily be detected and characterized, even if they are of unknown structure, as a result of partial degradation, and present in a complex biological sample, provided that mass spectrometry is performed at high resolution.

1.4. Pharmacokinetic and Metabolic Studies

Here we describe a bacterial expression system that produces proinsulin in fusion with a poly-histidine tag. It allows uniform labeling with stable isotopes at will, as demonstrated here with specific culture media using $^{15}NH_4Cl$ at 99.4% isotopic enrichment as the sole nitrogen source, deuterated water at 50% isotopic enrichment, or ^{12}C at 99.897% enrichment (^{13}C depleted, negative labeling). After appropriate purification, refolding, and transformation steps, proinsulin, insulin, and C-peptide analogs were obtained and fully characterized. The biological activity of the insulin analogs was found to be authentic as regards their ability to induce glycogen formation. Such nonradioactive analogs, otherwise of rigorously authentic structure, can easily be distinguished from their native form by mass spectrometry. They are thus a unique tool for in vivo pharmacokinetic studies and could be used as specific tracers if injected in humans. The striking appearance of the isotopic envelope provided by the negative labeling of insulin is demonstrated with a high-resolution mass spectrometric analysis; its nonnatural isotopic envelope provides a unique feature otherwise impossible to achieve without use of radioactivity. We therefore propose the use of negatively labeled analogs as a possible new kind of metabolic tracer and discuss the potential of analogs with such a clear isotopic signature for in vivo metabolic studies in humans.

2. Materials

2.1. Reagents and Associated Materials

1. Reagents and solvents (of analytical grade or better) were obtained from commercial sources and were used without further purification unless otherwise stated.
2. Water was purified using a Milli-Q system from Millipore (Bedford, MA).
3. The pRSV plasmid containing the gene encoding human pre-proinsulin was a gift of Prof. Philippe Halban and Dr. Jean-Claude Irminger (Louis Jeantet Laboratories, University Medical Center, Geneva, Switzerland).

4. Restriction enzymes *Bam*H1, *Eco*R1, and *Xba*1 were obtained from Boehringer (Mannheim, Germany).
5. Ultrapure agarose (BRL 5510UB) was obtained from Bethesda Research Laboratory, Bethesda, MD.
6. Stable isotope $^{15}NH_4Cl$ (99.4% atom ^{15}N) was obtained from Isotech, Miamisburg, OH (85-70203). D_2O 99.75% atom (Uvasol) was sterile filtered before use. Lyophilized Algal, enriched in ^{12}C to 99.897%, was from Cambridge Isotope Laboratory, Andover, MA.
7. Quiagen Ni-NTA agarose gel was obtained from Kontron, Milan, Italy (30210).
8. Proteolytic enzymes. TPCK-treated bovine trypsin was from Worthington, Lakewood, NJ (32A680), and porcine pancreatic carboxypeptidase B was from Boehringer Mannheim (103 233).
9. Monocomponent grade, zinc-form human insulin was the gift of Novo Nordisk, Bagsvaerd, Denmark. It was used in its zinc-free form after dialysis as described by Morihara et al. *(14)*.
10. Synthetic C-peptide of human sequence (31-residue form) was prepared by solid-phase synthesis on a Perkin Elmer Biosystems, Norwalk, CT, 430A peptide synthesizer using standard Fmoc [*N*-(9-fluorenyl)methoxycarbonyl] chemistry.

2.2. Separation and Extraction Techniques

1. Sodium dodecyl sulfate polyacrylamide gel electrophoresis (SDS-PAGE). Samples were loaded on a homogenous 20% SDS-PAGE gel and run using an automated Phast System (Pharmacia LKB Biotechnology, Uppsala, Sweden) with 80 V-h at 15°C. Gels were stained with Coomassie blue as recommended by the manufacturer.
2. Solid-phase extraction. Samples in acidic solution and with low organic phase content were loaded on a C_{18} Sep Pak RP-cartridge (Waters, Milford, MA) that had previously been washed with 20 mL ethanol and equilibrated with 20 mL 0.1% trifluoroacetic acid (TFA). After loading, the cartridge was washed with 20 mL 0.1% TFA and the required product eluted with 4–10 mL 0.1% TFA containing 80% acetonitrile.
3. Reversed-phase (RP) HPLC. Analytical and semipreparative RP-HPLC used a Beckman-Gold system (Beckman, Fullerton, CA) fitted with a model 126 gradient programmer and a model 166 detector. The manual injector (Beckman Altex 210A) was equipped with a 2-mL sample loop. The column was either 250×4 mm, packed with Nucleosil C_{18} 5 µm 300 Å particles (Macherey Nagel, Düren, Germany) and operated at a flow rate of 1 mL/min (analytical) or 250×10 mm, packed with the same particles and operated at a flow rate of 4 mL/min (semipreparative). Solvent A was 1 g TFA (Pierce, Rockford, IL) added to 1 L water. Solvent B was prepared by adding 1 g TFA to 100 mL water prior to making up to 1 L with acetonitrile (Lichrosolv, Merck, Darmstadt, Germany). Unless stated otherwise, the gradient used was from 20%B to 50%B over 30 min. The effluent was monitored at 214 nm and the system was operated under control of the Gold 3.0 software.

2.3. Mass Spectrometry

1. Electrospray mass spectrometry. Electrospray analyses were carried out using a VG-Trio 2000 mass spectrometer fitted with a 3000 mass unit RF generator (VG Biotech, Altrincham, UK). The mass spectrometer was equipped with an electrospray ion source (ESI-MS) and operated in the positive-ionization mode under control of the Lab-Base data system. 10 µL portions of the sample solutions (in 50% aqueous methanol containing 2% acetic acid or in 50% aqueous CH_3CN containing 0.2% formic acid) were injected at a flow rate of 2 µL/min. The quadrupole was scanned at 10 s/scan cycle, usually from m/z 500 to 1500, to cover the major peaks of the charge distribution. Data were processed using a six-point smoothing routine. External calibration of the mass scale was performed with a fresh 10-pmol/µL solution of horse myoglobin (Sigma, Buchs, Switzerland).
2. High-resolution matrix-assisted laser desorption/ionization (MALDI) spectra were recorded using a Fourier-transform ion cyclotron resonance (FT-ICR) mass spectrometer, equipped with an external ion source, as described by McIver et al. *(15)* and by Li et al. *(16,17)*. Commercial human insulin (Sigma), in-house recombinant ^{13}C-depleted human insulin, and in-house 99.4% ^{15}N-enriched human insulin were dissolved in 30% acetonitrile + water (0.1% TFA) and mixed with a matrix material (2,5-dihydroxybenzoic acid) and with fructose as co-matrix prior to MALDI-FTICR-MS analysis. Conditions used are given in the figure legends.

3. Methods
3.1. Gene Assembly

Two oligonucleotides were designed to extract, using polymerase chain reaction (PCR), the proinsulin cDNA from the pRSV plasmid and simultaneously insert two restriction enzyme cleavage sites on each end of the gene. An ATG sequence, encoding for a methionine, upstream from the B1 phenylalanine of proinsulin was also inserted (**Fig. 2**). These two single-stranded oligonucleotides were synthesized in an automated DNA/RNA synthesizer (Applied Biosystems 394), purified on a polyacrylamide gel, and extracted with a Sep Pak cartridge as described by the manufacturer.

The details of the PCR amplification procedure are as follows:

1. Carry out the PCR program using a DNA Thermal Cycler (Perkin Elmer 480) with 1 ng pRSV plasmid, 2.5 U Taq polymerase (Promega, Wallisellen, Switzerland M1861), 0.5 µL of 200 µM dNTP (Boehringer Mannheim, 1578 553), and 4 µg of each oligonucleotide in PCR buffer (Boehringer Mannheim) in the presence of 1.5 mM $MgCl_2$. Recover the final volume of 50 µL with 70 µL mineral oil (Sigma M-3516).
2. Submit the PCR reaction vessels to a 4-min denaturation step at 94°C, followed by 30 amplification cycles, each consisting of 30 s at 94°C (denaturation), 30 s at 60°C (hybridization), and 30 s at 72°C (polymerization).

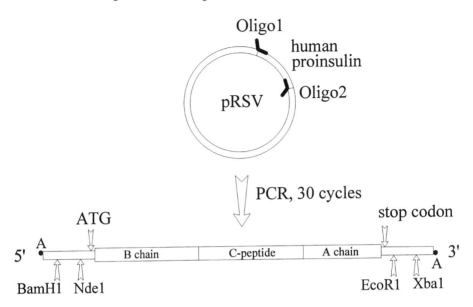

Fig. 2. Illustration of the PCR strategy. The pRSV plasmid containing the gene for the human sequence for proinsulin was submitted to 30 cycles of PCR reaction in the presence of two single-stranded oligonucleotides, which were partially complementary in the N- and C-terminal regions of the proinsulin sequence. The 5'-3' nucleotide sequence for Oligo1 is AA GGA TCC CAT ATG TTT GTG AAC CAA CAC CTG and that of Oligo2 is T GTC TAG AAT TCC CTA GTT GCA GTA GTT CTC. These oligonucleotides had been designed in order to insert specific sites for restriction enzymes at both extremities of the cDNA of interest for future cloning in an appropriate expression vector, and to insert an ATG sequence encoding methionine for future cleavage of the fusion protein with cyanogen bromide. The last PCR elongation cycle was deliberately prolonged so as to stimulate the addition, by Taq polymerase, of an adenosine at the 3' end of each strand to facilitate the insertion of our PCR product in the T-Vector.

3. Terminate the reaction and stimulate the addition, by the Taq polymerase, of an adenosine at the 3' end of each strand by leaving the reaction at 72°C for 10 min at the end of the last cycle.
4. Remove the oil and immediately submit 3 µL of the product to ligation in a final volume of 10 µL with the pCRII open plasmid (In Vitrogen, Carlsbad, CA) using the T-vector strategy *(18)*. Transform 3 µL of the ligation product using 50 µL of competent INVαF' cells (In Vitrogen) as described by Sambrook et al. *(19)*.

Positive colonies are isolated on agar plates using the α-complementation strategy *(20)* in the presence of kanamycin and the X-gal chromogenic agent *(21)*. The plasmids are extracted from an overnight culture in Luria Bertani

(LB) medium using a Quiagen Typ-20 (Kontron 12123) cartridge and an aliquot submitted to an enzymatic restriction mapping with EcoR1 *(19)*.

As indicated above, cDNA encoding human proinsulin is extracted and amplified from the pRSV plasmid using a PCR strategy requiring two synthetic oligonucleotides. Both of these synthetic products were found to give a single band on polyacrylamide gel. Again, analytical agarose-gel control of the PCR product gave a single band at the expected position of 0.3 kb, demonstrating that the amplification process was performed successfully (data not shown). Ligation of the PCR product to the pCRII plasmid using the T-vector strategy and the subsequent transformation in INVαF' cells led to 100 positive colonies that could be selected by α-complementation. Out of 12 positive colonies that were then analyzed by enzymatic restriction mapping with *Eco*R1, 8 were found to give the expected band at 0.3 kb on agarose gel.

3.1.1. Subcloning

The gene encoding for human proinsulin is extracted from the plasmid using the *Bam*H1 and *Xba*1 enzymes. The pDHFR expression vector *(22)* encoding for a hexahistidine tag in fusion with a 19-kDa derivative of murine dehydrofolate reductase is submitted to a similar enzymatic treatment. Processing then proceeds as follows:

1. Load both samples on a 2% ultrapure agarose gel. After migration and incubation in ethidium bromide, expose the gel briefly to a 254-nm UV light for visualization of the DNA.
2. Isolate the bands of interest (0.3 kb for proinsulin and 2.5 kb for the pDHFR plasmid) by excision and extract the DNA using Quick-Elute gel (Biofinex, Praroman-LeMouret, Switzerland, K 8730) according to Vogelstein and Gillespie *(23)*.
3. Ligate the two purified DNA fragments under conditions similar to those described in **Subheading 3.1.** and transform 5 µL of the ligation product *(19)* using 50 µL of a preparation of competent M15a *E. coli* cells containing the pREP4 repressor plasmid *(24)*.
4. Isolate positive colonies on agar plates in the presence of both kanamycin and ampicillin. Then perform a second screening for positive colonies by overexpression of the recombinant fusion protein in LB culture media in the presence of kanamycin (25 µg/mL) and ampicillin (100 µg/mL).
5. Dilute a saturated overnight culture of several colonies by 1:100 and leave at 37°C until the optical density (OD) at 600 nm is between 0.7 and 0.9 (after approx 4 h) and then add isopropyl-β-D-thiogalactopyranoside (IPTG) (2 mM final). After a further 5 h of incubation, take 0.5-mL aliquots and centrifuge at 4°C for 20 min at 1500 g in a bench centrifuge. Resuspend the pellets in 15 µL of loading buffer and apply a fraction to an SDS-PAGE gel.

Isolation of the expected bands at 2.5 kb for pDHFR and at 0.3 kb for proinsulin is realized during the above subcloning process. Of 24 positive colonies that were screened for overexpression of the fusion protein, 21 gave a clear overexpression band at 11 kDa on SDS-PAGE.

For characterization and further use, the fusion protein can be produced in larger amounts in fractions of 500-mL culture volume under similar expression conditions. After overexpression, the bacteria are centrifuged at 4°C at 1500g (Beckmann TJ-6, 3000 rpm) for 20 min. The cells are washed in 50 mL phosphate-buffered saline (PBS) buffer, centrifuged again, and the new pellets stored at −20°C.

3.2. Protein Labeling Procedures

3.2.1. Positive Labeling with 99.4% ^{15}N or 50% D_2O

1. Medium Preparation and Labeling
   a. Prepare 1 L synthetic culture medium (minimum essential medium [MEM]) containing $^{15}NH_4Cl$ (99.4% atom ^{15}N) as sole nitrogen source, from 686 mL sterile water by adding 200 mL of a sterilized stock solution made up from Na_2HPO_4 (30 g/L), KH_2PO_4 (15 g/L), $^{15}NH_4Cl$ (5 g/L), and $CaCl_2$ (15 mg/L) in water. To this mixture, add the following filter-sterilized solutions: 100 mL of phosphate buffer 0.5 M (pH 8.0), 10 mL of 20% (w/v) glucose, 2 mL of 1 M $MgSO_4$, 1 mL of 2 mg/mL vitamin B_1, and 3.4 mL of a solution that initially contains 30 mL of 1 M $MgCl_2$ and 3 mL of 1 mM $FeCl_3$.
   b. Carry out deuterium labeling in MEM medium, as described above, but with an unlabeled nitrogen source and in the presence of 50% (v/v) deuterated water (originally 99.75% atom).
2. Perform overexpression as described in **Subheading 3.1.1.**

3.2.2. Negative Labeling with 99.897% ^{12}C

1. Hydrolyze two samples (0.5 g each) of lyophilized Algal, enriched in ^{12}C to 99.897%, in HCl + H_2O (1:1 [v/v]) (50 mL each sample) at 108°C under a nitrogen atmosphere with a constant water reflux for 4 and 12 h, respectively.
2. Reduce both hydrolysates to a thick syrup in a Rotavapor (Büchi Labortechnik, Flawil, Switzerland) concentrator (10 mm Hg vacuum) and then further dry them for 14 h under high vacuum in a dessicator containing NaOH pellets.
3. Dissolve each sample in 20 mL water, bring to pH 7.0 with 0.8 M NaOH, and then centrifuge at 8000g for 30 min. Sterile-filter the supernatant and use it to replace the glucose as sole carbon source in the preparation of MEM (100 mL) (*see* **Note 1**).

The negatively labeled fusion protein is overexpressed in the presence of both antibiotics, as described in the earlier section, but using 1 mM IPTG for induction, which was found to be the lowest concentration allowing a high expression level, and therefore represented a compromise between maximized expression and minimized contamination from unlabeled IPTG.

3.2.3. Protein Extraction and Purification

The extraction procedure is performed, using the N-terminal hexahistidine tag, on high-affinity Ni-NTA agarose gel as recommended by the manufacturer (Quiagen, Kontron).

1. Dissolve pellets corresponding to a 500-mL culture by rotation for 2 h at 37°C in 50 mL of buffer A (6 M guanidinium chloride, 0.1 M NaH_2PO_4, 0.01 M Tris-HCl, adjusted to pH 8.0 with NaOH).
2. Centrifuge the mixtures at 10,000g for 20 min, dilute the supernatant with 3 volumes of buffer A prior to loading a 1/3 portion onto a 3 mL Ni-NTA agarose column equilibrated in the same buffer.
3. Wash the gel with 30 mL of buffer A, 50 mL of buffer B (8 M urea, 0.1 M NaH_2PO_4, 0.01 M Tris-HCl, adjusted to pH 8.0 with NaOH) and 50 mL of the same buffer adjusted to pH 6.3. Finally elute with 40 mL of the same buffer, but at pH 5.9. Regenerate the gel according to the manufacturer recommendation before loading the next 1/3 portion.
4. Dilute each eluate with 150 mL 0.1% TFA and purify using a double Sep-Pak system (two cartridges in series). Analyze aliquots by ESI-MS, SDS-PAGE, and RP-HPLC followed by ESI-MS of the collected peak.

3.3. Protein Modification

3.3.1. Formation of Disulphide Bridges in Proinsulin

The following steps, as well as those described below for the formation of insulin and C-peptide, are based on published work *(7,25)*.

1. Remove the histidine tag from the samples by 24-h treatment in 3 mL of 0.3 M BrCN in 75% formic acid (for a 1/3 portion of a 500-mL culture). This process causes cleavage at the methionine upstream of Phe-1 in the proinsulin sequence. The reaction can be followed by analytical HPLC and ESI-MS.

Disulphide bridges are formed by a first step of oxidative sulfitolysis followed by controlled β-mercaptoethanol treatment.

1. Dissolve proinsulin, obtained from 500-mL culture medium, in 2 mL of 6 M guanidinium-chloride (Microselect, Fluka 50935) and 0.1 M glycine adjusted to pH 9.0 with NaOH.
2. Add sodium sulphite (100 mg: 0.4 M final) and sodium tetrathionate (200 µL of a 1-M solution: 0.1 M final) and leave the reaction for 18 h at room temperature.
3. Dialyze (Visking 18/32 bags, Scientific Instruments, London, UK) the samples twice at 4°C against 2.5 L of a 50-mM glycine buffer (pH 9.0) containing 5 mM EDTA for 6 and 20 h, respectively.
4. Bring the pH of the dialysate to 10.5 with NaOH, and add β-mercaptoethanol to a final concentration of 750 µM. Leave the reaction on ice for 24 h and then acidify the solution with acetic acid.

5. Immediately purify the refolded proinsulin by semipreparative HPLC (25–45% solvent B in solvent A [*see* **Subheading 2.2.1.**] over 20 min). Analyze the product by ESI-MS.

3.3.2. Transformation into Insulin and C-peptide

1. Dissolve proinsulin obtained from **Subheading 3.2.1.** or **3.2.2.** in 2 mL of a 1% NH_4HCO_3 solution to which 5 µg of TPCK-treated trypsin (Worthington, 32A680) and 0.5 µg of carboxypeptidase B (Boehringer Mannheim, 103 233) have been added.
2. Leave the enzymatic reaction for 45 min at 37°C and then stop it by the addition of 100 µL of glacial acetic acid.
3. Immediately purify the products by HPLC as described under **Subheading 3.3.1.**, freeze-dry the two collected components, and analyze these by ESI-MS.
4. Further fractionate the insulin-containing component by RP-HPLC on the same system, but using 10% acetonitrile + 90% of a buffer composed of 100 mM H_3PO_4, 50 mM $NaClO_4$, and 20 mM triethylamine at pH 3.0 as solvent C and 50% acetonitrile + 50% of the same buffer as solvent D. Apply a gradient of 45–50% solvent D over 50 min.
5. Dilute the eluate with 5 vol of 0.1% TFA and purify using a Sep-Pak cartridge. Evaporate the acetonitrile and freeze-dry the samples prior to further characterization.

3.4. Discussion

3.4.1. Protein Synthesis

The use of LB, ^{15}N-enriched MEM, deuterium-enriched MEM, or ^{12}C-enriched Algal hydrolysate were all found to give the expected protein band at 11 kDa on SDS-PAGE gels. No obvious difference in intensity could be observed when the different culture media were compared, except for the Algal hydrolysate, in which the bacteria did not grow as well as in the other media. The affinity isolation using Ni-NTA agarose gel followed by Sep-Pak extraction was found to be extremely efficient, leading to a single band at 11 kDa on SDS-PAGE (with slight traces at 22 and 33 kDa, probably polymers) and almost no fusion protein remaining in the uncollected fractions. RP-HPLC analysis gave a single peak, although several shoulders were evident (not shown). ESI-MS analysis of the unlabeled protein gave a molecular ion envelope corresponding to a molecular weight of 11,056.0 Da, in agreement with the calculated value of 11,055.5 Da for the oxidised form of the fusion protein (**Fig. 3A**). As no significant impurities could be detected in the mass spectrum, the different shoulders observed on RP-HPLC were explained as various disulphide bonded isomers demonstrating the necessity of appropriate refolding. Typically, 12–15 mg of the fusion protein were obtained from 1 L of culture after Sep-Pak extraction.

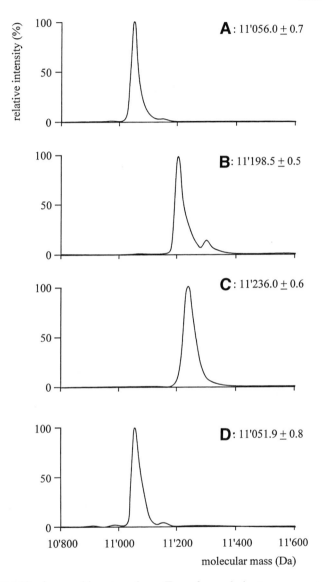

Fig. 3. ESI-MS of recombinant analogs. Transformed electrospray mass spectra of the recombinant protein in fusion with the hexahistidine tag following Ni-NTA affinity purification and solid-phase extraction. (**A**) Nonlabeled form (11,055.5 Daltons calculated). (**B**) 99.4% ^{15}N-enriched MEM. (**C**) 50% D_2O-enriched MEM. (**D**) 99.897% ^{12}C-enriched culture medium. Phosphoric acid or sulphuric acid adducts can be seen at higher mass, especially in **B**; this is commonly observed during mass spectrometric analysis. The resolution was not high enough to allow any special observation with the negatively labeled analog (apart from establishing a -4.1 mass unit difference from that of natural isotopic abundance), as illustrated in **D**.

3.4.2. Labeling

^{15}N enrichment led to a protein of molecular weight 11,198.5 Da (**Fig. 3B**), corresponding to an increase in molecular weight of 142.5 Daltons. As the fusion protein contains 145 nitrogen atoms, the final enrichment in ^{15}N was calculated as 98.6% and the labeling efficiency as 99.2% (the source was 99.4% ^{15}N and the natural abundance of ^{15}N is 0.37%). Deuterium labeling using a culture medium containing 50% deuterated water led to a compound of 11,236.0 Daltons molecular weight (**Fig. 3C**), corresponding to a 24.4% enrichment (+180.0 Daltons for 739 hydrogen atoms). This relatively low labeling efficiency was found to be sufficient for our studies and was not unexpected since the nutrients in the culture medium were of natural isotopic abundance, and non-carbon bound deuterium atoms may have been exchanged with hydrogen atoms during workup. No trials were performed to improve the deuterium labeling efficiency, because, on the one hand, the bacteria do not survive high deuterated water concentrations and, on the other hand, the physicochemical properties of a highly deuterated protein would probably be different. Negative labeling with ^{12}C-enriched Algal hydrolysate gave a measured molecular weight of 11,051.9 Daltons (**Fig. 3D**), corresponding, as expected, to a decrease in molecular weight of 4.1 Daltons. The final ^{12}C enrichment was calculated as 99.76, which should be compared with the natural abundance of 98.9%. This enrichment corresponds to a 79% labeling efficiency since the abundance of ^{12}C in the carbon source was 99.897%. The enrichment was confirmed by high-resolution mass spectrometry (*see* **Subheading 3.5.**). The fact that this efficiency is not closer to 100% is probably because of incorporation of nondepleted carbon from the IPTG overexpression stimulating agent, which is added at a crucial point. All these labeled proteins co-eluted with unlabeled material on RP-HPLC under the conditions described.

3.4.3. Analytical Control and Biological Assay

All the analytical controls performed during refolding, transformation, and conversion processes of the unlabeled and labeled forms of the recombinant protein gave the expected result. RP-HPLC always showed pure compounds although the retention times varied from step to step as would be expected from the physicochemical and sequence modifications. ESI-MS analysis routinely gave measured molecular weights within ±1 Daltons of the calculated values and did not reveal any unexpected contaminant. The last RP-HPLC purification step was found to be necessary and very efficient for separating human insulin from its des-ThrB30 form that is produced as a side product during the enzymatic conversion. Typically, 5 mg of pure refolded proinsulin or 3 mg of pure insulin could be obtained from a 1-L culture volume of LB.

Recoveries were found to be approximately 35% lower from MEM, and 50% lower from hydrolysed Algal media. Final products were found to have the expected molecular weight and showed no trace of contamination on RP-HPLC or ESI-MS analysis. The biosynthetically labeled forms of recombinant proinsulin, insulin, and C-peptide could not be distinguished from their unlabeled forms by RP-HPLC or by spectrophotometry. They were identical to materials from other available sources (commercial or C-peptide produced by solid-phase synthesis). Furthermore, our recombinant human insulin, and its analog labeled with 99.4% ^{15}N, induced glycogen formation in primary rat hepatocytes in culture at the same level as was achieved with commercial human insulin, within the 10% tolerance of our in vitro biological assay. This new assay has been described in detail elsewhere *(26)*.

3.4.4. Peptide Mapping

Comparison of our biosynthetic analogs with commercial insulin and synthetic C-peptide on a molecular level by peptide mapping confirmed their structure. The procedure for peptide mapping is as follows:

1. Dissolve 1 mg each of the insulin or C-peptide sample (commercial human insulin, in-house synthetic C-peptide, in-house recombinant unlabeled human insulin or C-peptide, and insulin or C-peptide analogs labeled with 99.4% ^{15}N) in 1 mL of 1% NH_4HCO_3 containing 10 µg of *Staph. aureus* V8 protease.
2. Stop the reaction with 100 µL glacial acetic acid after a 4-h incubation at 37°C and analyze a 15% fraction of each digest by RP-HPLC with an analytical column using a gradient of 10–40% solvent B in solvent A (*see* **Subheading 2.2.**) over 60 min.
3. Collect and freeze-dry the eluted components and analyze these by ESI-MS.

The RP-HPLC profiles were found to be identical for all the forms of insulin and C-peptide respectively (**Fig. 4**), and the molecular weights measured for each collected peak were all found to be within 1 Da of the expected values, each of them corresponding to a specific enzymatic digest fragment (**Tables 1 and 2**). It is interesting to note that the difference in molecular weight between the fragments of the unlabeled form of insulin and its analog uniformly labeled with ^{15}N exactly reflects the number of nitrogen atoms present in each fragment. Such strategies of labeling with stable isotopes can thus find applications in macromolecular composition analyses.

3.5. High-Resolution Mass Spectrometry

High-resolution mass spectra of native human insulin and the ^{15}N-enriched recombinant analog are shown in **Fig. 5A** and **B**, respectively. The spectrum of the enriched insulin (**Fig. 5B**) exhibits an almost natural isotopic envelope cen-

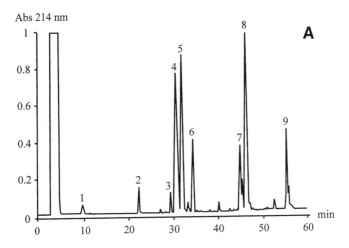

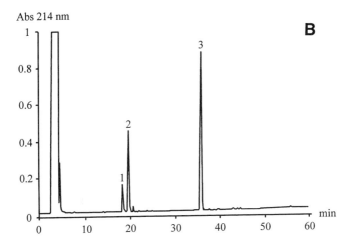

Fig. 4. Reversed-phase HPLC. HPLC chromatograms of (**A**) insulin and (**B**) C-peptide following *Staph. aureus* V8 protease treatment, as described in the text. The numbered peaks were collected and analyzed by ESI-MS (*see* data in **Tables 1** and **2**).

tered around 5871.5 Daltons, as had been previously observed by ESI-MS (**Table 1**), except for slightly higher relative intensities of the earlier signals (starting with the monoisotopic signal at m/z 5869.5) due to incomplete incorporation of the label. The high-resolution mass spectrum obtained with the negatively labeled recombinant insulin is shown in **Fig. 5C**; it corresponds to that obtained after calculation according to Werlen *(27)* and confirms the enrichment in ^{12}C to be approximately 99.7%. The difference between the natural and ^{13}C-depleted form (**Fig. 5A** and **C**) is obvious here, the isotopic ratios

Table 1
Peptide Mapping of Insulin[a]

Peak	Fragment	Calculated molecular weight (M_r)	Fragments of commercial insulin (M_r)	Fragments of recombinant natural insulin (M_r)	Fragments of ^{15}N-labeled insulin (M_r)	ΔM	Nr (N)
2	A13–17	664.8	665.7	664.7	670.8	6	6
3	A5–12_B1–13	2322.7	—	2322.8	2350.5	27.8	28
4	B22–30	1116.3	1116.8	1116.6	1129.2	12.9	13
5	A18–21_B14–21	1377.6	1377.8	1377.6	1392.2	14.6	14
6	A1–12_B1–13	2721.1	2721.7	2721.2	2752.5	31.4	32
7	A5–17_B1–13	2969.4	2969.7	2969.6	3002.7	33.3	34
8	A1–17_B1–13	3367.9	3368.1	3368.3	3405.5	37.6	38
9	Insulin	5807.7	5807.4	5808.6	5872.1	64.4	65

[a]Molecular weights measured by ESI-MS of the peaks collected from the HPLC chromatograms shown in **Fig. 4**. ΔM is the difference in Daltons between the calculated molecular weight of native insulin and the measured molecular weight of the ^{15}N analog, and Nr (N) is the number of nitrogen atoms present in each fragment.

Table 2
Peptide Mapping of C-Peptide[a]

Peak	Fragment	Calculated molecular weight (M_r)	Fragments of ^{15}N-labeled C-peptide (M_r)	ΔM	Nr (N)
1	4–11	887.0	897.3	10.3	10
2	1–11	1216.3	1229.4	13.1	13
3	12–27	1436.6	1453.9	17.3	17

[a]See **Table 1** footnote.

being quite distinct. The relative intensities of the isotopic signals at M+H values starting at 5804.7 Daltons and with 1-Dalton increments are successively 2, 25, 73, 100, 95, 75, 42, 29, and 11% for the natural form with the most intense signal in the fourth position. The corresponding intensities for the negatively labeled analog are successively 84, 100, 95, 62, 38, 27, 20, 8, and 3%, with the most intense signal in the second position. This experimental difference would be even more marked if a higher degree of labeling had been achieved, as, for example, with simultaneous depletion in ^{13}C, ^{15}N, ^{18}O, and, in particular, ^{34}S (which has a 4.4% natural abundance and effects a 2-Dalton increase in molecular weight) in the case of sulfur-containing proteins. Such a labeling would, in addition to an improvement of the specificity of the isotopic signature, also provide the advantage of an increased signal intensity and accordingly improve the sensitivity of the method. A fully monoisotopic analog, however, would be very expensive to produce and might give such a narrow signal (especially if multiply charged as in ESI-MS) that it might be confused with electronic noise.

3.6. Conclusions

Our results demonstrate that isotopic depletion can be used to provide a distinct isotopic signature for a protein. Such a tracer is nonradioactive, has very high similarity to the natural form, and is relatively easy to synthesize. High-resolution MALDI-FT-MS analysis further provides the advantage of being much less susceptible to contamination that might interfere in the mass spectrum; their atomic composition would be different, so the probability of contaminants giving isotopic signals at exactly the same m/z values is extremely low under the high-resolution experimental conditions used. For similar molecular weights, contaminants would be expected to give rise to resolved envelopes with signals that are distinct from those of the target compound and thus not interfering with it. Furthermore, in cases of mixtures of the native

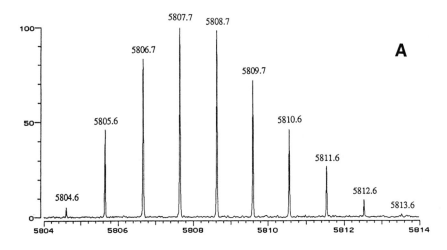
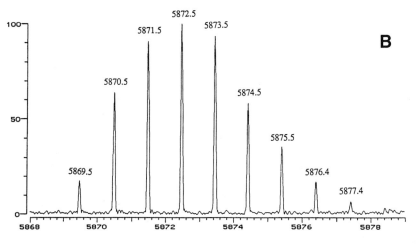
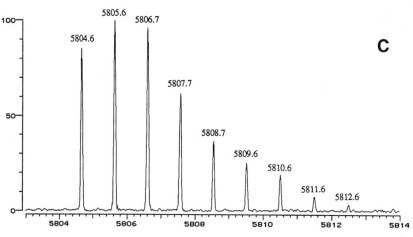

and negatively labeled form of insulin (as would happen with studies in humans after injection of the analog), a calculation of their relative amounts might be possible on the basis of the relative intensities of each isotopic signal.

Negative labeling is a new approach to protein metabolic studies that has not been described in this context so far. It is based on isotopic depletion rather than enrichment and should allow pharmacokinetic studies in humans or other complex biological environments independently of the natural form of the target compound. Such analogs, being authentic, can be injected in humans, allowing pharmacokinetic studies of the exogenous (injected) hormone independently of its native endogenous form. The specificity of the isotopic signature is also valid for degradation fragments from a labeled protein. Mass spectrometric analysis of a sample should therefore enable easy detection, with direct identification and characterization, of unknown degradation fragments, even in a complex biological extract. Negative labeling should therefore be more suited to the identification of metabolites than to the simple quantitation of a target protein. Although relatively expensive FTICR-MS instrumentation was used in the experiments described, it is possible to resolve the isotope clusters of the molecular ion regions of molecules up to about 6 kDa molecular weight using less expensive high-performance MALDI-TOF reflectron instruments.

4. Notes

Positive labeling with ^{15}N has already been used to provide an internal standard for quantitative studies of proteins based on isotope dilution mass spectrometry *(3,4)*. Such investigations are divided into three distinct experimental parts, which need to be distinguished when considering the use of our methodology: preparation of labeled analogs (**Notes 2–4**), sample handling, i.e., extraction and purification of the target compounds from blood samples (**Notes 5–7**), and mass spectrometric analyses (**Notes 8–12**).

1. Separate cultures made with each of the hydrolysates gave encouraging results, so the expressed proteins obtained in each case were finally pooled to obtain the mass spectra shown.

Fig. 5. MALDI-FTICR-MS. High-resolution mass spectra (vertical axis, relative intensity; horizontal axis, *m/z*) obtained with 10 pmol of (**A**) native human insulin, (**B**) ^{15}N-labeled human insulin (99.4%), and (**C**) negatively labeled recombinant human insulin prepared with a hydrolysate of 99.897% ^{12}C-enriched Algal. The signals observed on the *m/z* scale correspond to [M+H]$^+$ ions of the different isotopic forms of the hormone at resolving powers of 220,000, 90,000, and 92,000 for **A**, **B**, and **C**, respectively. The only difference between **A** and **B** is that the monoisotopic mass is shifted from *m/z* 5804.6 to *m/z* 5869.5 because there are 65 nitrogen atoms in the protein. The mass spectrum **C** shows a considerable shift toward the monoisotopic peak at 5804.6 with a different "nonnatural" isotopic envelope. (Reproduced with permission from **ref. *17***).

2. In labeled analog studies, a carrier protein (which is not injected into the animal or patient) is required to limit losses by adsorption during extraction and purification, and this is conveniently furnished through partial metabolic labeling of the analyte protein with deuterium. High deuterium enrichment is not required for such purposes; 25% is quite sufficient. Small variations in deuterium content do not affect the measurements.
3. The internal standard and the carrier are added in known amounts to the blood, serum or plasma sample before any further extraction procedure. The ^{15}N- and ^{15}N^{2}H-labeled recombinant analogs are usually a good choice to start with.
4. The internal standard must be added to the sample to be analyzed with meticulous control of the volume and concentration. The concentration of the stock solution must be determined accurately, ideally through repeated amino acid analyses.
5. For sample handling whenever feasible, we recommend the use of a strategy based on immunoaffinity chromatography, which has proved to be very successful for our insulin work and permits the necessary selective enrichment of proteins present at very low concentrations. The column retentate is analyzed by capillary HPLC prior to final mass spectrometric analysis.
6. Three different extraction procedures of the target compound from filtered plasma or serum samples (0.2–5 mL typically) have been developed and can be used either alone or in combination. Several variants of two of them have already been described in detail: immunoaffinity (with a single highly specific monoclonal antibody or using mixed-bed chromatography with several antibodies) and solid-phase extraction using C18 Sep-Pak cartridges *(3,4)*.
7. Following solid-phase extraction, ultrafiltration, carried out according to the following protocol, has been found to be an efficient technique for the second purification step:
   a. Resuspend the sample in 500 µL 0.1% TFA and load onto an Ultrafree UFC4LTK filtration unit (Millipore, 8302 Kloten, Switzerland) which has a 30-kDa cutoff and has previously been washed with 500 µL 0.1% TFA by centrifugation.
   b. Place a 1.5-mL Eppendorf tube with the cap removed between the filter and the collecting chamber to recover the filtrate in a smaller tube.
   c. Centrifuge the Ultrafree unit for 20 min on a bench centrifuge (Hettich Universal II) at full speed (5000 rpm). Wash the sample tube with 500 µL 0.1% TFA and then load this volume onto the Ultrafree. Centrifuge the ultrafilter again for 20 min and then repeat the process with a further 250 µL 0.1% TFA. Finally, freeze-dry the pooled ultrafiltrate directly in the Eppendorf tube.
8. Electrospray ionization mass spectrometry with a quadrupole analyzer is suitable for quantitative IDA studies, and expensive high-resolution FTICR-MS instrumentation is not required.
9. Sensitivity and resolution are limiting factors with ESI-MS using a quadrupole analyzer, and a compromise has to be found in each case to optimize the signal-to-noise ratio. The quadrupole analyzer is scanned over a narrow *m/z* range to cover only the signals corresponding to the major peaks of the charge distribution and thus maximize sensitivity.

10. We have also been able to use MALDI-TOFMS for IDA measurements of insulin levels with success (data not shown). However, although they are extremely sensitive, the low dynamic range of these instruments is a drawback for IDA measurements.
11. In preliminary experiments, we have collected samples directly on a MALDI-TOFMS plate for analysis and have obtained excellent spectra and partial sequence information (up to eight residues) by carboxypeptidase Y digestion and by postsource decay at the 500-fmol level (data not shown).
12. Preliminary experiments with nanoelectrospray ionization and iontrap (LCQ, Finnigan, San Jose, CA) or orthogonal time-of-flight (QTOF, Micromass, Manchester, England) instrumentation have shown that much higher sensitivities are possible, together with extensive sequence information from fragments. A concentration of 1 fmol/µL and a sample volume < 1 µL gave excellent signal-to-noise ratios, even when recording ranges of 100 *m/z* units.

Acknowledgments

We thank in particular Prof. Philippe Halban and Dr. Jean-Claude Irminger from the Louis Jeantet Research Laboratories in Geneva for gifts of plasmids and for interesting and stimulating discussions. We thank Dr. Yunzhi Li and Prof. Robert T. McIver from the Department of Chemistry of the University of California in Irvine for the high-resolution MALDI-FT-MS analyses. We are grateful to Prof. Jean-Claude Jaton of our Department, to Dr. Reto Crameri from the Schweizer Institut für Asthma und Allergische Forschung in Davos, to Prof. Shige Harayama and Dr. Raymond Werlen, formerly of our Department, and to Dr. Daniel Scherly and Dr. Philippe Calain from the Department of Genetics and Microbiology of the Faculty of Medicine in Geneva for helpful discussions and support. We thank the Swiss National Science Foundation and the Sandoz Research fund for generously supporting this project.

References

1. Halban, P. A. and Irminger, J.-C. (1994) Sorting and processing of secretory proteins. *Biochem. J.* **299**, 1–18.
2. Stöcklin, R., Rose, K., Green, B. N., and Offord, R. E. (1994) The semisynthesis of [octadeutero-PheB1–octadeutero-ValB2]-porcine insulin and its characterization by mass spectrometry. *Protein Eng.* **7**, 285–289.
3. Stöcklin, R., Vu, L., Vadas, L., Cerini, F., Kippen, A. D., Offord, R. E., et al. (1997) A stable isotope dilution assay for the in vivo determination of insulin levels in humans by mass spectrometry. *Diabetes* **46**, 44–50.
4. Kippen, A. D., Cerini, F., Vadas, L., Stöcklin, R., Vu, L., Offord, R. E., et al. (1997) Development of an isotope dilution assay for precise determination of insulin, C-peptide, and proinsulin levels in non-diabetic and type II diabetic individuals with comparison to immunoassay. *J. Biol. Chem.* **272**, 12,513–12,522.

5. Berg, H., Walter, M., Mauch, L., Seissler, J., and Northemann, W. (1993) Recombinant human preproinsulin. Expression, purification and reaction with insulin autoantibodies in sera from patients with insulin-dependent diabetes mellitus. *J. Immunol. Methods* **164,** 221–231.
6. Chance, R. E., Hoffmann, J. A., Kroeff, E. P., Johnson, M. G., Schmirmer, E. W., Bromer, W. W., et al. (1981) *Peptides: Synthesis-Structure-Function,* (Rich, D. H. and Gross, E., eds.), Pierce Chemical, Rockford, IL, pp. 721–728.
7. Heath, W. F., Belagaje, R. M., Brooke, G. S., Chance, R. E., Hoffmann, J. A., Long, H. B., et al. (1992) (A-C-B) human proinsulin, a novel insulin agonist and intermediate in the synthesis of biosynthetic human insulin. *J. Biol. Chem.* **267,** 419–425.
8. Offord, R. E. (1980) *Semisynthetic Proteins,* John Wiley & Sons, Chichester, UK.
9. Tang, J. and Hu, M. (1993) Production of human proinsulin in *E. coli* in a nonfusion form. *Biotechnol. Lett.* **15,** 661–666.
10. Couprie, J., Remerowski, M. L., Bailleul, A., Courcon, M., Gilles, N., Quemeneur, E., et al. (1999) Differences between the electronic environments of reduced and oxidized *Escherichia coli* DsbA inferred from heteronuclear magnetic resonance spectroscopy. *Protein Sci.* **7,** 2065–2080.
11. Hansen, A. P., Petros, A. M., Mazar, A. P., Pederson, T. M., Rueter, A., and Fesik, S. W. (1992) A practical method for uniform isotopic labeling of recombinant proteins in mammalian cells. *Biochemistry* **31,** 12,713–12,718.
12. Marshall, A. G., Senko, M. W., Li, W., Li, M., Dillon, S., Guan, S., et al. (1997) *J. Am. Chem. Soc.* **119,** 433–434.
13. Zubarev, R. A. and Demirev, P. A. (1998) *J. Am. Soc. Mass Spectrom.* **9,** 149–156.
14. Morihara, K., Oka, T., Tsuzuki, H., Tochino, Y., and Kanaya, T. (1980) Achromobacter protease I-catalyzed conversion of porcine insulin into human insulin. *Biochem. Biophys. Res. Commun.* **92,** 396–402.
15. McIver, R. T. J., Li, Y., and Hunter, R. L. (1994) High-resolution laser desorption mass spectrometry of peptides and small proteins. *Proc. Natl. Acad. Sci. USA* **91,** 4801–4805.
16. Li, Y., Hunter, R. L., and McIver, R. T. J. (1994) High-resolution mass spectrometer for protein chemistry. *Nature* **370,** 393–395.
17. Li, Y., Hunter, R., and McIver, R. T. (1996) *Int. J. Mass Spectrom. Ion Process* **157/158,** 175–188.
18. Marschuk, D., Drumm, M., Saulino, A., and Collins, F. S. (1990) Construction of T-vectors, a rapid general system for direct cloning of unmodified PCR products. *Nucleic Acids Res.* **19,** 1154–1154.
19. Sambrook, J., Fritsch, E. F., and Maniatis, T. (1989) *Molecular Cloning: A Laboratory Manual,* 2 ed, Cold Spring Harbor Laboratory, Cold Spring Harbor, NY.
20. Ullmann, A., Jacob, F., and Monod, J. (1967) Characterization by in vitro complementation of a peptide corresponding to an operator-proximal segment of the beta-galactosidase structural gene of Escherichia coli. *J. Mol. Biol.* **23,** 339–345.
21. Horwitz, J. P., Chua, J., Curby, R. J., Tomson, A. J., DaRooge, M. A., Fischer, B. E., Mauricio, J., and Klundt, I. (1964) Substrate for cytochemical demonstration

of enzyme activity. I. Some substituted 3-indolyl-beta-D-glycopyranosides. *J. Med. Chem.* **7,** 574–579.
22. Stüber, D., Matile, H., and Garotta, G. (1990) *Immunological methods* 4th ed. (Lefkovits, I. and Pernis, B., eds.), pp. 121–152.
23. Vogelstein, B. and Gillepsie, D. (1979) Preparative and analytical purification of DNA from agarose. *Proc. Natl. Acad. Sci. USA* **76,** 615–619.
24. Farabaugh, P. J. (1978) Sequence of the lacI gene. *Nature* **274,** 765–769.
25. Semple, J. W., Cockle, S. A., and Delovitch, T. (1988) Purification and characterization of radiolabelled biosynthetic human insulin from *Escherichia coli.* Kinetics of processing by antigen presenting cells. *Mol. Immunol.* **25,** 1291–1298.
26. Vu, L., Pralong, W. F., Cerini, F., Gjinovci, A., Stocklin, R., Rose, K., et al. (1998) Short-term insulin-induced glycogen formation in primary hepatocytes as a screening bioassay for insulin action. *Anal. Biochem.* **262,** 17–22.
27. Werlen, R. (1994) *Rapid Commun. Mass Spectrom.* **8,** 976–980.

18

Identification of Snake Species by Toxin Mass Fingerprinting of Their Venoms

Reto Stöcklin, Dietrich Mebs, Jean-Claude Boulain, Pierre-Alain Panchaud, Henri Virelizier, and Cécile Gillard-Factor

1. Introduction

A new method is proposed to identify venomous snakes, which is based on electrospray ionization mass spectrometry (ESI-MS) detection and is demonstrated with Asiatic snake venoms from the controversial genus *Naja* (cobras). Appropriate combinations of chromatographic techniques and ESI-MS are used to analyze the crude venom of single specimens. Highly specific toxin mass maps, which can be used as a unique fingerprint for the systematic classification of the snake, are obtained; these results are compared with those obtained using standard samples and with the calculated molecular weights of characterized toxins. By off-line ESI-MS analysis of high-performance liquid chromatography (HPLC) fractions of two venom samples, one from Vietnam (undefined *Naja sp.*) and the other from Thailand (*Naja kaouthia*), it was found that both snakes belong to the same species, namely, *Naja kaouthia*. Using on-line liquid chromatography (LC)/ESI-MS, a direct analysis of crude venom from a single specimen of an unidentified white cobra from Thailand was performed. Two standard venom samples of *Naja naja* and *Naja kaouthia* were also analyzed using this improved strategy. By this approach, a peptide mass map of these three samples was obtained within a day and, in addition, an unambiguous systematic classification of the white cobra as *Naja kaouthia* was obtained. This method is able to identify clearly the origin and purity of crude or partially fractionated venom, which is an important advantage for medical use or in antivenom production.

From: *Methods in Molecular Biology, vol. 146:*
Protein and Peptide Analysis: New Mass Spectrometric Applications
Edited by: J. R. Chapman © Humana Press Inc., Totowa, NJ

1.1. Snake Venoms

Snake venoms *(1–3)* are complex mixtures, typically made of more than 50 different peptides and proteins that exhibit specific activity. Neurotoxins, cytotoxins, and phospholipases, as well as proteases that cause hemorrhages or affect homeostasis or fibrinolysis, are some of the more common venom components. They are applicable to a wide range of applications including basic research, clinical diagnosis, use as therapeutic agents, and the production of antiserum. However, despite their biomedical importance, the origin and precise identification of both crude venoms and purified toxins are often not clear. This phenomenon is closely linked to the classification of snakes, which is far from being solved. Most systematic studies rely either on geographic and morphological characters or on DNA sequencing of the mitochondrial cytochrome B oxidase subunit I gene *(4,5)*. However, neither of these two methods, either alone or in combination, provides information on venom composition, which is of major interest for clinicians and in biomedical research.

For characterizing crude venoms, chromatographic techniques have been used in combination with screening assays for biological activities such as neurotoxicity or phospholipase A2 *(6,7)*. Isoelectrofocusing, capillary isotachophoresis, sodium dodecyl sulfate-polyacrylamide gel electrophoresis (SDS-PAGE), and two-dimensional electrophoretic studies have been performed on crude venoms *(8–13)*. However, these methods are not able to distinguish closely related venom constituents present in a sample; moreover, chromatographic or electrophoretic profiles of crude venom cannot be used for the identification of snake species. Therefore, the combination of chromatography and mass spectrometry has been applied off-line and on-line for direct analysis of crude snake venoms.

1.2. Strategy

As a soft ionization technique, ESI-MS plays an important role in the analysis of intact polypeptides *(14)* and has been widely used for their characterization by molecular weight determination, disulfide bridge assignment, glycosylation site determination, sequencing, and so forth *(15–17)*. Its main advantages are high accuracy, high sensitivity (pmol level), and high resolution in molecular weight determination (up to 150 kDa). The samples in solution are introduced into the ionization source in a continuous liquid flow. Chromatographic systems such as reversed-phase HPLC can thus be coupled on-line to the mass spectrometer (LC/ESI-MS), which is used as a mass detector for the eluate (**Fig. 1**). Such strategies are currently used for the study of protein digests in sequence and structure analyses *(18–21)*.

For the analysis of individual toxins *(22–24)* or crude venoms *(25–30)*, or even for the detection of specific toxins in blood samples *(31)*, mass spectrometry has been used on several occasions. Capillary electrophoresis (CE) in con-

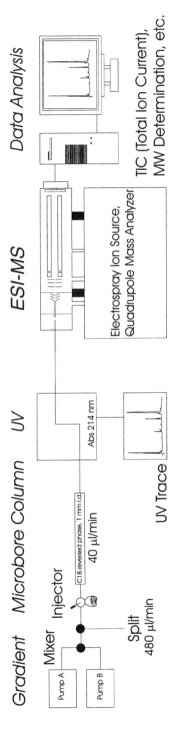

Fig. 1. On-line LC/ESI-MS instrumentation: following UV detection at 214 nm, the eluate of the microbore reversed-phase HPLC column is directly introduced into the electrospray instrument, which is scanned over a wide mass range every 2.4 s during the whole analytical process (about 45 min). Measurement of the ion current (at all m/z values) reaching the mass spectrometer detector results in a TIC chromatogram. For each separated fraction, mass spectra can be averaged and deconvoluted to obtain molecular weight information.

junction with ESI-MS has also been successfully applied to the analysis of crude snake venoms *(28,29)*. However, these studies exhibited some experimental and instrumental restrictions. The mass spectrometer used had a limited mass range (*m/z* 1200), restricting the detection to only one multiply charged ion for most toxins and causing problems of accuracy in molecular weight assignment. The extremely small amounts used in CE prevented full-scan data acquisition; most of the analyses had to be performed in the static single-ion monitoring mode so that repeated experiments had to be performed for a full screening of the content of a venom.

Our preliminary investigations *(27,30)* using on-line LC/ESI-MS for the direct analysis of crude snake venoms and the results obtained using MALDI-TOFMS with tarantula venoms *(25,26)* have shown that accurate molecular weight information on many different compounds can be obtained in a single experiment. In the present chapter we propose the use of liquid chromatography followed by mass spectrometric detection for the direct analysis of crude venoms, the data being summarized as toxin mass maps, which can be used as unique fingerprints for each snake. This approach is based on the principle that snakes of the same species will essentially express toxins with the same amino acid sequences (thus of the same molecular weights) in their venom glands, and that these will be different from those present in other species, i.e., they will have different molecular weights. This analytical methodology should not be affected by variations in relative amounts of the constituents of different venom samples, which is the major cause of intraspecific variation in venom composition.

Venoms from elapid snakes of the genus *Naja* were used as a model in this study, and the methodology was developed off-line in a first stage and then further improved to on-line LC/ESI-MS analysis of crude venom samples, allowing toxin mass fingerprinting in a single experiment.

2. Materials

2.1. Chemical Reagents

1. Reagents and solvents (acetonitrile, trifluoroacetic acid [TFA] and acetic acid) were all of analytical grade or better. Water was purified on a Milli-Q system (Millipore, Bedford, MA).
2. 2% Acetic acid as diluent for the crude venom (0.1% TFA may also be used).
3. 0.1% aqueous TFA (HPLC solvent A).
4. 90% aqueous CH_3CN + 0.1% TFA (HPLC solvent B).
5. 50% aqueous methanol + 2% acetic acid for off-line ESI-MS analyses (50% aqueous CH_3CN + 0.2% formic acid may also be used).
6. A fresh, 10 pmol/µL solution of a standard toxin or porcine insulin (Novo-Nordisk, Bagsvaerd, Denmark) in 35% aqueous CH_3CN + 0.1% TFA to tune the

mass spectrometer for on-line LC/ESI-MS analyses. Stock solutions of 1 mg/mL can be stored at –20°C.
7. A fresh, 10 pmol/µL solution of horse heart myoglobin (Sigma, Buchs, Switzerland) in 35% aqueous CH_3CN + 0.1% TFA to calibrate the mass spectrometer for on-line LC/ESI-MS analyses. Stock solutions of 1 mg/mL can be stored at –20°C. A mixture of insulin and angiotensin may also be used for calibration.

2.2. Venom Samples

1. For off-line analysis, crude venom from *Naja sp.* originating from the region of Ho-Chi-Minh City (Vietnam) was kindly provided by Prof. Trinh Kim Anh, and *Naja kaouthia* (monocled cobra) venom was purchased from the Queen Saovabha Institute, Bangkok (Thailand). For on-line analysis, venom from a single captive white cobra originating from the region of Chiang Mai (North Thailand) and belonging to a rare endemic population, which has not yet been described in the literature, was obtained by milking and freeze-drying. This specimen is neither an albino, nor related to the yellowish colored cobra from central Thailand described as *Naja kaouthia suphanensis (32)* and identified as *Naja kaouthia* by DNA sequencing *(5)*. This white cobra has no homology to the Indochinese spitting cobra (*Naja siamensis*), which can be found in the same area. It shows some morphological similarities with the distant species *Naja naja,* but is better linked to the closer species *Naja kaouthia* that can be found to the south and west of its habitat. *Naja naja* (Indian cobra) and *Naja kaouthia* standard crude venoms, which were also used for on-line analysis, were obtained from captive-born snakes originating from India and Thailand, respectively, by milking followed by freeze-drying.
2. Membrane filter (0.45 µm) to remove insoluble macromolecules and other particles present from milking prior to HPLC injection of the crude venom samples.

2.3. HPLC Systems *(see* **Note 1***)*

Overall, the HPLC systems used consisted of a gradient controller, C_8 or C_{18} reversed-phase column, an injector with a 30–200-µL sample loop and an ultraviolet (UV)-absorbance detector. The systems were capable of providing either a low (40 µL/min) pulse-free flow with reproducible gradient that could be passed directly, after UV detection, into the electrospray ion source of a quadrupole mass spectrometer, or a 0.6 mL/min flow which, with an appropriate splitter, allowed 10–40 µL/min to enter the ion source while the remainder was collected immediately after UV detection.

A typical low-flow (40 µL/min) system, as used in this work, comprised a 150 × 1 mm ID reversed-phase column packed with Nucleosil C_{18} 5-µm, 300-Å particles (Macherey Nagel, Düren, Germany), two PU980 pumps (Jasco, Cremella, Italy), an SUS mixer (Shimadzu, Kyoto, Japan), and a manual injector (Valco, Schenkon, Switzerland). UV monitoring at 214 nm used a Kontron 432 UV detector (Milano, Italy) fitted with a 90-nL capillary Z-flow cell (LC Packings, Amsterdam, the Netherlands).

A typical higher flow (0.6-mL/min) system, as used in this work, comprised two M6000A pumps, a model 720 system controller, a model 490E absorbance detector (all from Waters, Bedford, MA), and a manual injector (Valco) together with a column (250 × 4 mm ID) packed with Nucleosil C_8 5-µm, 300-Å particles (Macherey Nagel).

2.4. Mass Spectrometry (see Note 2)

1. Off-line analyses used a VG-Trio 2000 instrument fitted with a 3000 mass units radiofrequency (RF) generator (VG Biotech, Altrincham, UK). The mass spectrometer was equipped with an electrospray ion source and was operated in the positive-ionization mode under control of the Lab-Base data system.
2. On-line LC/MS analyses used a Nermag R1010 quadrupole mass analyzer with a mass range of 2000 (Quad Service, Poissy, France). The instrument was equipped with an electrospray ion source (Analytica of Brandford, Brandford, CT) and was operated in the positive-ionization mode. The data system used the HP Chem software (Hewlett-Packard, Palo Alto, CA) (*see* **Note 3**).

2.5. Databases (see Note 4)

1. The measured molecular weights of venom constituents were compared with those calculated from previously described toxins, which can be found in the VENOMS *(33)* and SWISS-PROT *(34)* databases, which cover the whole literature in this field.

3. Methods
3.1. Off-Line Analyses
3.1.1. Reversed-Phase HPLC

The *Naja sp.* and *Naja kaouthia* venoms, from Vietnam and Thailand, respectively, were fractionated by reversed-phase HPLC, prior to mass spectrometric analysis, as follows.

1. Carry out a chromatographic separation using the higher flow HPLC system (*see* **Subheading 2.3.**) with the 250 × 4 mm ID column packed with Nucleosil C_8 5-µm, 300-Å particles operated at a flow rate of 0.6 mL/min. Solvent A was 0.1% TFA in water (v/v), and solvent B was 0.1% TFA in water + 90% acetonitrile (v/v).
2. Equilibrate the column with 100% A prior to injection of the crude venom sample (50 µg of a 1 mg/mL stock solution in 2% acetic acid filtered through a 0.45-µm membrane).
3. Maintain 100% A for 5 min after injection and then start a linear gradient (1% B/min) to 60% B, monitoring the effluent at 214 nm.
4. Collect the major fractions and freeze-dry these prior to ESI-MS analysis.

3.1.2. Mass Spectrometry (see **Notes 2 and 5**)

ESI-MS analysis of the purified fractions was carried out on the VG-Trio 2000 electrospray instrument (*see* **Subheading 2.4.**) as follows.

1. Dissolve the collected fractions (see **Subheading 3.1.1.**), at an approximate concentration of 10 pmol/µL, in a mixture of methanol and water 1:1 (v/v) containing 1% acetic acid.
2. Inject 10-µL portions of the fraction solutions into the electrospray system at a flow rate of 2 µL/min.
3. Scan the quadrupole with a 10 s/scan cycle, usually from m/z 500 to 1500, to include all the major molecular ion peaks (see **Notes 6–8**). Data were processed using a six-point smoothing-routine. External calibration of the mass scale was performed with a solution of horse heart myoglobin (see **Note 9**).

3.2. On-Line LC-ESI-MS Analyses

3.2.1. Microbore HPLC

The venom samples of white cobra, as well as samples of *Naja naja* and *Naja kaouthia* standards, were analyzed by on-line LC/ESI-MS as follows.

1. Carry out a microbore HPLC separation using the low-flow HPLC system (see **Subheading 2.3.**) with the 150 × 1 mm ID reversed-phase column packed with Nucleosil C_{18} 5-µm, 300-Å particles operated at a flow rate of 40 µL/min. Solvents (see **Note 10**) are the same as those used for off-line separation (see **Subheading 3.1.1.**).
2. Equilibrate the column with 20% B prior to injection of the crude venom sample (30 µg of a 2 mg/mL stock solution in 0.1% TFA filtered through a 0.45-µm membrane).
3. Maintain 20% B for 5 min after injection and then initiate a linear gradient (1% B/min) to 60% B.
4. Following UV detection, introduce the eluate directly into the mass spectrometer.

3.2.2. Mass Spectrometry (see **Notes 2 and 5**)

On-line mass spectrometric analyses were carried out using the Nermag R1010 quadrupole electrospray instrument. Experiments were performed by scanning from m/z 500 to 1800 with a step size of 0.1 mass unit and a rate of 0.42 scan/s Mass spectra corresponding to each component shown by the total-ion current (TIC) chromatogram were averaged, allowing an accurate molecular weight determination. External calibration of the mass scale was performed with a mixture of porcine insulin and neurotensin (Sigma-Aldrich, St. Louis, MO) (see **Note 9**).

3.3. Discussion

3.3.1. Off-Line Analyses

By reversed-phase HPLC fractionation of the two venom samples (*Naja sp.* from Vietnam and *Naja kaouthia* from Thailand), 13 fractions were collected. ESI-MS analysis of each fraction resulted in the determination of 26 molecular

Table 1
Off-Line Analysis of Venom Samples from *Naja sp.* and *Naja kaouthia* (Vietnam and Thailand, respectively)[a]

Peak no.	Measured mol wt (Daltons)	
	Naja kaouthia (Thailand)	*Naja sp.* (Vietnam)
1	6880.2; 6850.0	6880.3; 6850.4
2	6965.1; 7062.7	6964.8; 7062.4
3	**6975.6**; 7073.2	**6974.9**; 7072.5
4	**7820.4**	**7820.9**
5	6844.5; 6827.2; 6860.2	6845.0; 6827.3; 6860.2
6	7710.7	7711.4
7	7612.8	7612.7
8	7594.7; 7356.1	7595.0; 7356.3
9	**6708.8; 6736.7**	**6709.1; 6736.9**
10	**6986.3**; 7083.2	**6986.9**; 7083.5
11	**6645.6; 6692.8**; 6984.7	**6646.0; 6692.5**; 6984.6
12	**6739.0**; 6835.7	**6738.5**; 6835.8
13	6926.7; 6786.9; 24934.0	6926.3; 6786.2; 24933.0

[a]Lists are of molecular weights measured after HPLC separation. Both samples gave extremely similar results. Peak numbering refers to **Fig. 2**; molecular weights corresponding to described toxins are in bold characters and can be found again in **Table 2**.

weights, as shown in **Table 1**. The UV chromatogram was similar for both venom samples and is illustrated in **Fig. 2A**. An example of the mass spectra obtained is shown in **Fig. 2B**. Both samples gave nearly identical results, and the molecular weights measured correlated within ±1 Dalton, indicating that both snakes belonged to the same species. This was confirmed by a comparison with the data obtained for the two standard samples in the on-line LC/ESI-MS experiment (**Table 2**), as no correlation could be found between the fingerprints of the *Naja sp.* and that of the *Naja naja* standard sample. On the basis of these results, *Naja sp.* from Vietnam can be classified as *Naja kaouthia*.

3.3.2. On-Line LC/ESI-MS

The UV and TIC chromatograms obtained with the three *Naja* venom samples are shown in **Figs. 3** and **4**, where a close similarity of the elution patterns for the venoms from the white cobra and *Naja kaouthia* is observed. More than 30 molecular masses were detected in each sample; the most significant of these are summarized in **Table 2**. For the two standard samples, various molecular weights of polypeptides ranging between 6 and 8 kDa could be assigned to neurotoxins, cytotoxins, or cytotoxin homologs already described in the literature and listed in the VENOMS *(33)* and SWISS-PROT *(34)* data-

Toxin Mass Fingerprints of Crude Venoms 325

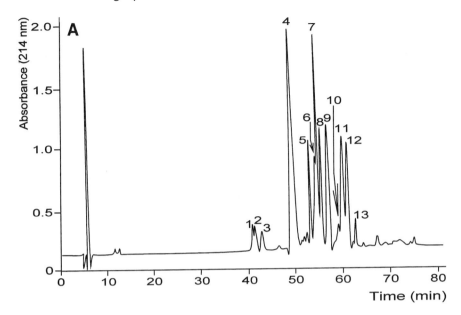

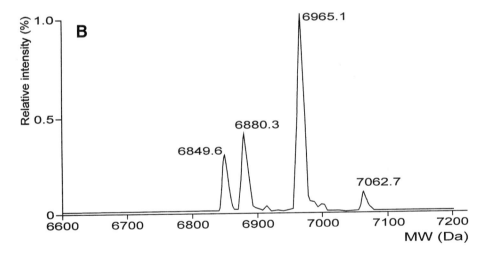

Fig. 2. Off-line analysis of venom samples from *Naja sp.* and *Naja kaouthia* (Vietnam and Thailand, respectively). (**A**) Microbore HPLC chromatogram at 214 nm of crude *Naja kaouthia* venom from Thailand. (**B**) Transformed mass spectrum (mass rather than *m/z* scale) of fraction 2 showing the presence of four compounds with different molecular weights. The compounds with molecular weights of 6965.1 and 7062.7 Daltons are two polypeptide toxins that co-elute in this fraction. The molecular weights of 6880.3 and 6849.6 Daltons correspond to remaining traces of material that eluted mainly in fraction 1. The *Naja sp.* sample from Vietnam exhibited a nearly identical UV chromatogram (data not shown).

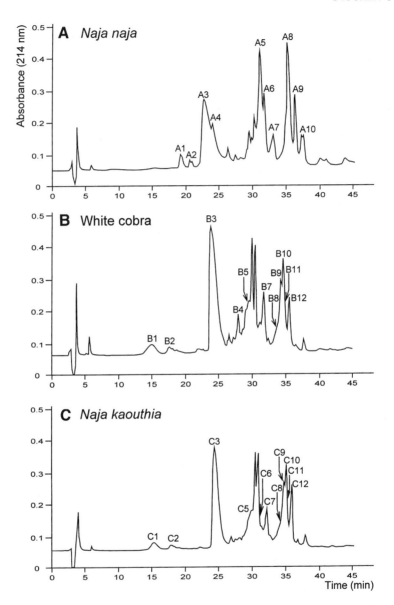

Fig. 3. Elution pattern, with UV detection at 214 nm, of venom samples from (**A**) *Naja naja* standard, (**B**) the white cobra, and (**C**) *Naja kaouthia* standard. A similarity between traces (**B**) and (**C**) can be observed.

bases. Among 10 of the toxins described for *Naja naja*, seven compounds with corresponding molecular weights could be determined, and all the 9 toxins belonging to these three toxin families (neurotoxins, cytotoxins, and cytotoxin

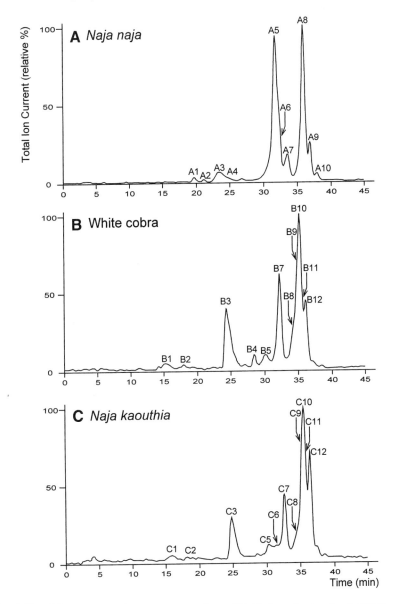

Fig. 4. Total-ion current chromatograms from (**A**) *Naja naja* standard, (**B**) the white cobra, and (**C**) *Naja kaouthia* standard. Again, a similarity between traces (**B**) and (**C**) can be observed. The TIC relative intensities are not directly quantitative as they essentially depend on the ease with which each compound is ionized in the electrospray ion source; peak A3, for example, shows strong UV absorption (**Fig. 3**), but a very weak total-ion current.

Table 2
Molecular Weights Measured by On-Line LC-ESI-MS
Analysis of *Naja naja,* the White Cobra, and *Naja kaouthia*[a]

Peak no.	mol wt (Daltons)		
	Measured	Calculated	Toxin
Naja naja			
A1	6817.3		
A2	7534.3		
A3	7836.4	7837.0	Long neurotoxin 1 *(40)*
A4	7824.2	7823.0	Long neurotoxin 3 *(41)*
	7878.5	7879.0	Long neurotoxin 4 *(41)*
A5	6781.6	6783.2	Cytotoxin 1 *(42)*
A6	6736.8	6737.2	Cytotoxin 3 *(43)*
A7	7006.0	7006.4	Cytotoxin homolog *(44)*
A8	6754.5	6755.3	Cytotoxin 2 *(45)*
A9	6672.7		
A10	6806.3		
White *Naja*			
B1	6849.3; 6879.9; 6965.7		
B2	6975.9		
B3	7822.1		
B4	6845.2		
B5	7616.0		
B7	6736.0; 6708.9		
B8	6986.5		
B9	6646.9		
B10	6693.3		
B11	6764.8		
B12	6738.4		
Naja kaouthia			
C1	6850.5; 6878.3; 6965.5		
C2	6976.0	6974.7	Short neurotoxin 1 *(46)*
C3	7822.0	7821.0	Long neurotoxin 1 *(47)*
C5	7615.2		
C6	6709.0	6709.2	Cytotoxin 3 *(48)*
C7	6736.2	6737.2	Cytotoxin 2 *(48)*
	6709.4	6709.2	Cytotoxin 3 *(48)*
C8	6985.8	6986.4	Cytotoxin homolog *(49)*
C9	6646.8	6646.2	Cytotoxin 5 *(46)*
C10	6692.9	6693.2	Cytotoxin 1 *(46)*
C11	6764.6		
C12	6738.3	6739.3	Cytotoxin 4 *(46)*

[a]Included is a comparison with the calculated molecular weights of characterized toxins. There is high similarity between the measured molecular weights of the white cobra venom and those of *Naja kaouthia*. The peak numbering refers to **Figs. 3** and **4**.

homologs) that have been described for *Naja kaouthia* were identified (**Table 2**). The maximum difference between the measured and calculated molecular weights of the toxins was 1.6 Daltons for *Naja naja* and 1.3 Daltons for the *Naja kaouthia* sample, respectively, and this allowed an unambiguous distinction and identification of the two snake species. The toxin mass fingerprint of the white *Naja* corresponds exactly to that of *Naja kaouthia*, but shows no similarity to the *Naja naja* map. With an accuracy of ±0.8 Daltons, most of the measured white *Naja* molecular masses correspond exactly to those of the standard *Naja kaouthia* sample. The white *Naja* thus clearly belongs to the *Naja kaouthia* complex.

Cytotoxin 4 *(35)*, long neurotoxin 2 *(36)*, and long neurotoxin 5 *(37)* are the three toxins from *Naja naja* that could not be detected. Apart from possible sequence errors or mistakes in the determination of the original snake venom from which the described toxins were sequenced, other possibilities such as the fact that a toxin may be expressed in extremely low amounts in a single specimen, which renders its detection impossible, may explain why some expected molecular masses could not be found by on-line LC/ESI-MS. Moreover, some compounds may carry too many negatively charged side chains to allow proper ionization in the positive-ion mode and would be better detected in the negative-ion mode. Mass spectrometric analysis may also suffer from the suppressing effect of another compound that coelutes. The chromatographic conditions (reversed-phase at low pH) may also not be appropriate in certain cases. Another possible explanation would be the presence of posttranslational modifications, such as amidation, that alter the molecular weight but are not detectable by Edman degradation sequencing.

3.4. Conclusions

To identify the origin of snake venoms, a new method based on UV and mass spectrometric detection is proposed and demonstrated using snake venoms of the controversial *Naja* genus, viz. *Naja naja* and *Naja kaouthia*. A venom sample from Vietnam and an unknown white *Naja* from Thailand could both be assigned to *Naja kaouthia*. Use of the first, and simplest, approach (off-line analysis) shows that high technology on-line microbore LC/ESI-MS, even if it is the most efficient approach, is not indispensable, and essential information can also be obtained within hours by the simpler method on partially purified fractions. Both analytical approaches allow a definite classification of venomous snakes, even when morphological and anatomic characteristics are inefficient since animals can exhibit relatively high variations depending on both phylogeny (evolutionary context) and ecogenesis (environmental context).

The results demonstrate that on-line LC/ES-MS is a very fast and efficient method for preliminary studies of crude venom mixtures. This methodology, however, is not restricted to investigations of venomous snakes. It could, of course, easily be extended to other venomous animals or even any kind of tissue or biological extract. It also gives access to a structural investigation of individual compounds that is otherwise impossible to achieve. The various molecular weights measured, which could not be assigned to known compounds, illustrate the fact that many compounds have still not been described and confirm the complexity of snake venoms.

Toxin families can, for example, be defined on the basis of the molecular weight range and the relative retention times on the chromatographic system used to separate their individual forms. It seems that short-chain neurotoxins (60–62 residues, 6600–7300 Daltons) tend to elute first in the chromatographic system described, followed by the long-chain neurotoxins (66–74 residues, 7250–8200 Daltons) and then the most basic cytotoxins (60–62 residues, 6600–6900 Daltons). This kind of approach should facilitate the classification and understanding of the potential biological activity of previously undescribed compounds, and more important, could possibly be of great help in the search for new families of bioactive compounds with or without toxic or enzymatic activities. It is interesting to note here, for example, the presence of a 24,934-Daltons protein, a molecular weight that does not correspond to any protein family previously described in cobra venom. It could consist of polymers of smaller single-chain peptides, but might also be a new protein with a so far undefined enzymatic activity.

Furthermore, as only a fraction of the sample is introduced into the mass spectrometer, the remainder can be collected for further structural and physiological studies. Mass spectrometry can also play an essential role in such investigations, as it is not restricted to molecular weight assessment. A combination of chemical or enzymatic digestions, followed by a reductive treatment, is currently used for disulphide bridge assignment or for investigation of posttranslational modification. Recent developments have shown that mass spectrometry can also be used for amino acid sequencing of polypeptides (by tandem mass spectrometric analysis or by progressive treatment with a carboxypeptidase for example), or for noncovalent interaction studies as well as for in vivo pharmacokinetic studies *(16,38,39)*.

Finally, the technique of toxin mass fingerprinting can also be extremely useful for quality control of crude venom batches, which is essential in venom and antivenom production, for the identification of new compounds and for eliciting structural information on individual toxins.

4. Notes

1. Narrow-bore HPLC with a split is recommended in those cases were collection of fractions is of interest, whereas microbore HPLC without a split and with direct

introduction into the ion source gives better analytical results. High pressure syringe pumps working at 40 µL/min with a 1-mm ID reversed phase column followed by a split allowing 2–5 µL/min to enter the ion source with the major remaining fraction passing through the UV cell followed by fraction collection is an excellent compromise, providing sufficient material for further biochemical studies such as sequencing or amino acid analysis.

2. Although some of the terms are specific to the instruments used here, these details apply essentially to all electrospray instruments and are the result of compromises that need to be made in each specific case.

3. A good computer with enough disk space is important for data analysis, as the files may easily reach 100 Mb in size depending on run duration, scan time, and mass range.

4. When calculating the molecular masses of sequences of the individual toxins as found in some databases, it is important to be aware that (1) any cysteine residues are usually disulphide bonded, (2) average molecular masses (not monoisotopic) have been measured under these conditions and (3) the signal peptides have to be removed from the sequence.

5. A reduction in resolution, which results in a gain in sensitivity, is recommended. A peak-width at half-height of 1.0 mass unit for a signal at m/z 1000 is usually sufficient.

6. A minimum m/z scan range of 500–1500 is recommended to cover most of the toxin signals; m/z 400–1800 may be better in some cases, but will reduce sensitivity and accuracy of the results.

7. The scan time should ideally be between 4 and 6 s to maintain good accuracy and to accumulate enough scans for deconvolution.

8. In case of smaller, lower molecular weight, compounds present in venoms of other animal species (scorpions, cones, spiders), the scan range should be increased to cover lower m/z values and the ion source cone voltage reduced to minimize the risk of in-source fragmentation.

9. To optimize the signal, tuning should be performed prior to mass calibration using a solution of a model toxin in the solvents used for the chromatographic step (30–40% acetonitrile, corresponding to the major elution zone). Insulin may also be used for tuning if no toxin is available.

10. Using acetic acid or formic acid instead of TFA in the HPLC solvent may typically increase the mass spectrometer signal intensity by a factor of 10 on many instruments, but will also reduce the quality of the chromatographic separation. This may result in peak broadening, a corresponding sample dilution, and therefore reduced sensitivity in the mass spectrometric analysis.

Acknowledgments

We thank Michel Guillod (Reptiles du Monde, Switzerland) for providing the venom and for his herpetological expertise. We are grateful to Prof. Robin E. Offord and Dr. Keith Rose from the Department of Medical Biochemistry, University of Geneva, Switzerland and the High Technology Program of the

Medical Faculty of the University of Geneva for stimulating discussions and access to the VG-Trio 2000 equipment. We thank the Lyonnaise des Eaux (France) for financial support.

References

1. Harvey, A. L., ed. (1990) *Snake Toxins*. Pergamon, New York.
2. Shier, W. T. and Mebs, D., eds. (1990) *Handbook of Toxinology*, Marcel Dekker, New York.
3. Stocker, K. F., ed. (1990) *Medical Use of Snake Venom Proteins*, CRC Press, Boca Raton, FL.
4. Wüster, W. and Thorpe, R. S. (1991) Asiatic cobras: systematics and snakebite. *Experientia* **47,** 205–209.
5. Wüster, W., Thorpe, R. S., Cox, M. J., Jintakune, P., and Nabhitabhata, J. (1996) Population systematics of the snake genus *Naja* (Reptilia: Serpentes: Elapidae) in Indochina: Multivariate morphometrics and comparative mitochondrial DNA sequencing (cytochrome oxidase I). *J. Evol. Biol.* **8,** 493–510.
6. Bougis, P. E., Marchot, P., and Rochat, H. (1986) Characterization of Elapidae snake venom components using optimized reverse-phase high-performance liquid chromatographic conditions and screening assays for alpha-neurotoxin and phospholipase A2 activities. *Biochemistry.* **25,** 7235–7243.
7. Da Silva, N. J., Jr., Griffin, P. R., and Aird, S. D. (1991) Comparative chromatography of Brazilian coral snake (*Micrurus*) venoms. *Comp. Biochem. Physiol.* **100B,** 117–126.
8. Daltry, J. C., Ponnudurai, G., Shin, C. K., Tan, N. H., Thorpe, R. S., and Wuster, W. (1996) Electrophoretic profiles and biological activities: intraspecific variation in the venom of the Malayan pit viper (*Calloselasma rhodostoma*). *Toxicon* **34,** 67–79.
9. Daltry, J. C., Wuester, W., and Thorpe, R. S. (1996) Diet and snake venom evolution. *Nature* **379,** 537–540.
10. Kent, C. G., Tu, A. T., and Geren, C. R. (1984) Isotachophoretic and immunological analysis of venoms from sea snakes (*Laticauda semifasciata*) and brown recluse spiders (*Loxosceles reclusa*) of different morphology, locality, sex, and developmental stages. *Comp. Biochem. Physiol. B.* **77,** 303–311.
11. Marshall, T. and Williams, K. M. (1994) Analysis of snake venoms by sodium dodecyl sulfate-polyacrylamide gel electrophoresis and two-dimensional electrophoresis. *Appl. Theor. Electrophor.* **4,** 25–31.
12. Mendoza, C. E., Bhatti, T., and Bhatti, A. R. (1992) Electrophoretic analysis of snake venoms. *J. Chromatogr.* **580,** 355–363.
13. Tu, A. T., Stermitz, J., Ishizaki, H., and Nonaka, S. (1980) Comparative study of pit viper venoms of genera *Trimeresurus* from Asia and *Bothrops* from America: an immunological and isotachophoretic study. *Comp. Biochem. Physiol.* **66B,** 249–254.
14. Fenn, J. B., Mann, M., Meng, C. K., Wong, S. F., and Whitehouse, C. M. (1989) Electrospray ionization for mass spectrometry of large biomolecules. *Science* **246,** 64–71.
15. Fenselau, C., Vestling, M. M., and Cotter, R. J. (1993) Mass spectrometric analysis of proteins. *Curr. Opin. Biotechnol.* **4,** 14–19.

16. Loo, J. A. (1995) Bioanalytical mass spectrometry: many flavors to choose. *Bioconjugate Chem.* **6**, 644–665.
17. Siuzdak, G. (1994) The emergence of mass spectrometry in biochemical research. *Proc. Natl. Acad. Sci. USA* **91**, 11,290–11,297.
18. James, P., Quadroni, M., Carafoli, E., and Gonnet, G. (1993) Protein identification by mass profile fingerprinting. *Biochem. Biophys. Res. Commun.* **195**, 58–64.
19. Morris, H. R. and Pucci, P. (1985) A new method for rapid assignment of S-S bridges in proteins. *Biochem. Biophys. Res. Commun.* **126**, 1122–1128.
20. Pappin, D. J. C., Hojrup, P., and Bleasby, A. J. (1993) Rapid identification of proteins by peptide-mass fingerprinting. *Curr. Biolo.* **3**, 327–332.
21. Yates III, J. R., Speicher, S., Griffin, P., and Hunkapiller, T. (1993) Peptide mass maps: a highly informative approach to protein identification. *Anal. Biochem.* **214**, 397–408.
22. Castaneda, O., Sotolongo, V., Amor, A. M., Stöcklin, R., Anderson, A. J., Harvey, A. L., et al. (1995) Characterization of a potassium channel toxin from the Caribbean Sea anemone *Stichodactyla helianthus*. *Toxicon* **33**, 603–613.
23. McDowell, R. S., Dennis, M. S., Louie, A., Shuster, M., Mulkerrin, M. G., and Lazarus, R. A. (1992) Mambin, a potent glycoprotein IIb-IIIa antagonist and platelet aggregation inhibitor structurally related to the short neurotoxins. *Biochemistry* **31**, 4766–4772.
24. Tyler, M. I., Retson-Yip, K. V., Gibson, M. K., Barnett, D., Howe, E., Stöcklin, R., et al. (1997) Isolation and amino acid sequence of a new long-chain neurotoxin with two chromatographic isoforms from the venom of the australian death adder (Acanthophis antarcticus). *Toxicon.* **35**, 555–562.
25. Escoubas, P., Celerier, M. L., and Nakajima, T. (1997) High-performance liquid chromatography matrix-assisted laser desorption/ionization time-of-flight mass spectrometry peptide fingerprinting of tarantula venoms in the genus *Brachypelma*: chemotaxonomic and biochemical applications. *Rapid Commun. Mass Spectrom.* **11**, 1891–1899.
26. Escoubas, P., Whiteley, B. J., Kristensen, C. P., Célérier, M.-L., Corzo, G., and Nakajima, T. (1998) Multidimensional peptide fingerprinting by high performance liquid chromatography, capillary zone electrophoresis and matrix-assisted laser desorption/ionization time-of-flight mass spectrometry for the identification of tarantula venom samples. *Rapid Commun. Mass Spectrom.* **12**, 1075–1084.
27. Gillard, C., Virelizier, H., Arpino, P., and Stöcklin, R. (1996), Classification of the white *Naja* by on-line LC-ES-MS, in *Eighteenth International Symposium on Capillary Chromatography*, vol. III, ISSC, Riva del Garda, Italy, pp. 2197–2197.
28. Perkins, J. R., Parker, C. E., and Tomer, K. B. (1993) The characterization of snake venoms using capillary electrophoresis in conjunction with electrospray mass spectrometry: Black Mambas. *Electrophoresis* **14**, 458–468.
29. Perkins, J. R. and Tomer, K. B. (1995) Characterization of the lower-molecular-mass fraction of venoms from *Dendroaspis jamesoni kaimosae* and *Micrurus fulvius* using capillary-electrophoresis electrospray mass spectrometry. *Eur. J. Biochem.* **233**, 815–827.

30. Stöcklin, R. and Savoy, L.-A. (1994) On-line LC-ES-MS: a new method for direct analysis of crude venom. *Toxicon* **32**, 408–408 (abstract).
31. Nelson, R. W., Krone, J. R., Bieber, A. L., and Williams, P. (1995) Mass spectrometric immunoassay. *Anal. Chem.* **67**, 1153–1158.
32. Nutaphand, W. (1986) *Cobra*, Pata Zoo Publication, Bangkok, unnumbered and unpaginated (in Thai).
33. Stöcklin, R. and Cretton, G. (1999) *VENOMS: The ultimate Database on Venomous Animals and their Venoms,* Atheris Laboratories, Bernex-Geneva, Switzerland. Professional edition, computer program, Ver. **1.0**, CD-Rom.
34. Bairoch, A. and Apweiler, R. (1996) The SWISS-PROT protein sequence data bank and its new supplement TREMBL. *Nucleic Acids Res.* **24**, 21–25.
35. Takechi, M. et al., unpublished results (1973), cited by: Dufton, M. J. and Hider, R. C. (1983) Conformational properties of the neurotoxins and cytotoxins isolated from Elapid snake venoms. *CRC Crit. Rev. Biochem.* **14**, 113–171.
36. Ohta, M., Sasaki, T., and Hayashi, K. (1976) The primary structure of toxin B from the venom of the Indian cobra Naja naja. *FEBS Lett.* **72**, 161–166.
37. Hayashi, K., Takechi, M., and Sasaki, T. (1971) Amino acid sequence of cytotoxin I from the venom of the Indian cobra (*Naja naja*). *Biochem. Biophys. Res. Commun.* **45**, 1357–1362.
38. Chait, B. T., Wang, R., Beavis, R. C., and Kent, S. B. H. (1993) Protein ladder sequencing. *Science* **262**, 89–92.
39. Stöcklin, R., Vu, L., Vadas, L., Cerini, F., Kippen, A. D., Offord, R. E., et al. (1997) A stable isotope dilution assay for in vivo determination of insulin levels in man by mass spectrometry. *Diabetes* **46**, 44–50.
40. Nakai, K., Sasaki, T., and Hayashi, K. (1971) Amino acid sequence of toxin A from the venom of the Indian cobra (*Naja naja*). *Biochem. Biophys. Res. Commun.* **44**, 893–897.
41. Ohta, M., Sasaki, T., and Hayashi, K. (1981) The amino acid sequence of toxin D isolated from the venom of Indian cobra (*Naja naja*). *Biochim. Biophys. Acta* **671**, 123–128.
42. Endo, T. and Tamiya, N. (1987) Current view on the structure-function relationship of postsynaptic neurotoxins from snake venoms. *Pharmacol. Ther.* **34**, 403–451.
43. Kaneda, N., Takechi, M., Sasaki, T. and Hayashi, K. (1984) Amino acid sequence of cytotoxin IIa isolated from the venom of the Indian cobra (*Naja naja*). *Biochem. Int.* **9**, 603–610.
44. Takechi, M., Tanaka, Y., and Hayashi, K. (1987) Amino acid sequence of a less-cytotoxic basic polypeptide (LCBP) isolated from the venom of the Indian cobra (*Naja naja*). *Biochem. Int.* **14**, 145–152.
45. Takechi, M., Hayashi, K., and Sasaki, T. (1972) The amino acid sequence of cytotoxin II from the venom of the Indian cobra (*Naja naja*). *Mol. Pharmacol.* **8**, 446–451.
46. Ohkura, K., Inoue, S., Ikeda, K., and Hayashi, K. (1988) Amino-acid sequences of four cytotoxins (cytotoxins I, II, III and IV) purified from the venom of the Thailand cobra, *Naja naja siamensis*. *Biochim. Biophys. Acta* **954**, 148–153.

47. Arnberg, H., Eaker, D., and Karlsson E., unpublished results, cited by: Karlsson, E. (1973) Chemistry of some potent animal toxins. *Experientia* **29,** 1319–1327.
48. Joubert, F. J. and Taljaard, N. (1980) The complete primary structures of three cytotoxins (CM-6, CM-7 and CM-7A) from *Naja naja kaouthia* (Siamese cobra) snake venom. *Toxicon.* **18,** 455–467.
49. Inoue, S., Ohkura, K., Ikeda, K., and Hayashi, K. (1987) Amino acid sequence of a cytotoxin-like basic protein with low cytotoxic activity from the venom of the Thailand cobra *Naja naja siamensis. FEBS Lett.* **218,** 17–21.

19

Mass Spectrometric Characterization of the β-Subunit of Human Chorionic Gonadotropin

Roderick S. Black and Larry D. Bowers

1. Introduction

Human chorionic gonadotropin (hCG) is a member of the family of glycoprotein hormones, which includes human luteinizing hormone (hLH), thyrotropin (TSH), and follicle-stimulating hormone (FSH). Each of these is heterodimeric, consisting of a noncovalently bound α- and β-subunit. The α-subunits are identical within the glycoprotein hormone family, whereas the β-subunits are unique, conferring specific biochemical properties to each hormone.

Physiologically, hCG is produced by the placenta and stimulates the steroid production of the *corpus leuteum* necessary for the maintenance of pregnancy. This results in the excretion of a high concentration of hCG in the urine of pregnant women. Home pregnancy test kits detect the presence of hCG in urine by a simple immunoassay. hCG and its degradation products are also present in the plasma and urine of some individuals with trophoblastic disease and hydatidiform mole. Accordingly, immunoassays are also used in clinical monitoring of these conditions (*1*). There are correlations between specific medical states and the chemical structure of the hCG produced. For example, nicking of the hCG molecule is believed to occur at a different location in the peptide chain in normal pregnancies vs. abnormal pregnancies or choriocarcinoma (*2,3*). Typical methods for characterization of nicking and other structural variations can be labor intensive, and are usually impractical for routine clinical use. Presently available immunoassay kits are simple to use but do not provide the analytical specificity required to differentiate among variant hCGs produced by the body during different physiological or pathologic states such as normal pregnancy or choriocarcinoma. Mass spectrometry and liquid chromatography/mass spectrometry (LC/MS) methods potentially

From: *Methods in Molecular Biology, vol. 146:
Protein and Peptide Analysis: New Mass Spectrometric Applications*
Edited by: J. R. Chapman © Humana Press Inc., Totowa, NJ

offer both the speed and specificity necessary for characterization of these molecular variations.

Pharmaceutical hCG can be administered to promote steroid production in the treatment of certain health problems including infertility and testicular nondescent. Unfortunately, healthy athletes sometimes exploit the steroid-producing activity of hCG in an effort to gain a competitive advantage *(4)*. The International Olympic Committee (IOC) and other athletic federations have banned the administration of hCG for this purpose. Thus, the analysis of hCG by mass spectrometry is also motivated by an interest in confirming cases of hCG abuse. The specificity of a simple immunoassay is generally not sufficient for a forensic confirmation such as an athletic drug test. Our laboratory has developed a highly discriminating hCG confirmation method that employs immunoaffinity trapping in conjunction with the LC/MS detection methods outlined in this chapter *(5)*. A complementary method employing matrix-assisted laser desorption/ionization time-of-flight mass spectrometry (MALDI-TOFMS) has also been reported *(6)*.

In addition to routine analysis or confirmation of hCG and structurally related molecules in biological samples, mass spectrometry will also be essential in the complete characterization of hCG analytical standards. Virtually any preparation of hCG is expected to be heterogeneous due to variation in carbohydrate structure, nicking, and other molecular variations. The World Health Organization (WHO) hCG standards are not well characterized with regard to structural heterogeneity. Mass spectrometric data would provide a much more definitive description of candidate standards than immunoaffinity- or chromatography-based methods. Furthermore, immunoassays for specific types of variant hCGs and hCG-related molecules produced by tumors could be developed more efficiently if mass spectrometric methods were available. Although the thorough characterization of hCG standards by mass spectrometry has yet to be undertaken, the procedures provided here constitute a useful framework for both structural confirmation and characterization of novel hCG preparations. Both the protein backbone and attached carbohydrate can be characterized using a combination of electrospray LC/MS and microelectrospray mass spectrometry.

Our experimental strategies concern specifically the β-subunit of hCG, or hCG-β, because it is unique to hCG. Normal hCG-β has a molecular weight of about 23,000 Daltons, and reportedly bears O-linked glycosylations at Ser121, Ser127, Ser132, and Ser138, as well as N-linked glycosylations at Asn13 and Asn30 *(7–9)*. The mass spectrometric confirmation of the structure requires dissociation of the intact hCG heterodimer followed by isolation of hCG-β by high-performance liquid chromatography (HPLC). The hCG-β is then reduced and pyridylethylated to protect the 12 cysteine residues from becoming reoxidized, as shown in **Fig. 1**. Pyridylethylation is chosen for alkylation

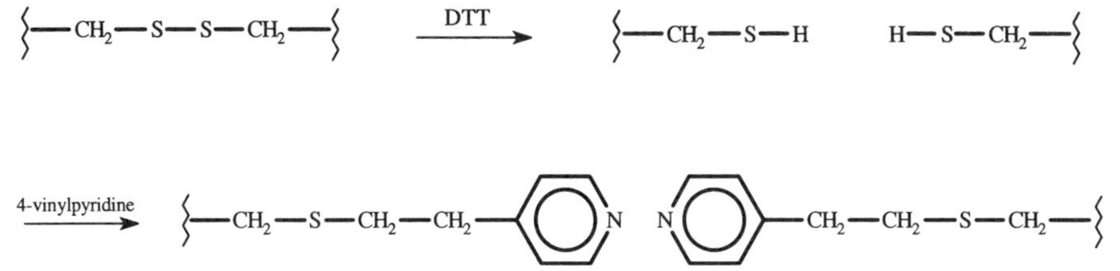

Fig. 1. Reduction and pyridylethylation of disulfides using DTT and 4-vinylpyridine.

because it introduces a potential positive charge site at each cysteine. The charge sites are essential for positive-ion electrospray mass spectrometric analysis (*see* **Note 1**). hCG-β is then digested with trypsin to generate peptides and glycopeptides for LC/MS analysis. The primary structure of hCG-β, including the tryptic cleavage sites, is shown in **Fig. 2**. The peaks in the LC/MS chromatogram are assigned as shown in **Fig. 3**. The published structures of the attached carbohydrates, shown in **Tables 1** and **2**, facilitate the identification of the glycosylated tryptic fragments.

A method for determining the extent of O-glycosylation at each Ser-linked site is also provided *(10)*. The carboxy-terminal fragment corresponding to residues 109–145 is released by α-chymotryptic digestion of a reduced and carboxymethylated sample of hCG-β; all four of the reported O-glycosylation sites of hCG-β are contained within this fragment. The glycopeptide is isolated by HPLC (**Fig. 4**) and treated with a β-elimination solution of 25 mM KOH in 5:4:1 DMSO + water + ethanol (v/v/v). This promotes efficient β-elimination of the serine-linked carbohydrates of hCG-β (*see* **Note 2**). Unglycosylated serine residues are not affected in this reaction; however, each serine residue bearing a carbohydrate is replaced with a dehydroalanine (Dha or Δ) residue. Dha has a distinct mass, and mass spectrometry can be used to distinguish formerly glycosylated fragments from fragments containing free serines, as shown in **Fig. 5**.

2. Materials

See **Note 3**.

2.1. Peptide Backbone and Carbohydrate Structure Analysis

2.1.1. Isolation of hCG-β

1. 0.1% trifluoroacetic acid (TFA) in water (v/v).
2. HPLC system with ultraviolet (UV) detector (set at 220 nm) and reversed-phase column (*see* **Note 4**).
3. Mobile phase A: HPLC-grade water containing 0.1% TFA (v/v), filtered and degassed.

NH2-Ser-Lys – Glu-Pro-Leu-Arg-Pro-Arg – Cys-Arg-Pro-Ile-**Asn**-Ala-Thr-Leu-Ala-Val-Glu-Lys – Glu-Gly-Cys-
T1   T2   T3

Pro-Val-Cys-Ile-Thr-Val-**Asn**-Thr-Thr-Ile-Cys-Ala-Gly-Tyr-Cys-Pro-Thr-Met-Thr-Arg – Val-Leu-Gln-Gly-Val-Leu-
T4

Pro-Ala-Leu-Pro-Gln-Val-Val-Cys-Asn-Tyr-Arg – Asp-Val-Arg – Phe-Glu-Ser-Ile-Arg – Leu-Pro-Gly-Cys-Pro-
T5   T6   T7   T8

Arg – Gly-Val-Asn-Pro-Val-Val-Ser-Tyr-Ala-Val-Ala-Leu-Ser-Cys-Gln-Cys-Ala-Leu-Cys-Arg-Arg – Ser-Thr-Thr-
T(9 + 10)

Asp-Cys-Gly-Gly-Pro-Lys – Asp-His-Pro-Leu-Thr-Cys-Asp-Asp-Pro-Arg – Phe-Gln-Asp-Ser-Ser-**Ser**-Lys –
T11   T12   T13

Ala-Pro-Pro-**Ser**-Leu-Pro-Ser-Pro-**Ser**-Arg – Leu-Pro-Gly-Pro-**Ser**-Asp-Thr-Pro-Ile-Leu-Pro-Gln-COOH
T14   T15

Fig. 2. The primary sequence of hCG-β. Reported glycosylation sites are indicated in boldface. Tryptic cleavage sites are indicated by a long dash (—). Tryptic fragments are indicated as T1, T2, and so forth, as indicated in **Tables 3** and **4**.

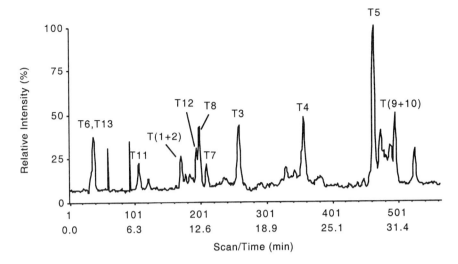

Fig. 3. LC/MS total-ion chromatogram for reduced and pyridylethylated hCG-β, acquired using the conditions given in **Subheading 3.1.4.** Tryptic fragments are labeled as T1, T2, and so forth, as indicated in **Tables 3** and **4**.

4. Mobile phase B: HPLC-grade acetonitrile containing 0.1% TFA (v/v), filtered and degassed.
5. Vacuum centrifuge (e.g., Savant SpeedVac, Farmingdale, NY).

2.1.2. Reduction and Pyridylethylation

1. 5-mL reaction vial. (*See* **Note 5**.)
2. Nitrogen source.
3. Nitrogen-purged reduction buffer, freshly prepared: 6 M guanidine HCl, 0.5 M Tris, 2 mM EDTA, pH 8.3 (adjusted with HCl), containing 1.2 mg/mL dithiothreitol (DTT; *see* **Note 6**). (Solution should be nitrogen-purged for at least 5 min before DTT is added.)
4. 37°C incubator.
5. 4-Vinylpyridine.
6. Ice bath.
7. HPLC system with UV/Vis detector and reversed-phase column (*see* **Notes 4** and **7**).
8. Mobile phase A: HPLC-grade water containing 45 mM ammonium acetate, pH 5.0 (adjusted with acetic acid), filtered and degassed.
9. Mobile phase B: 70% acetonitrile + 30% water (both HPLC-grade) containing 45 mM ammonium acetate and the same volume of acetic acid as added to mobile phase A, filtered and degassed.
10. Vacuum centrifuge.

2.1.3. Digestion with Trypsin

1. 50 mM ammonium bicarbonate solution, pH 8.0 (adjusted with ammonium hydroxide).
2. pH indicator paper.

Table 1
Reported Structures of N-linked Carbohydrates of hCG-β

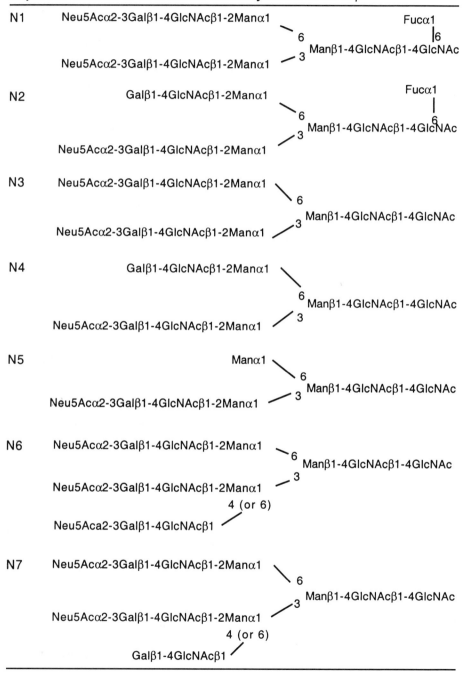

Table 2
Reported Structures of O-linked Carbohydrates of hCG-β

O1	NeuAcα2-3Galβ1-3GalNac — 　　　　　　　　　　　| β1-6 NeuAcα2-3Galβ1-4GlcNAc
O2	NeuAcα2-3Galβ1-3GalNAc — 　　　　　　　　　　　| α2-6 　　　　　　　　　　NeuAc
O3	NeuAcα2-3Galβ1-3GalNAc —
O4	NeuAcα2-6GalNAc —
O5	Galβ1-3GalNac — 　　　　　　| β1-6 Galβ1-4GlcNAc
O6	Galβ1-3GalNAc —

3. 0.5 mg/mL trypsin solution in 50 mM ammonium bicarbonate, pH 8.0.
4. 37°C incubator.
5. 1.0 M ammonium hydroxide.

2.1.4. LC/MS Analysis

See **Note 8**.

1. Microbore LC/MS instrument, e.g., Beckman (Fullerton, CA) Model 126 solvent delivery system plumbed for low-flow gradient operation, with a Deltabond (Keystone Scientific, State College, PA) 1 × 150 mm C_{18} column (particle size, 5

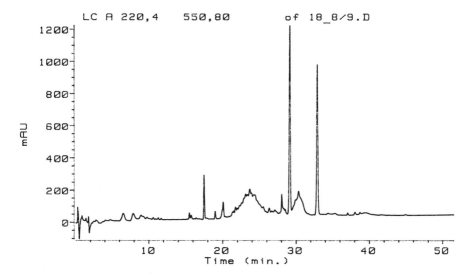

Fig. 4. LC/UV chromatogram of reduced and carboxymethylated hCG-β, acquired using the conditions described in **Subheading 3.2.2.** The broad cluster of peaks at 20–26 min corresponds to the 109–145 fragment, which is collected preparatively for determination of O-linked glycosylation sites.

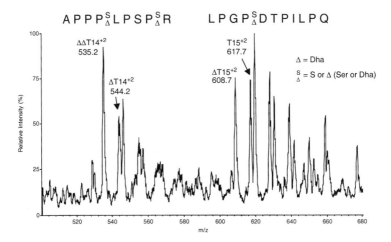

Fig. 5. The extent of O-glycosylation of hCG-β can be determined using the mass spectrum of tryptic β-eliminated 109–145. The large population of $\Delta T15^{+2}$, in which Ser has been substituted with Dha, is produced by β-elimination of glycosylated material, whereas $T15^{+2}$ corresponds to material that was not glycosylated. T14 containing one Dha ($\Delta T14^{+2}$) and two Dha ($\Delta\Delta T14^{+2}$) is also observed. The predicted location of T14 with no Dha is m/z 553.2 and, if present, it is within a cluster that includes sodiated and ammoniated adducts of $\Delta\Delta T14^{+2}$ and $\Delta T14^{+2}$.

μm; pore size, 300 Å) operated at 50 μL/min and a PE-Sciex (Concord, Ontario, Canada) API III⁺ triple-quadrupole mass spectrometer (see **Note 9**).
2. Mobile phase A: HPLC-grade water containing 0.1% TFA (v/v), filtered and degassed.
3. Mobile phase B: HPLC-grade acetonitrile containing 0.1% TFA (v/v), filtered and degassed.

2.2. Glycosylation States of O-Linked Sites

2.2.1. Reduction and Carboxymethylation

1. 5-mL reaction vial. (See **Note 5**.)
2. Nitrogen source.
3. Nitrogen-purged reduction buffer, freshly prepared: 6 M guanidine HCl, 0.5 M Tris, 2 mM EDTA, pH 8.3 (adjusted with HCl), containing 0.37 mg/mL DTT (see **Note 10**). (Solution should nitrogen-purged for at least 5 min before DTT is added.)
4. 37°C incubator.
5. Iodoacetic acid solution, freshly prepared: 6 M guanidine HCl, 0.5 M Tris, 2 mM EDTA, and 50 mg/mL iodoacetic acid.
6. Ice bath.
7. Glacial acetic acid.
8. HPLC system with UV/Vis detector and reversed-phase column (see **Note 4**).
9. Mobile phase A: HPLC-grade water containing 45 mM ammonium acetate, pH 5.0 (adjusted with acetic acid), filtered and degassed.
10. Mobile phase B: 70% acetonitrile + 30% water (both HPLC-grade) containing 45 mM ammonium acetate and the same volume of acetic acid as added to mobile phase A, filtered and degassed.
11. Vacuum centrifuge.

2.2.2. Isolation of 109–145 Glycopeptide

1. 100 mM ammonium hydrogen carbonate, pH 8.7 (adjusted with ammonium hydroxide).
2. 0.5 mg/mL α-chymotrypsin solution in 100 mM ammonium hydrogen carbonate, pH 8.7.
3. pH indicator paper.
4. Glacial acetic acid.
5. HPLC system with UV detector (set at 220 nm) and reversed-phase column (see **Note 4**); e.g., Vydac C_{18} (Hesperia, CA), 4.6 × 150 mm (particle size, 5 μm; pore size, 300 Å; cat. no. 218TP5415), maintained at 40°C with a flow rate of 1.0 mL/min.
6. Mobile phase A: HPLC-grade water containing 45 mM ammonium acetate, pH 5.0 (adjusted with acetic acid), filtered and degassed.
7. Mobile phase B: 70% acetonitrile + 30% water (both HPLC-grade), containing 45 mM ammonium acetate and the same volume of acetic acid as added to mobile phase A, filtered and degassed.
8. Vacuum centrifuge.

2.2.3. β-Elimination

1. β-Elimination solution: 25 mM KOH in DMSO + water + ethanol (5:4:1 v/v/v).
2. 45°C incubator.
3. 1.0 M HCl.
4. pH indicator paper.
5. HPLC system with column and mobile phases specified in **Subheading 2.2.2.**
6. Vacuum centrifuge.

2.2.4. Tryptic Digestion

1. 50 mM ammonium bicarbonate solution, pH 8.0 (pH adjusted with sodium hydroxide).
2. 0.5 mg/mL trypsin solution in 50 mM ammonium bicarbonate, pH 8.0.
3. 37°C incubator.

2.2.5. Microelectrospray Mass Spectrometric Analysis (See **Note 11**)

1. Vacuum centrifuge.
2. 1:1 methanol + water (v/v, both HPLC-grade) solution containing 0.05% formic acid (v/v); (*see* **Note 3**).
3. Microelectrospray mass spectrometer.

3. Methods
3.1. Peptide Backbone and Carbohydrate Structure Analysis
3.1.1. Isolation of hCG-β

1. Dissolve hCG in aqueous 0.1% TFA (v/v) to a concentration of 1–5 mg/mL. hCG is from Sigma (St. Louis, MO; cat. no. C-5297) or Organon (W. Orange, NJ; sold as Pregnyl).
2. Incubate 0.5 h at room temperature to dissociate the α- and β-subunits.
3. Separate the α- and β-subunits by reversed-phase gradient HPLC; for example, with the Vydac C_{18} column (*see* **Note 4**), begin with a 5-min hold at 95:5 A-B followed by a linear ramp over 60 min to 30:70 A-B.
4. The subunits will elute as two very broad "peaks." The second peak is the β-subunit; collect it preparatively from the HPLC.
5. Dry the β-subunit by vacuum centrifugation.
6. Estimate the mass of hCG-β obtained by assuming a yield of about 0.6 g hCG-β per 1.0 g of intact hCG initially dissolved. (*See* **Note 12**.)

3.1.2. Reduction and Pyridylethylation

See **Notes 1** and **6**.

1. Dilute hCG-β to a concentration of about 3 mg/mL and transfer to a 5-mL reaction vial.
2. Purge 5 min with nitrogen.
3. Add freshly prepared, nitrogen-purged reduction buffer at a 1:1 volume ratio.
4. Incubate 30 min at 37°C, continuously purging the vial headspace with nitrogen during incubation.

Table 3
Unglycosylated Tryptic Fragments of Reduced and Pyridylethylated hCG-β

Tryptic fragment	Amino acid residues	Peptide mass	Retention time (min)
T6	61–63	389.2	2.3
T11	96–104	970.4	6.7
T1, 2	1–8	982.6	10.7
T8	69–74	747.4	12.3
T12	105–114	1273.6	12.4
T7	64–68	651.4	13.1
T5	44–60	1974.2	29.0
T9	75–94	2368.2	31.0

5. Add 4-vinylpyridine at a ratio of 1.1 µL per 1 mg of hCG-β.
6. Wrap the vial with aluminum foil to minimize light exposure and incubate for 30 min at room temperature, purging the headspace with nitrogen during incubation.
7. Inject on reversed-phase gradient HPLC; for example, with the Vydac C_{18} column, begin with a 5-min hold at 95:5 A-B followed by a linear ramp over 60 min to 30:70 A-B (*see* **Notes 4** and **7**).
8. Collect the eluting peak preparatively and evaporate the solvent by vacuum centrifugation.

3.1.3. Digestion with Trypsin

1. Reconstitute the reduced and pyridylethylated hCG-β in 50 mM ammonium bicarbonate buffer (pH 8.0) using a ratio of 2 mL buffer per 1 mg of hCG-β.
2. Adjust pH to 8.0 with 1.0 M ammonium hydroxide if necessary (*see* **Note 13**).
3. Add 0.5 mg/mL trypsin solution at a ratio of 40 µL of trypsin solution per 1 mg of hCG-β.
4. Incubate overnight at 37°C.

3.1.4. LC/MS Analysis

1. Inject 5 µL of the tryptic digest of reduced and pyridylethylated hCG-β, which corresponds to about 100 pmol, on the LC/MS. Use a linear gradient of 95:5–50:50 A-B over 60 min. Operate with a flow rate of 50 µL/min at 40°C. If a Sciex API III+ mass spectrometer is employed, *see* **Note 9** for additional conditions.
2. Assign unglycosylated peptide peaks in the total-ion chromatogram (TIC, e.g. **Fig. 3**), using the masses and retention times shown in **Table 3** as a guide. Check for ions in all possible charge states within the m/z range of the instrument.
3. Assign glycosylated peptide peaks in the total-ion chromatogram (TIC), using the masses and retention times shown in **Table 4** as a guide (*see* **Note 14**). Check for ions in all possible charge states within the m/z range of the instrument. Note

Table 4
Glycosylated Tryptic Fragments of Reduced and Pyridylethylated hCG-β

Tryptic fragment	Amino acid residues	Peptide mass	Retention time (min)	Carbohydrate attached	Glycopeptide Mass (theory)
T13	115–122	884.4	2.3	O1	2198.4
				O3	1542.4
				O5	1616.4
				O6	1251.4
T3	9–20	1418.8	16.2	N1	3769.8
				N2	3478.8
				N3	3623.7
				N4	3332.7
				N5	2967.9
				N6	4279.7
				N7	3988.7
T4	21–43	2852.3	22.4	N1	5203.2
				N2	4911.5
T14	123–133	1104.6	—[a]	O6	1469.6
T15	134–145	1233.7	—[a]	O5	1963.8

[a]Fragments T14 and T15 are observed only by mass spectrometry infusion of the tryptic digest after treatment with neuraminidase and *O*-glycanase (*see* **Note 14**).

that **Table 4** includes only those glycopeptides observed by our laboratory. Because of carbohydrate heterogeneity, the specific tryptic glycopeptides observed will vary among hCG preparations. The "collision-dissociation scanning" method of Huddleston et al. (*11*) can be used to aid in distinguishing glycopeptides from unglycosylated peptides in the chromatogram (*see* **Note 9**).

3.2. Glycosylation States of O-Linked Sites

3.2.1. Reduction and Carboxymethylation

1. Dilute hCG-β to a concentration of about 3 mg/mL and transfer to a 5-mL reaction vial.
2. Purge 5 min with nitrogen.
3. Add freshly prepared, nitrogen-purged reduction buffer at a 1:1 volume ratio.
4. Incubate 30 min at 37°C, continuously purging the vial headspace with nitrogen during incubation.
5. Add freshly prepared, nitrogen-purged iodoacetic acid solution at a ratio of 50 µL per 1 mg of hCG-β.
6. Wrap the vial with aluminum foil to minimize light exposure and incubate for 30 min at room temperature, purging the headspace with nitrogen during incubation.
7. Transfer the reaction vial to an ice bath and cool for several minutes.
8. Quench the reaction by adding glacial acetic acid dropwise to adjust the pH to 4 (*see* **Note 15**), while agitating with a gentle stream of nitrogen.

9. Inject the mixture on reversed-phase gradient HPLC; for example, begin with a 5-min hold at 95:5 A-B followed by a linear ramp over 60 min to 30:70 A-B.
10. Collect the eluting peak preparatively and evaporate the solvent by vacuum centrifugation.

3.2.2. Isolation of 109–145 Glycopeptide

1. Add 100 mM pH 8.7 ammonium bicarbonate solution to the dried, reduced, and carboxymethylated hCG-β to give a concentration of 1–3 mg/mL.
2. Add 0.5 mg/mL α-chymotrypsin solution at a ratio of 20 μL α-chymotrypsin solution per 1 mg of hCG-β.
3. Incubate 5 min at 37°C.
4. Quench the digest by adjusting the pH to about 5 using glacial acetic acid (*see* **Note 15**).
5. Inject on reversed-phase gradient HPLC; for the Vydac C_{18} column, begin with a 5-min hold at 95:5 A-B followed by a linear ramp over 60 min to 30:70 A-B (*see* **Note 4**).
6. Collect the 109–145 fragment, which, using the conditions given here, elutes from 20 to 26 min (**Fig. 4**). The fragment is a heterogeneous glycopeptide and elutes as a broad cluster of chromatographic peaks.
7. Dry the collected 109–145 fragment using vacuum centrifugation.

3.2.3. β-Elimination

See **Notes 16** and **17**.

1. Reconstitute the dry 109–145 fragment using 40 μL of β-elimination solution per 1 mg of hCG-β initially employed (*see* **Notes 2** and **13**).
2. Incubate 2 h at 45°C.
3. Stop the reaction by adjusting the pH to approximately 5 using 1.0 M HCl.
4. Desalt the solution by injecting it on reversed-phase gradient HPLC; use the gradient given in **Subheading 3.2.2.**
5. Collect the β-eliminated 109–145 fragment and dry it using a vacuum centrifuge.

3.2.4. Tryptic Digestion

1. Reconstitute the β-eliminated 109–145 fragment in pH 8.0 ammonium bicarbonate solution using a ratio of 40 μL of buffer per 1 mg of hCG-β initially employed. If necessary, adjust the pH to 8 using 1.0 M aqueous ammonium bicarbonate.
2. Add trypsin solution at a ratio of 5 μL of trypsin solution per 1 mg of hCG-β digested.
3. Incubate at 37°C for 8 h before analysis by microelectrospray mass spectrometry.

3.2.5. Microelectrospray Mass Spectrometric Analysis

1. Dry a 10 μL aliquot of the tryptic digest solution on the vacuum centrifuge.
2. Reconstitute the sample with 10 μL of 1:1 methanol + water containing 0.05% formic acid.

3. Load the sample on the microelectrospray mass spectrometer.
4. Acquire the mass spectrum. **Figure 5** shows an example in which the doubly protonated pseudomolecular ions predominate.
5. Assign any peaks corresponding to initially unglycosylated species, which contain free Ser. The doubly protonated tryptic fragments are T13, m/z 443.2; T14, m/z 553.3; and T15, m/z 617.9.
6. Assign peaks corresponding to initially glycosylated species, which contain Dha in place of each formerly glycosylated Ser residue (see **Note 18**). The doubly protonated tryptic fragments of species bearing former glycosylations are T13 (1 Dha), m/z 434.2; T14 (1 Dha), m/z 544.3; T14 (2 Dha), m/z 535.3; and T15 (1 Dha), m/z 608.9. Note that the substitution of Ser by Dha results in a mass that is 18 u less than that of the Ser-containing analog.

4. Notes

1. For analysis of tryptic hCG-β, pyridylethylation of cysteine has a distinct advantage over carboxymethylation, because it introduces an additional potential positive charge site at each cysteine (**Fig. 1**). This lowers the molecular mass-to-charge ratio, effectively extending the mass range of the analysis. For example, glycosylated T4, which has a carbohydrate attached at Asn30, cannot be detected using a typical quadrupole mass spectrometer if carboxymethylation is used for alkylation of cysteines. However, because T4 has four cysteine residues, pyridylethylation increases the predicted maximum charge state from +2 to +6, and glycopeptide ions are readily observed *(12)*.
2. The traditional β-elimination solution, NaOH in purely aqueous solvent, produces extensive degradation of the 109–145 peptide backbone. The KOH/DMSO-containing solvent system *(14)* minimizes this destruction, and considerably enhances the relative reaction rate.
3. Trypsin should be L-1-tosylamide-2-phenylethylchloromethyl ketone (TPCK)-treated (Sigma cat. no. T-8642), and α-chymotrypsin should be 1-chloro-3-tosylamido-7-amino-2-heptanone (TLCK)-treated (Sigma cat. no. C-3142). The proteolytic specificity of the corresponding untreated proteases is less reliable, especially during long-term incubations. Chemicals should be HPLC grade or better. For all analyses involving mass spectrometry, we employ Burdick and Jackson Brand solvents (VWR Scientific, Chicago, IL) designed specifically for mass spectrometric analysis. TFA for LC and LC/MS mobile phases should be peptide sequencing grade, conveniently purchased in individual 1-mL ampoules (e.g., Sigma cat. no. T-1647). Formic acid for mobile phases and mass spectrometric infusion should also be of the highest grade available (e.g., Sigma catalog number F-4636, supplied in a glass bottle). These precautions are important to minimize interferences from metal ions such as sodium, as well as phthalate ester impurities.
4. All preparative HPLC separations employ a Vydac reversed-phase C_{18} column, 4.6 × 150 mm (particle size, 5 μm; pore size, 300 Å; cat. no. 218TP5415), maintained at 40°C with a flow rate of 1.0 mL/min.

5. We routinely use glass vials to minimize contamination of samples with phthalate esters, which are present in many plastics. However, the introduction of sodium ions from glass vials leads to sodium adduct formation on ESI-MS, as seen here in **Fig. 5**, and irreversible adsorption of proteins and peptides to glass may be responsible for significant sample loss. Plastic vials can be used if phthalate contamination does not become a problem.
6. The amount of DTT added prior to pyridylethylation corresponds to an approximately 10-fold molar excess over hCG-β disulfides (*see also* **Subheading 3.1.2.**). After reduction, 4-vinylpyridine is added in a twofold molar excess over DTT sulfhydryls. Use of a much larger excess of 4-vinylpyridine should be avoided, as it leads to formation of unknown pyridylethyl adducts in addition to the desired pyridylethylation of cysteine *(12)*.
7. The large quantity of 4-vinylpyridine injected during isolation of hCG-β from the reduction and pyridylethylation mixture may rapidly diminish the efficiency of the HPLC column. It is advisable to designate a column to be used only for the 4-vinylpyridine cleanup. This column can be used repeatedly for this purpose, as the chromatographic efficiency of the cleanup step itself is not critical. Alternatively, solid-phase extraction (SPE) can be employed.
8. For analyses of hCG-β that demand the highest mass spectrometer sensitivity, we employ 0.05% formic acid (v/v) as the mobile phase additive in place of 0.1% TFA. The chromatographic signal-to-noise is increased by about fivefold *(5)*. This substitution does not alter the appearance of the chromatogram or mass spectra in any other significant way.
9. For microbore LC/MS analysis (*see also* **Subheading 3.1.4.**), we employ a Beckman model 126 solvent delivery system HPLC, plumbed for low-flow gradient operation with a Deltabond 1 × 150 mm C_{18} column (particle size, 5 µm; pore size, 300 Å), and a PE-Sciex API III$^+$ mass spectrometer. The LC is operated at 50 µL/min at 45°C. A Valco-type "tee" is used to divert approximately 90% of the LC effluent through poly(ether ether ketone) (PEEK) tubing to the UV detector, with the remainder (approximately 5 µL/min) flowing through the 50-µm ID fused-silica tubing to the electrospray ionization interface of the mass spectrometer. This provides a parallel detector and facilitates system troubleshooting, but does not diminish the mass spectrometer response. The mass spectrometer is tuned and calibrated using ammoniated adducts of propylene glycol, according to the manufacturer's instructions. Other settings are as follows: ionspray voltage, 4500 V; nebulizer zero-grade air, 0.6 L/min; nebulizer pressure, 40–50 psi; curtain gas, 1.2 L/min. The orifice potential should be held at 65 V during scanning from *m/z* 150 to 370 and at 120 V during scanning from *m/z* 370 to 1900. This orifice stepping routine, called collision-excitation scanning, allows glycopeptides to be distinguished from unglycosylated molecules based on the presence of characteristic carbohydrate fragment ions at *m/z* 204, 274, and 366 *(11)*.
10. The reduction and carboxymethylation procedure employs an approximately 4-fold molar excess of DTT over cystine disulfides, followed by a 2.5-fold excess of iodoacetic acid over DTT sulfhydryls. Our procedure is similar to that of

Rademaker and Thomas-Oates *(13)*. We found that this relatively small excess of reagents restricts the carboxymethylation to the single cysteine residue of the 109–145 fragment. By contrast, a gross excess of iodoacetic acid resulted in partial carboxymethylation of other chemical moieties, which complicated interpretation of the mass spectrum.

11. The microelectrospray apparatus can be purchased commercially or constructed in-house. We routinely employ an inexpensive sprayer design based on that of Covey *(15)*. A PE-Sciex API III$^+$ triple-quadrupole instrument is employed. Samples are infused from a 10 µL syringe at a rate of approximately 200 nL/min using a Harvard Apparatus model 22 syringe pump (South Natick, MA). Spectra are acquired using the Tune program (PE-Sciex version 2.5). Samples are infused using an electrospray ionization potential of 5000 V and an orifice potential of 70 V. The curtain gas (nitrogen 99.999%) flow rate is 1.2 L/min. Mass spectrometer experiments are conducted by scanning the first quadrupole analyzer using a mass step size of 0.1 u and a dwell time of 10 ms.

12. Several steps in the procedures given in this chapter require an estimate of the mass of hCG-β treated. The initial quantity of hCG-β isolated can be estimated based the starting quantity of intact hCG as described in **Subheading 3.1.1.** To compute the amount of hCG-β employed in each subsequent step, assume that the mass changes accompanying HPLC cleanup, reduction, alkylation, and so forth are negligible.

13. Depending on the completeness of vacuum centrifugation, there may be residual ammonium acetate remaining after evaporation of the HPLC mobile phases. It is advisable to measure and, if necessary, adjust the pH of reconstituted solutions before proteolytic digestion. Also, measure the apparent pH of the β-elimination solution both before and after it is added to the 109–145 glycopeptide. If the apparent pH drops, due to residual ammonium acetate, the pH can be raised by dropwise addition of 0.25 M KOH in DMSO + water + ethanol (5:4:1 v/v/v).

14. Two of the O-glycosylated tryptic fragments of reduced and pyridylethylated hCG-β are not observed using the LC/MS procedure described herein. To detect these fragments, follow the protocol for isolation, β-elimination, and tryptic digestion of the 109–145 fragment. Alternatively, the carbohydrates can be partially removed by treatment with glycosidases such as neuraminidase or *O*-glycanase. The T14 and T15 glycopeptides noted in **Table 4** are observed after glycosidase treatment *(12)*.

15. To avoid consuming significant quantities of sample during pH measurement, deliver 1–2 µL of the sample to broad-range pH indicator paper from the tip of an Oxford-type pipette. This provides adequate measurement for all pH adjustment steps described in this chapter, with the exception of mobile phase pH adjustments, which should be performed using a digital pH meter.

16. In addition to converting the Ser residues bearing carbohydrates to Dha, we have evidence that the β-elimination conditions also convert the carboxymethyl-cysteine (Cmc) at position 112 of hCG-β 109–145 to Dha. Interpretation of the mass spectrum of intact β-eliminated 109–145 could be complicated by this reaction. However, our procedure calls for tryptic digestion of the β-eliminated 109–145 peptide (*see* **Subheading 3.2.4.**). The state of Cys112 subsequently becomes

unimportant, as the tryptic fragments containing the potential glycosylation sites do not contain position 112.

17. Using our β-elimination procedure, we find no evidence that the peptide backbone undergoes specific cleavage at Dha. However, it has been noted that Dha can undergo acid-catalyzed addition of water at Dha to form α-hydroxyalanine, followed by base-catalyzed cleavage of the peptide backbone *(16)*. To avoid these reactions, the pH should be adjusted during each step as described in **Subheading 3.2.3.**, and care should be taken not to overshoot the pH.

18. The T14 tryptic fragment contains two potential glycosylation sites. If the mass spectrum indicates glycosylation of T14 is incomplete, it is possible to determine the specific glycosylation state of each site by collision-induced dissociation (CID) with tandem mass spectrometry. The specific experimental conditions are beyond the scope of this chapter.

References

1. Cole, L. A. (1997) Immunoassay of human chorionic gonadotropin, its free subunits, and metabolites. *Clin. Chem.* **43,** 2233–2243.
2. Birken, S., Gawinowicz, M. A., Kardana, A., and Cole, L. A. (1991) The Heterogeneity of human chorionic gonadotropin (hCG). II. Characteristics and origins of nicks in hCG reference standards. *Endocrinology* **129,** 1551–1558.
3. Liu, C. and Bowers, L. D. (1997) Mass spectrometric characterization of nicked fragments of the β-subunit of human chorionic gonadotropin. *Clin. Chem.* **43,** 1–9.
4. Bowers, L. D. (1997) Analytical advances in detection of performance-enhancing compounds. *Clin. Chem.* **43,** 1299–1304.
5. Liu, C. and Bowers, L. D. (1996) Immunoaffinity trapping of human chorionic gonadotropin and its high-performance liquid chromatographic-mass spectrometric confirmation. *J. Chromatogr. B* **687,** 213–220.
6. Laidler, P., Cowan, D. A., Hider, R. C., Keane, A., and Kicman, A. T. (1995) Tryptic mapping of human chorionic gonadotropin by matrix-assisted laser desorption/ionization mass spectrometry. *Rapid Commun. Mass Spectrom.* **9,** 1021–1026.
7. Morgan, F. J., Birken, S., and Canfield, R. E. (1975) The amino acid sequence of human chorionic gonadotropin. *J. Biol. Chem.* **250,** 5247–5258.
8. Birken, S. and Canfield, R. E. (1977) Isolation of amino acid sequence of COOH-terminal fragments from the β-subunit of human choriogonadotropin. *J. Biol. Chem.* **252,** 5386–5392.
9. Keutmann, H. T. and Williams, R. M. (1977) Human chorionic gonadotropin. Amino acid sequence of the hormone-specific COOH-terminal region. *J. Biol. Chem.* **252,** 5393–5397.
10. Black, R. S. and Bowers, L. D. (1998) Determination of O-glycosylation sites: a β-subunit of human chorionic gonadotropin model system, in *Proceedings of the 46th ASMS Conference on Mass Spectrometry and Allied Topics,* Orlando, FL, p. 1302.
11. Huddleston, M. J., Bean, M. F., and Carr, S. A. (1993) Collisional fragmentation of glycopeptides by electrospray ionization LC/MS and LC/MS/MS: methods for selective detection of glycopeptides in protein digests. *Anal. Chem.* **65,** 877–884.

12. Liu, C. and Bowers, L. D. (1997) Mass spectrometric characterization of the β-subunit of human chorionic gonadotropin. *J. Mass Spectrom.* **32,** 33–42.
13. Rademaker, G. J. and Thomas-Oates, J. (1996) Analysis of glycoproteins and glycopeptides using fast-atom bombardment, in *Methods in Molecular Biology,* vol. 61, *Protein and Peptide Analysis by Mass Spectrometry* (Chapman, J. R., ed.), Humana, Totowa, NJ.
14. Downs, F., Herp, A., Moschera, J., and Pigman, W. (1973) β-Elimination and reduction reactions and some applications of dimethylsulfoxide on submaxillary glycoproteins. *Biochim. Biophys. Acta* **328,** 182–192.
15. Covey, T. (1995) Characterization and implementation of micro-flow electrospray nebulizers, in *Proceedings of the 43rd ASMS Conference on Mass Spectrometry and Allied Topics,* Atlanta, GA, p. 669.
16. Rollema, H. S., Metzger, J. W., Both, P., Kuipers, O. P., and Siezen, R. J. (1996) Structure and biological activity of chemically modified nisin A species. *Eur. J. Biochem.* **241,** 716–722.

20

Analysis of Gluten in Foods by MALDI-TOFMS

Enrique Méndez, Israel Valdés, and Emilio Camafeita

1. Introduction

Celiac disease (CD), gluten-sensitive enteropathy, is a permanent intolerance to gluten prolamins from wheat (gliadins), barley (hordeins), rye (secalins), and oats (avenins). The only treatment for celiac patients is a strict diet of foods without gluten. To control the gluten content in the diet of celiac patients, several conventional immunologic procedures comprising immunoblotting and home-made or commercial enzyme-linked immunosorbent assay (ELISA) methods using different monoclonal or polyclonal antibodies against a variety of gliadin components are commonly used (*1–4*). However, comparison of ELISA data, especially in the low gluten content range close to toxic levels, reveals inconsistencies since these are epitope-dependent methods. Therefore these systems are not reliable as sensitive methods for the measurement of gluten content in foods.

In addition, one of the main problems remaining to be solved in gluten analysis is the occasional occurrence of possible false-positive results arising from the immunologic systems routinely employed, e.g., ELISA. This question is difficult to address since no alternative nonimmunologic methods have been used for gluten analysis.

For this reason, the development of nonimmunologic alternative procedures to determine the four types of toxic gluten in foods is of great interest. Matrix-assisted laser desorption/ionization with time-of-flight mass spectrometry (MALDI-TOFMS) has been demonstrated to be efficacious and powerful in the analysis of wheat gliadin, barley hordein, rye secalin, and oat avenin protein components (*5,6*). We recently reported the use of MALDI-TOFMS as the first nonimmunologic method to quantify wheat gliadins and oat avenins in foods based on the direct observation of the gliadin and avenin mass spectral patterns *(7,8)*, as well as using the method to demonstrate the existence of false-positive ELISA results.

2. Materials
2.1. Proteins
1. Bovine serum albumin (Sigma, St. Louis, MO).

2.2. Cereal Cultivars
1. Wheat (*Triticum durum* L cv. *Spelta*).
2. Barley (*Hordeum vulgare* L cv. *Distica cameo*).
3. Rye (*Secale cereale* L cv. *Petkus*).
4. Oat (*Avena sativa* L cv. *PA-101*).

2.3. Chemical Reagents
1. Acetonitrile (Merck, Darmstadt, Germany).
2. Trifluoroacetic acid (TFA) (Merck).
3. Ethanol (Scharlau, Barcelona, Spain).
4. Sinapinic (*trans*-3,5-dimethoxy-4-hydroxycinnamic) acid (Fluka, Bochs, Switzerland).
5. *n*-Octyl-β-D-glucopyranoside (Fluka).

2.4. Solvents and Solutions
1. 60% aqueous ethanol.
2. 60% aqueous ethanol containing 0.1% TFA.
3. 50 mM *n*-octyl-β-D-glucopyranoside detergent solution.
4. Saturated sinapinic acid in 30% aqueous acetonitrile containing 0.1% TFA.
5. 0.4, 1.0, 4.0, and 10.0 µg of gliadin standard each in 1 mL of 60% aqueous ethanol (*see* **Subheadings 3.1.** and **3.2.**).
6. 0.4, 1.0, and 4.0 µg of avenin standard each in 1 mL of 60% aqueous ethanol (*see* **Subheadings 3.1.** and **3.3.**).

2.5. Mass Spectrometry
The MALDI-TOF mass spectrometer used for analyzing the wheat gliadin, barley hordein, rye secalin, and oat avenin glutens extracted from food samples was a Bruker (Bremen, Germany) Reflex II MALDI-TOF mass spectrometer equipped with an ion source with visualization optics and an N_2 laser (337 nm).

3. Methods
3.1. Extraction of Gliadin and Avenin Standards
1. Homogenize 1.0 g of wheat flour (*Spelta* or another cultivar) or oat flour (*PA-101*) for 1 h in polypropylene tubes with 10 mL of 60% ethanol at room temperature using a rotary shaker.

Gluten in Foods by MALDI-TOFMS 357

2. Centrifuge the solutions for 5 min at 4500*g* also at room temperature.
3. Transfer the supernatant to clean tubes and determine the gliadin or avenin concentration by amino acid analysis (2970 µg/mL for gliadins and 1560 µg/mL for avenins).

3.2. Preparation of Gliadin Standard Solutions for Calibration

1. Take a 10-µL aliquot of 2970 µg of gliadin standard in 60% aqueous ethanol (**step 3, Subheading 3.1.**) and dilute 10-fold with 90 µL of 60% aqueous ethanol to obtain 100 µL at 297 µg/mL. Then take a 10-µL aliquot of this 297 µg/mL solution (in 60% aqueous ethanol) and mix with 287-µL of 60% aqueous ethanol to give a standard gliadin solution of 10.0 µg/mL.
2. Take a 100-µL aliquot of the 10–µg/mL standard from **step 1** and mix with 150 µL of 60% aqueous ethanol to give a standard gliadin solution of 4.0 µg/mL.
3. Take a 100-µL aliquot of the 4.0-µg/mL standard from **step 2** and mix with 300 µL of 60% aqueous ethanol to give a gliadin solution of 1.0 µg/mL.
4. Take a 100-µL aliquot of 1.0–µg/mL standard from **step 3** and mix with 150 µL of 60% aqueous ethanol to give a gliadin solution of 0.4 µg/mL.

3.3. Preparation of Avenin Standard Solutions for Calibration

1. Take a 10-µL aliquot of 1560 µg of avenin standard in 60% aqueous ethanol (**step 3, Subheading 3.1.**) and dilute 10-fold with 90 µL of 60% aqueous ethanol to obtain 100 µL at 156 µg/mL. Then take a 10-µL aliquot of this 156-µg/mL solution (in 60% aqueous ethanol) and mix with 146 µL of 60% aqueous ethanol to give a standard avenin solution of 10.0 µg/mL.
2. Take a 100-µL aliquot of the 10–µg/mL standard from **step 1** and mix with 150 µL of 60% aqueous ethanol to give a standard avenin solution of 4.0 µg/mL.
3. Take a 100-µL aliquot of 4.0 µg/mL in 60% aqueous ethanol and mix with 300 µL of 60% aqueous ethanol to give a standard avenin solution of 1.0 µg/mL.
4. Take a 100-µL aliquot of 1.0 µg/mL in 60% aqueous ethanol and mix with 150 µL of 60% aqueous ethanol to make a standard avenin solution of 0.4 µg/mL.

3.4. Sample Preparation for the Analysis of High Gluten Content Food Samples by MALDI-TOFMS

The following procedure is recommended for the analyses of gliadins, hordeins, secalins, and avenins from wheat, barley, rye, and oat flours, and starches or foods with high gluten content

1. Homogenize 0.1 g of gliadins, hordeins, secalins, or avenins from wheat, barley, rye or oat flours, or starches with high gluten content for 1 h in polypropylene tubes with 10 mL of 60% ethanol at room temperature using a rotary shaker.
2. Centrifuge the homogenized samples for 5 min at 4500*g* also at room temperature.
3. Transfer the supernatant to clean tubes to be analyzed within 24 h (*see* **Note 1**) and discard the pellet.

4. Mix 10 µL of the ethanol extract with 2 µL of 50 mM n-octyl-β-D-glucopyranoside detergent solution and a matrix solution consisting of 100 µL of saturated sinapinic acid solution in 30% aqueous acetonitrile 0.1% TFA.
5. Deposit 0.5 µL of the mixture from **step 4** onto a stainless steel MALDI probe tip and allow to dry for 5 min at room temperature.
6. Introduce the sample target into the vacuum system of the mass spectrometer.
7. Record mass spectra in the linear positive-ion mode at 20 kV acceleration voltage and with 1.75 kV on the linear detector by accumulating 100 spectra of single laser shots at threshold irradiance. Only highly intense, well-resolved mass signals arising from five to seven selected target spots are considered further.
8. For mass calibration, a second target with a protein of known mass, usually bovine serum albumin ([M+H]$^+$ at m/z 66431), is prepared. The equipment is then externally calibrated employing singly, doubly, and triply charged signals from the bovine serum albumin.

Figure 1 displays the typical mass profile of wheat gliadins (30–40 kDa), rye secalins (32 and 39 kDa), and oat avenins (20–30 kDa). Barley hordein mass patterns differ markedly depending on the particular barley cultivar *(9)* (*see* **Notes 2** and **3**).

3.5. Sample Preparation for the Analysis of Gluten-Free Food Samples by MALDI-TOFMS

We reported the use of MALDI-TOFMS as a rapid screening system to determine the presence of gliadins in food samples (*see* **Note 4**) by directly monitoring the occurrence of the protonated gliadin mass pattern *(7)*. This experimental strategy is illustrated in **Fig. 2**. The complete procedure is as follows.

1. Allow wet food samples to dry in an oven at 37°C overnight and then mill as in **step 2**. The water content is determined by weighting the sample before and after the drying procedure. Quantification results should be corrected to take this water content into account. Proceed directly to **step 2** with dry food samples.
2. Weight 50 g of dry food sample and mill to obtain a fine powder.
3. Homogenize 0.1 g of the powder from **step 2** for 1 h in polypropylene tubes with 10 mL of 60% ethanol at room temperature using a rotary shaker.
4. Centrifuge the homogenized samples for 5 min at 4500g also at room temperature.
5. Transfer the supernatant to clean tubes to be analyzed within 24 h (*see* **Note 1**) and discard the pellet.
6. Mix a 40-µL aliquot of the extract in an Eppendorf tube with 8 µL of 50 mM n-octyl-β-D-glucopyranoside detergent solution and 25 µL of a matrix solution consisting of saturated sinapinic acid in 30% aqueous acetonitrile containing 0.1% TFA.
7. Dry the mixture in a Speedi-Vac centrifuge (40–50 min) and then redissolve in 6 µL of 60% aqueous ethanol containing 0.1% TFA.

Gluten in Foods by MALDI-TOFMS

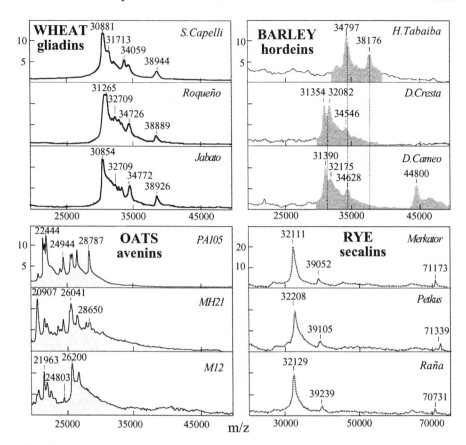

Fig. 1. MALDI-TOF mass spectra of the prolamin extracts from three different wheat (top left), barley (top right), oat (bottom left), and rye (bottom right) cultivars. Molecular masses are provided only for some specific peaks within the selected gliadin, hordein, avenin, or secalin mass range.

8. Deposit a 0.5-µL volume of this solution on a stainless steel MALDI probe tip and allow to dry at room temperature for 5 min.
9. Introduce the sample target into the vacuum system of the mass spectrometer.
10. Record mass spectra in the linear positive-ion mode at 20 kV acceleration voltage and with 1.75 kV on the linear detector by accumulating 100 spectra from single laser shots at threshold irradiance. Only highly intense, well-resolved mass signals arising from five to seven selected target spots are considered further.
11. For mass calibration, a second target with a protein of known mass, usually bovine serum albumin ([M+H]$^+$ at m/z 66431 Daltons), is prepared. The equipment is then externally calibrated employing singly, doubly, and triply charged signals from the bovine serum albumin.

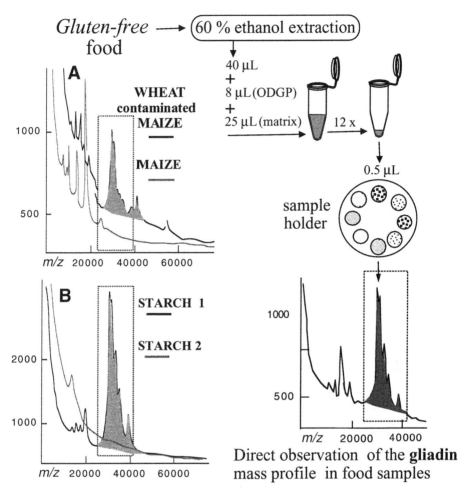

Fig. 2. Schematic representation of the procedure employed to analyze gluten in food samples by MALDI-TOF mass spectrometry. After the ethanol extraction, the sample + detergent + matrix mixture is dried in a Speedi-Vac centrifuge, redissolved in a 12 X smaller volume, and then mass analyzed to allow direct observation of the gliadin mass profile in the food sample. This is illustrated by the mass analysis of wheat-contaminated and noncontaminated maize (**A**) and wheat starch (**B**) samples. The gliadin mass range (30–40 kDa) has been highlighted.

We routinely use this procedure as a rapid screening test for the presence of gliadins in food samples by visualizing the corresponding gliadin mass pattern. This is illustrated in **Fig. 2** for two wheat-contaminated samples (maize flour and wheat starch) in which gliadins are found as contaminants, compared with a pure maize flour and another wheat starch where they are absent.

Gluten in Foods by MALDI-TOFMS

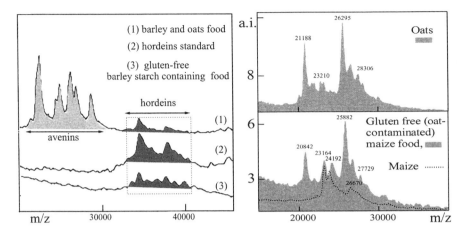

Fig. 3. MALDI-TOF mass spectra of barley-containing foods (left) and an oat-contaminated gluten-free maize food (right bottom) together with those of oat avenins (top) and a maize (bottom, dashed) cultivar.

3.6. Detection of Barley and Oat Contamination in Gluten-Free Foods by MALDI-TOFMS

Since all ELISA systems utilized to control the gluten content in food samples fail to recognize avenins and hordeins, MALDI-TOFMS is an ideal tool for the detection of oat or barley contamination in gluten-free foods *(4)*. This is exemplified in **Fig. 3** by the analysis of two gluten-free foods together with barley- and oat-containing samples in which barley and oats, unambiguously revealed by MALDI-TOFMS, remain undetected by commercial ELISA kits.

3.7. Comparative Analysis of Complex Formula Gluten-Free Foods Extracted at Different Concentrations in 60% Aqueous Ethanol

Even though the gliadin, hordein, secalin, and avenin fractions are selectively extracted by means of an ethanol extraction, accompanying nonprolamin components (e.g., salts, sugars, colorants, fats, noncereal proteins) are often co-extracted. These components may produce a partial or complete lack of prolamin mass signals (**Fig. 4**) if the extraction is performed at 1 g/10 mL of 60% ethanol as originally described *(7)*. As can be seen, the relative recovery is much lower when using 1 g/10 mL than with 0.1 g/10 mL. Good mass signals are also obtained by simply diluting the extract (**Fig. 4**) 1:10 or as much as is necessary. To avoid this inconvenience, we recommend that all types of gluten-free samples be routinely extracted at 0.1 g/10 mL of aqueous ethanol (*see* **Note 5**).

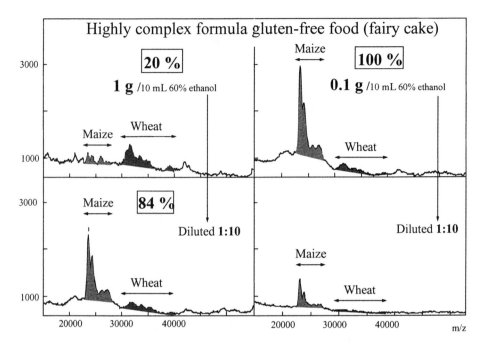

Fig. 4. Comparison of mass spectra of the prolamins extracted with alcohol from a highly complex formula gluten-free food extracted at two concentrations (1 and 0.1 g/10 mL). Analyses after dilution of the sample 1:10 in 60% ethanol are also displayed. The estimated percentage recovery as deduced from the gliadin mass pattern area is indicated in a box.

3.8. Assessment of ELISA False-Positive Results by MALDI-TOFMS

MALDI-TOFMS permits a proper assessment of ELISA false-positive results in the analysis of gluten-free foods. **Figure 5** displays two gluten-free samples in which no gliadin, hordein, secalin, or avenin mass signals are observed using MALDI-TOFMS, in contrast to the unexpected high-gluten content values obtained using commercial ELISA kits.

3.9. Quantification of Gliadins and Avenins Based on Mass Area Measurement

The gliadin or avenin content of each sample is calculated by comparing mass areas measured from food samples against the appropriate calibration graph prepared from mass areas obtained with standard gliadin or avenin solutions. Mass area measurements for gliadins (in the m/z 25,000–45,000 region) or avenins (in the m/z 19,000–33,000 region) were determined as follows (*see* **Note 6**):

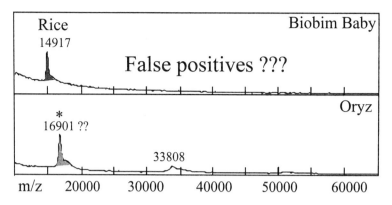

Fig. 5. ELISA false-positive results revealed by MALDI-TOFMS, i.e., the absence of gliadins, hordeins, secalins, and avenins in food samples in contrast to commercial ELISA kit data. *Unidentified mass peak of a co-extracted protein component.

1. Convert the intensity and m/z values from mass spectra to ASCII files and use a spreadsheet program, e.g., Microsoft Excel, to display all spectra on the same intensity scale and measure gliadin or avenin mass areas by means of Simpson's method. Alternatively, mass areas can be determined by cutting out and weighing mass spectra plotted on paper.
2. Obtain a calibration graph by plotting the gliadin or avenin mass areas against the standard concentration at 0.4, 1,0 4.0, and 10.0 µg/mL (**Fig. 6**).
3. Directly mass analyze alcohol extracts from gluten-free samples according to the procedure described in **Subheading 3.5**. Results are expressed as concentration (µg/mL) of gliadins or avenins.
4. Recalculate these values into milligrams of gliadins or avenins per 100 g of food or (ppm) by taking into account the sample extraction procedure (i.e., 0.1–1 g in 10 mL 60% aqueous ethanol). Gluten values are expressed as twice the gliadin or avenin content.

Results obtained from the quantification of gliadin- and avenin-containing and gluten-free food samples by this mass spectrometric procedure are shown in **Tables 1** and **2**. We presently routinely employ mass area measurement for the quantification (*see* **Notes 7** and **8**) since these peak area measurement correlate better with data obtained by ELISA than do results from gliadin peak height measurement as originally described *(7)*. Results obtained by each method are summarized in **Tables 1** and **2**.

4. Notes

1. As a general recommendation, analysis of foods by MALDI-TOFMS should be carried out immediately after the ethanol extraction. This can be especially critical for complex formula gluten-free foods, in which co-extracted material some-

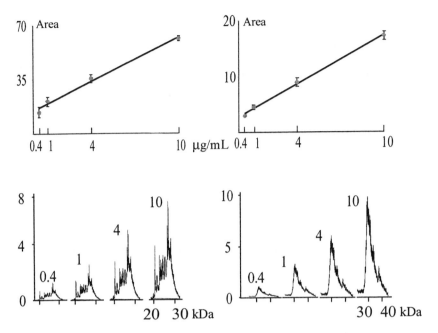

Fig. 6. **(Top)** Calibration graph from 0.4 to 10 µg/mL using the avenin standard from *Avena sativa* L. cv. *PA-101* (left) and the gliadin standard from *Triticum durum* L. cv. *Spelta* (right). **(Bottom)** MALDI-TOF mass spectra corresponding to the above concentrations showing the 20–30 kDa avenin range (left) and the 30–40 kDa gliadin range (right) employed to measure peak areas. Error bars represent the standard deviation of three independent duplicated measurements from three to five selected target spots.

times precipitates a few hours after the ethanol extraction and may alter the mass spectra.

2. It should be noted that while the mass profiles of wheat gliadins (30–40 kDa), rye secalins (32 and 39 kDa), and oat avenins (20–30 kDa) appear to be very similar in different cultivars of the same cereal, in contrast, barley hordein mass patterns differ markedly depending on the particular barley cultivar.
3. The fact that these typical mass ranges are different permits a selective identification of the specific prolamin type in foods, mainly when made with one or two of the above cereals. Difficulties arise when analyzing foods made with the four cereals wheat, barley, rye, and oats at varying ratios. Nevertheless, oat avenins and wheat gliadins are easily discernible from the remaining components, whereas hordeins and secalins can be identified, although more ambiguously, from the corresponding increased intensity in the gliadin mass pattern.
4. It is noteworthy that MALDI-TOFMS is a highly sensitive procedure for the quantification of gluten in foods. The detection limit for gliadins and avenins is

Table 1
Analysis of Gliadins (ppm) in Gluten-Free Foods and Starch Samples by ELISA and Mass Spectrometry

Sample	ELISA^a	Mass spectrometry	
Starch 1	0.9	2.1^b	(1.0)^c
Starch 2	8.1	11.8	(5.8)
Starch 3	11.0	9.7	(10.8)
Starch 5	4.6	4.3	(2.6)
1155	7.2	12.0	
1190	1.3	ND	
2098	4.4	6.4	

^{a}Cocktail Sandwich ELISA *(4)*.
^{b}Area measurement.
^{c}Height measurement.

Table 2
Analysis of Avenins (ppm) in Oat-Containing Samples by Competitive ELISA and Mass Spectrometry

Sample	Competitive ELISA (mg/100 g)	Mass spectrometry (mg/100 g)
Control grain^a	< 4*	ND
Biscuits (2031)^b	23	< 8
Maize food (1020)^b	40	14
Oat starch (2027)^b	55	60
Rolled oats (2166)	230	810
Rolled oats (2170)	880	835
Rolled oats (2169)	960	590

^{a}Oat-free samples as controls.
^{b}Oat-contaminated gluten-free samples.

0.4 µg/mL with a working linear range of 0.4–10.0 µg/mL. The sensitivity of MALDI-TOFMS is similar to that of the most sensitive ELISA systems designed for gluten analysis. However, the advantage of employing MALDI-TOFMS is that it reveals the characteristic prolamin mass profile, thus allowing identification of the specific cereal type used to make the food subjected to analysis, such as oat avenins and barley hordeins, which cannot be determined by using commercial ELISA kits.

5. We have improved the mass analysis of gluten-free samples by routinely extracting samples at 0.1 g/10 mL instead of at 1 g/10 mL which we originally employed, since this is the concentration conventionally utilized for ELISA. While differences between extraction at 1 and 0.1 mg/10 mL may not be observed for simple

formula, gluten-free foods, extraction of complex-formula foods at 1 g/10 mL often leads to a partial or complete lack of prolamin mass signals. If this is the case, a 1:10 dilution is encouraged.

6. One of the greatest problems found in the quantification of gliadins or avenins arises when analyzing food samples made with cereal mixtures, in which the presence of singly and doubly protonated mass signals from these complex prolamin protein mixtures can make the measurement of mass areas difficult. Gliadin and avenin quantification procedures are reliable provided that wheat or oats are the only cereals present in the gluten-free food. However, if these samples have a complex formula, co-extracted components can again affect the mass pattern intensity.
7. We presently routinely employ mass area measurement for gliadin and avenin quantification since these data correlate better with ELISA measurements than do measurements from the quantification procedure based on measuring the height of the most intense α-gliadin peak (at $m/z \approx 31,000$) as originally described in **ref. 7**.
8. Originally the use of internal standards added to the alcohol extracts was intended. Unfortunately, the very high numbers of protonated and doubly protonated mass signals in these highly complex protein mixtures overlap those of the internal standards, mainly when attempting to analyze foods containing mixtures of gliadins, secalins, hordeins, and avenins.

References

1. Skerritt, J. H. and Hill, A. S. (1991) Enzyme-immunoassay for determination of gluten in food; collaborative study. *Assoc. Off. Anal. Chem.* **74,** 257–264.
2. Friis, S. U. (1988) Enzyme-linked immunosorbent assay for quantification of cereal proteins in coeliac disease. *Clin. Chim. Acta* **178,** 261–270.
3. Ellis, H., Rosen-Bronson, S., O'Reilly, N., and Ciclitira, P. J. (1998) Measurement of gluten using a monoclonal antibody to a celiac toxic peptide of A gliadin. *Gut* **43,** 190–195.
4. Sorell, L., López, J. A., Valdés, I., Alfonso, P., Camafeita, E., Acevedo, B., et al. (1998) An innovative sandwich ELISA system based on an antibody cocktail for gluten analysis. *FEBS Lett.* **439,** 46–50.
5. Méndez, E., Camafeita, E., S.-Sebastián, J., Valle, Y., Solís, J., Mayer-Posner, F. J., et al. (1995) Screening of gluten avenins in foods by matrix-assisted laser desorption/ionization time-of-flight mass spectrometry. *Rapid Commun. Mass Spectrom.* S123–128.
6. Camafeita, E., Alfonso, P., Acevedo, B., and Méndez, E. (1997) Sample preparation optimization for the analysis of gliadins in food by matrix-assisted laser desorption/ionization time-of-flight mass spectrometry. *J. Mass Spectrom.* **32,** 444–449.
7. Camafeita, E., Alfonso, P., Mothes, T., and Méndez E. (1997) MALDI-TOF mass spectrometric micro-analysis: the first non-immunological alternative attempt to quantify gluten gliadins in food samples. *J. Mass Spectrom.* **32,** 940–947.

8. Camafeita, E. and Méndez, E. (1998) Screening of gluten avenins in food by matrix-assisted laser desorption/ionozation time-of-flight mass spectrometry. *J. Mass Spectrom.* **33,** 1023–1028.
9. Camafeita, E., Solís, J., Alfonso, P., López, J. A., Sorell, L., and Méndez, E. (1998) Selective identification by matrix-assisted laser desorption/ionization time-of-flight mass spectrometry of different types of gluten in foods containing cereal mixtures. *J. Chromatogr. A* **823,** 299–306.

21

Quantitation of Nucleotidyl Cyclase and Cyclic Nucleotide-Sensitive Protein Kinase Activities by Fast-Atom Bombardment Mass Spectrometry

A Paradigm for Multiple Component Monitoring in Enzyme Incubations by Quantitative Mass Spectrometry

Russell P. Newton

1. Introduction

1.1. Nucleotidyl Cyclases and Cyclic Nucleotide-Responsive Kinases

Two cyclic nucleotides, adenosine 3',5'-cyclic monophosphate (cyclic AMP) and guanosine 3',5'-cyclic monophosphate (cyclic GMP), are established components of biological signal transduction processes. Two sets of enzymes are identified as determinants of cyclic nucleotide activity in such regulatory processes; the nucleotidyl cyclases, which catalyse the synthesis of the cyclic nucleotide from the corresponding nucleoside triphosphate, and the cyclic nucleotide-responsive protein kinases, which are stimulated by the cyclic nucleotides to phosphorylate target proteins, thereby modifying their activity (**Scheme 1**). This cyclic nucleotide signal is switched off by hydrolysis to a mononucleotide by phosphodiesterase. Whereas the cyclic AMP and cyclic GMP second messenger systems are major targets for pharmacologic manipulation, other cyclic nucleotides (cytidine-, inosine-, uridine-, and deoxythymidine-3',5'-cyclic monophosphate) also occur naturally, together with enzymes capable of their synthesis and hydrolysis, but no precise physiological functions have yet been elucidated for these compounds.

From: *Methods in Molecular Biology, vol. 146:
Protein and Peptide Analysis: New Mass Spectrometric Applications*
Edited by: J. R. Chapman © Humana Press Inc., Totowa, NJ

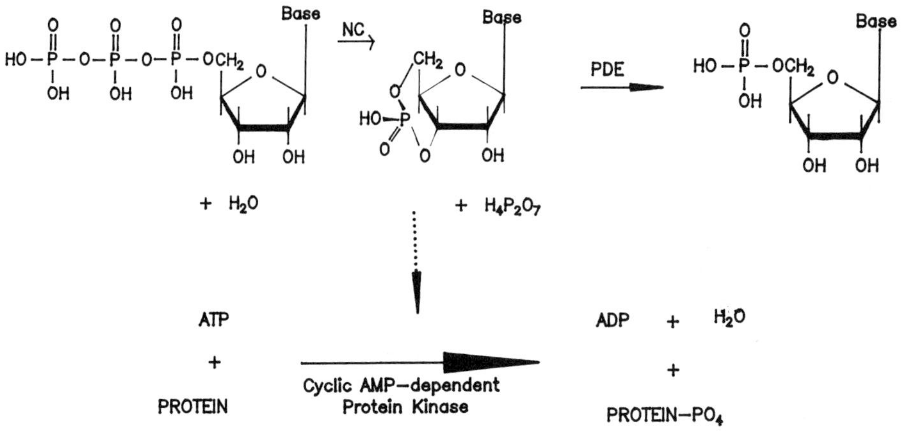

Scheme 1.

1.2. Mass Spectrometric Analyses of Cyclic Nucleotides

Prior to the advent of soft-ionization techniques, mass spectrometric analyses of cyclic nucleotides were only possible after derivatization, so that mass spectrometric quantitation of enzymes involved in cyclic nucleotide action was not then practicable *(1,2)*. Fast-atom bombardment mass spectrometry (FABMS) provided a means of soft ionization that enabled the acquisition, without derivatization, of cyclic nucleotide mass spectra containing a pseudomolecular ion. Although the FAB mass spectra do not provide extensive fragmentation data, selection of a specific or ion in the mass spectrum and subjecting this to a second fragmentation (collisionally-induced dissociation, [CID]) enables valuable structural information to be obtained by tandem mass spectrometry (MS/MS). Thus, in a qualitative context, FABMS, in conjunction with CID and mass-analyzed ion kinetic energy spectrum (MIKES) scanning, has proved invaluable in the unequivocal identification of putative cyclic nucleotides in tissue extracts *(3)* and enzyme incubates *(4)*, in the differentiation of cyclic nucleotide isomers *(5)*, and in the structural elucidation of analogs *(6–8)*. In quantitative studies of the nucleotidyl cyclases and cyclic nucleotide-responsive protein kinases, the conventional routine assay protocols are based on the use of a radiolabeled substrate and, following chromatographic or other separation of substrate and product, determination of label present in the product, i.e., a cyclic nucleotide in the cyclase assay and the target protein in the kinase assay *(9)*. The disadvantage of these procedures, in addition to the requirement for a separation step, is that only one datum is available from each incubation. The use of quantitative FABMS provides a facility (multiple component monitoring) that enables the assay of changes in concentration of several components of the enzyme reaction simultaneously. For example, with a cyclase, it is possible to monitor the appearance of more than one product, the consumption of alternative substrates, binding of effectors, further metabolism of products, and the appearance of side products, all from the same set of incubations *(10,11)*.

1.3. Quantitation by FABMS

Although some enzyme systems can be assayed directly by monitoring a single peak in the FAB mass spectrum *(12)*, there is a comparatively narrow range of analyte concentration within which peak intensity is directly proportional to concentration for cyclic nucleotides and nucleotides *(13)*. The problem is compounded when a complex mixture of components is present as in an enzyme incubation mixture. In many quantitative FABMS studies the solution to these problems is the inclusion of a deuterated or other heavy isotope-labeled internal standard; for several of the cyclic nucleotides this option is not financially viable and for the majority such labeled compounds are not commercially available. This problem is overcome by determining the peak heights of the protonated molecules, their adducts and characteristic fragments of the compounds, together with the analogous peaks derived from the glycerol sample matrix; quantitation is then carried out by a proportionation of the sums of the relative peak intensities. It is necessary to carry out parallel series of incubations containing active and inactivated enzyme preparations and to utilize the differences between them in order to obtain valid kinetic data *(10,11,14,15)*.

2. Materials
2.1. Chemicals

All chemicals are high-performance liquid chromatography (HPLC) analytical grade or the highest purity commercially available.

1. Buffers. Tris-HCl (pH 7.4) (*see* **Note 1**) is used for the enzyme incubations. In the standard radiometric assay the concentration is 50 mM; in the mass spectrometric assay this concentration is reduced, as detailed in **Subheading 3.1.3.**
2. Glycerol + water (1:1 v/v) matrix is used to apply the sample to the FAB probe. Water is HPLC grade double deionized.
3. Cyclic nucleotides and nucleotides (Sigma, Poole, Dorset, UK or Boehringer-Mannheim, Lewes, UK). Where possible free acids or magnesium salts are used in preference to the sodium salts.
4. Nucleotide analogs (Sigma, Boehringer-Mannheim, or Biolog, Bremen, Germany).
5. Radiolabeled nucleoside triphosphates (Amersham International, Amersham, UK).

2.2. Equipment

1. A freeze drier capable of rapidly reducing at least 50 enzyme incubates to dryness simultaneously. Rotary film evaporation is too time consuming and does

not provide reproducible data. Samples are best freeze dried in tapered vessels such as Ependorf tubes or silylation vials.
2. A mass spectrometer with a fast-atom bombardment ion source (*see* **Notes 2** and **3**) and tandem MS facility; the use of two such instruments is detailed below. These instruments are a reversed-geometry VG ZAB-2F mass spectrometer (Micromass, Wythenshawe, Manchester, UK) and a Finnegan 8400 reversed-geometry double-focusing mass spectrometer (Finnegan MAT, Bremen, Germany). The use of these instruments to record positive-ion spectra (*see* **Note 4**) is described in **Subheadings 3.2.1.** and **3.2.2.**, respectively.
3. Equipment for the purification and radiometric assay of the cyclase and kinase.

3. Methods
3.1. Pre-Mass Spectrometry Enzyme Procedures
3.1.1. Enzyme Preparation

The nucleotidyl cyclase or cyclic nucleotide-responsive protein kinase preparation to be assayed must be purified to the greatest practicable extent. The purer the enzyme preparation the less its effect on the complexity of the mass spectra. The purification protocol used depends on the enzyme and tissue source; if required, the reader is referred to **ref. 16** for guidance.

3.1.2. Radiometric Determination of Enzyme Activity

To carry out the ensuing steps it is necessary to determine the cyclase or kinase activity by the conventional radiometric assays; if required, the reader is referred to **refs. 16** and **17** for guidance.

3.1.3. Modification of Assay Buffer Concentration

To minimize the contribution of buffer-derived peaks to the mass spectra, the buffer concentration must be reduced to the minimum concentration that still permits the reaction to proceed at the original rate for the duration of the incubation period. This is determined as follows:

1. Carry out the standard radiometric assay for the enzyme at the highest substrate concentration with a series of dilutions of the assay buffer.
2. Select the lowest buffer concentration that is still effective (*see* **Note 5**).

3.1.4. Enzyme Incubation for Mass Spectrometric Quantitation

1. Carry out enzyme incubations under the same conditions as for the radiometric assay (*see* **Subheading 3.1.2.**) with the exception that the substrate is unlabeled and the buffer concentration is reduced as determined in **Subheading 3.1.3., step 2**.
2. Stop the reaction by heating at 90°C for 90 s.
3. Centrifuge off the denatured protein and freeze dry the supernatant.

Multiple Component Monitoring Paradigm

3.2. Acquisition of Mass Spectra

Mass spectra for cyclase and kinase quantitation have been obtained on two instruments in our laboratories (*see* **Subheadings 3.2.1.** and **3.2.2.**).

3.2.1. Acquisition of Data with the VG ZAB-2F

1. Apply samples directly to the FAB probe tip in 3–5 µL of glycerol/water matrix.
2. Set the following ion source conditions (*see* **Notes 6** and **7**). Typically operate the FAB atom gun with 8-kV accelerating potential and 1–2-mA discharge current, using xenon (or argon) gas. Set the ion source accelerating potential to 8 kV to record positive-ion mass spectra.
3. Set the following MS/MS conditions. Generate CID spectra by using nitrogen as a collision gas in the second field-free region gas cell at a pressure that gives a reading of 800 µPa on the nearby ion gauge. Record MIKE spectra by using the magnetic sector to select the ion to be collisionally dissociated and then scanning the electric sector voltage under data system control (*see* **Note 8**).

3.2.2. Acquisition of Data with the Finnegan MAT 8400

1. Apply samples directly to the FAB probe tip in 3–5 µL of glycerol/water matrix.
2. Set the following ion source conditions Typically operate the FAB atom gun with 9-kV accelerating potential and 0.5-mA discharge current. Set the ion source accelerating potential to record positive-ion mass spectra.
3. Set the following MS/MS conditions. Generate CID spectra by using nitrogen in a collision gas cell located close to the focal plane between the magnetic and electric sector. Set a collision energy of 3 keV. Record MIKE (and normal) spectra using a Finnegan INCOS computer, which also controls the scans of magnetic and electric sector fields (*see* **Note 8**).

3.2.3. Reference Spectra

1. Prepare standard solutions of the substrate(s), product(s), and other incubation components to be monitored at a known concentration of 2–5 µg/µL.
2. Mix the solutions from **step 1** with an equal volume of glycerol and record reference FAB mass spectra as in **Subheadings 3.2.1., step 2** or **3.2.2., step 2**.
3. Record CID/MIKE spectra from the protonated molecule and the sodium adduct ion for each compound as in **Subheadings 3.2.1., step 3** or **3.2.2., step 3**.

3.2.4. Mass Spectrometry of Enzyme Incubates

1. Dissolve the freeze-dried enzyme incubates in the minimum volume of water. It is essential that the pairs of denatured control and active enzyme incubate be dissolved in the same volume.
2. Mix a few microliters with an equal volume of glycerol and obtain the mass spectrum as described in **Subheadings 3.2.1., step 2** or **3.2.2., step 2**.
3. If the cyclic nucleotide or nucleotide peaks have a higher relative abundance than the glycerol matrix-derived peaks at either m/z 277 or 299, then dilute the enzyme

incubate sample further and repeat the spectrum acquisition until a dilution is obtained that gives a spectrum with the major cyclic nucleotide or nucleotide-derived peak at 80–85% of the height of the major glycerol peak at m/z 277 or 299 (see **Note 9**).

3.3. Quantitation

The change in concentration of a component X during the course of the enzymic reaction is obtained by monitoring the intensities of a series of diagnostic peaks of X in both active (experimental) and inactive (control) samples and expressing them relative to matrix-derived peaks in both cases:

$$\Delta[X] = ([\Sigma I_n X/\Sigma I_n Gro]_{cont} - [\Sigma I_n X/\Sigma I_n Gro]_{exp})/ [\Sigma I_n X/\Sigma I_n Gro]_{cont}$$

where $I_n X$ and $I_n Gro$ are the sums of the relative peak intensities of the characteristic ions from the analyte X and glycerol matrix respectively, and $\Delta[X]$ is the enzyme-catalyzed change in the concentration of X. For cyclic nucleotides and nucleotides, the peak intensities monitored are those of the protonated molecule, its cation, its glycerol and buffer adducts, plus any adduct formed with any other incubation component, and each of these ions minus 17 and 18 mass units, corresponding to losses of NH_3 and H_2O, respectively. The matrix peaks monitored are the protonated glycerol trimer, tetramer, and pentamer, their cation and buffer adducts, together with each of these ions minus 18 mass units, corresponding to the loss of H_2O.

In the proportionation equation $\Delta[X]$ is a relative term. To quantitate the enzyme activity in absolute terms, the proportional change in analyte concentration is multiplied by the original concentration of substrate to provide the rate of disappearance of substrate. If product formation is to be measured, e.g., as in the cyclase reaction, then control samples must be spiked with known amounts of cyclic nucleotide and the $\Delta[X]$ value calculated for the known amount of product. From this relationship, the value of $\Delta[X]$ for the reaction can then be converted into absolute units for product formed or released.

3.4. Representative Data

3.4.1. General Appearance of Spectra

The positive-ion FAB mass spectra of ATP and cyclic AMP are shown in **Fig. 1A** and **B**. Both show the characteristic peaks of $[M + H]^+$ and $[MH + Gro]^+$ at m/z 508 and 600 and their sodium adducts (see **Note 1**) at 530, 552, 574, 596, and 618 $[MH+ nNa - nH]^+$ and 622 $[MH + Gro + Na - H]^+$ for ATP and at m/z 330, 422, 352, 374, 396, and 444 for the corresponding compositions for cyclic AMP. Characteristic glycerol (mol wt 92) matrix peaks at m/z 277 $[3Gro + H]^+$, 299, 321, 369, 391, 413, 461, 483, 553, 575, and 645 are also seen. These spectra were obtained with the magnesium salts of the nucleotide

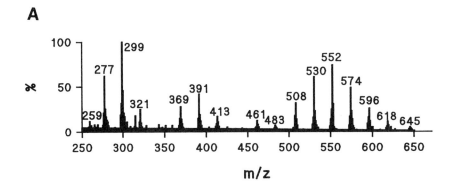

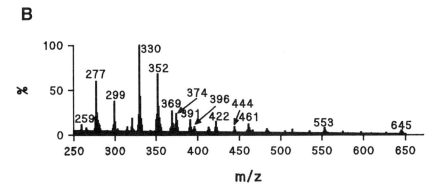

Fig. 1. Positive-ion FAB mass spectra of (**A**) adenosine 5'-triphosphate and (**B**) adenosine 3',5'-cyclic monophosphate. (Reproduced with permission from **ref. 11**).

dissolved in the reduced concentration assay buffer and are essentially identical to those obtained with the free acids dissolved in deionized water, indicating that neither the presence of buffer nor magnesium ions substantially contributes to the spectrum in the region of interest (see **Note 10**).

With highly purified preparations of adenylyl cyclase, the differences between the spectrum from the adenylyl cyclase incubation containing the active enzyme and that from the inactivated control enzyme incubation clearly indicate the changes in concentration of the substrate, ATP, and the product, cyclic AMP. In the denatured control preparation (**Fig. 2A**) there are peaks corresponding to the molecular ion of ATP and its sodium adducts at m/z 508, 530, 552, 574, 596, and 618. On the other hand, no peaks at m/z 330 or 352 are apparent, indicating the absence of cyclic AMP, nor are there substantial peaks at m/z 428, 450, or 472, which would have indicated nonenzymic breakdown of ATP to ADP. In the spectrum of the adenylyl cyclase incubate containing active

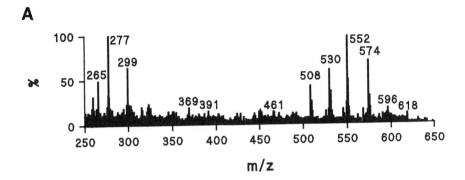

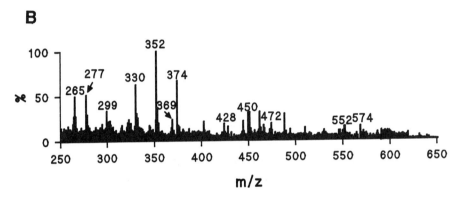

Fig. 2. Positive-ion FAB mass spectra of (**A**) incubate containing inactivated (control) adenylyl cyclase, and (**B**) incubate containing active adenylyl cyclase. (Reproduced with permission from **ref. 11**).

enzyme (**Fig. 2B**), the ATP-derived peaks have significantly diminished, with only m/z 552 and 574 still evident; peaks at m/z 428, 450, and 472 indicate the conversion of ATP to ADP by ATPase in the active enzyme preparation. The peaks at m/z 330, 352, and 374 correspond to $[M + H]^+$, $[M + Na]^+$ and $[MNa + Na - H]^+$ for cyclic AMP. The identity of the product as cyclic AMP can be unequivocally established by CID/MIKE analysis of the peak at m/z 330 (**Fig. 3**) when the spectrum obtained is essentially identical to that of standard cyclic AMP and contains diagnostic peaks at m/z 136, 164, and 178, corresponding to $[BH_2]^+$, $[BH_2 + 28]^+$, and $[BH_2 + 42]^+$, where B represents the nucleotide base.

3.4.2. Quantitative Data

The usefulness of quantitative FABMS assays of adenylyl cyclase can be gauged from the extra information obtainable compared with that from the

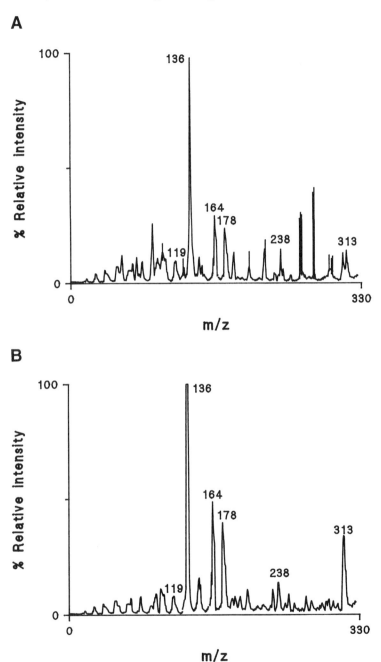

Fig. 3. CID/MIKE spectra of *m/z* 330 in the FAB mass spectrum of **(A)** incubate containing active adenylyl cyclase and **(B)** standard cyclic AMP. (Reproduced with permission from **ref. *11***).

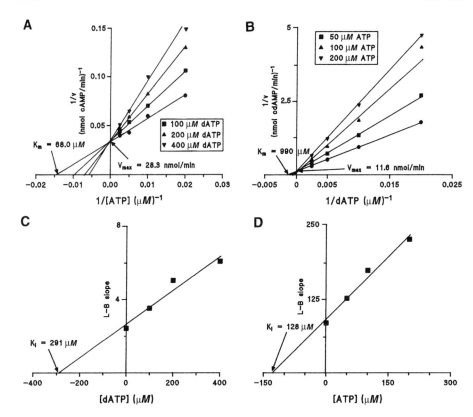

Fig. 4. Plots of kinetic data for adenylyl cyclase obtained by quantitative mass spectrometry. (**A**) Lineweaver-Burk plot for cyclic AMP synthesis from ATP in the presence of increasing concentrations of dATP. (**B**) Lineweaver-Burk plot for cyclic dAMP synthesis from dATP in the presence of increasing concentrations of ATP. (**C**) secondary plots of Lineweaver-Burk slopes against inhibitor concentration for (**C**) cyclic AMP synthesis and (**D**) cyclic dAMP synthesis. (Reproduced with permission from **ref. 11**).

radiometric assay *(11)*. By monitoring the appearance of ADP-derived peaks at m/z 428, 450, and 470 it is possible to determine the efficiency of ATP-regenerating systems employed in standard adenylyl cyclase assay protocols. By monitoring the cyclic AMP- and AMP-derived peaks it is possible to determine cyclic AMP/AMP ratios obtained with a variety of different phosphodiesterase inhibitors present, thereby determining their efficiencies. Alternative substrates such as adenylyl methylene diphosphate (AMP-PCP), adenylyl imidodiphosphate (AMP-PNP), and deoxyadenosine triphosphate (dATP) can be used in combination with the natural substrate, ATP, and kinetic data for the effects of one substrate upon the other obtained, data not accessible by the routine radiometric assay (**Fig. 4**). With a second nucleotidyl cyclase, cytidylyl cyclase, the multiple component monitoring facility has enabled monitoring of side product formation as well as synthesis of cyclic CMP from enzyme preparations from different subcellular fractions, the resultant kinetic data being consistent with current theories of more than one isoform of the enzyme being present *(10)*.

With a highly purified preparation of cyclic AMP-dependent protein kinase, the kinase activity, together with other significant data inaccessible by the routine radiometric assay, can be obtained from quantitative FABMS of a series of kinase incubations *(15)* (**Fig. 5**). In **Fig. 5A** and **B**, spectra from incubations containing ATP, cyclic AMP, and the denatured kinase and from incubations containing the two nucleotides plus the active kinase, respectively, are shown. A difference in the relative intensities of the cyclic AMP-derived peaks at m/z 330 and 352 is evident between the experimental and control spectra, indicating the binding of cyclic AMP to the active protein kinase preparation. Peaks at m/z 508, 530, and 552 correspond to $[M + H]^+$, $[M + Na]^+$, and $[MNa+Na - H]^+$ for ATP, with peaks at m/z 428 and 450 corresponding to $[M + H]^+$ and $[M + Na]^+$ for ADP; the appearance of the latter in the spectrum from the inactivated kinase is indicative of nonenzymic breakdown of ATP (**Fig. 5A**), whereas the diminution of ATP-derived and the increase of ADP-derived peak heights on comparing **Fig. 5A** with **5B** are indications of either ATPase activity or cyclic AMP-induced autophosphorylation of the protein kinase. The effect of addition of histone, a protein substrate for phosphorylation, is seen in **Fig. 5C** and **5D**. The cyclic AMP-derived peaks at m/z 330, 352, and 374 are weaker in the spectrum from the active kinase (**Fig. 5D**) compared with the spectrum from the denatured kinase (**Fig. 5C**), indicating the binding of cyclic AMP to the active enzyme. In comparison with the change in ATP-derived peaks between **Fig. 5A** and **5B**, these ATP-derived peaks at m/z 508, 530, and 552 show a much larger decrease in relative intensity from **Fig. 5C** to **5D**, reflecting the larger consumption of ATP for phophorylation of the histone by the active kinase. This decrease in ATP-derived peaks is paralleled by an increase in the ADP-derived peaks at m/z 428 and 450 from **Fig. 5C** to **5D**.

With protein kinase assays by quantitative FABMS, in addition to the extra facility, illustrated above, of monitoring cyclic AMP binding with assay of the kinase activity by ATP consumption, the multiple component monitoring facility allows the effect of the inclusion of other cyclic nucleotides to be determined. When cyclic GMP is included instead of cyclic AMP (**Fig. 5E** and **5F**), the decrease in the height of the cyclic GMP-derived peaks at m/z 346, 368, 390, and 438 from the spectrum in **Fig. 5E** to that in **5F** indicates the binding of cyclic GMP to the active cyclic AMP-dependent protein kinase. The relative diminution in ATP-derived peaks at m/z 508, 530, and 552 and relative increase in ADP-derived peaks at m/z 428 and 550 from **Fig. 5E** to **5F**

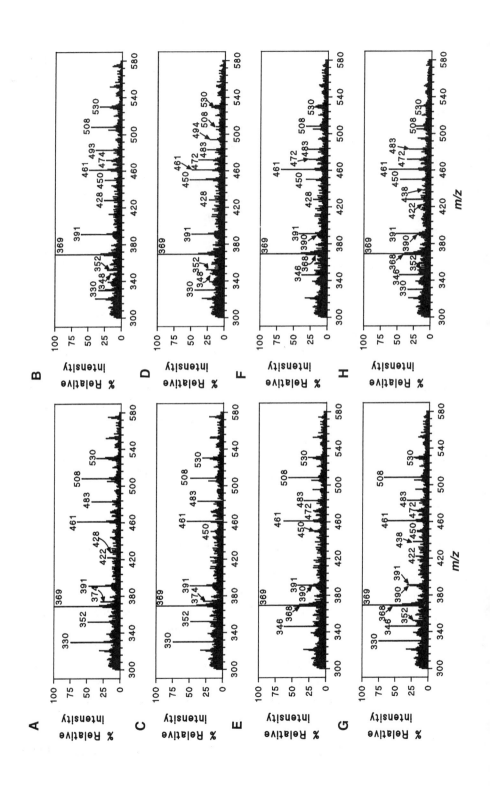

is smaller than the analogous change from **Fig. 5C** to **5D**, indicating that cyclic GMP stimulates the kinase, but to a lesser extent than cyclic AMP. The effect of simultaneous inclusion of cyclic AMP and cyclic GMP in the kinase assay is shown in **Fig. 5G** and **5H**. The diminution in cyclic GMP-derived peaks at *m/z* 346, 368, 390, and 438 between **Fig. 5G** and **5H** is less than that between **Fig. 5E** and **5F**, indicating that cyclic AMP is reducing the binding of cyclic GMP. The decrease in ATP-derived peaks at *m/z* 508, 530, and 552 and the increase in ADP-derived peaks from **Fig. 5G** to **5H** are greater than that between **Fig. 5E** and **5F** but less than that between **Fig. 5C** and **5D**, indicating that kinase activity in the presence of cyclic AMP and cyclic GMP together is greater than that with cyclic GMP alone but less than that with cyclic AMP alone. The cyclic nucleotide specificities obtained by quantitative FABMS and by radiometric kinase assay show good correlation between the data obtained (**Fig. 6A** and **6B**). The advantage conferred by quantitative FABMS is that the binding of other agonists (other cyclic nucleotides in the example given) in the presence of increasing cyclic AMP concentrations (**Fig. 7A**) and of cyclic AMP in the presence of competing cyclic nucleotides (**Fig. 7B**) can be determined simultaneously with the kinase activity and also any intrinsic phosphodiesterase activity *(15)*. In addition to the study of cyclic AMP-dependent protein kinase, this approach has also been used successfully in kinetic studies of cyclic CMP-responsive protein kinase *(14)*.

3.5. Summary

The procedures described above have been widely and successfully applied by the author and his collaborators in studies of cyclic nucleotide biochemistry. However the principle of multiple component monitoring by quantitative mass spectrometry is potentially applicable to any enzyme system in which the substrates, products, and effectors are of M_r >1 kDa and have similar mass spectrometric ranges of sensitivity.

Fig. 5. Positive-ion FAB mass spectra of protein kinase incubates. (**A**) Denatured protein kinase preparation incubated with cyclic AMP and ATP. (**B**) Active protein kinase preparation incubated with cyclic AMP and ATP. (**C**) Denatured protein kinase preparation incubated with cyclic AMP, ATP, and histone. (**D**) Active protein kinase preparation incubated with cyclic AMP, ATP and histone. (**E**) Denatured protein kinase preparation incubated with cyclic GMP, ATP, and histone. (**F**) Active protein kinase preparation incubated with cyclic GMP, ATP, and histone. (**G**) Denatured protein kinase preparation incubated with cyclic AMP, cyclic GMP, ATP, and histone. (**H**) Active protein kinase preparation incubated with cyclic AMP, cyclic GMP, ATP, and histone. (Reproduced with permission from **ref. *15***).

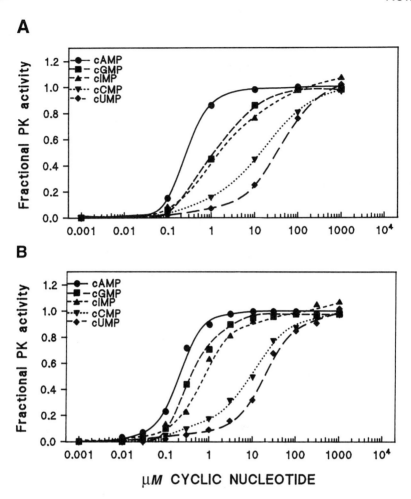

Fig. 6. Cyclic AMP-dependent protein kinase activity, in the presence of increasing concentrations of cyclic nucleotides, calculated in (**A**) by quantitative mass spectrometry and in (**B**) by the conventional radiometric assay. (Reproduced with permission from **ref. 15**).

4. Notes

1. The presence of multiple cation adducts in the spectra renders interpretation, peak assignment, and consequently quantitation much more difficult. It is virtually impossible to totally exclude sodium ions from biological samples, but every effort must be made to keep these to a minimum by using high-grade chemicals and water. The presence of high levels of potassium salts will, however, compound the problem and lead to mixed adducts, which makes quantitation even more difficult. Consequently, as far as pos-

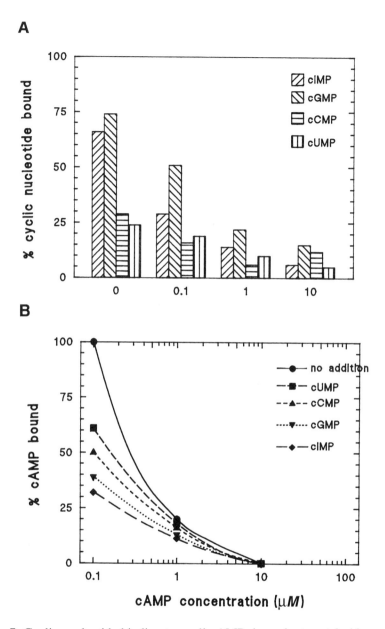

Fig. 7. Cyclic nucleotide binding to cyclic AMP-dependent protein kinase determined by quantitative FABMS. (**A**) Binding of other cyclic nucleotides in the presence of increasing cyclic AMP concentrations. (**B**) Binding of cyclic AMP in the presence of other cyclic nucleotides. (Reproduced with permission from **ref. 15**).

sible, sodium rather than potassium salts should be used if such salts cannot be avoided as one of the assay buffer components. Phosphate buffers should not be used.

2. Continuous flow- or dynamic-FABMS, in which the sample is pumped directly onto the FAB tip, offers a potential increase in sensitivity and the facility to remove aliquots for quantitation directly from the enzyme reaction vessel *(18)*. In our hands these potential advantages are, however, offset by the frequent periods of downtime as a result of blockages in the system.
3. Electrospray mass spectrometry offers increased sensitivity over FABMS and has been used successfully in cyclic nucleotide analysis *(19)*; the benefits of increased sensitivity are currently offset by greater problems with salt and phosphate contamination and reduced dynamic range but the advent of nanoelectrospray and pre-column concentration *(20)* are now combining to increase sensitivity dramatically.
4. Negative-ion FABMS removes the problems of spectrum complexity caused by cation adducts; however, different problems ensue from phosphate anions and, at least on the instrumentation we have used, positive-ion FAB is at least an order of magnitude more sensitive than negative-ion FAB for cyclase and kinase analyses.
5. Typically for a 10–15-min incubation with a highly purified enzyme it is possible to reduce the buffer concentration from 50 mM to 500 µM; ideally the buffer should be within the same order of magnitude or below that of the substrate concentration.
6. Sample lifetime under these operating conditions is typically 3–5 min.
7. Tight control of the FAB sample probe and power supply discharge current are required to maintain a steady precursor ion current throughout the acquisition of MIKE spectra.
8. Most modern mass spectrometer data systems will acquire and store the requisite data for processing in a spreadsheet. In the absence of such a data system, or where usage is saturated, two software packages, Ungraph (Biosoft, Cambridge, UK) or Un-Scan-It (Silk Scientific, Orem, UT) can be used to transpose the spectra into x, y data, which can then be processed by a package designed for life science computations such as Fig. P. (Biosoft).
9. In regard to sensitivity for cyclic nucleotides it is possible to obtain a useful positive-ion mass spectrum from 50–100 ng of sample.
10. If difficulty is experienced in assigning peaks to assay components, the complexity of the incubate composition should be sequentially constructed by the addition of one component at a time in a series of incubations. If difficulty persists, a second mass spectrum should be obtained after the sample has been spiked with one or more analyte standards. A further useful stratagem is to replace the glycerol matrix with thioglycerol and rerun; matrix-derived peaks will now be displaced by 16 mass units. It can also be useful to use a Li^+ salt of a component to be monitored and identify the ion clusters separated by 6 mass units, which contain Li^+ adducts.

References

1. Newton, R. P. (1996) Mass spectrometric analysis of cyclic nucleotides and related enzymes, in *Applications of Modern Mass Spectrometry in Plant Science*

Research, (Newton, R. P. and Walton, T. J., eds.), Oxford University Press, Oxford, pp. 159–181.
2. Newton, R. P. (1997) Applications of mass spectrometry in biochemical studies of nucleosides, nucleotides and nucleic acids, in *Mass Spectrometry in the Biomolecular Sciences* (Caprioli R. M., Malorni, A., and Sindona G., eds.), Kluwer, Dordrecht, pp. 427–454.
3. Kingston, E. E., Beynon, J. H., and Newton, R. P. (1984) The identification of cyclic nucleotides in living systems using collision-induced dissociation of ions generated by fast atom bombardment. *Biomed. Mass Spectrom.* **11,** 367–374.
4. Newton, R. P., Hakeem, N. A., Salvage, B. J., Wassenaar, G., and Kingston, E. E. (1988) Cytidylate cyclase activity: identification of cytidine 3',5'-cyclic monophosphate and four novel cytidine cyclic phosphates as biosynthetic products from cytidine triphosphate. *Rapid Commun. Mass Spectrom.* **2,** 118–126.
5. Kingston, E. E., Beynon, J. H., Newton, R. P. , and Liehr, J. G. (1985) The differentiation of isomeric biological compounds using collision-induced dissociation of ions generated by fast atom bombardment. *Biomed. Mass Spectrom.* **12,** 525–535.
6. Newton, R. P. Evans, A. M., Hassan, H. G., Walton, T. J., Brenton, A. G., and Walton, T. J. (1993) Identification of cyclic nucleotide derivatives synthesized for radioimmunoassay development by fast-atom bombardment with collision-induced dissociation and mass-analysed ion kinetic energy spectroscopy. *Organic Mass Spectrom.* **28,** 899–906.
7. Langridge, J. I., Evans, A. M., Walton, T. J., Harris, F. M., Brenton, A. G., and Newton, R. P. (1993) Fast-atom bombardment mass spectrometric analysis of cyclic nucleotide and nucleoside triphosphate analogues used in studies of cyclic nucleotide-related enzymes. *Rapid Commun. Mass Spectrom.* **7,** 725–753.
8. Newton, R. P., Bayliss, M. A., Wilkins, A. C. R., Games, D. E., Brenton, A. G., Langridge, J. I., et al. (1995) Fast-atom bombardment mass spectrometric analysis of cyclic-nucleotide analogues used in studies of cyclic nucleotide-dependent protein kinases. *J. Mass Spectrom. Rapid Commun. Mass Spectrom.* S107–110.
9. Newton, R. P. (1992) Cyclic nucleotides, in *Cell Biology Labfax* (Dealtry, G. and Rickwood, D., eds.), Bios, Oxford, pp. 153–166.
10. Newton, R. P., Groot, N., van Geyschem J., Diffley, P. E., Walton, T. J., Bayliss, M. A., et al. (1997) Estimation of cytidylyl cyclase activity and monitoring of side-product formation by fast-atom bombardment mass spectrometry. *Rapid Commun. Mass Spectrom.* **11,** 189–194.
11. Newton, R. P., Bayliss, A. M., van Geyschem, J., Harris, F. M., Games, D. E., Brenton, A. G., et al. (1997) Kinetic analysis and multiple component monitoring of effectors of adenylyl cyclase activity by quantitative fast-atom bombardment mass spectrometry. *Rapid Commun. Mass Spectrom.* **11,** 1060–1066.
12. Caprioli, R. M. (1987) Enzymes and mass spectrometry: a dynamic combination. *Mass Spectrom. Rev.* **6,** 237–287.
13. Newton, R. P. (1993) Contributions of fast-atom bombardment mass spectrometry to studies of cyclic nucleotide biochemistry. *Rapid Commun. Mass Spectrom.* **7,** 528–537.

14. Newton, R. P., Khan, J. A., Brenton, A. G., Langridge, J. I., Harris, F. M., and Walton, T. J. (1992) Quantitation by fast atom bombardment mass spectrometry: assay of cytidine 3',5'-cyclic monophosphate-responsive protein kinase. *Rapid Commun. Mass Spectrom.* **6,** 601–607.
15. Newton, R. P., Evans, A. M., Langridge, J. I., Walton, T. J., Harris, F. M., and Brenton, A. G. (1995) Assay of cyclic AMP-dependent protein kinase by quantitative fast-atom bombardment mass spectrometry. *Anal. Biochem.* **224,** 32–38.
16. Corbin, J. D. and Johnson, R. A. (1988) Initiation and termination of cyclic nucleotide action. *Methods in Enzymology*, Vol. **159**.
17. Corbin, J. D. and Johnson, R. A. (1991) Adenylyl cyclase, G-proteins, and guanylyl cyclase. *Methods in Enzymology* Vol. **195**.
18. Langridge, J. I., Brenton, A. G., Walton, T. J., Harris, F. M., and Newton, R. P. (1993) Analysis of cyclic nucleotide-related enzymes by continuous-flow fast-atom bombardment mass spectrometry. *Rapid Commun. Mass Spectrom.* **7,** 293–303.
19. Witters, E., Roef, L.,Newton, R. P., van Dongen, W., Esmans, E. L., and van Onckelen, H. A. (1996) Quantitation of cyclic nucleotides in biological samples by negative electrospray tandem mass spectrometry coupled to ion suppression liquid chromatography. *Rapid Commun. Mass Spectrom.* **10,** 225–231.
20. Vanhoutte, K., van Dongen, W., and Esmans, E. (1998) On-line nanoscale liquid chromatography nano-electrospray mass spectrometry: effect of the mobile phase composition and the electrospray tip design on the performance of a nanoflow electrospray probe. *Rapid Commun. Mass Spectrom.* **12,** 15–24.

22

Influence of Salts, Buffers, Detergents, Solvents, and Matrices on MALDI-MS Protein Analysis in Complex Mixtures

K. Olaf Börnsen

1. Introduction

Matrix-assisted laser desorption/ionization mass spectrometry, (MALDI-MS) was initially used to analyze pure peptide and protein solutions (1–3). Later, MALDI was applied to oligonucleotides, carbohydrates, polymers and other chemical classes (4–9). Concomitantly, other mass spectrometric methods, such as electrospray ionization (ESI-MS), increased in utility. Unlike ESI-MS, the main feature of MALDI-MS is the potential to measure mixtures of complex samples directly. The lack of nearly any fragmentation during the desorption and ionization steps, along with the high sensitivity of the technique, means that all peaks in a MALDI spectrum (except matrix-related peaks) can be correlated to a single component of a given mixture. In this way it is clear that MALDI-MS is also an excellent tool for complex samples such as cell culture media, cell lysates, blood plasma, serum, or even milk (10–14). Without any chromatographic separation method, MALDI-MS can rapidly provide an analytical fingerprint of these rather complex mixtures. Today, not only can these measurements be carried out qualitatively, but quantitative measurements are also possible using an internal standard (15–17). The main features of the MALDI method are its relative ease and speed of analysis.

Currently, MALDI-MS-based analyses are entering the field of biotechnology, in which extremely complex systems like proteins, immunoglobulins in cultivation media, metabolic profiles, or protein-protein associates or agglomerates are to be analyzed (13,18). Detergents, as well as inorganic and organic buffers, play a major role in biocultivations to stabilize the solution, avoid degradation of labile proteins, maintain the functionality of biopolymers, and suppress adsorption losses *(19–21)*. It is well known that such complex systems pose special demands on a MALDI investigation, especially when the simplicity of the approach is to be maintained. As a consequence it is necessary to optimize the MALDI matrix preparation (matrix, co-matrix, solvent, buffer, and detergent) and often necessary to remove excessive salt levels via a sample pretreatment step. With an optimized method, quantitative data, including reaction monitoring, can be recorded with a low deviation.

2. Materials

1. Ferulic acid (4-hydroxy-3-methoxycinnamic acid), 2,6-dihydroxyacetophenone (DHAP), diammonium hydrogen citrate (DAHC), Tween 80, HEPES, and 2-propanol (all from Fluka, Buchs, Switzerland).
2. Phospholipase D (white cabbage), cytochrome C (horse), ovalbumin, γ-globulin and β-galactosidase (all from Fluka).
3. Triton X-100 (Merck, Darmstadt, Germany).
4. Cell media FMX-8 (Gibco, Gaithersburg, MD),
5. Membrane for droplet dialysis, type MV, 0.025 mm; 25 mm diameter; (VSWP 02500; Millipore, Bedford, MA).
6. Cut-off filters: Microcon-30 microconcentrators, 1.5 mL, 30-kDa cut-off (Amicon, Beverly, MA).
7. Whista blood plasma (Novartis, Basel, Switzerland).

3. Methods

3.1. Signal Optimization

In working with biological samples, it often becomes necessary to improve the sensitivity or selectivity of an analysis to make the desired proteins visible in a MALDI spectrum (*see* **Note 1**). With these complex sample matrices, several optimization steps have to be carried out. First, the right matrix has to be selected. For example, ferulic acid (FA) and DHAP have a broad application range *(22–26)*. This is necessary because a matrix ideally should make all proteins and peptides equally visible in a given mixture. FA has better performance in the upper mass range, whereas DHAP is optimal in the lower mass range. However, selecting the matrix is but one step. Another step is to use the right solvent system, as the sample molecules have to be embedded into the matrix by a co-crystallization process *(21)*. If a given sample molecule can be built into the crystal lattice of the matrix, the best analyte signals will be obtained in the MALDI spectrum. Also, the method of producing crystals is important. Fast evaporation of the matrix and sample solution on the probe tip is necessary to produce reproducible and highly sensitive spectra *(27–32)*. One possibility for fast solvent evaporation is vacuum drying of the probe tip. Besides this, for biological samples, the right buffer system and/or detergent is necessary for sample stability *(21)*. All these components have to be tested for matrix compatibility, because these molecules and ions also have to be embedded into the matrix crystal lattice. However, despite this understanding of the behavior of the matrix/sample system, if these additives are more than 50%, the resulting MALDI spectrum will be poor, because the matrix cannot build up a crystalline lattice. In this case, it is the matrix that would be distributed in a crystalline lattice formed from the additives. Therefore it becomes necessary to use a small-sample pretreatment step, like drop dialysis for desalting of the sample or a cut-off filter to separate the sample into high and low molecular weight fractions *(34–37)*.

3.1.1. The Matrix

1. Solubilize 30 mg FA in 1 mL solvent (*see* **Subheading 3.1.2.**).
2. Solubilize 60 mg DHAP and 85 mg DAHC in 1 mL solvent (*see* **Subheading 3.1.2.**).

FA was found to be suitable for large proteins. It gives highest peak intensities in the upper mass range (> 100 kDa) compared with other matrices. For example, the optimization necessary for analyzing a cell lysate with MALDI-MS was tested with a mixture of cytochrome c (12,360 Daltons), ovalbumin (45 kDa), and γ-globulin (147 kDa) proteins mixed with cell media solution. Just enough cell medium was added so that the protein peaks did not totally disappear from the spectrum. With this synthetic mixture different matrices can be rapidly tested. Some matrices give signals for only one or a few proteins. Others show no signal at all. However, FA gives the best intensities for all proteins *(21)*.

For applications below 15 kDa, DHAP matrix gives the best results. For example, complex mixtures like milk or blood plasma can be measured at these lower masses with this matrix. The addition of DAHC to the matrix always improves the MALDI spectrum, because nearly all alkali metal adduct ions are suppressed very effectively. This treatment is absolutely necessary for oligonucleotides *(5,24,25)*.

3.1.2. The Solvent

In order to optimize the solvent 0.5 mL 2-propanol was first mixed with 0.5 mL pure water (*see* **Note 2**).

Then, in the next step, the matrix solvent was optimized under the same experimental conditions with the test mixture described in **Subheading 3.1.1.** It was found that 2-propanol as the matrix solvent tolerates high salt concentrations and improves all signal intensities *(19)*. This effect is also seen in **Fig. 1**. **Figure 1A** shows 2.3×10^{-6} M β-galactosidase (117 kDa) in FA matrix with 2-propanol and water (1:1) as the matrix solvent. With a sinapinic acid matrix

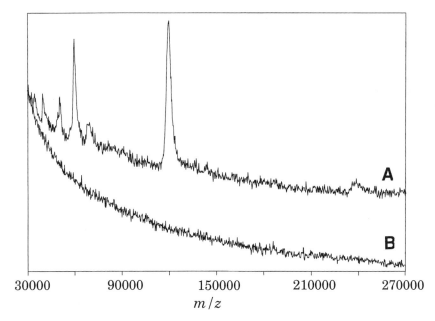

Fig. 1. Comparison of different matrix systems with β-galactosidase as analyte. (**A**) ferulic acid with 2-propanol and water; (**B**) sinapinic acid with acetonitrile.

and acetonitrile solvent no signals could be observed (**Fig. 1B**), because of the high salt load of the sample. Furthermore, it was found that 2-propanol improves signal intensities with many other samples *(13,21)*. Another advantage of this solvent is that it allows the use of much higher protein concentrations than does the commonly used acetonitrile solvent. Often the ratio of matrix to analyte molecules is crucial, e.g., too high an analyte concentration can result in suppression of any analyte ions, so that the sample has to be diluted with more matrix. In the case of FA and 2-propanol and water, a 5 times higher sample concentration could be tolerated compared with other systems. It should also be noted that oligonucleotides can be measured with better results (DHAP and DAHC as matrix) using 2-propanol and water as solvent *(21)*.

The reason for these improvements might be found in the solvation behavior. If the solubility of the sample and the matrix component are balanced out, excess of sample and impurities are precipitated at the beginning of the crystallization process. Then an optimal co-crystallization can occur avoiding local high concentrations and allowing ideal embedding conditions. It seems that proteins and oligonucleotides can now be embedded and distributed much more effectively into the matrix crystals. This assumption is also supported by the known protein precipitation behavior of 2-propanol *(21)*. This effect can also be observed in the matrix/sample solution. Often, several minutes after the matrix solution and a highly concentrated sample solution are mixed, a small precipitation can be observed.

It is well known that high salt concentrations generally suppress any signals in the MALDI spectrum. However, by using 2-propanol, these salts can often be tolerated. This solvent behavior might be explained by the salts being more soluble in water, so they remain dissolved longer and do not disturb the co-crystallization process of matrix and sample. However, a full experimental basis still has to be established for a final conclusion in this case.

3.1.3. The Buffer

1. Add 20 mM HEPES, mol wt, 238.31 Daltons to the sample solution as necessary. Alternatively, add a similar strength solution of MOPS (pH 6.5–7.9) to the sample solution (*see* **Note 3**).
2. Adjust the HEPES solution to a pH between 7.0 and 8.0 by adding NaOH solution.

Buffers are often very important in keeping biological molecules soluble *(33,38)*. If the sample starts to fall out of solution, no MALDI spectrum or only a poor spectrum will be obtained. It was found that the use of HEPES or MOPS buffer can improve, for example, a bovine albumin MALDI spectrum by a factor of two in comparison with pure water *(39,40)*. Phosphate buffer should be avoided, because the same amount of phosphate buffer discriminates against all signals in the mass spectrum. Also, buffers and salt systems that are hygroscopic, such as sodium acetate, should not be used because of the resultant wet surface.

3.1.4. Detergents

A general procedure is to add Triton X-100 or Tween 80 at a concentration of 0.1–0.6% to the sample as required (*see* **Note 4**).

When working with complex biological materials in MALDI-mass spectrometry it is necessary to use such detergents, otherwise the protein, especially in low concentrations ($< 5 \times 10^{-7}$ M) will be rapidly adsorbed on accessible surfaces. Furthermore, detergents generally have to be used in protein analysis to prevent proteins from agglomeration and from adsorption on tubes or membranes, and to enhance their solubility (especially of hydrophobic proteins, like membrane proteins) *(19–21)*. If no detergent is used in a MALDI-MS sample preparation, agglomeration and adsorption can dramatically suppress protein peaks in MALDI spectra *(21)*. However, in MALDI analysis, the detergent has to be compatible with the matrix itself. Thus, it is known that detergents such as sodium dodecyl sulfate (SDS) adversely affect MALDI measurements, resulting in poor and nonreproducible spectra *(41)*.

The use of Tween 80 as detergent is promising, because all compounds can be detected in the given test mixture (*see* **Subheading 3.1.1.**). However, Tween

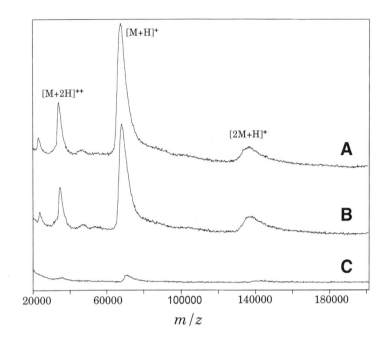

Fig. 2. Positive-ion spectra of phospholipase D. The concentration of protein was always 25 g/mL. The upper mass spectrum (**A**) shows the measurement of phospholipase D in pure water. For spectra **B** and **C**, silica powder was added to the protein solution. In spectrum **C** most of the protein is adsorbed at the surface of the silica powder. In spectrum **B**, by adding Triton X-100 (0.1%), the signal has increased dramatically because the detergent prevents adsorption of phospholipase D on the silica powder.

80 has a very low critical micelle concentration (CMC), which could be important in some cases. Similar results were also obtained with Triton X-100 as detergent. All protein peaks were detected with greater intensities when using detergents, most likely because of reduced adsorption during the sample preparation steps. It should be recognized, however, that any spectrum will show peaks from the detergent itself between m/z 400 and 900 (Triton X-100) or m/z 1300 and 2000 (Tween 80). Triton X-100 is a small molecule and with concentrations up to 0.6% no disadvantages in MALDI measurements could be found *(13,21)*.

To demonstrate the effect of adsorption of large proteins, a phospholipase D solution was treated with silica powder. The adsorption phenomenon is shown in **Fig. 2**. **Figure 2A** shows the spectrum is of phospholipase D in a standard solution (25 mg/mL i.e. 3.7×10^{-7} M) with 148 fmol on the probe tip. This protein solution was then mixed with silica powder and incubated for 10 min.

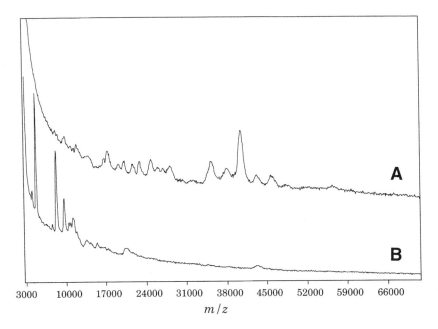

Fig. 3. Effect of detergent addition to a clear solution of CHOSSF3 cell lysate. (**A**) positive-ion spectrum after adding 0.61% Triton X-100; (**B**) pure untreated cell lysate solution for comparison.

The solid phase was spun down, the supernatant mixed with ferulic acid, and the MALDI spectrum shown in **Fig. 2C** obtained. The mass spectrum shows the poor result with almost 96% of the protein molecules stuck to the surface of the silica. In a control experiment, a 0.1% Triton X-100 solution was used instead of pure water and then mixed with the silica powder (**Fig. 2B**). Here the peak intensity of phospholipase D is significantly increased, being 81% of the signal intensity obtained in pure water. This experiment nicely demonstrates the effect of adsorption on large surface areas and the remedial effects of detergents. However, it also shows that the amount of detergent must be relatively high to minimize protein losses on a surface *(21)*.

Another important effect of detergents in MALDI spectra should also be noted. This is that hydrophobic protein samples become visible in MALDI spectra when a detergent is used. For example, a cell lysate solution was measured under normal conditions without any detergent (**Fig. 3B**). After addition of 0.6% of Triton X-100 to the same clear sample solution, the resulting spectrum reveals higher molecular weight proteins, as shown in **Fig. 3A**. It is interesting to note that after removing the detergent (with adsorbing beads) from this sample solution, the resulting spectrum is the same as that in **Fig. 3B** *(21)*.

3.1.5. Crystallization

1. If required, pretreat the sample probe surface, by use of some tiny matrix crystals or fine glass powder (4 μm) to provide a rough surface which can help to initiate a fast crystallization process.
2. When high salt concentrations may be present, alternatively apply a dry layer of DAHC crystals (from 0.3 μL 100 mM aqueous DAHC solution) before the sample/matrix solution since this can improve the fast crystallization process and the quality of the subsequent MALDI spectra in this situation (*see* **Note 5**).
3. Dry the matrix sample solution on the sample probe tip as quickly as possible (5–20 s) to provide a fine homogeneous crystal layer for MALDI-MS measurements.

MALDI is based on the behavior of the matrix. The mass spectral analysis is not the critical part of the process. The success of determining molecular weights by MALDI-MS is based strongly on the preparation of the matrix so that special care is required in this area. Without a correct embedding of the sample molecules into the matrix crystal lattice, no results or only poor results can be expected. Slow crystallization of matrix and analyte produces different kinds of crystals. At some places on the probe, matrix/salt crystals appear, whereas at other places matrix/analyte crystals are seen. A problem occurs when mixtures are to be analyzed when there will be some crystals where peptide A is highly concentrated and other crystals that show only peptide B. In these cases only qualitative information, often requiring time-consuming measurements and resulting in a poor sensitivity, is obtained. Vacuum drying is a powerful crystallization method for the matrix/sample solution *(27–32)* in which the matrix/analyte system has no time to separate on the probe tip. Fast crystallization can, however, also be effected by the use of microdrops. A homogeneous crystal layer on the probe tip is always important. It should be noted that the matrix-analyte, solvent, and additives, together with the crystallization process, should be seen as one system in which all the components have to be adjusted to each other.

3.2. Sample Pretreatment

3.2.1. Drop Dialysis

1. Place 40 mL of a 200-mM DAHC solution containing 0.5% Triton X-100 into a small glass flask.
2. Using tweezers, place the membrane (shiny side up) on the solution surface and allow it to float.
3. Now put 10–40 μL of the analyte solution to be desalted onto the floating membrane by means of a micropipet.
4. After 10–50 min remove the droplet from the membrane completely (*see* **Note 6** and **Fig. 4**).

Influence of MALDI-MS Matrix Preparations

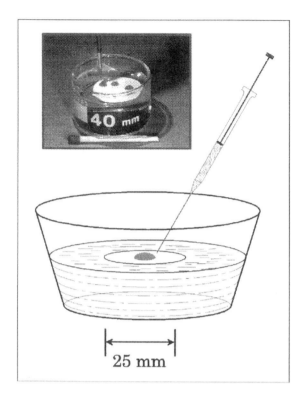

Fig. 4. Schematic of drop dialysis procedure.

Normally in mass spectrometry high salt concentrations are problematic. However, salt-free samples can be easily measured. MALDI-MS has shown its tolerance towards many buffers in low concentrations (< 10 mM) and an independence of pH and solvent choice, which combine to make it a strong analytical method. However, even this technology is limited when higher salt or buffer concentrations (≥ 5%) are found so that, for a given sample, extremely broad peaks and low sensitivities result or only a number of salt adduct ions are visible.

Görisch *(34,35)* originally described the desalting method in **steps 1–4** of this section, as droplet dialysis (with pure water for the reservoir). This method perfectly fits the needs of MALDI-MS since small amounts of sample can easily and rapidly be desalted. Detergent is used to avoid adsorption loss of the sample on the membrane surface, whereas DAHC is used to prevent the sample droplet from dilution by the osmotic flow of pure water. The dialysis time depends on the amount of salt in the sample. It should be noted that with a constant buffer concentration in the reservoir and a constant dialysis time this process is highly reproducible and can be also used for quantitative measure-

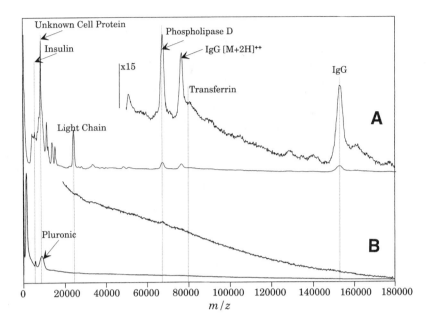

Fig. 5. Positive-ion mass spectrum of γ-globulin and phospholipase D using cell culture medium as solvent (**A**) after 30 min. of drop dialysis and (**B**) without any sample pretreatment.

ments. *(13,31,42)* However, this desalting method is based only on diffusion. **Figure 5** shows an example of a cell culture medium spiked with phospholipase D. Without sample pretreatment, only a small amount of insulin (5230 Daltons) is visible, whereas IgG and phospholipase D are suppressed by the high salt load (5%), as well as by the high level of amino acids and glucose in the cell culture medium (**Fig. 5B**). After 45 min of desalting by drop dialysis, the spectrum in **Fig. 5A** could be obtained. Desalting rendered all components of the sample visible *(11)*.

3.2.2. Cut-off Filter

1. Add 0.1% Triton X-100 and 0.1% trifluoroacetic acid (TFA) solution to the sample (plasma) to avoid adsorption loss of proteins or peptides on the cut-off membrane.
2. Fill a 30-kDa Microcon microconcentrator with 150 µL sample solution and centrifuge at 9000*g* for 6 min.
3. Use the supernatant directly for MALDI-MS measurements (*see* **Note 7**).

For fast fractionation of a given complex sample, microconcentrators are quite helpful. For example, when looking for a specific 2000-Dalton peptide in blood plasma, only the low molecular weight components are of interest. Here best results were observed by using a 30-kDa cut-off microconcentrator, because the excess of large proteins and other particles will be forced and adsorbed into the channels by centrifugation. This action dramatically reduces the effective diameter of the channels, with the result that only molecules below 5000 Daltons can still pass the membrane. An example is shown in **Fig. 6**. In contrast, use of larger cut-off values will increase the number of additional components and will reduce the sensitivity of the MALDI-MS measurement.

3.3. Quantitation

1. Define a reference peptide (or protein), which has a similar crystallization behavior, using the given matrix, to the component to be measured (analyte). This reference should also have similar basicity to the analyte. Phospholipase D is a very good internal standard for the upper mass range (> 40 kDa), because it is available in high purity, behaves well under MALDI conditions, and is cost effective (*see* **Note 8**).
2. Mix the reference peptide with the sample, so that analyte and reference have a similar concentration, resulting in similar peak heights.
3. Use a fast crystallization method (16–30 s). The resulting matrix should be like a fine powder.
4. Carry out the MALDI-MS measurements by taking 50–100 laser shots from different spots. Choose these spots on an automatic basis; do not use only "sweet spots."
5. Measure the peak heights, from the baseline, for the reference and sample molecules. Normalize all data to the reference.
6. Perform a calibration measurement.

With MALDI-MS it is not possible to determine the absolute concentration of a given sample directly since signal intensity depends strongly on laser power. Another reason is the variation in matrix crystallization since different crystals can produce different signal intensities from a given sample molecule. Again, signal intensity changes from one laser shot to the next as a result of different matrix crystals. In all cases, however, MALDI-MS produces *relative* intensity values that are quite reproducible. Thus, an additional molecule, included in the sample at a known constant concentration, can be used as an internal reference, and different sample signal intensities can be normalized to the intensity of the signal from this additional molecule *(13,15,31,42)*.

Different measurements of the sample molecule can be compared with each other directly so long as all spectra are normalized to the same intensity of the reference molecule (**Fig. 6**). This method, however, does not work if sample and reference molecule are chemically different. If, as a result of the chemical structure, the crystallization behavior is not the same, sample and reference molecule would not exhibit the same relative intensity to each other. On the other hand, different carbohydrates are quite similar in their behavior. Therefore, the molecules will be embedded in the matrix in the same way and with the same crystallization time. As a result, the reproducibility of the relative signal

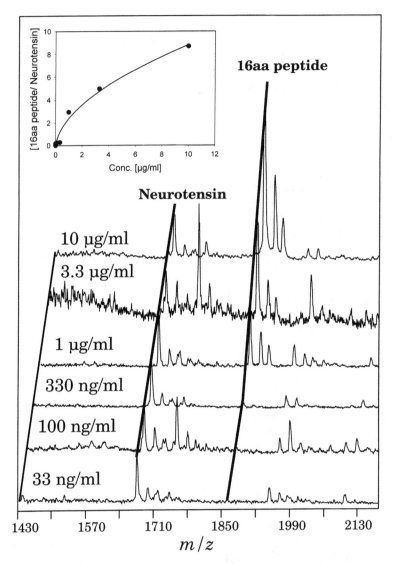

Fig. 6. MALDI spectra of blood plasma spiked with a 16-amino acid peptide at different concentrations (indicated as µg peptide/mL plasma). All samples have been filtered using a 30-kDa cut-off membrane. The spectra are normalized to the same peak height of neurotensin used as internal standard. The inset shows the resulting calibration curve.

intensities will be excellent, as shown in **Fig. 7** for the analysis of γ-globulin in cell culture media.

Industrial polymers provide another good example. Thus, within a polymer distribution, all molecules have approximately the same physical con-

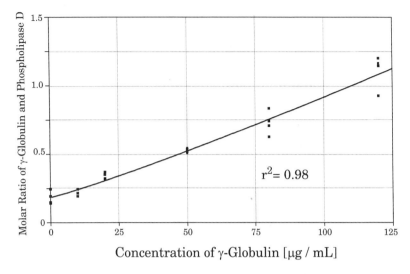

Fig. 7. Calibration curve for γ-globulin in cell culture medium. All measurements are normalized to the peak height of phospholipase D used as internal standard.

stants so that if the polymer chain looses one repeating unit, the physical behavior will still be identical. This behavior can be observed in the MALDI spectra of many polymer distributions, and the same regularity is also observed in the field of peptides. For example, a peptide of 20 amino acids will not change its crystallization behavior if one or even more amino acids are missing.

3.4. Conclusions

The crystallization behavior, the solubility of the sample and matrix, and the effects of buffer and detergent described here provide a useful picture to explain many differences in MALDI spectra. Often this kind of understanding helps in making samples amenable to MALDI analysis. Therefore these findings should also be understood as a helpful tool.

4. Notes

1. When the use of MALDI shows no signal for a given sample, this could be due to several reasons:
    a. The sample is too highly concentrated—dilute with matrix.
    b. Unsuitable matrix—change the matrix.
    c. Unsuitable solvent system—change the solvent (sample and matrix should always be dissolved in the same solvent).
    d. The sample requires the addition of a detergent, especially if the sample has a hydrophobic character.

e. The sample contains a high salt load—desalt the sample (*see* **Subheading 3.2.1.**)
f. There are too many different components in the sample—fractionate the sample (*see* **Subheading 3.2.2.**)
g. Peptide is sulfated or contains other acidic groups—record a spectrum in the negative-ion mode. It should be noted that that MALDI-MS can, in general, directly ionize only molecules that contain polar functional groups. For some nonpolar compounds cationization is appropriate (e.g., add Ag^+ or K^+ ions to the solution).

2. The low surface tension of some solvents with a high vapor pressure can cause the droplet to spread over a larger area than a water droplet would do. A low crystal density will result. The subsequent MALDI desorption process will then produce only spectra of low intensity or permit only one shot on a selected location on the probe tip. Therefore the surface tension of the solvent has to be taken into account together with the construction of the probe tip used. The tip should be constructed in such a way that the droplet will stay within a relatively small area.

3. Avoid buffer concentrations of more than 100 mM. Basic buffer systems (solutions with pH > 8) are rather difficult and often impossible to measure by MALDI-MS. In some of these cases, 3-aminoquinoline was found to be useful as a basic matrix *(43)*.

4. In a given sample, the hydrophilic protein signal can be suppressed relative to that from the hydrophobic proteins by the use of a detergent. With high detergent concentrations ($\approx$ 1 mM), micelles can begin to build up, and further improvement in the MALDI spectrum is not seen.

5. Only a given number of sample molecules can be embedded in the matrix and therefore provide a signal under MALDI-MS conditions. On this basis, investigations with a single component sample always result in the highest sensitivity whereas a large number of components will reduce sensitivity.

6. Dialysis on the membrane should not take too much time. With longer times, the sample will be partially lost by diffusion through the membrane or by adsorption on the membrane itself. The reservoir should not be stirred.

7. With small quantities (< 40 µL) or low concentrations (< 10^{-6} M) of proteins or peptides, sample can be lost by adsorption on the cut-off membrane of the filter. Unfortunately, this effect can be rather pronounced. The use of buffers and/or detergents, as well as 0.1% TFA can help.

8. It must be stated that this method based on exact relative intensities becomes difficult when important functional chemical groups are different in the sample and reference molecules. A small molecule with and without a sulfate group will show an enormous difference in signal intensity. In general, a sulfated molecule will give the highest intensity in the form of a negative ion $[M-H]^-$. On the other hand, a molecule without this group cannot produce negative ions at all and will give a good signal intensity only as a positive ion $[M+H]^+$. It should also be noted that the use of another matrix, solvent, or crystallization method will require a new calibration curve.

Acknowledgments

The author thanks Ute Kallweit, Marion Gass, Horst Conzelmann, and Derek Brandt for their work with buffers, solvents, detergents, and desalting methods; Joke van Adrichem and Augusta van Steijn for helpful discussions of biotechnology and drug aspects; Christian Leist for supporting the project; and John Chakel for critical reading of the manuscript.

References

1. Karas, M., Ingendoh, A., Bahr, U., and Hillenkamp, F. (1989) Ultraviolet-laser desorption/ionization mass spectrometry of femtomolar amounts of large proteins. *Biomed. Environ. Mass Spectrom.* **18**, 841–843.
2. Widmer, H. M., Börnsen, K. O., and Schär M. (1990) Recent progress in laser analytics. *Chimia* **44**, 417–424.
3. Börnsen, K. O., Schär, M. and Widmer, H. M., Matrix-assisted laser desorption and ionization mass spectrometry and its applications in chemistry. *Chimia* **44**, 412–417.
4. Börnsen, K.O, Schär, M., Gassmann, E., and Steiner, V. (1991) Analytical applications of matrix-assisted laser desorption and ionization mass spectrometry. *Biol. Mass Spectrom.* **20**, 471–478.
5. Pieles, U., Zürcher, W., Schär, M., and Moser, E. H. (1993) Matrix-assisted laser desorption ionization time-of-flight mass spectrometry: a powerful tool for the mass and sequence analysis of natural and modified oligonucleotides. *Nucleic Acids Res.* **14**, 3191–3196.
6. Steiner, V., Schär, M., Börnsen, K. O., and Gassmann E. (1993) Characterization of carbohydrate heterogeneity of glycoprotein. *Anal. Methods Instrum.* **1**, 124–126.
7. Mohr, M. D., Börnsen, K. O., and Widmer, H. M. (1995) Matrix-assisted laser desorption and ionization mass spectrometry: improved matrix for oligosaccharides. *Rapid. Commun. Mass Spectrom.* **9**, 809–814.
8. Bürger, H. M., Müller H-M., Seebach, D., Börnsen, K. O., Schär, M., and Widmer H. M. (1993) Matrix-assisted laser desorption and ionization mass spectrometry matrix-assisted laser desorption and ionization as a mass spectrometric tool for the analysis of poly[(R)-3-hydroxybutanoates]. Comparison with gel permeation chromatography. *Macromolecules* **26**, 4783–4790.
9. Börnsen, K. O. and Mohr, M. D. (1995) High mass resolution of small molecules with MALDI on a linear time of flight instrument. *Anal. Methods Instrum.* **2**, 202–205.
10. Beavis, R. C. and Chait B. T. (1990) Rapid, sensitive analysis of protein mixtures (milk) by mass spectrometry. *Proc. Natl. Acad. Sci. USA* **87**, 6873–6877.
11. Marsilio, R., Catinella, S., Seraglia, R., and Traldi, P. (1995) Matrix-assisted laser desorption/ionization mass spectrometry for the rapid evaluation of thermal damage in milk. *Rapid Commun. Mass Spectrom.* **9**, 550–552.
12. Stahl, B., Thurl, S., Zeng, J., Karas, M., Hillenkamp, F., Steup, M., and et al. (1994) Oligosaccharides from human milk as revealed by matrix-assisted laser desorption/ionization mass spectrometry. *Anal. Biochem.* **223**, 218–226.

13. Van Adrichem, J. H. M., Börnsen, K. O., Conzelmann, H., Gass, M. A. S., Eppenberger, H., Kresbach, G. M., et al. (1998) Investigation of protein patterns in mammalian cells and culture supernatants by MALDI-MS. *Anal. Chem.* **70**, 923–930.
14. Harvey, D.J, Rudd, P. M., Bateman, R. H., Bordoli, R. S., Howes, K., Hoyes, J. B., et al. (1994) Examination of complex oligosaccharides by matrix-assisted laser desorption/ionization mass spectrometry on time-of-flight and magnetic sector instruments. *Org. Mass Spectrom.* **29**, 753–765.
15. Börnsen, K. O. and Mohr M. D. (1995) Are quantitative measurements possible with MALDI-MS? *Anal. Methods Instrum.* **2**, 158–160.
16. Gusev, A. I., Wilkinson, W. R., Proctor, A., and Hercules, D.M (1993) Quantitative analysis of peptides by matrix-assisted laser desorption/ionization time-of-flight mass spectrometry. *Appl. Spectrosc.* **47**, 1091–1092.
17. Duncan, M. W., Matanovic, G., and Cerpa-Poljak, A. (1993) Quantitative analysis of low molecular weight compounds of biological interest by matrix-assisted laser desorption ionization. *Rapid Commun. Mass Spectrom.* **7**, 1090–1094.
18. Liang, X., Zheng, K. Z., Qian M. G., and Lubman D. M. (1996) Determination of bacterial protein profiles by MALDI-MS with high performance liquid chromatography. *Rapid. Commun. Mass Spectrom.* **10**, 1219–1226.
19. Rosinke, B., Strupat, K., Hillenkamp, F., Rosenbusch, J., Dencher, N., Krüger, U., et al. (1995) MALDI-MS of membrane proteins and non-covalent complexes. *J. Mass Spectrom.* **30**, 1462–1468.
20. Gharahdaghi, F.,Kirchner, M., Fernandez, J., and Mische, S. M. (1994) Peptide-mass profiles of polyvinylidene difluoride-bound proteins by MALDI TOF-MS in the presence of nonionic detergents. *Anal. Biochem.* **233**, 94–99.
21. Cohen S. L. and Chait B. T. (1996) Influence of matrix solution conditions on the MALDI-MS analysis of peptides and proteins. *Anal. Chem.* **68**, 31–37.
22. Börnsen, K. O., Gass, M. A. S., Bruin, G. J. M., van Adrichem, J. H. M., Biro M. C., Kresbach, G. M., et al. (1997) Influence of solvents and detergents on matrix-assisted laser desorption/ionization mass spectrometry measurements of proteins and oligonucleotides. *Rapid Commun. Mass Spectrom.* **11**, 603–609.
23. Beavis, R. C. and Chait B. T. (1989) Cinnamic acid derivates as matrixes for ultraviolet laser desorption mass spectrometry of proteins. *Rapid Commun. Mass Spectrom.* **3**, 432–435.
24. Wunderlich, M. (1993) Matrixunterstützte Laserdesorptions-Massenspektrometrie an Oligonukleotiden. Fachhochschule für Technik und Wirtschaft, diploma thesis, Reutlingen, Germany.
25. Schuette, J. M., Pieles, U., Maleknia, S. D., Srivatsa, G. S., Cole, D. L., Moser, H. E., et al. (1995) Sequence analysis of phosphorothioate oligonucleotides via matrix-assisted laser desorption ionization time-of-flight mass spectrometry. *J. Pharm. Biomed. Anal.* **13**, 1195–1203.
26. Gorman, J. J., Ferguson, B. L. and Nguyen, T. B. (1996) Use of 2,6-dihydroxy-acetophenone for analysis of fragile peptides, disulfide bonding and small proteins by MALDI. *Rapid Commun. Mass Spectrom.* **10**, 529–536.

27. Weinberger, S. R., Börnsen, K. O., Finnchy, J. W., Robertson, V., and Musselman, B. D. (1993) An evaluation of crystallization methods for matrix assisted laser desorption/ionization of proteins. *Procedings of the 41st ASMS Conference on Mass Spectrometry*, pp. 775a-775b.
28. Nicola, A. J., Gusev, A. I., Proctor, A., Jackson, E. K., and Hercules, D. M. (1995) Application of the fast-evaporation sample preparation method for improving quantification of angiotensin II by matrix-assisted laser desorption/ionization. *Rapid Commun. Mass Spectrom.* **9,** 1164–1171.
29. Vorm, O., Roepstorff, P., and Mann, M. (1994) Improved resolution and very high sensitivity in MALDI TOF of matrix surfaces made by fast evaporation. *Anal. Chem.* **66,** 3281–3287.
30. Axelsson J., Hoberg A.-M., Waterson, C., Myatt, P., Shield, G. L., Varney, J., et al. (1997) Improved reproducibility and increased signal desorption/ionization as a result of electrospray sample preparation. *Rapid Commun. Mass Spectrom.* **11,** 209–213.
31. Nicola, A. J., Gusev, A. I., Proctor, A., Jackson, E. K., and Hercules, D. M. (1995) Application of the fast-evaporation sample preparation method for improving quantification of angiotensin II by MALDI. *Rapid Commun. Mass Spectrom.* **9,** 1164–1171.
32. Kussmann, M., Nordhoff, E., Rahbek-Nielsen, H., Haebel, S., Rossel-Larsen, M., Jakobsen, L., et al. (1997) MALDI-MS sample preparation techniques designed for various peptide and protein analytes. *J. Mass Spectrom.* **32,** 593–601.
33. Ferguson, W. J., Braunschweiger, K. I., Braunschweiger, W. R., Smith, J. R., McCormick, J. J., Wasmann C. C., et al. (1980) Hydrogen ion buffers for biological research. *Anal. Biochem.* **104,** 300–310.
34. Marazyk, R. and Sergeant A. (1980) A simple method for dialysis of small-volume samples. *Anal. Biochem.* **105,** 403.
35. Görisch, H. (1988) Drop dialysis: rime course of salt and protein exchange. *Anal. Biochem.* **173,** 393–398.
36. Brandt, D. K., Dioploma Thesis: Einsatz von MALDI-TOF Massenspektrometric zur Annalytik von Peptiden in Blutserum (1998) Fachhockschulebeider Basel, Basel, Switzerland.
37. Yanase, H., Cahill, S., Martin De Llano, J. J., Manning, L. R., Schneider, K., Chait, B. T., et al. (1994) Properties of a recombinant human haemoglobin with aspartic acid 99(β), an important intersubunit contact site, substituted by lysine. *Protein Sci.* **3,** 1213–1223.
38. Good, N. E., Winget, G. D., Winter, W., Connnolly, T. N., Izawa, S., and Singh, R. M. M (1966) Hydrogen ion buffers for biological research. *Biochemistry* **5,** 467–477.
39. Kallweit, U. (1995) Analysen komplexer biologischer Systeme mit MALDI-TOF-Massenspektrometrie, diploma thesis. Fachhochschule, Reutlingen, Germany.
40. Kallweit, U., Börnsen, K. O., Kresbach, G. M., and Widmer, H. M. (1996) Matrix compatible buffers for analysis of proteins with matrix-assisted laser desorption ionization mass spectrometry. *Rapid Commun. Mass Spectrom.* **10,** 845–849.

41. Amado, F. M. L., Santana-Marques, M. G., Ferer-Correira, A. J., and Tomer, K. B. (1997) Analysis of peptide and protein samples containing surfactants by MALDI-MS. *Anal. Chem.* **69,** 1102–1106.
42. Conzelmann, H. (1997) Quantifizierung von IgG in Zellkulturmedien mit Hilfe der matrixunterstützten Laser Desorptions/Ionisations-Massenspektrometrie, unter Anwendung einer optimierten Probenvorbereitung, diploma thesis, Fachhochschule für Technik, Mannheim, Germany.
43. Fitzgerald, M. C., Parr, G. R., and Smith, L. M. (1993) Basic matrices for the MALDI mass spectrometry of proteins and oligonucleotides. *Anal. Chem.* **65,** 3204–3211.

23

Sample Preparation Techniques for Peptides and Proteins Analyzed by MALDI-MS

Martin Kussmann and Peter Roepstorff

1. Introduction

Protein mass analysis and peptide mapping by matrix-assisted laser desorption/ionization time-of-flight mass spectrometry (MALDI-TOFMS) has emerged as a powerful tool for identification and characterization of proteins. The method involves (1) the mass spectrometric analysis of the intact protein, and, in particular, (2) the proteolytic cleavage of the latter, followed by mass determination of the generated peptides. The subsequent identification of the protein is performed by matching the set of measured peptide masses against the calculated masses of theoretical digest, based on protein or cDNA sequence databases *(1–3)*. The detailed characterization of primary structure and covalent modifications of the protein is performed by differential peptide mapping, i.e., by comparative analysis of different proteolytic digests prior to and after application of, for example, phosphatases *(4)* or glycosidases *(5)*.

Sample preparation is known to be the crucial procedure in MALDI-MS analysis of peptides and proteins. This procedure encompasses two steps: first is the isolation and purification of a single component or a mixture, free of contaminants such as buffers, salts, detergents, or denaturants. The second step encompasses sample processing on the MALDI target, i.e., choice of matrix, matrix and analyte concentration, pH adjustment, crystallization conditions, use of additives, and on-target sample clean-up.

The detection sensitivity and the accuracy of mass determination of protein and peptide analytes is, apart from instrumental parameters, not only limited by the absolute amount of analyte available but rather by the presence of contaminants, such as salts, ionic detergents, and high molecular weight surfactants, present in the sample. These contaminants increase the chemical noise in the mass spectrum and can cause suppression of analyte ions. This circumstance is particularly pronounced when one is analyzing proteins derived from gel electrophoretic separations *(6)*, an approach that has recently gained importance in the context of proteome analysis.

To date, a variety of MALDI sample preparation recipes has been reported, and several strategies have been presented to overcome the aforementioned limitations. The choice of matrix and the use or presence of matrix additives, e.g., detergents, has been demonstrated to influence significantly the quality of MALDI-MS spectra of peptides and proteins *(7–9)*. Moreover, the influence of choice of solvent for matrix and sample, sample pH, and crystal growth time on the quality of MALDI-MS analyses has been systematically studied in two recent publications *(10,11)*.

The most frequently used on-target preparation is the dried-droplet method *(12)* often followed by washing of the prepared sample with aliquots of acidified water *(13,14)*. The accessibility of the contaminants, their solubility in the washing solution, and the acceptable loss of sample material limit the efficiency of this washing procedure. In one approach, the tolerance of MALDI toward involatile contaminants has been improved by a slow matrix crystallization procedure *(14)*. Later, the preparation of a thin, homogeneous layer of sample + matrix co-crystals was shown to improve sensitivity and mass accuracy in MALDI-MS analysis of peptides and proved to be more compatible with on-target sample clean-up *(15,16)*.

All on-target sample clean-up procedures require that the matrix is insoluble in the washing solution. The use of water-soluble matrices such as 2,5-dihydroxybenzoic acid (DHB) is therefore excluded. Furthermore, the success of the different on-target washing protocols is highly dependent on the nature of the contaminants as well as the nature of the analyte molecules. Therefore, on-target immobilization of the analyte molecules and miniaturized pretarget chromatography have been developed. A recent paper *(17)* describes coating of the target with self-assembled monomolecular layers of octadecyl mercaptan (C_{18}). This approach relies on the well-characterized hydrophobic interactions of peptides with a reversed-phase surface. Once the analyte molecules are bound, efficient washing can be performed prior to the addition of the matrix solution.

A natural limitation in terms of binding capacity arises from the minimal active surface area. Furthermore, hydrophobic contaminants compete for binding sites with the analyte molecules and are not removed in the washing steps. The lack of binding capacity can be overcome by the use of porous hydrophobic membranes such as poly(vinylidenedifluoride) (PVDF) or polyethylene

(PE) as a sample support *(18)*. Regenerated cellulose dialysis membranes coated with α-cyano-4-hydroxy-cinnamic acid (HCCA) crystals are an alternative sample support *(19,20)*. The obvious limitation of these target-coating methods is that most of the active surface to which binding occurs is not accessible to the laser.

A more versatile strategy uses nanoscale reversed-phase columns designed for direct sample elution onto the MALDI target. Zhang et al. *(21)* described a sample preparation device consisting of a fused-silica capillary packed with C_{18} coated particles and, connected to three syringe pumps via a multiport injector valve. The sample was eluted from the column with matrix solution, and aliquots as small as 5 nL were fractionated onto the MALDI target. Kalkum and Gauss *(22)* described an instrumental set-up for deposition of sample droplets with a volume as low as 65 pL onto a thin layer of HCCA, using piezo-jet dispensers. A feature common to all these reports is the use of sophisticated custom-built instrumentation.

A simple microcolumn purification and sample preparation technique using exclusively standard laboratory materials was introduced independently by Annan et al. *(23)* and by our group *(24,25)*. A long, narrow pipet tip serves as the column body, in which a small reversed-phase column is packed. A disposable plastic syringe provides the drive for all liquid transport. After washing, the sample is eluted onto the MALDI target followed by addition of matrix solution. An obvious advantage of this approach is that the risk of memory effects can be completely excluded.

As a supplement to and summary of these earlier investigations, we describe here several established sample preparation techniques and variants of miniaturized chromatography, all of which are applicable to any laboratory and which are designed for peptide and protein analytes of different origin, nature, and purity. The methods tolerate the presence of salts, contaminants, and additives and enable further sample processing on the target, e.g., disulfide bond reduction, proteolytic digestion, and efficient washing steps. Due to its increasing importance in the context of proteome analysis, the preparation of proteolytic digests from proteins separated by gel electrophoresis is also described. The present collection of recipes should be considered a "cook book" for MALDI sample preparation and is largely based on previous in-house studies offering a summary of sample preparation procedures *(24,26)* and reporting on microchromatographic techniques *(25,27)*.

2. Materials
2.1. Solvents

All solvents were of the highest purity available, i.e., p.a. quality or high-performance liquid chromatography (HPLC) grade. Trifluoroacetic acid (TFA)

Table 1
Preference (1–3: high to low priority) of Matrix Selection for Different Analytes and their Compatibility with the Sample Preparation Techniques in Subheading 3.2.

Matrix	Peptide mapping	Small proteins	Large proteins	Glyco-peptides	Glyco-proteins	Compatibility
HCCA	1	3		2		Dried-droplet, thin/thick-layer, sandwich
SA		1	1	3	2	Dried-droplet, sandwich
DHB	2	2	1	1	1	Dried-droplet
THAP	2					Dried-droplet, thin-layer

and HPLC-grade acetonitrile (ACN) were obtained from Rathburn. Millipore water was used for all solutions

2.2. Matrices

The following matrices were used: HCCA *(28)* (Sigma, St. Louis, MO); sinapinic acid *(29)* (SA, Fluka, Buchs, Switzerland); 2,5-dihydroxybenzoic acid *(30)* (DHB; Aldrich or Hewlett Packard, Palo Alto, CA); a 9:1 mixture *(31)* of DHB and 2-hydroxy-5-methoxybenzoic acid (Aldrich, Milwaukee, WI); 2,4,6-trihydroxyacetophenone (THAP, Aldrich). The compatibility of these matrices with different analytes and with different sample preparation techniques is summarized in **Table 1**.

The HCCA used in these experiments was recrystallized from methanol before use. The DHB matrix solution obtained from Hewlett Packard was found to be sufficiently pure for direct use. The solvent was, however, replaced by 30% (v/v) ACN containing 0.1% TFA (v/v). Nitrocellulose (NC) membrane was purchased from BioRad.

2.2.1. Matrix Solutions

1. **HCCA:** (I) 20 µg/µL in ACN + 0.1% TFA (70:30, v/v); (II) 20 µg/µL in acetone + water (99:1, v/v).
2. **SA:** (I) 20 µg/µL in ACN + 0.1% TFA (40:60, v/v); (II) 20 µg/µL in acetone + water (99:1, v/v).
3. **DHB:** (I) DHB (Aldrich): 20 µg/µL in ACN + 0.1% TFA (20:80–0:100, v/v); (II) 9:1 mixture of DHB and 2-hydroxy-5-methoxybenzoic acid (Aldrich): 20 µg/µL in ACN + 0.1% TFA (20:80–0:100, v/v); (III) DHB matrix solution from Hewlett Packard, no. HP G2056A. Replace solvent by an equal amount of ACN + 0.1% TFA (30:70, v/v).
4. **THAP:** (I) 15 µg/µL in ACN + water (70:30, v/v); (II) 20 µg/µL in 100% MeOH.

2.3. Reagents for Protein Chemistry

Reagents for disulfide bond reduction and alkylation, viz., dithiothreitol, iodoacetic acid, iodoacetamide, and 4-vinylpyridine, were purchased from Sigma/Aldrich in p.a. quality.

2.4. Chromatography

GELoader tips and Combitips (1.25-µL capacity, no. 0030 048.083) were purchased from Eppendorf. Poros 10 R2 medium for reversed-phase chromatography, Porozyme immobilized trypsin medium, and Poros 50 R1 and 50 R2 media were all obtained from PerSeptive Biosystems (Framingham, MA).

2.5. Mass Spectrometry

MALDI mass spectra were recorded on a Bruker Reflex mass spectrometer or on a PerSeptive Voyager Elite instrument, both equipped with delayed-ion extraction technology. The spectra were acquired in positive-ion linear or positive-ion reflectron mode. Typically, 5–20 laser shots were added per spectrum.

For subsequent data processing, the software packages PerSeptive-Grams (Galactic, Salem, NH) and GPMAW (Lighthouse Data, Odense, Denmark) were used. Protein identification was achieved with the program Peptide Search (EMBL, Heidelberg, Germany).

3. Methods

The protocols and remarks that make up **Subheading 3.** refer to established sample preparation techniques, including variants of miniaturized chromatography, which have been designed for the analysis, by MALDI-MS, of a wide range, in type and purity, of peptide and protein analytes.

3.1. Preparation of Proteolytic Digests from PAGE-Separated Proteins

Protein spots were excised from the gel and subjected to *in situ* digestion with trypsin according to previously described procedures *(32)*, with modifications reported by Shevchenko et al. *(33)*. The modified protocol is as follows:

3.1.1. Excision and Washing

1. Wash the gel piece twice for 10 min with water and excise the gel band as closely as possible with a clean scalpel (*see* **Note 1**).
2. Cut the gel band into approx 1 × 1-mm pieces and transfer them into a 1.5-mL Eppendorf tube.
3. Wash the gel pieces twice for 10 min with water + ACN 50:50 (v/v) and vortex several times (*see* **Note 2**).

4. Pipet off the supernatant, then add approx 20 µL ACN, and let the gel pieces shrink.
5. Pipet off the supernatant and rehydrate the gel pieces with 100 mM NH$_4$HCO$_3$ (pH 7.8).
6. After 5 min, add an equal volume of ACN and then, after 15 min, pipet off the supernatant and dry the gel pieces in a vacuum centrifuge (20–30 min) (*see* **Note 3**).

3.1.2. Reduction and Alkylation

1. Rehydrate the gel particles in 10 mM DTT in 100 mM NH$_4$HCO$_3$ (pH 7.8) and incubate them for 45 min at 56°C (*see* **Note 4**).
2. Chill the tubes to room temperature, pipet off the supernatant, and replace it by the same volume of fresh 55 mM iodoacetamide in 100 mM NH$_4$HCO$_3$ (pH 7.8).
3. Incubate for 30 min at room temperature in the dark.
4. Pipet off the supernatant and wash the gel pieces with 100 mM NH$_4$HCO$_3$ (pH 7.8) and ACN as described in **Subheading 3.1.1., steps 5,6** (and *see* **Note 5**).

3.1.3. In-gel Digestion

See **Note 6**.

1. Dry the gel pieces in a vacuum centrifuge (shrink large gel pieces with ACN before drying) and add 12.5 ng/µL trypsin in chilled 100 mM NH$_4$HCO$_3$ (pH 7.8). The volume used should be enough to cover the gel pieces.
2. Incubate for 45 min on ice and add more digestion buffer, if necessary.
3. After 45 min, pipet off the supernatant and add 5–20 µL 100 mM NH$_4$HCO$_3$ (pH 7.8) without trypsin.
4. Digest for 4 h or overnight at 37°C.

3.1.4. Extraction of Peptides and Analysis by MALDI-MS

1. Pipet off supernatant from **Subheading 3.1.3., step 4** and analyze by MALDI-MS.
2. Add 25 mM NH$_4$HCO$_3$ (pH 7.8; enough to cover the gel pieces) and incubate for 10 min.
3. Add the same volume of ACN and incubate for another 10 min. Pipet off and save the supernatant.
4. Repeat the extraction 2 times with 5% HCOOH + ACN (50:50). Pipet off and save the supernatants, pool all extracts, and dry them in a vacuum centrifuge.
5. Redissolve extracts in 5–30 µL 5% HCOOH and analyze by MALDI-MS.

3.2. On-Target Sample Preparation

3.2.1. Dried-Droplet Method

Mix 0.5–2 µL of sample and 0.5–1 µL of matrix solution [(HCCA (I), SA (I), DHB (I, II, III), THAP (I, II)] on the target and allow to dry in the ambient air or, optionally, in a gentle stream of forced air or argon (*see* **Notes 7** and **8**).

Table 2
TFA Concentrations Needed for On-Target Acidification to pH < 2 with 0.5–1 µL TFA Solution[a]

Buffer (mM)	5	25	50	75	100	500
NH$_4$HCO$_3$	0[b]	0.5	0.5	1.0	1.0	5.0
NaH$_2$PO$_4$	0.5	0.5	0.5	1.0	1.0	5.0
Na$_2$HPO$_4$	0.5	1.0	1.0	2.0	2.0	10.0
Tris-HCl	0.5	0.5	1.0	1.0	1.0	5.0

[a]The adjustment is given for different buffers and buffer concentrations.
[b]The TFA concentration is given as a % figure.

3.2.2. Methods Based on a Preformed Matrix Layer

3.2.2.1. HCCA OR THAP AS MATRIX

1. Prepare a thin layer of small, homogeneous matrix crystals on the target by placing either 0.5–1 µL of HCCA (II) or 0.5–1 µL of THAP (II) onto the target and allowing the droplet to spread and dry.
2. In the case of acidic analyte solutions (pH < 2), place 0.5 µL of this solution on top of the matrix layer. Otherwise, deposit 0.5 µL of aqueous TFA solution, according to **Table 2**, onto the matrix layer followed by the addition of 0.5 µL of analyte solution. After solvent evaporation, wash the sample 1–3 times by adding 5–10 µL 0.1% TFA onto the sample and removing after a few seconds using a forced air flow.

3.2.2.2. HCCA PLUS NITROCELLULOSE AS MATRIX ADDITIVE

See also **Subheading 3.5.5.**

1. Dissolve two parts HCCA and one part NC in acetone + isopropanol (4:1) to final concentrations of 20 µg/µL HCCA and 10 µg/µL NC.
2. Deposit 0.5 µL of this solution onto the target and allow to spread and dry.
3. Perform **step 2** of **Subheading 3.2.2.1.**

3.2.2.3. THICK-LAYER METHOD WITH NITROCELLULOSE AS MATRIX ADDITIVE (SPIN-DRY TECHNIQUE)

See **Note 9**, **Subheading 3.5.5.**, and **refs. 34** and **35**.

1. Dissolve equal amounts of NC membrane and HCCA in acetone (40 µg/µL each) and subsequently dilute this solution with isopropanol to 20 µg/µL.
2. Add 1% (v/v) 0.1% TFA and then apply 2 × 10 µL of the resulting solution onto a rotating target so that the solution is immediately spin-dried and yields a uniform NC/matrix layer.
3. Place 0.5 µL 2% TFA and 0.5 µL sample solution on top of the NC/matrix layer and allow to dry.

4. If the sample is highly contaminated, wash 1–3 times with 10 µL 0.1% TFA.
5. After a few seconds of incubation, remove the TFA droplet by spinning.
6. Finally, deposit another 0.5 µL of 2% TFA and 0.5 µL of HCCA (I) and dry.
7. If indicated by poor spectrum quality, perform one to two further washes with 0.1% TFA as described in **step 2** of **Subheading 3.2.2.1.**

3.2.2.4. SANDWICH METHOD

Also *see* **Subheading 3.5.3.**

1. Prepare a thin layer of matrix crystals as in **Subheading 3.2.2.1., step 1**.
2. Add droplets of (1) 1–2 µL aqueous TFA solution (**Table 2**), (2) 0.5 µL of sample solution, and (3) 0.5 µL matrix solution (HCCA [I], SA [I]) and allow this mixture to dry. Optionally, wash the sample 1–3 times with 5–10 µL 0.1% TFA by placing the droplet onto the target, incubating for a few seconds, and removing it with a pipet (*see* **Note 10**).

3.3. On-Target Reactions

Also *see* **Subheading 3.5.4.**

3.3.1. On-Target DTT Reduction

1. After a first analysis of a sample prepared with the matrix SA according to the dried-droplet method, carry out reduction of disulfide bonds by placing a 5-µL droplet of DTT solution (150 mM in 100 mM ammonium bicarbonate, pH 7.8) on top of the matrix/sample crystals.
2. After reaction for 10 min at ambient temperature, acidify the moist sample with 5 µL 2% TFA.
3. Pipet off the resulting droplet and wash the sample 1–3 times with 5–10 µL 0.1% TFA according to **Note 8**.
4. Finally, add 0.5 µL of matrix solution (HCCA [I], SA [I]) and allow the product to dry.

3.3.2. On-Target Proteolytic Digestion

See **Subheading 3.3.3.** for an alternative procedure. A protein sample preparation according to the dried-droplet method with SA as the matrix can be followed by proteolytic digestion on-target.

1. Dissolve a lyophilized aliquot of 50–100 ng trypsin in 2–5 µL of 50 mM ammonium bicarbonate (pH 6–8) or 50 mM ammonium acetate (pH 6–8) (enzyme to substrate ratio [E/S] approx 1:10).
2. Place the protease solution on top of the sample/SA layer.
3. Allow the mixture to react for 30 min at 37°C in a moist environment.
4. Remove the buffer droplet and use it as the sample solution in a subsequent sandwich sample preparation (**Subheading 3.2.2.4.**) or proceed to **step 5** of this section.
5. Acidify the moist sample, dry, and wash according to **Note 8**.

3.3.3. On-Target Proteolytic Digestion Using n-Octyl-Glucopyranoside

n-Octyl-glucopyranoside (OGP) may be used as an alternative for on-target digestions under denaturing conditions as follows:

1. Add 1 µL sample solution (freshly prepared in buffer or a dried-droplet sample preparation redissolved in 50 mM bicarbonate) followed by 0.5 µL of a 10 mM solution of OGP in water.
2. Add 1 µL of 50 mM bicarbonate buffer (pH 7.8) and 0.5 µL trypsin solution (E/S approx 1:10) onto the target.
3. Allow the mixture to react for 1 h at 37°C in a moist environment.
4. Add 1 µL 2% TFA and 0.5 µL HCCA (I), allow the sample to dry, and wash according to **Note 8**.

3.4. Miniaturized Pretarget Chromatography

See **Subheading 3.5.7**.

3.4.1. Micro-Scale Sample Purification

See **refs. 24** and **25**.

3.4.1.1. PREPARATION AND EQUILIBRATION OF THE MICROSCALE COLUMN

1. Squeeze a GELoader tip (Eppendorf) carefully at the lower end of the extended outlet with a pair of flat-nosed pliers. Doing this, the inner diameter is reduced to < 50 µm. This procedure is needed to avoid losses of stationary phase material (Poros R1/50 µm or Poros R2/50 µm) during column packing and operation.
2. Fill 50 µL of methanol into the GELoader tip using a 200-µL pipet tip.
3. Pipet 2–3 µL of a methanol suspension of Poros 50 R1 or Poros 50 R2 material, for proteins or peptides, respectively, into the methanol layer with a 1–10-µL tip, taking care to exclude any air bubbles.
4. When the Poros material has settled (prepacked column), mount the GELoader tip onto a 1.25-mL Eppendorf Combitip, which is full of air, and push the methanol quickly through the packing material. During this procedure, the Poros material forms a small column with a volume of approx 1 µL in the GELoader tip outlet above the squeezed section with the reduced diameter.
5. Load 20 µL 0.1% TFA with a 1–10-µL pipet tip and pass about 80% of this liquid over the stationary phase as described in the previous step. In this state, the columns can then be stored for least for 1 d, at 4°C, if they have been sealed.

3.4.1.2. LOADING, WASHING, AND ELUTION OF THE SAMPLE

1. Load the sample solution onto the column (*see* **Note 11**). With sample volumes < 40 µL, place the solution above the microcolumn using a second GELoader tip and push it through as described in **Subheading 3.4.1.1., step 4**. With volumes > 40 µL, load the sample solution into a second Eppendorf Combitip, and use this to push the solution through the column.

2. Put 100–200 µL of washing solvent ([100-X]% of 0.1% TFA + X% ACN [v/v] where X% ACN is in the sample solution) into the syringe and push this through the column.
3. Place 2 µL of elution solvent ([100-Y]% 0.1% TFA + Y% ACN (v/v) where Y% ACN ensures complete elution) on top of the stationary phase using a third GELoader tip and elute the sample directly onto the target for subsequent MALDI-MS sample preparation. When Y% is difficult to predict, use 0.1% TFA + ACN (20:80, v/v) for the elution of both peptides and proteins.

3.4.1.3. STEPWISE ELUTION ALTERNATIVE FOR COMPLEX PEPTIDE MIXTURES

1. The column is loaded and washed as in **Subheading 3.4.1.2.**
2. Elute the peptides onto the MALDI target with 5-µL portions of solvent containing 0.1% TFA and an increasing concentration of ACN (5% increment per 5-µL portion) up to 35%.

3.4.2. Nanoscale Sample Purification

See **ref. 27**.

3.4.2.1. PREPARATION OF NANOSCALE REVERSED-PHASE COLUMNS

1. Carefully flatten the column body (GELoader tip, Eppendorf; see **Subheading 3.4.1.1.**) near the end of the outlet using fine tweezers. By heating the tweezers briefly in a flame prior to flattening the pipet tip, the outlet is permanently narrowed. This is recommended when using Poros R1/2 material of 10- or 20-µm particle size.
2. To ensure that the column is not blocked, dispense isopropanol (10 µL) into the column body, from the wide end, and push it gently down to the outlet using a syringe (Eppendorf Combitip, 1.25 mL).
3. Deposit a suspension of chromatography medium in isopropanol into the loaded solution and push it gently through, so that the chromatography medium forms a column of approx 1.5 mm in length, near the outlet.

3.4.2.2. LOADING, WASHING, AND ELUTION OF THE SAMPLE

1. Prior to using the column, perform a washing step with 15 µL of ACN + 0.1% TFA, 80:20 (v/v), followed by an equilibration step with 5 µL 0.1% TFA (v/v). These solvents and all samples should be loaded onto the column with GELoader tips and pushed through the reversed-phase material using a plastic syringe (Eppendorf Combitip, 1.25 mL).
2. Acidify the peptide sample with TFA to a concentration of approx 0.5% (v/v), load an aliquot of the sample onto the column, and push it slowly through the column material.
3. After loading the sample, perform a washing step with 10 µL 0.1% TFA (v/v).
4. Run the column completely dry by pusing air through it for a few seconds. This is imperative for smooth and continuous liquid flow in the next step of the procedure.
5. Load 0.5 µL eluent, also containing the MALDI matrix (saturated solution of HCCA in ACN + 0.1% TFA, 45:55 [v/v] or DHB: 5 mg/mL, ACN + 0.1% TFA, 3:7 [v/v]), and pass this slowly through the column.

6. Dot the eluted material onto the MALDI target in fractions of 50–100 nL. The major part of the analyte will elute in the first fraction, but hydrophobic peptides may elute in the second fraction (*see* **Note 12**).

3.4.3. Microscale Tryptic On-Column Digestion

See ref. 25.

3.4.3.1. PREPARATION OF TRYPSIN MICROCOLUMNS

1. Flatten a GELoader pipette-tip near the end of the outlet using flat-nosed pliers as described for microcolumn preparation (**Subheading 3.4.1.1., step 1**).
2. Dispense a suspension of Porozyme immobilized trypsin medium (1 µL) in the column body from the wide end, followed by 5 µL 5 mM NH_4HCO_3 (pH 7.8) + ACN 95:5 (v/v) (digestion buffer).
3. Fit the column body to a plastic syringe (Eppendorf Combitip, 1.25 mL) filled with air, and press down to expel liquid from the suspension to form a column with a volume of approximately 0.3–0.5 µL.
4. Remove the column body from the syringe and fill with 15 µL 5 mM NH_4HCO_3 (pH 7.8) + ACN 70:30 (v/v) (elution buffer) from another GELoader tip. Push this elution buffer through the column in the same way using an air-filled syringe.
5. Equilibrate the column with 15 µL digestion buffer following the same procedure.

3.4.3.2. ON-COLUMN DIGESTION

1. Dissolve the protein sample in digestion buffer, which also contains 5 mM n-octylglucopyranoside (OGP), and apply 1–2 µL of this mixture to the top of the column using a GELoader tip.
2. Push the sample slowly through the column, using the air-filled syringe, and collect the eluate onto a MALDI target.
3. Pass a further 1–2 µL elution buffer through the column to ensure complete elution of peptides.

3.5. Supplementary Comments and Examples

3.5.1 Matrix selection

The choice of matrix must be adapted to the properties of the analyte. The most common choices of matrix, dependent on the analyte and on compatibility of the matrix with the on-target sample preparation techniques described here, are summarized in **Table 1**.

HCCA is the first choice for peptide mapping analyses and is compatible with all sample preparation procedures described in this chapter. It is mostly used in the sandwich technique because this combination yields high peptide ion abundances (even with highly contaminated samples), good protein sequence coverage, and little methionine or tryptophan side chain oxidation. The HCCA/sandwich combination has, with respect to the latter, proved superior to the use of HCCA in the dried-droplet or thin-layer method. However, for peptide sample amounts of < 100 fmol, the thin-layer method, preferably with NC as additive, exhibits a higher detection sensitivity.

The choice of matrix can considerably influence the sequence coverage in peptide mapping. According to our experience, peptide-mapping results obtained with the matrix THAP can complement those obtained with HCCA in terms of protein sequence coverage. The same effect is usually seen with DHB as a complementary matrix to HCCA. DHB and THAP both yield less chemical noise in the low-mass range and hence improve the detection of small peptides. This effect is especially pronounced when using DHB in the nanoscale purification technique *(27)*, as exemplified in **Fig. 1**. **Figure 1** shows tryptic peptide maps of a silver-stained protein-gel spot obtained using **(A)** the thin-layer method and **(B,C)** the nanocolumn method, using HCCA **(B)** and DHB **(C)** as matrices, respectively *(27)*. In each case, 0.5 µL of digestion supernatant was used. To improve the mass accuracy, 10 fmol of angiotensin III were loaded onto the nanocolumn prior to the sample and the corresponding signal used for internal mass calibration of spectra. With the thin-layer method, relatively few peptide signals were detected, and the database search resulted in a number of equally possible proteins. By contrast, with the nanocolumn method, mitochondrial aconitase was unequivocally identified with HCCA, as well as with DHB *(27)*. It was observed that the peptide signals derived from the gel-separated protein were, in general, relatively more abundant with DHB as the matrix than with HCCA. In addition, a number of peptide signals were exclusively detected with one or the other matrix. However, the use of either of the two matrices resulted in the correct protein identification.

Sinapinic acid is the favored matrix for protein analysis, whereas for glycopeptides and glycoproteins DHB (I–III) is preferred (*see* **Table 1**). As reported previously *(36,37)*, the quality of glycopeptide mass spectra obtained with DHB in the dried-droplet sample preparation is highly dependent on the morphology of the sample/matrix crystals. The large rim crystals almost exclusively yield protonated glycopeptides, whereas the smaller crystals in the central sample area preferentially exhibit sodium and potassium adduct ions. Thus, use of the on-target crystallization process to a certain extent affects the separation of alkali salts. This sample heterogeneity can be overcome by the application of DHB in the nanoscale sample clean-up, yielding a small and homogeneous sample spot, which is almost entirely irradiated by the laser beam *(27)*. HCCA with the sandwich method sometimes yields better results in glycopeptide analysis and is therefore often used as the matrix in a second sample preparation.

MALDI-MS Sample Preparation Techniques

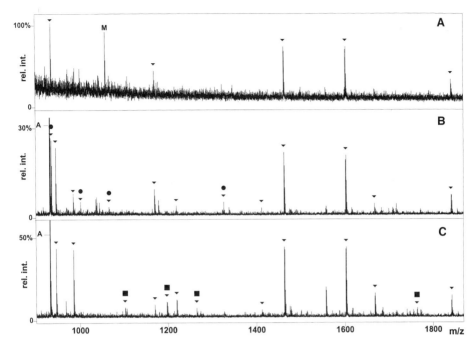

Fig. 1. Delayed extraction (DE) MALDI RE-TOF spectra obtained from a 0.5-μL digestion supernatant from a medium silver-stained gel spot. **(A)** Thin-layer sample preparation: the crystalline sample was washed on the MALDI target with 5 μL 5% formic acid. A commonly observed matrix signal is indicated by 'M'. **(B,C)** Nanocolumn sample preparation using **(B)** HCCA and **(C)** DHB as the matrices. Triangles indicate signals matching tryptic fragments of human mitochondrial aconitase. Circles signify matching peptides detected with HCCA, but not with DHB. Squares indicate matching peptides exclusively detected with DHB. The signal of the added internal calibrant, angiotensin III, is indicated by the 'A'. (Adapted with permission from **ref. 27**).

3.5.2. Influence of pH

Whereas pH 3 has been reported as the upper limit for sensitive MALDI-MS peptide detection with HCCA as the matrix *(10)*, acidification to a pH below 2.0 is essential for samples containing considerable amounts of salts, buffers, and detergents. **Table 2** lists the TFA concentrations needed for on-target acidification to pH < 2 for a variety of buffers and buffer concentrations *(11)*.

3.5.3. Features of the Sandwich Sample Preparation

The sandwich technique is compatible with HCCA and SA matrices and often yields significantly better peptide mapping results than are obtained with

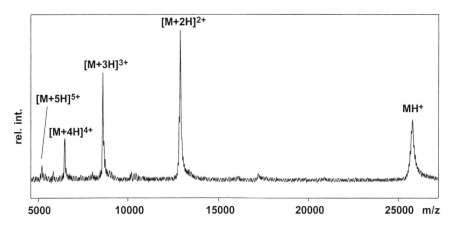

Fig. 2. Linear-mode DE MALDI mass spectrum of the protein csgA, showing an abundant series of singly and multiply protonated molecules. The analyte solution contained 10-20 pmol/μL protein, 8 M urea, and various buffers and salts. This solution (5 μL) was subjected to a sandwich sample preparation including extensive washing according to **Subheading 3.2.2.4.** (Reproduced with permission from **ref.** *24*).

the dried-droplet method. The sandwich method has, in particular, proved to be superior in terms of tolerance toward high amounts of impurities (denaturants, detergents, salts, and buffers), especially when the sample is not allowed to dry before washing. The spectrum in **Fig. 2** shows an abundant series of singly and multiply protonated molecules of the His-tagged C-signaling A protein (csgA; from *Myxococcus xanthus*) *(24)*. The analyte solution contained 10–20 pmol/μL of protein in 20 mM Tris-HCl (pH 8.0), 100 mM NaCl, 6 mM imidazole, and 8 M urea. This solution (5 μL) was used directly in a sandwich sample preparation with HCCA, and three washing steps (3 × 10 μL 0.1% TFA) were performed 1 min after addition of HCCA (I). By contrast, none of the other matrices and on-target sample preparations described in this chapter afforded abundant protein molecular ions with this sample.

The thin-layer sample preparation often results in considerable oxidation of methionine and tryptophan side chains. The sandwich method is an alternative that can circumvent this amino acid modification during sample preparation.

3.5.4. On-Target Reactions

The on-target DTT reduction of peptides and peptide mixtures has, for instance, been successfully applied to disulfide bond determination of the major cat allergen Fel d 1 *(38)*. Sometimes, however, particularly if the on-target reduction is applied to complex peptide mixtures, not all the expected reduced cysteinyl peptides are observed.

Proteolytic digestion can be performed on-target as an alternative to conventional digestion in solution, for instance, in the case of very low sample amounts or when a quick protein identification is needed *(39)*. If a protein is particularly resistant to proteolysis, even in 8 M urea or 6 M guanidinium hydrochloride, proteolytic degradation can sometimes be achieved by addition of OGP to the (on-target) digestion mixture. OGP is a nonionic detergent, known to aid in protein solubilization, which often improves the quality of MALDI mass spectra of proteins or peptide mixtures. The latter finding is in contrast to the use of ionic detergents, such as sodium dodecyl sulphate (SDS), which drastically lower the quality of MALDI mass spectra.

3.5.5. Nitrocellulose as Matrix Additive

The addition of nitrocellulose to the matrix solution is especially valuable for the analysis of highly contaminated samples, e.g., protein digests derived from gel samples. Nitrocellulose significantly reduces, if not eliminates, alkali adduct ions as well as signals deriving from polymers such as PEG. The capacity for preventing contaminants from desorption is most pronounced in the thick-layer method, whereas the thin-layer preparation with NC as additive provides the best compromise between "spectrum clean-up" and sensitivity (*see* **Subheading 3.5.6.**).

The preparation of a homogeneous matrix/nitrocellulose layer is important. In addition, particularly if the modified NC preparation on a Scotch tape (*see* **Note 9**) is used, calibration is crucial. It is recommended that the sample be spiked with an internal standard. Alternatively, an external calibration, obtained from a protein or peptide standard also prepared by the thick-layer method, can be used.

3.5.6. Sensitivity Comparison of Different Sample Preparations

A decision as to which method provides the highest detection sensitivity for a given analyte is predominantly made on the basis of the amount and composition of contaminants in the sample rather than just on the absolute amount of analyte. In the case of samples containing moderate amounts of buffers and little detergent, denaturant, or synthetic polymer, the thin-layer HCCA/NC method is more sensitive than the thick-layer technique for peptide detection, and the latter is more sensitive than the sandwich method. However, if large amounts of contaminants, e.g., urea (**Fig. 2**), guanidinium hydrochloride, or PEG are present, the sandwich or the thick-layer methods are advantageous because the contaminants can be washed off on-target (sandwich) or are efficiently adsorbed by the matrix (thick-layer).

3.5.7. Miniaturized Pretarget Chromatography

If contaminants prevent the detection of peptide or protein molecular ions after application of the sample preparation and on-target purification methods mentioned so far, or, if an analyte concentration step is needed prior to sample preparation, a micropurification procedure is called for. Again, because the presence of nonprotein contaminants often limits the detection sensitivity and the spectrum quality, a pretarget micro- or nanoscale sample clean-up is almost always advantageous. Features of this technique are efficient purification and concentration of pmol to low fmol amounts of peptides and proteins, and circumvention of memory effects since a separate column is used for each sample. The microscale sample purification offers a larger capacity for contaminant removal, whereas the nanoscale technique is optimized to deal with minute sample amounts.

3.5. Concluding Remarks

There is no universal sample preparation method that yields good results for a broad variety of peptide and protein analytes, rather, a number of alternatives should be considered. These include different matrix compounds combined with various sample preparation techniques. Based on specific information on the analyte properties, it is possible to select a promising first method (**Table 1**), but subsequent variations often are needed to optimize the results. The combinations of matrix compounds, sample preparation procedures, and use of additives described in this study are frequently used in our laboratory. However, this selection does not claim completeness since new matrices, additives, and techniques will continually emerge.

Acknowledgments

Johan Gobom is acknowledged for scientific discussion. The Danish Biotechnology Programme is acknowledged for financial support.

4. Notes

1. Optionally, excise a blank gel piece of the same size as a control.
2. All solvent volumes should be approximately twice the volume of the excised gel plug.
3. An option for silver-stained gels is to just shrink these with ACN.
4. Reduction is still recommended, if reducing sample and running buffers were used in the PAGE operation.
5. Coomassie-stained bands should be transparent by now; if > 10 pmol protein is present, remove residual staining by further washing.
6. This *in situ* digestion protocol is compatible with the following electrophoretic parameters: 1/2-D SDS-PAGE; gel thickness: 0.5–1 mm; acrylamide concentration: 7.5–18%; sample buffer: Laemmli, 0.1% SDS; staining: Coomassie brilliant blue R-250/G-250, silver staining (extensive washing of the gel with ACN can be omitted), reverse staining (zinc-imidazole), or detection of ^{35}S-labeled proteins by autoradiography.

7. Roman numerals refer to the matrix solution compositions listed in **Subheading 2.2.1.**
8. In the case of on-target sample acidification, add 0.5 µL of aqueous TFA according to **Table 2**. If the analyte solution contains buffers, urea, guanidinium hydrochloride, or other involatile contaminants and if HCCA, SA, or THAP are used as the matrix, wash the sample by placing 5–10 µL ice-cold 0.1% TFA 1–3 times onto the target and pipeting it off after a few seconds.
9. Alternatively, if using a multisample target (Voyager Elite MS) that is difficult to spin, prepare the NC/matrix layer on a small piece of Scotch tape (adhesive on both sides) glued onto the rotator, and wash subsequently with 100 µL 0.1% TFA. Transfer this tape onto the multisample target and carry out **steps 3–7** of **Subheading 3.2.2.3.**
10. Whenever large amounts of contaminants such as urea or guanidinium hydrochloride are present, add 5–20 µL of 0.1% TFA approx 1–2 min after addition of HCCA (I) or SA (I), i.e., shortly after crystallization had commenced. After another 1–2 min, pipet off the solvent and perform 1–3 washing steps as in **Note 8**.
11. Prior to the clean-up procedure documented in **Subheading 3.4.1.2.**, it is necessary to acidify the sample solution with 0.1–2% TFA to pH < 2 and also to ensure that the amount of organic solvent (e.g., ACN, methanol) in the sample solution still allows efficient retention of the analyte molecules on the stationary phase. If the latter is difficult to predict, the sample solution should not contain any organic solvent.
12. Internal calibrants can optionally be loaded onto the column prior to sample loading. For example, 5–10 fmol angiotensin III in 0.1% TFA (10–20 fmol/mL) can be used.

References

1. Pappin, D. J. C., Højrup, P., and Bleasby, A. J. (1993) Protein identification by peptide mass fingerprinting. *Curr. Opin. Biol.* **3**, 327–332.
2. James, P., Quadroni, M., Carafoli, E., and Gonnet, G. (1993) Protein identification by mass profile fingerprinting. *Biochem. Biophys. Res. Commun.* **195**, 58–64.
3. Mann, M., Højrup, P., and Roepstorff, P. (1993) Use of mass spectrometric molecular weight information to identify proteins in sequence databases. *Biol. Mass Spectrom.* **22**, 338–345.
4. Resing, K. A., Mansour, S. J., Hermann, A. S., Johnson, R. S., Candia, J. M., Fukusawa, K., et al. (1995) Determination of v-Mos-catalyzed phosphorylation sites and autophosphorylation sites on MAP kinase kinase by ESI/MS. *Biochemistry* **34**, 2610–2620.
5. Burlingame, A. L. (1996) Characterization of protein glycosylation by mass spectrometry. *Curr. Opin. Biotechnol.* **7**, 4–10.
6. Henzel, W. J., Billeci, T. M., Stults, J. T., Wong, S. C., Grimley, C., and Watanabe, C., (1993) Identifying proteins from two-dimensional gels by molecular mass searching of peptide fragments in protein sequence data bases. *Proc. Natl. Acad. Sci. USA* **90**, 5011–5015.

7. Chou, J. Z., Kreek, M. J., and Chait, B. T. (1994) Matrix-assisted laser desorption mass spectrometry of biotransformation products of dynorphin A in vitro. *J. Am. Soc. Mass Spectrom.* **5,** 10–16.
8. Gusev, A. L., Wilkinson, W. R., Proctor, A., and Hercules, D. M. (1995) Improvement of signal reproducibility and matrix/co-matrix effects in MALDI analysis. *Anal. Chem.* **67,** 1034–1041.
9. Billeci, T. M. and Stults, J. T. (1993) Tryptic mapping of recombinant proteins by matrix-assisted laser desorption/iomization mass spectrometry. *Anal. Chem.* **65,** 1709–1716.
10. Cohen, S. L. and Chait, B. T. (1996) Influence of matrix solution conditions on the MALDI-MS analysis of peptides and proteins. *Anal. Chem.* **68,** 31–37.
11. Jensen, C., Haebel, S., Andersen, S. O., and Roepstorff, P. (1997) Towards monitoring of protein purification by matrix-assisted laser desorption ionization mass spectrometry. *Int. J. Mass Spectrom. Ion Processes* **160,** 339–356.
12. Karas, M. and Hillenkamp, F. (1988) Correspondence: laser desorption ionization of proteins with molecular masses exceeding 10.000 Daltons. *Anal. Chem.* **60,** 2299–2301.
13. Beavis, R. C. and Chait, B. T. (1990) High-accuracy molecular mass determination of proteins using matrix-assisted laser desorption mass spectrometry. *Anal. Chem.* **62,** 1836–1840.
14. Xiang, F. and Beavis, R. C. (1993) Growing protein-doped sinapic acid crystals for laser desorption: an alternative preparation method for difficult samples. *Org. Mass Spectrom.* **28,** 1424–1429.
15. Xiang, F. and Beavis, R. C. (1994) A method to increase contaminant tolerance in protein matrix-assisted laser desorption/ionization by the fabrication of thin protein-doped polycrystalline films. *Rapid Commun. Mass Spectrom.* **8,** 199–204.
16. Vorm, O., Roepstorff, P., and Mann, M. (1994) Improved resolution and very high sensitivity in MALDI TOF of matrix surfaces made by fast evaporation. *Anal. Chem.* **66,** 3281–3287.
17. Brockman, A. H., Dodd, B. S., and Orlando, R. (1997) A desalting approach for MALDI-MS using on-probe hydrophobic self-assembled monolayers. *Anal. Chem.* **69,** 4716–4720.
18. Blackledge, J. A. and Alexander, A. J. (1995) Polyethylene membrane as a sample support for direct matrix-assisted laser desorption/ionization mass spectrometric analysis of high mass proteins. *Anal. Chem.* **67,** 843–848.
19. Zhang, H. and Caprioli, R. M. (1996) Direct analysis of aqueous samples by matrix-assisted laser desorption ionization mass spectrometry using membrane targets pre-coated with matrix. *J. Mass Spectrom.* **31,** 690–692.
20. Worrall, T. A., Cotter, R. J., and Woods, A. S. (1998) Purification of contaminated peptides and proteins on synthetic membrane surfaces for matrix-assisted laser desorption/ionization mass spectrometry. *Anal. Chem.* **70,** 750–756.
21. Zhang, H., Andren, P. E., and Caprioli, R. M. (1995) Micro-preparation procedure for high-sensitivity matrix-assisted laser desorption ionization mass spectrometry. *J. Mass Spectrom.* **30,** 1768–1771.

22. Kalkum, M. and Gauss, C. (1997) Piezo-jet pipetting system for automated microscale preparation of high density MALDI-MS sample plates, in *Proceedings of the 45th ASMS Conference on Mass Spectrometry and Allied Topics*, Palm Springs, CA, p. 324.
23. Annan, R. S., McNulty, D. E., and Carr, S. A. (1996) High sensitivity identification of proteins from SDS-PAGE using in-gel digestion and MALDI-PSD, in *Proceedings of the 44th ASMS Conference on Mass Spectrometry and Allied Topics*, Portland, OR, p. 702.
24. Kussmann, M., Nordhoff, E., Rahbek-Nielsen, H., Haebel, S., Rossel-Larsen, M., Jakobsen, L., et al. (1997) MALDI-MS sample preparation techniques designed for various peptide and protein analytes. *J. Mass Spectrom.* **32**, 593–601.
25. Gobom, J., Nordhoff, E., Ekman, R., and Roepstorff, P. (1997) Rapid micro-scale proteolysis of proteins for MALDI-MS peptide mapping using immobilised trypsin. *Int. J. Mass Spectrom. Ion. Processes* **169/170**, 153–163.
26. Kussmann, M. and Roepstorff, P. (1998) Characterisation of the covalent structure of proteins from biological material by MALDI mass spectrometry—possibilities and limitations. *Spectroscopy* **14**, 1–27.
27. Gobom, J., Nordhoff, E., Mirgorodskaya, E., Ekman, R., and Roepstorff, P. (1999) A sample purification and preparation technique based on nano-scale RP-columns for the sensitive analysis of complex peptide mixtures by MALDI-MS. *J. Mass Spectrom.* **24**, 105–116.
28. Beavis, R. C., Chaudhary, T., and Chait, B. T. (1992) OMS Letters: alpha-cyano-4-hydroxy cinnamic acid as a matrix for matrix-assisted laser desorption mass spectrometry. *Org. Mass Spectrom.* **27**, 156–158.
29. Beavis, R. C. and Chait, B. T. (1989) Matrix-assisted laser-desorption mass spectrometry using 355 nm radiation. *Rapid Commun. Mass Spectrom.* **3**, 432–435.
30. Strupat, K., Karas, M., and Hillenkamp, F. (1991) 2,5-Dihydroxybenzoic acid: a new matrix for laser desorption/ionization mass spectrometry. *Int. J. Mass Spectrom. Ion Physics* **111**, 89–102.
31. Karas, M., Ehring, H., Nordhoff, E., Stahl, B., Strupat, K., Grehl, M., et al. (1993) Matrix-assisted laser desorption mass spectrometry with additives to 2.5-dihydroxy benzoic acid. *Org. Mass Spectrom* **28**, 1476–1481.
32. Rosenfeldt, J., Capdevielle, J., Guillemot, J. C., and Ferrara, P. (1992) In-gel digestion of proteins for internal sequence analysis after one- or two-dimensional gel electrophoresis. *Anal. Biochem.* **203**, 173–179.
33. Shevchenko, A., Wilm, M., Vorm, O., and Mann, M. (1996) Mass spectrometric sequencing of proteins from silver-stained polyacrylamide gels. *Anal. Chem.* **68**, 850–858.
34. Roepstorff, P. (1994) Plasma desorption mass spectrometry of peptides and proteins, in *Cell Biology: A Laboratory Handbook* (Celis, J., ed.), Academic, Orlando, FL, pp. 399–404.
35. Perera, I. K., Perkins, J., and Knatartzoglou, S. (1995) Spin-coated samples for high-resolution matrix-assisted laser desorption/ionization time-of-flight mass spectrometry of large proteins. *Rapid Commun. Mass Spectrom.* **9**, 180–187.

36. Stahl, B., Steup, M., Karas, M., and Hillenkamp, F. (1991) Analysis of neutral oligosaccharides by matrix-assisted laser desorption/ionization mass spectrometry. *Anal. Chem.* **63,** 1463–1466.
37. Westman, A., Huth-Fehre, T., Demirev, P., and Sundqvist, B. U. R. (1995) Sample morphology effects in matrix-assisted laser desorption/ionization mass spectrometry of proteins. *J. Mass Spectrom.* **30,** 206–211.
38. Kristensen, A. K., Schou, C., and Roepstorff, P. (1997) Determination of isoforms, N-linked glycan structure and disulfide bond linkages of the major cat allergen Fel d 1 by a mass spectrometric approach. *Biol. Chem.* **378,** 899–908.
39. Ásgeirsson, B., Haebel, S., Thorsteinsson, L., Helgason, E., Gudmundsson, K. O., Gudmundsson, G., et al. (1998) Hereditary cystatin C amyloid angiopathy; monitoring the presence of the Leu68 → Gln cystatin C variant in cerebrospinal fluids and monocyte cultures by MS. *Biochem. J.* **329,** 497–503.

24

Analysis of Hydrophobic Proteins and Peptides by Mass Spectrometry

Johann Schaller

1. Introduction

The introduction of electrospray ionization mass spectrometry (ESI-MS) *(1)* and matrix-assisted laser desorption/ionization mass spectrometry (MALDI-MS) *(2)* in the late eighties has had a tremendous influence on the analysis of biomolecules at the molecular level in biochemistry and biology. The vast number of publications dealing with many different aspects of biomolecules has, however, focused primarily on water-soluble compounds *(3)*. The poor solubility and scarcity of hydrophobic peptides and proteins has hampered extensive analysis of these materials by ESI-MS or by MALDI-MS.

1.1. MALDI-MS Analysis of Hydrophobic Peptides and Proteins

Some reasonably accurate MALDI-MS data are available for rhodopsin and bacteriorhodopsin *(4–6)*, porin, and cholesterol esterase *(5)*. The 75-kDa major surface layer protein from *Clostridium thermosaccharolyticum* was determined with a mass accuracy of ±0.2% *(7)*. Very recently, the subunits of several integral membrane proteins, namely, cytochrome bo_3 oxidase and cytochrome bd oxidase from *E. coli* as well as cytochrome bc_1 and cytochrome c oxidase from *Rhodobacter sphaeroides* were analyzed by MALDI-MS with an accuracy of better than 0.5% in most cases *(8)*. Good data were obtained for the subunits of bovine cytochrome c oxidase with accuracies better than 0.05% *(9)*. A recent article in this series described sample preparation protocols for the analysis of hydrophobic peptides and proteins by MALDI-MS *(10)*. The analysis of membrane channel proteins (porins) in the presence of detergents was recently described *(11)*.

1.2. ESI-MS Analysis of Hydrophobic Peptides and Proteins

Some data with rather good accuracies are available from ESI-MS measurements. Bacteriorhodopsin, fragments of bacteriorhodopsin, and genetically and chemically modified bacteriorhodopsin were analyzed with accuracies close to 0.01% *(12–16)*. Bovine cytochrome c oxidase subunits *(12,17)*, subunit 6 of bovine ATP synthase *(12,17)*, the three subunits from the mannose transporter complex in *E. coli (14)*, and the subunits from the spinach and pea photosystem II reaction center *(16,18,19)* were successfully analyzed with good mass accuracies. In addition, a method was described for the analysis of hydrophobic proteins and peptides electroeluted from sodium dodecyl sulfate-polyacrylamide gel electrophoresis (SDS-PAGE) *(20)*. Valuable data on the influence of various detergents on ESI-MS measurements in general are also available *(21)*.

2. Materials
2.1. Reagents

1. 10 mM Tris-HCl pH 7.0.
2. 5 mM HEPES pH 7.0.
3. n-Dodecyl-β-D-maltoside (DDM), cyclohexyl-hexyl-β-D-maltoside (Cymal-6), SDS, and dodecyltrimethylammonium chloride (DTAC) (Anatrace, Maumee, OH).
4. Trifluoracetic acid (TFA), acetonitrile, water, chloroform, methanol and 1-propanol, all high-performance liquid chromatography (HPLC) grade (Merck, Dietikon, Switzerland or Fluka, Buchs, Switzerland).
5. Hexafluoro-2-propanol (HFIP) purum (Fluka).
6. Other reagents and solvents (all analytical grade), from Merck or Fluka.

2.2. Analytes

The mannose transporter complex from *E. coli* was isolated according to Erni et al. *(22)*. Bacterioopsin, together with V8 protease fragments of bacterioopsin, from *Halobacterium halobium* were prepared according to Wuethrich and Sigrist *(23)*. The glucose-specific IIBC component of the phosphotransferase system in *E. coli* and the outer membrane protein (OMPA) from *E. coli* were prepared according to Buhr et al. *(24)* and Garavito et al. *(25)*, respectively.

2.3. Instrumentation

1. VG Platform single-stage quadrupole mass spectrometer with an atmospheric pressure electrospray ion source and MassLynx software (Micromass, Manchester, UK). Upper mass limit m/z 3000 (*see* **Note 1**).
2. Harvard syringe pump, model 22 (Harvard Apparatus, South Natick, MA).

Hydrophobic Proteins and Peptides

3. Dual-syringe solvent delivery system 140B (Applied Biosystems, Foster City, CA).
4. Rheodyne injection valve, model 7225 (Rheodyne, Inc., Cotati, CA) with a 10-µL injection loop.
5. LiChroCART column (4.0 × 25 mm × 7 µm) with Lichrosorb RP-8 (reversed phase) packing (Merck, Darmstadt, Germany).
6. Z-shaped flow cell; 8-mm path length, 35-nL cell volume (LC Packings, Amsterdam, The Netherlands).
7. Variable wavelength absorbance detector 759A (Applied Biosystems).

3. Methods
3.1. Sample Purification
3.1.1. Mannose Transporter Complex

1. Solubilize the mannose transporter complex in 10 mM Tris-HCl buffer (pH 7.0) containing 250 mM imidazole and one of 0.02% DDM, 0.5% SDS, or dodecyltrimethylammonium chloride (concentration unknown).
2. Purify the solubilized complex on the Lichrosorb RP-8 column using a linear gradient from 0 to 100% B over 30 min at a flow rate of 50 µL/min with either of the following solvent systems:
   a. A: 0.1% TFA in water, B: neat formic acid.
   b. A: 0.1% TFA in water, B: 0.1% TFA in acetonitrile.
   c. A: 0.1% TFA in 2:1 water + methanol (v/v), B: 0.1% TFA in chloroform.

3.1.2. Outer Membrane Protein (OMPA)

After the final preparation step, the OMPA solution (concentration ≈1 µg/µL) contains 20 mM sodium phosphate buffer (pH 7.3) with 100 mM NaCl, 0.5 mM DTT, and 1.2 mM Cymal-6 (nonionic detergent).

1. Purify the preparation of outer membrane protein (OMPA) from *E. coli* on the same column as in **Subheading 3.1.1., step 2**, using a separation system of solvent A (10 min isocratic), step to solvent B, then solvent B (30 min isocratic).

3.2. Electrospray Mass Spectrometry

See **Note 1**.

3.2.1. Mass Spectrometer Operation

1. Typically, apply 3500 V on the spraying capillary, 500 V on the counter electrode and 50–100 V on the sample cone. Keep the source temperature at 60°C (any changes to these values are mentioned in the text).
2. Scan the mass spectrometer over an appropriate mass range for a total time of about 1 min for loop-injected samples. Acquire 10–20 scans and then average these to obtain the final spectrum. Use the MassLynx software to obtain the average molecular mass of the samples.

3.2.2. Loop Injection of Samples

Normally, inject 10 µL of sample solution by means of the Rheodyne injection valve. Introduce the sample into the ion source with the Harvard syringe pump at a constant flow rate of 10 µL/min using various carrier solvents, depending on the solubility properties of the proteins (*see* **Subheadings 3.3.** and **3.4.** for details).

3.2.3. LC/MS Analysis of Samples

1. Connect the dual-syringe solvent delivery system, reversed-phase HPLC column, Z-shaped flow cell, and variable wavelength absorbance detector, via fused-silica capillary tubing (75-µm internal diameter), to the mass spectrometer.
2. Scan the mass spectrometer at intervals of 2.5 s and record the total-ion current of all masses in the range m/z 600–2200 together with the UV chromatogram at 280 nm from the HPLC separation.
3. Accumulate and combine the mass scans and determine the average molecular mass with the aid of the MassLynx software.

3.3. Bacterioopsin and V8 Protease Fragments of Bacterioopsin from Halobacterium halobium Analyzed by Loop Injection ESI-MS

1. Dissolve bacterioopsin (BO) from *Halobacterium halobium* in hexafluoro-2-propanol (10 pmol/µL) or use the BO already contained in the solvent from the final purification by HPLC/fast protein liquid chromatography (FPLC) (0.1% TFA in acetonitrile + 1-propanol + water) *(23)* and inject into the carrier solvent flow (*see* following text for details).

The most accurate results with BO were obtained with neat formic acid as carrier solvent at an increased ion source temperature (*see* **Note 2**) of 100°C (**Fig. 1**). The measured mass of 26,781.5 ± 3.6 Daltons agrees very well with the calculated value of 26,783.6 Daltons. Alternatively, 0.5% acetic acid in hexafluoro-2-propanol + methanol + water (2:5:2, v/v/v) or 2% formic acid in chloroform + methanol + water (4:4:1 or 2:5:2, v/v/v) can be used as carrier solvent, but with a loss of accuracy (26,790.5 ± 7.0 Daltons and 26,960.0 ± 7.5 Daltons, respectively). In addition, these solvent mixtures have a more or less pronounced tendency to form adducts with BO.

The V8 protease fragments V2A (Ser162/Glu232) and V2B (Val167/Glu232) of BO, contained in the effluent from HPLC/FPLC separation (*see* **Subheading 3.3., step 1**), were analyzed using 2% acetic acid in chloroform + methanol + water (2:5:2, v/v/v) as carrier solvent. The MassLynx transformed masses of 7748.0 ± 0.4 Daltons and 7147.1 ± 0.5 Daltons agree very well with the calculated values of 7748.3 Daltons and 7147.6 Daltons, respectively (spectra not shown). The pronounced tendency of neat formic acid to form adducts with V2A and V2B is in contrast to the measurements of BO itself, where no substantial adduct formation was observed.

Hydrophobic Proteins and Peptides

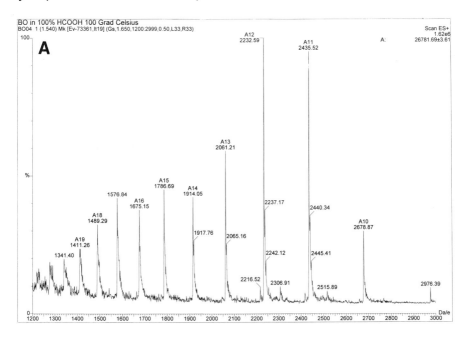

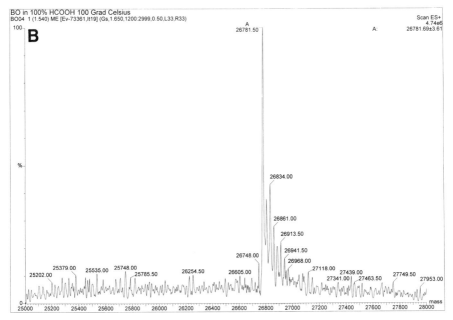

Fig. 1. (**A**) Electrospray mass spectrum and (**B**) transformed mass spectrum of BO (calculated mass: 26,783.6 Daltons). The protein was dissolved in HFIP and analyzed with neat formic acid as the carrier solvent at an elevated ion source temperature of 100°C (10 µL injected at 10 pmol/µL).

3.4. Glucose-Specific IIBC Component of the Phosphotransferase System in E. coli Analyzed by Loop Injection ESI-MS

The glucose-specific IIBC component (IICB^{Glc}), carrying a C-terminal hexahistidine tag (Met1/His477)Phe+Gln-Ser-Arg-Ser-(His)$_6$ to facilitate an efficient isolation by affinity chromatography, was obtained following recombinant expression.

1. Prepare a very high concentration (150–200 pmol/µL) solution of IICB^{Glc} in 5 mM HEPES buffer (pH 7.0) containing 0.4 mM DDM (*see* **Note 3**).
2. Dilute the freshly prepared IICB^{Glc} solution (dilution range: 1:10–1:100) into acetonitrile + water (1:1, v/v) containing 10% formic acid to give final concentrations of 2–20 pmol/µL and inject into the carrier solvent flow (see following text for details).

Using acetonitrile + water (1:1, v/v) as carrier solvent, a very accurate mass determination was obtained; 51,987.0 ± 5.4 Daltons against a calculated value of 51,985.9 Daltons (**Fig. 2**). A fragment of IICB^{Glc} (Met1/Asp392), generated by cleavage with proteinase K, was also successfully analyzed applying the same conditions as described above. The measured mass was 41,946.0 ± 3.8 Daltons against a calculated value of 41,945.6 Daltons (spectra not shown).

3.5. LC/MS

3.5.1. The Mannose Transporter Complex from E. coli

The three subunits IIAB^{Man} (one-third of these molecules contain an N-terminal Met residue according to Edman degradation), IIC^{Man}, and IID^{Man} (containing an additional octapeptide Met-Asn-(His)$_6$ at the N-terminus to facilitate efficient isolation by affinity chromatography) were obtained following recombinant expression.

1. Separate the mannose transporter complex and remove the detergent by RP-HPLC prior to mass spectrometric analysis. Use a step-gradient HPLC solvent system from 0.1% TFA in water to neat formic acid (details as in **Subheading 3.1.1., step 2**, solvent system **a**).

The most accurate results were obtained using the detergents DDM (uncharged) and DTAC (positively charged). The two subunits IIAB^{Man} and IID^{Man} could be separated by RP-HPLC (**Fig. 3A** and **B**) and identified by mass spectrometry (**Fig. 3C** and **D**). The main masses determined for IIAB^{Man} and IID^{Man} are 34,933.9 ± 11.1 Daltons (**Fig. 3C**) and 31,950.6 ± 15.8 Daltons (**Fig. 3D**), respectively, for the DDM-containing sample and 34,920.7 ± 6.7 Daltons and 31,909.4 ± 18.2 Daltons, respectively, for the DTAC-containing sample (*see* **Note 4**). The corresponding calculated values are 34,916.4 Daltons (for IIAB^{Man} without Met), 35,047.6 Daltons with Met, and 31,892.5 Daltons

Hydrophobic Proteins and Peptides

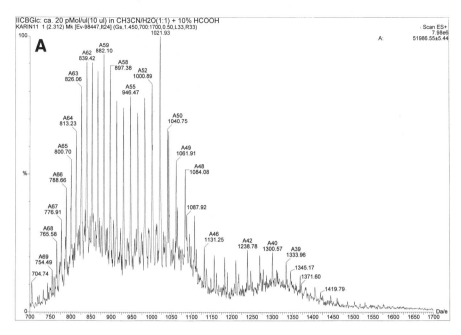

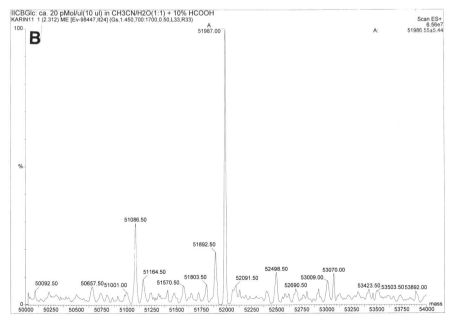

Fig. 2. (**A**) Electrospray mass spectrum and (**B**) transformed mass spectrum of IICB^{Glc} (calculated mass: 51,985.9 Daltons). IICB^{Glc} was freshly prepared at a high concentration (≈ 200 pmol/µL) containing DDM and diluted to the appropriate concentration immediately prior to mass spectrometric analysis.

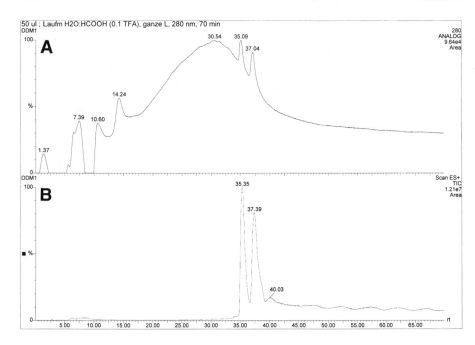

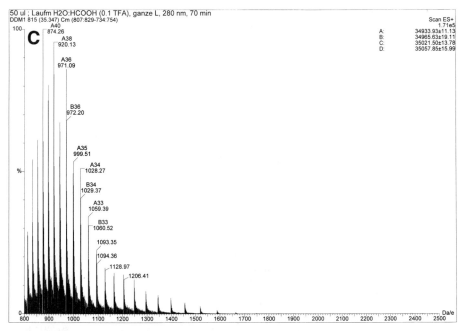

Fig. 3. *(continued on opposite page)* RP-HPLC separtion of the mannose transporter complex consisting of the three subunits IIAB^{Man}, IIC^{Man}, and IID^{Man}. **(A)** UV absorption at 280 nm and **(B)** the total-ion current from all masses in the range *m/z* 600

Hydrophobic Proteins and Peptides 433

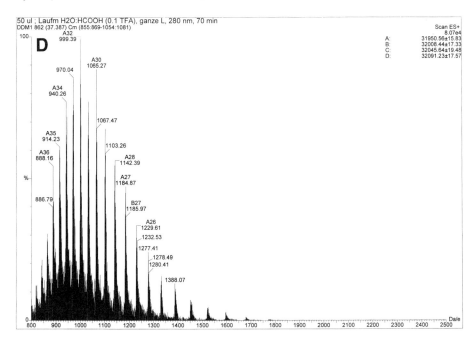

to 2200. The subunits were separated on a Lichrosorb RP-8 column (Lichrocart, 4.0 × 25 mm × 7 µm) using a linear gradient from 0 to 100% B in 30 min at a flow rate of 50 µL/min. Solvent A is 0.1% TFA in water; solvent B is neat formic acid. (**C** and **D**) Electrospray mass spectra of the subunits IIAB^{Man} and IID^{Man}, respectively, of the mannose transporter system (calculated masses: 34,916.4 Daltons (without Met) and 35,047.6 Daltons (with Met) for IIAB^{Man} and 31,892.5 Daltons for IID^{Man}). The spectra correspond to the two peaks in the total-ion current trace in **Fig. 3B**.

for IID^{Man}. Although the mass accuracy and the standard deviation are certainly not as good as in the case of water-soluble proteins, emphasis has to be put on the fact that this was the first time that such precise masses of IIAB^{Man} and IID^{Man} had been obtained experimentally. Unfortunately, it was not possible to determine an exact mass for the very hydrophobic subunit IIC^{Man} (calculated mass 27,636 Daltons), although it seems that IIC^{Man} is contained in the descending slope of the IID^{Man} peak.

3.5.2. The Outer Membrane Protein OMPA from E. coli

1. Further purify the OMPA, removing the salt and detergent (Cymal-6) prior to mass analysis, by RP-HPLC. Use a step-gradient HPLC solvent system from 0.1% TFA in water to neat formic acid (details as in **Subheading 3.1.1., step 2**, solvent system 1).

The OMPA protein could be separated by RP-HPLC from salt and detergent (**Fig. 4A** and **B**). The OMPA protein was contained in the main peak and iden-

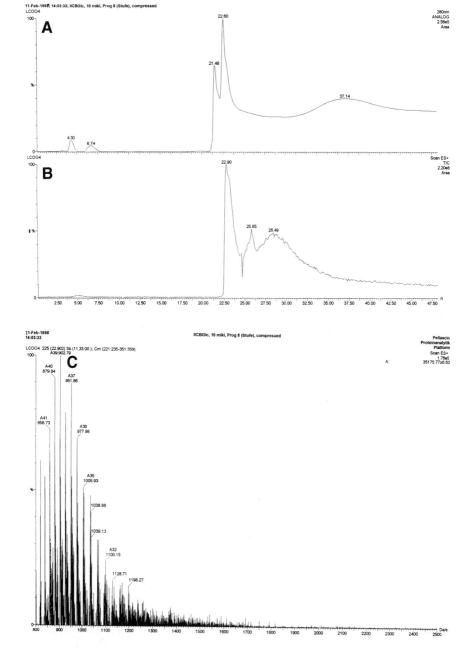

Fig. 4. RP-HPLC purification of the OMPA protein on a Lichrosorb RP-8 column (Lichrocart, 4.0 × 25 mm × 7 µm) using a step gradient at a flow rate of 50 µL/min. Solvent A is 0.1% TFA in water; solvent B is neat formic acid. (**A**) UV absorption at 280 nm and (**B**) the total-ion current from all masses in the range m/z 800 to 2500. (**C**) Electrospray mass spectrum of OMPA (calculated mass: 35,172.3 Daltons). The spectrum corresponds to the total-ion current trace in **Fig. 4B**.

tified by an accurate mass measurement (**Fig. 4C**). The experimentally determined main mass of 35,175.8 ± 6.6 Daltons is in good agreement with the theoretical value of 35,172.3 Daltons. The other peaks primarily contain the nonionic detergent (Cymal-6).

4. Notes

1. Molecular weights were calculated from the electrospray spectra using the MassLynx software incorporating the MaxEnt algorithm.
2. Lowering the ion source temperature from 100°C to 60°C (the recommended temperature for the instrument used) led to a reproducible, although not dramatic, loss of mass accuracy (26,780.0 ± 5.0 Daltons).
3. A high protein concentration for subsequent dilution and a nonionic detergent (0.01% DDM), with a low CMC are important prerequisites for a successful measurement by loop injection. It is essential to have access to a freshly prepared stock solution of IICB^{Glc}, and immediate analysis after dilution is recommended to obtain optimal results.
4. The chromatograms and the spectra of the DDM-and the DTAC-containing samples are alike. Therefore, only data for the DDM sample are shown in **Fig. 3**.

Acknowledgments

Without the generous supply of the various membrane proteins these measurements would not have been possible. Therefore, I would like to thank D. Kaesermann for the BO samples, F. Huber for the mannose transporter complex, and B. Erni and K. Flükiger for the IIBC^{Glc} component and the OMPA protein. Special thanks go to B.C. Pellascio for performing many of the measurements. All contributions were from the Department of Chemistry and Biochemistry, University of Bern.

References

1. Fenn, J. B., Mann, M., Meng, C. K., Wong, S. F., and Whitehouse, C. M. (1989) Electrospray ionization for mass spectrometry of large biomolecules. *Science* **246**, 64–71.
2. Karas, M. and Hillenkamp, F. (1988) Laser desorption/ionization of proteins with molecular masses exceeding 10,000 daltons. *Anal. Chem.* **60**, 2299–2301.
3. Burlingame, A. L., Boyd, R. K., and Gaskell, S. J. (1998) Mass spectrometry. *Anal. Chem.* **70**, 647R–716R.
4. Schey, K. L., Papac, D. I., Knapp, D. R., and Crouch, R. K. (1992) Matrix-assisted laser desorption mass spectrometry of rhodopsin and bacteriorhodopsin. *Biophys. J.* **63**, 1240–1243.
5. Rosinke, B., Strupat, K., Hillenkamp, F., Rosenbusch, J., Dencher, N., Kruger, U., et al. (1995) Matrix-assisted laser desorption/ionization mass spectrometry (MALDI-MS) of membrane proteins and non-covalent complexes. *J. Mass Spectrom.* **30**, 1462–1468.

6. Barnidge, D. R., Dratz, E. A., Sunner, J., and Jesaitis, A. J. (1997) Identification of transmembrane tryptic peptides of rhodopsin using matrix-assisted laser desorption/ionization time-of-flight mass spectrometry. *Protein Sci.* **6,** 816–824.
7. Allmaier, G., Schaffer, C., Messner, P., Rapp, U., and Mayer-Posner, F. J. (1995) Accurate determination of the molecular weight of the major surface layer protein isolated from *Chlostridium thermosaccharolyticum* by time-of-flight mass spectrometry. *J. Bacteriol.* **177,** 1402–1404.
8. Ghaim, J. B., Tsatsos, P. H., Katsonouri, A., Mitchell, D. M., Salcedo-Hernandez, R., and Gennis, R. B. (1997) Matrix-assisted laser desorption/ionization mass spectrometry of membrane proteins: demonstration of a simple method to determine subunit molecular weights of hydrophobic subunits. *Biochim. Biophys. Acta* **1330,** 113–120.
9. Marx, M. K., Mayer-Posner, F., Soulimane, T., and Buse, G. (1998) Matrix-assisted laser desorption/ionization mass spectrometry analysis and thiol-group determination of isoforms of bovine cytochrome c oxidase, a hydrophobic multi-subunit membrane protein. *Anal. Biochem.* **256,** 192–199.
10. Schey, K. L. (1996) Hydrophobic proteins and peptides analyzed by matrix-assisted laser desorption/ionization, in *Methods in Molecular Biology, vol. 61* (Chapman, J. R., ed.), Humana, Totowa, NJ, pp. 227–230.
11. Schnaible, V., Michels, J., Zeth, K., Freigang, J., Welte, W., Buhler, S., et al. (1997) Approaches to the characterization of membrane channel proteins (porins) by UV MALDI-MS. *Int. J. Mass Spectrom. Ion Processes* **169/170,** 165–177.
12. Schindler, P. A., Van Dorsselaer, A., and Falick, A. M. (1993) Analysis of hydrophobic proteins and peptides by electrospray ionization mass spectrometry. *Anal. Biochem.* **213,** 256–263.
13. Hufnagel, P., Schweiger, U., Eckerskorn, C., and Oesterhelt, D. (1996) Electrospray ionization mass spectrometry of genetically and chemically modified bacteriorhodopsin. *Anal. Biochem.* **243,** 46–54.
14. Schaller, J., Pellascio, B. C., and Schlunegger, U. P. (1997) Analysis of hydrophobic proteins and peptides by electrospray ionization mass spectrometry. *Rapid Commun. Mass Spectrom.* **11,** 418–426.
15. Ball, L. E., Oatis, J. E., Jr., Dharmasiri. K., Busman, M., Wang, J., Cowden L. B., et al. (1998) Mass spectrometric analysis of integral membrane proteins: application to complete mapping of bacteriorhodopsin and rhodopsin. *Protein Science* **7,** 758–764.
16. Whitelegge, J. P., Gundersen, C. B., and Faull, K. F. (1998) Electrospray-ionization mass spectrometry of intact intrinsic membrane proteins. *Protein Sci.* **7,** 1423–1430.
17. Fearnley, I. M. and Walker, J. E. (1996) Analysis of hydrophobic proteins and peptides by electrospray ionization MS. *Biochem. Soc. Trans.* **24,** 912–917.
18. Sharma, J., Panico, M., Barber, J., and Morris, H. R. (1997) Characterization of the low molecular weight photosystem II reaction center subunits and their light-induced modifications by mass spectrometry. *J. Biol. Chem.* **272,** 3935–3943.
19. Sharma, J., Panico, M., Barber, J., and Morris, H. R. (1997) Purification and determination of intact molecular mass by electrospray ionization mass spectrometry of the photosystem II reaction center subunits. *J. Biol. Chem.* **272,** 33,153–33,157.

20. Le Maire, M., Deschamps, S., Møller, J. V., Le Caer, J.-P., and Rossier, J. (1993) Electrospray ionization mass spectrometry on hydrophobic peptides electroeluted from sodium dodecyl sulfate—polyacrylamide gel electrophoresis application to the topology of the sarcoplasmic reticulum Ca^{2+} ATPase. *Anal. Biochem.* **214,** 50–57.
21. Ogorzalek Loo, R. R., Dales, N., and Andrews, P. C. (1994) Surfactant effects on protein structure examined by electrospray ionization mass spectrometry. *Protein Sci.* **3,** 1975–1983.
22. Erni, B., Zanolari, B., and Kocher, H. P. (1987) The mannose permease of *E. coli* consists of three different proteins. Amino acid sequence and function in sugar transport, sugar phosphorylation and penetration of phage (DNA. *J. Biol. Chem.* **262,** 5238–5247.
23. Wuethrich, M. and Sigrist, H. (1990) Peptide building blocks from bacteriorhodopsin: isolation and physicochemical characterization of two individual transmembrane segments. *J. Protein Chem.* **9,** 201–207.
24. Buhr, A., Flükiger, K., and Erni, B. (1994) The glucose transporter of *E. coli.* Overexpression, purification and characterization of functional domains. *J. Biol. Chem.* **269,** 23,437–23,443.
25. Garavito, R. M., Hinz, U., and Neuhaus, J. M. (1984) The crystallization of outer membrane proteins from *E. coli.* Studies on *lamB* and *ompA* gene products. *J. Biol. Chem.* **259,** 4254–4257.

25

Analysis of Proteins and Peptides Directly from Biological Fluids by Immunoprecipitation/Mass Spectrometry

Sacha N. Uljon, Louis Mazzarelli, Brian T. Chait, and Rong Wang

1. Introduction

We have developed a simple and sensitive method for characterizing amyloid β-peptide (Aβ) and its variants in biological media and clinical specimens in Alzheimer's disease research (1). This method, termed immunoprecipitation/mass spectrometry (IP/MS), utilizes anti-Aβ-specific monoclonal antibodies, 4G8 and 6E10 (2,3), to capture Aβ and its related peptides directly from complex biological fluids. The molecular masses of these captured Aβ peptides are measured by matrix-assisted laser desorption/ionization time-of-flight mass spectrometry (MALDI-TOFMS) (4) and used to determine the identity of these peptides. The signal intensity of a given Aβ peptide relative to that of an internal standard is used, together with its standard quantitation curve, to determine the concentration of the peptide.

This method is sensitive and has high-specificity resulting from the combination of high quality antibodies and mass spectrometry. Immunoprecipitation allows for the purification of target proteins or peptides without complex and costly column purification. Low concentration proteins or peptides can be selectively concentrated by the selected antibodies during the immunoprecipitation and then directly analyzed by MALDI-TOFMS. MALDI-TOFMS was selected in preference to other mass spectrometry techniques because of its simplicity and ease of interpretation of the resulting mass spectra as well as the high tolerance of MALDI to impurities and nonpeptide or nonprotein contaminants. The accurate molecular mass measurement and the high mass resolution make mass spectrometry an excellent means for studying protein and peptide mixtures. We have applied IP/MS to the characterization of Aβ peptides from a variety of biological media in studies of Alzheimer's disease *(5–9)*. The present methodology is one example of many methods that utilize affinity techniques together with mass spectrometry for the analysis of peptides and proteins *(10–23)*.

The procedure described here was developed for the analysis of Aβ peptides from biological media by IP/MS. The described procedure can be made suitable for the analysis of other specific proteins and peptides through a judicious choice of appropriate specific antibodies. Some important lessons that we have learned from our experiments will be discussed.

2. Materials

2.1. Immunoprecipitation Reagents

1. Monoclonal antibodies 4G8 (Seneteck PLC Senescence Technology, Napa, CA) (*see* **Note 1**).
2. Protein G-Plus/Protein A agarose beads (Oncogene Science, Uniondale, NY) (*see* **Note 2**) or
3. M-280 sheep anti-mouse IgG Dynabeads (Dynal, Lake Success, NY) (*see* **Note 3**).

2.2. Protease Inhibitors

See **Note 4**.

1. Ethylenediaminetetraacetic acid, disodium salt (EDTA).
2. Leupeptin.
3. Pepstatin A.
4. Phenylmethylsulfonyl fluoride (PMSF).
5. N-α-p-tosyl-L-lysine chloromethyl ketone (TLCK).
6. N-tosyl-L-phenylalanine chloromethyl ketone (TPCK).

2.3. Chemicals

1. α-Cyano-4-hydroxycinnamic acid (4HCCA), (Aldrich, Milwaukee, WI) (*see* **Note 5**).
2. Sodium chloride (NaCl) (Fisher Scientific, Springfield, NJ).
3. Sodium azide (NaAz) (Sigma, St. Louis, MO).
4. Trifluroacetic acid (sequencing grade, Pierce, Rockford, IL).
5. Trizma hydrochloride (Tri-HCl) (Sigma).

2.4. Detergents

See **Note 6**.

1. N-Octylglucoside (NOG) (Boehringer Mannheim, Indianapolis, IN).
2. CHAPS (Boehringer Mannheim).

2.5. Proteins and Peptides

1. Bovine insulin (Sigma) (see **Note 7**).
2. Bovine serum albumin (BSA; Sigma).
3. Amyloid peptide fragment 12–28 (Aβ12–28) (Sigma) (see **Note 8**).

2.6. Solvents

See **Note 9**.

1. Acetonitrile (Pierce).
2. Ethanol (Fisher Scientific).
3. Formic acid (Fisher Scientific).
4. Isopropanol (Mallinckrodt, Paris, KY).
5. Methanol (Fisher Scientific).
6. Water (distilled and deionized, prepared in house with a Milli-Q^{UV} Plus ultrapure water system (Millipore).

2.7. Laboratory Equipment

1. Eppendorf model 5415C centrifuge (Brinkmann, Westbury, NY).
2. Eppendorf model 5417R centrifuge (Brinkmann).
3. Bench top vortex (VWR Scientific, West Chester, PA).
4. Labquake shaker (Barnstead/Thermolyne, Dubuque, IO).
5. Dynal magnetic particle concentrator (Dynal MPC®-M) (Dynal, Lake Success, NY).
6. Laboratory vacuum system with liquid collection trap.
7. Mass spectrometer (MALDI-TOF; PE PerSeptive Biosystems).

3. Methods

3.1. Reagent Preparation

3.1.1. Stock Solutions

1. For TBS (10×): dilute NaCl (43.8 g), Tris-HCl (7.9 g), and NaAz (0.13 g) in water (450 mL). Titrate to pH 7.5 and add water to 500 mL. The final working buffer concentration is 150 mM NaCl and 10 mM Tris-HCl (pH 7.5).
2. For dilution buffer (10×): dilute NaCl (40.9 g), Tris-HCl (7.9 g), NOG (5.0 g), and NaAz (0.13 g) in water (450 mL). Titrate to pH 8.0 and add water to 500 mL. The final working buffer concentration is 140 mM NaCl, 0.1% NOG and 10 mM Tris-HCl (pH 8).
3. For Tris-HCl buffer (10×): dilute Tris-HCl (7.9 g) and NaAz (0.13 g) in water (450 mL). Titrate to pH 8.0 and add water to 500 mL. The final working buffer concentration is 10 mM Tris-HCl (pH 8).

3.1.2. Protein Solutions

1. BSA solution (2%, w/v).
2. Insulin solution (500 μM) (internal mass calibrant solution; see **Note 7**).

Table 1
Properties of Some Commonly Used Protease Inhibitors and their Solution Preparation

Name	Class	Mol. wt.	Conc.	Stock conc. (Fold)	Solvent	Stability
EDTA-Na	Metallo	372.2	2.0 mM	186 mg/mL (250x)	Water, pH 8–9	+4°C (6 mo)
Leupeptin	Thiol	475.6	10.0 µM	1 mg/mL (200x)	Water	−20°C (6 mo)
Pepstatin A	Aspartyl	685.9	1.0 µM	0.35 mg/mL (500x)	Methanol	−20°C (1 mo)
PMSF	Serine	174.2	1.0 mM	8.71 mg/mL (50x)	Isopropanol	+25°C (9 mo)
TLCK	Serine	369.3	0.1 mM	7.4 mg/mL (200x)	0.1% TFA	+25°C pH < 6
TPCK	Serine (including chymo)	351.8	0.2 mM	14 mg/mL (200x)	Ethanol	

3.1.3. Protease Inhibitor Solutions

Refer to **Table 1** for the preparation of protease stock solution.

3.1.4. MALDI-Matrix Solution

1. Prepare matrix solvent by mixing formic acid, water, and isopropanol in the ratio of 1:4:4 (v/v/v).
2. Add 5 mg of 4HCCA into an Eppendorf tube.
3. Add 200 µL of matrix solvent and thoroughly vortex for ≈ 2 min to form a milky suspension.
4. Centrifuge at 16,000g for 2 min.
5. Transfer the upper saturated matrix solution into a separate Eppendorf tube and keep the tube in a dark place at room temperature for later use.

3.1.5. Extraction Solution

See **Note 10**.

1. Aliquot 100 µL of MALDI-matrix solution.
2. Add 1–2 µL of internal mass calibrant solution (*see* **Note 7**).

3.2. Precipitation Beads Preparation

3.2.1. Protein G-Plus/Protein A Agarose Beads

1. Suspend beads by gently vortexing (*see* **Note 11**).
2. Aliquot the required volume of beads to a 1.5-mL Eppendorf tube using a wide-mouthed pipet tip (normally 3 µL of bead suspension per sample).

3. Add an equal volume of 2% BSA solution.
4. Incubate at 4°C by rotating for 1 h.
5. Spin the Eppendorf tubes at 10,000g for 1 min to bring down the beads.
6. Aspirate supernatant carefully using a syringe needle (30 G1/2) with laboratory vacuum.
7. Add 500 µL of ice-cold TBS and gently vortex for 2 s.
8. Spin the Eppendorf tubes at 10,000g for 1 min to bring down the beads.
9. Repeat **steps 6–8**.
10. Suspend beads in TBS to give the same volume as the initial bead volume in **step 2**.

3.2.2. Dynabeads

1. Suspend beads by gently vortexing (*see* **Note 12**).
2. Aliquot chosen volume of beads to a 1.5-mL Eppendorf tube using a wide-mouthed pipet tip (normally 20 µL of bead suspension per sample).
3. Collect beads by placing the tube on a Dynal magnetic particle concentrator for 1 min and pipet out supernatant.
4. Remove the tube from the magnetic particle concentrator and suspend beads in 250–500 µL of TBS.
5. Aspirate supernatant carefully as in **step 3**.
6. Repeat **steps 4** and **5**.
7. Suspend beads with 200 µL of TBS (if more beads are prepared, the volume of TBS needs to be justified).
8. Add the desired amount of antibody (normally 1 µL antibody per sample).
9. Incubate at 4°C by rotating for 2 h.
10. Add one-tenth of volume of 2% BSA and continue incubation for 1 h.
11. Collect beads as in **step 3**.
12. Wash beads with TBS as in **steps 4** and **5**.
13. Suspend beads in TBS to give the same volume as the initial bead volume in **step 2**.

3.3. Immunoprecipitation

Unless otherwise specified, all steps are carried out on ice or at 4°C.

3.3.1. Precipitation with Protein G-Plus/Protein A Agarose Beads

1. Aliquot samples (0.5–1.0 mL) into 1.5-mL Eppendorf tubes (*see* **Note 13**).
2. Add protease inhibitor solutions.
3. Add internal standard (*see* **Note 14**).
4. Add CHAPS (1%, w/v) (*see* **Note 6**).
5. Add 1.0 µL antibody.
6. Add 3.0 µL of Protein G-Plus/Protein A agarose beads (50% suspended slurry) with a wide-mouthed pipet tip.
7. Place sample tubes on a rotator and incubate at 4°C for 4–12 h (*see* **Note 15**).

3.3.2. Precipitation with Dynabeads

1. Aliquot samples (0.5–1.0 mL) into 1.5-mL Eppendorf tubes (*see* **Note 13**).
2. Add protease inhibitor solutions.

3. Add internal standard (*see* **Note 14**).
4. Add 10.0 µL of dynabeads with a wide-mouthed pipet tip.
5. Place sample tubes on a rotator and incubate at 4°C for 4–12 h (*see* **Note 15**).

3.4. Washing

3.4.1. Agarose Beads as Immunoprecipitation Reagent

1. Spin the Eppendorf tubes at 10,000g for 1 min to bring down the beads.
2. Aspirate supernatant carefully with a syringe needle (30 G1/2) using laboratory vacuum.
3. Add 500 µL of ice-cold dilution buffer and gently vortex for 2 s.
4. Spin the Eppendorf tubes at 10,000g for 1 min.
5. Aspirate the washing buffer carefully as in **step 2**.
6. Repeat **steps 3–5**.
7. Add 500 µL of ice-cold Tris buffer and gently vortex for 2 s.
8. Spin and aspirate the washing buffer as in **steps 4** and **5**.
9. Add 500 µL of ice-cold distilled deionized water and gently vortex for 2 s.
10. Spin and aspirate the washing buffer as **steps 4** and **5**.
11. Spin the Eppendorf tubes one last time and aspirate any remaining liquid very carefully (*see* **Note 16**).

3.4.2. Dynabeads as Immunoprecipitation Reagent

1. Place the Eppendorf tubes on a Dynal magnetic particle concentrator for 1 min (or wait for all of the beads to be attracted to the magnet).
2. Aspirate supernatant carefully with a syringe needle (30 G1/2) using laboratory vacuum.
3. Add 500 µL of ice-cold dilution buffer and gently vortex for 2 s.
4. Place the Eppendorf tubes on a Dynal magnetic particle concentrator for 1 min as in **step 1**.
5. Aspirate the washing buffer carefully as in **step 2**.
6. Repeat **steps 3–5**.
7. Add 500 µL of ice-cold Tris buffer and gently vortex for 2 s.
8. Collect the beads and aspirate the washing buffer as in **steps 4** and **5**.
9. Add 500 µL of ice-cold distilled deionized water and gently vortex for 2 s.
10. Collect the beads and aspirate the washing buffer as in **steps 4** and **5** (*see* **Note 16**).

3.5. Sample Preparation for Mass Spectrometric Analysis

3.5.1. Preparation of Matrix Thin Layer Base

See **Note 17**.

1. Apply 20 µL of saturated 4HCCA-matrix solution onto a sample plate (or probe) with a gel-loading tip and smear matrix solution evenly over the sample loading area.

2. Wait until the matrix solution dries at room temperature and then continue drying the sample plate under vacuum (for example, in a Speed-Vacuum Dryer or mass spectrometer vacuum) for ≈ 10 min.
3. Wipe the surface with a dry Kimwipe to remove excess matrix, leaving a very thin layer of matrix (not visible to the naked eye).

3.5.2. MALDI Sample Preparation

1. Add 3.0 µL of extraction solution, containing internal standard mass calibrant, to the beads in the Eppendorf tube and close the cover of the tubes (*see* **Note 18**).
2. Incubate at room temperature for 1 min.
3. Remove the extraction solution from the beads using a fine-bore Gel-loading tip.
4. Spot the extraction solution onto the matrix thin layer and dry at room temperature (*see* **Note 19**).

3.6. Mass Spectrometric Analysis

Follow the instructions in the operation manual provided by the mass spectrometer manufacturer. The TOF mass spectrometer is operated in the linear mode with delayed extraction (*see* **Note 20**).

3.7. Quantitation

Quantitative measurement of proteins or peptides by IP/MS is done by using internal standard (*see* **Note 8**) and standard quantitation curves.

3.7.1. Standard Quantitation Curve

1. Prepare a series of concentrations of proteins or peptides of interest using appropriate media (*see* **Note 21**).
2. Add internal standard(s) to the different concentration solutions to give a fixed concentration of internal standard (*see* **Note 22**).
3. Conduct the immunoprecipitation and the mass spectrometric analysis as described above.
4. Measure the peak heights from the proteins or peptides ($H_{Protein}$) and from the internal standard ($H_{Standard}$) using the same mass spectrum.
5. Calculate the normalized relative peak intensities ($I_{Protein}$) as $I_{Protein} = H_{Protein}/H_{Standard}$.
6. Plot the standard quantitation curves using these normalized relative peak intensities against the known concentrations of the proteins or peptides.

3.7.2. Quantitative Measurement

1. Add internal quantitation standard as in the preparation of the standard curve
2. Immunoprecipitation (*see* **Subheadings 3.2.–3.4.**).
3. Mass spectrometric analysis (*see* **Subheadings 3.5.** and **3.6.**).
4. Measure the peak heights and calculate the relative peak intensities ($I'_{Protein}$) as described in **Subheading 3.7.1**.
5. Determine concentration by comparing ($I'_{Protein}$) with the standard quantitation curve.

4. Notes

1. Monoclonal anti-Aβ specific antibodies 4G8 (IgG$_{2b}$ mouse, 2.5 mg/mL, anti-Aβ [17–24]) and 6E10 (IgG$_1$ mouse, 3.3 mg/mL, anti-Aβ[5–11]) were used in our Aβ peptide analyses *(2,3)*. Both monoclonal and polyclonal antibodies can be used for immunoprecipitation. It is important to note the immunoglobulin isotype of the antibody because different immunoglobulin isotypes require the use of different immunoprecipitation reagents to collect the immunocomplex. In some cases, it is necessary to use a secondary antibody to increase the precipitation efficiency. Attention should be paid to the concentration of the antibody, which will determine the total binding capacity of the IP/MS assay.
2. Protein G-Plus/Protein A agarose beads were used together with antibodies 4G8 and 6E10 in our Aβ peptide analyses and are given here as an example. Protein A and Protein G are cell surface bacterial proteins that bind to the Fc region of IgG molecules. Both Protein A and Protein G can bind to immunoglobulins from many species. Protein A agarose is recommended for use in immunoprecipitation using affinity-purified rabbit polyclonal antibodies or unmodified mouse monoclonal antibodies. The binding of immunoglobulin to protein A is species and IgG subclass dependent. Protein A binds poorly to mouse IgG$_1$ or IgG$_3$ and to all rat antibodies. Protein G has a wider spectrum of binding to immunoglobulins from different species and subclasses than does protein A agarose. Protein G also has multiple immunoglobulin binding sites for higher binding capacity of mouse and rat IgG. Protein G-Plus is produced from recombinant protein G, which lacks the serum albumin binding sites as well as the normal protease cleavage sites. Therefore, protein G-Plus agarose is more protease resistant and does not bind serum albumin. The binding specificities of Protein A, Protein G, Protein G-Plus and Protein G-Plus/Protein A agarose beads are listed in **Tables 2** and **3**. We have also used other precipitation reagents (for example, magnetic beads, *see* **Note 3**) for the IP/MS assay. Immobilized antibodies can be directly used in the IP/MS protein assay. It is preferable to affinity-purify the antibodies prior to immobilization *(24)*.
3. Dynabeads are uniform superparamagnetic polystyrene spheres *(25)*. They consist of magnetic material (Fe_2O_3) and are coated with a thin polystyrene shell. Many Dynabead products can be used for immunoprecipitation of protein and peptides. For example, M-280 sheep anti-mouse IgG (not recommended for use with mouse IgG$_3$ subclass antibodies) and M-280 sheep anti-rabbit IgG beads have been used successfully in our laboratory.
4. It is necessary to add protease inhibitors to the biological sample solutions to prevent the proteins from being degraded. The protease inhibitors listed here and in **Table 1** are commonly used in our laboratory. Although some of the protease inhibitor solutions are stable for a long period, it is strongly recommended to prepare them freshly for each application. All the protease inhibitors were purchased from Sigma.
5. It may be necessary to purify 4HCCA before use because impurities in some commercial 4HCCA preparations will seriously reduce the sensitivity of the mass spectrometric measurement. The following recrystallization procedure is recommended:

Table 2
Binding Specificity of Protein A, Protein G, Protein G-Plus and Protein G-Plus/Protein A to IgG Species (29)[a]

Species	Protein A	Protein G	Protein G-Plus	Protein G-Plus/ Protein A
Cat	+	–	–	+
Chicken	–	–	–	–
Cow	–	+	+	+
Dog	±	–	–	±
Goat	±	+	+	+
Guinea pig	+	+	+	+
Horse	–	+	+	+
Human	+	+	+	+
Pig	+	+	+	+
Rabbit	+	+	+	+
Rat	–	±	+	+
Sheep	–	+	+	+

[a] +, strong binding; ±, weak binding; –, no binding.

Table 3
Binding Specificity of Protein A, Protein G, Protein G-Plus and Protein G-Plus/Protein A to Subtype of Mouse and Human Immunoglobulins (29)[a]

Subclass (Mouse)	Protein A	Protein G	Protein G-Plus	Protein G-Plus/A	Subclass (Human)	Protein A	Protein G	Protein G-Plus	Protein G-Plus/A
IgA	–	–	–	–	IgA	±	–	–	±
IgD	–	–	–	–	IgD	–	–	–	–
IgE	–	–	–	–	IgE	–	–	–	–
IgG$_1$	±	+	+	+	IgG$_1$	+	+	+	+
IgG$_{2a}$	+	±	±	+	IgG$_2$	+	+	+	+
IgG$_{2b}$	+	±	±	+	IgG$_3$	–	+	+	+
IgG$_3$	–	+	+	+	IgG$_4$	±	±	±	±
IgM	–	–	–	–	IgM	+	–	–	+
Serum Albumin	–	–	–	–	Serum Albumin	±	–	–	±

[a] +, strong binding; ±, weak binding; –, no binding.

  a. Dissolve 4HCCA (≈ 0.5 g) in 10 mL of 5% ammonium hydroxide (NH$_4$OH).
  b. Centrifuge at 5,000g for 10 min and transfer the supernatant to a new 15-mL conical centrifuge tube.

c. Titrate the 4HCCA solution with 5 M HCl to pH 3–4. (To prevent the precipitation of impurities, do not reduce the pH below 3).
d. Centrifuge at 5,000g for 10 min and aspirate the supernatant with care.
e. Wash the 4HCCA pellet 3 times with acidified water (pH 3–4) by vortexing; centrifuge and aspirate the supernatant.
f. Wash the 4HCCA pellet 3 times with deionized water as in **step 5e**.
g. Dissolve the 4HCCA pellet in 5 mL of acetonitrile (HPLC grade).
h. Centrifuge at 5,000g for 10 min.
i. Aliquot the 4HCCA solution into a 1.5-mL Eppendorf centrifuge tube.
j. Dry the 4HCCA solution in the tube using a Speed-Vac concentrator.

6. Detergents were used to reduce nonspecific binding in immunoprecipitation. CHAPS (1%) was added before immunoprecipitation for cell lysates and human tissue specimens. NOG (0.1%) was used in the washing buffer (26).

7. Although bovine insulin was used as a mass calibrant in our Aβ assay, the mass calibrant should be selected according to the molecular masses of the proteins or peptides under study. For the best accuracy, the masses to be determined should be bracketed between the calibrant masses (either two different mass calibrants or one mass calibrant having both singly and doubly charged ions).

8. Synthetic Aβ12–28 peptide was used as an internal standard for quantitation. Internal standards should be selected to have the following properties: (1) they do not occur naturally; (2) they have different molecular masses from the proteins or peptides to be analyzed; and (3) they contain the epitope that is recognized by the antibody used in the immunoprecipitation. For quantitative analysis, the best internal standard is the stable isotope-labeled protein/peptide of interest.

9. Solvents should be HPLC grade or the highest grade possible.

10. If there is a desire to analyze the immunoprecipitated proteins or peptides directly by methods other than mass spectrometry, do not add the matrix to the elution solvent mixtures of formic acid + water + isopropanol (1:4:4, v/v/v) or 0.1% TFA in water + acetonitrile (1:1, v/v). Matrix can be added later to a portion of this eluent for subsequent mass spectrometric analysis. However, it is very important to notice that the use of different solvents will result in different relative peak intensities for proteins or peptides (27). For quantitative measurements, it is important to keep the solvent composition constant. Bovine insulin (0.2 μM) was used as the internal mass calibrant in our Aβ peptide analyses. The 4HCCA matrix-containing extraction solution should be prepared no more than 15 min prior to elution and should be kept at room temperature to prevent possible precipitation of the saturated matrix solution. Low temperature will cause the matrix to precipitate.

11. Before transferring a chosen amount of beads, they must be thoroughly suspended to give a homogeneous suspension.

12. Dynabeads M-280 secondary coated products are supplied at 6.7 × 10^8 beads/mL in PBS with 0.1% BSA and 0.02% sodium azide (pH 7.4). Beads should be stored at 4–8°C. Dynabeads should be washed before use. Before transferring the desired amount of beads, they must be thoroughly suspended to give a homogeneous suspension.

13. Immunoprecipitation can be carried out using a variety of sample volumes. We have successfully performed IP/MS assays from a volume as large as 10 mL of cell culture medium. When using these larger volumes, the amount of antibody and immunoprecipitation reagents (beads) should be increased. For example, immunoprecipitations of Aβ peptide from 10 mL of cell culture media were conducted by adding 2.0 µL of antibody 4G8 and 10.0 µL of Protein G-Plus/Protein A agarose beads. It is necessary to increase the amount of antibody because immunoprecipitation is a reversible binding process, and the binding equilibrium depends on the concentrations of both antibody and antigen *(28)*.
14. For immunoprecipitation of Aβ peptide from 1.0 mL media, use 10.0 µL of Aβ12–28 peptide solution (2.0 µM, freshly prepared from a 100-µM stock solution diluted with dilution buffer). This results in a final Aβ12–28 concentration of 20 nM.
15. Longer incubation times may increase the sensitivity but may also increase nonspecific binding (resulting in a high background). In addition, the probability of oxidation and peptidase hydrolysis reactions increases as a function of incubation time.
16. It is extremely important to remove any excess water from the beads at this stage. If there is aqueous solution left, it will dilute the organic portion in the extraction solution. This dilution will seriously reduce the elution efficiency and cause precipitation of the MALDI matrix.
17. The MALDI-MS sample preparation procedure described here is the thin polycrystalline film (or thin-layer) method *(4)*, which produces a more homogenous sample spot than the more widely used dried droplet method. The dried droplet method can also be used for the present application. For general principles of successful MALDI-MS sample preparation, it is strongly recommended to read the chapter "Matrix-Assisted Laser Desorption Ionization Mass-Spectrometry of Proteins" in *Methods in Enzymology,* vol. 270 *(4)*.
18. The extraction solution must be a saturated 4HCCA-matrix solution. Otherwise, the matrix thin layer will dissolve. The extraction solution must also be free of matrix precipitate when it is added to the immunoprecipitation beads. Small, undissolved matrix particles will induce further matrix molecule crystallization in the Eppendorf tube, which will pull down protein/peptide molecules, resulting in serious sample loss.
19. If it is proves difficult to obtain a useful mass spectrum, the sample probe can be washed with ice-cold 0.1% aqueous TFA to remove salts and other water-soluble contaminants. Withdraw the sample plate from the mass spectrometer and wash by placing ≈ 2–3 µL of 0.1% TFA on top of the sample spot. Wait for a few seconds and then remove the liquid by vacuum suction *(4)*.
20. Occasionally, precipitation beads will be transferred onto the MALDI-MS sample plate, which produces a nonhomogeneous sample and consequently results in peak broadening. The use of delayed-extraction in the TOF mass spectrometer can minimize this peak broadening. We find that the linear mode TOF mass spectrometer provides higher sensitivity and less mass discrimination than does the reflectron mode. We always use the linear mode because sensitivity is of paramount importance in most of our studies.

21. In our Aβ peptide assay, concentrations of 0.05, 0.1, 1.0, 10.0, and 100 nM of Aβ(1–40) and Aβ(1–42) were prepared in the same media used for cell culture or in TBS buffer for human (or experimental animal) specimens.
22. In our Aβ peptide assay, Aβ(12–28), 20.0 nM, was added to each of the above standard solutions (*see* **Note 21**) prior to immunoprecipitation.

Acknowledgments

The work was supported by Alzheimer's Association grants of PRG 94-131, Mrs. Florence Martin Pilot Research grant, RG1-96-070 to R.W., and NIH RR00862 to B.T.C. The authors are grateful to Drs. Yingming Zhao and Martine Cadene for useful discussions.

References

1. Wang, R., Sweeney, D., Gandy, S., Sisodia, S. S., and Gandy, S. E. (1996) The profile of soluble amyloid β protein in cultured cell media: detection and quantification of amyloid β protein and variants by immunoprecipitation-mass spectrometry. *J. Biol. Chem.* **271**, 31,894–31,902.
2. Kim, K. S., Miller, D. L., Sapienza, V. J., Chen, C.-M. J., Bai, C., Iqbal, I. G., et al. (1988) Production and characterization of monoclonal antibodies reactive to synthetic cerebrovascular amyloid peptide. *Neurosci. Res. Commun.* **2**, 121–130.
3. Kim, K. S., Wen, G. Y., Bancher, C., Chen, C.-M. J., Sapienza, V. J., Hong, H. et al. (1990) Detection and quantitation of amyloid β-peptide with 2 monoclonal antibodies. *Neurosci. Res. Commun.* **7**, 113–122.
4. Beavis, R. C. and Chait, B. T. (1996) Matrix-assisted laser desorption ionization mass spectrometry of proteins, in *Methods in Enzymology*, (Academic, San Diego, CA, pp. 519–551.
5. Xu, H., Sweeney, D., Wang, R., Thinakaran, G., Lo, A. C., Sisodia, S. S., et al. (1997) Generation of Alzheimer beta-amyloid protein in the trans-Golgi network in the apparent absence of vesicle formation. *Proc. Natl. Acad. Sci. USA* **94**, 3748–3752.
6. Xu, H. X., Gouras, G. K., Greenfield, J. P., Vincent, B., Naslund, J., Mazzarelli, L., et al. (1998) Estrogen reduces neuronal generation of Alzheimer beta-amyloid peptides. *Nature Med.* **4**, 447–451.
7. Sudoh, S., Kawamura, Y., Sato, C., Wang, R., Saido, T. C., Oyama, F., et al. (1998) Presenilin 1 mutations linked to familial Alzheimer's disease increase the intracellular levels of amyloid-protein 1-42 and its N-terminally truncated variant(s) which are generated at distinct sites. *J. Neurochem.* **71**, 1535–1543.
8. Gouras, G. K., Xu, H., Jovanovic, J. N., Buxbaum, J. D., Wang, R., Greengard, P., et al. (1998) Generation and regulation of β-amyloid peptide variants in neurons. *J. Neurochem.* **71**, 1920–1925.
9. McGowan, E., Sanders, S., Iwatsubo, T., Takeuchi, A., Saido, T., Zehr, C., Yu, X., et al. (1998) Amyloid phenotype characterization of transgenic mice over-expressing both mutant Amyloid precursor protein and mutant presenilin 1 transgenes. *Neurobiol. Dis.* **6**, 231–244.

10. Hutchens, T. W. and Yip, T. T. (1993) New desorption strategies for the mass spectrometric analysis of macromolecules. *Rapid Commun. Mass Spectrom.* **7,** 576–580.
11. Papac, D. I., Hoyes, J., and Tomer, K. B. (1994) Direct analysis of affinity-bound analytes by MALDI/TOF MS. *Anal. Chem.* **66,** 2609–2613.
12. Nelson, R. W., Krone, J. R., Bieber, A. L., and Williams, P. (1995) Mass spectrometric immunoassay. *Anal. Chem.* **67,** 1153–1158.
13. Zhao, Y., Muir, T. W., Kent, S. B., Tischer, E., Scardina, J. M., and Chait, B. T. (1996) Mapping protein-protein interactions by affinity-directed mass spectrometry. *Proc. Natl. Acad. Sci. USA* **93,** 4020–4024.
14. Hsieh, Y. L., Wang, H., Elicone, C., Mark, J., Martin, S. A., and Regnier, F. (1996) Automated analytical system for the examination of protein primary structure. *Anal. Chem.* **68,** 455–462.
15. Harmon, B. J., Gu, X., and Wang, D. I. (1996) Rapid monitoring of site-specific glycosylation microheterogeneity of recombinant human interferon-γ. *Anal. Chem.* **68,** 1465–1473.
16. Tomlinson, A. J., Jameson, S., and Naylor, S. (1996) Strategy for isolating and sequencing biologically derived MHC class I peptides. *J. Chromatogr. A.* **744,** 273–278.
17. Brockman, A. H. and Orlando, R. (1996) New immobilization chemistry for probe affinity mass spectrometry. *Rapid Commun. Mass Spectrom.* **10,** 1688–1692.
18. Krone, J. R., Nelson, R. W., Dogruel, D., Williams, P., and Granzow, R. (1997) BIA/MS: interfacing biomolecular interaction analysis with mass spectrometry. *Anal. Biochem.* **244,** 124–132.
19. Kaur, S., McGuire, L., Tang, D., Dollinger, G. and Huebner, V. (1997) Affinity selection and mass spectrometry-based strategies to identify lead compounds in combinatorial libraries. *J. Protein Chem.* **16,** 505–511.
20. Cai, J. and Henion, J. (1997) Quantitative multi-residue determination of β-agonists in bovine urine using on-line immunoaffinity extraction-coupled column packed capillary liquid chromatography-tandem mass spectrometry. *J. Chromatogr. B.* **691,** 357–370.
21. Liang, X., Lubman, D. M., Rossi, D. T., Nordblom, G. D., and Barksdale, C. M. (1998) On-probe immunoaffinity extraction by matrix-assisted laser desorption/ionization mass spectrometry. *Anal. Chem.* **70,** 498–503.
22. Volz, J., Bosch, F. U., Wunderlin, M., Schuhmacher, M., Melchers, K., Bensch, K., et al. (1998) Molecular characterization of metal-binding polypeptide domains by electrospray ionization mass spectrometry and metal chelate affinity chromatography. *J. Chromatogr. A.* **800,** 29–37.
23. Kuwata, H., Yip, T. T., Yip, C. L., Tomita, M., and Hutchens, T. W. (1998) Bactericidal domain of lactoferrin: detection, quantitation, and characterization of lactoferricin in serum by SELDI affinity mass spectrometry. *Biochem. Biophys. Res. Commun.* **245,** 764–773.
24. Hermanson, G. T., Mallia, A. K. and Smith, P. K. (1992) *Immobilized Affinity Ligand Techniques*. Academic, San Diego, CA.

25. *Biomagnetic Applications in Cellular Immunology* (1998) Dynal, Oslo, Norway.
26. Zhao, Y. and Chait, B. T. (1994) Protein epitope mapping by mass spectrometry. *Anal. Chem.* **66,** 3723–3726.
27. Cohen, S. L. and Chait, B. T. (1996) Influence of matrix solution conditions on the MALDI-MS analysis of peptides and proteins. *Anal. Chem.* **68,** 31–37.
28. Harlow, E. and Lane, D. (1988) *Antibodies: A Laboratory Manual.* Cold Spring Harbor Laboratory, Cold Spring Harbor, NY.
29. Discovery Tools for Biomedical Research (1992) Oncogene Science, Uniondale, NY.

26

Detection of Molecular Determinants in Complex Biological Systems Using MALDI-TOF Affinity Mass Spectrometry

Judy Van de Water and M. Eric Gershwin

1. Introduction

Attempts to characterize antigens present in low concentration or with small amounts of available tissue can be difficult using standard methodology. For example, the molecule in question may be too small to resolve on a standard sodium dodecyl sulfate polyacrylamide gel electrophoresis (SDS-PAGE) gel or may not be detectable by immunoblotting due to loss of antigenicity following gel electrophoresis. Clearly, this type of analysis would be aided by enabling new technologies or by the implementation of alternative strategies. Hutchens et al. *(1)* have previously outlined new strategies for the enhanced detection and structural characterization of biopolymers by laser desorption/ionization time-of-flight mass spectrometry, an approach that has been termed affinity mass spectrometry (AMS). This technique can be used with enzyme-linked signal amplification to determine the relative amount of a particular antigen present in a complex mixture. We have used AMS with enzyme-linked signal amplification to search for small amounts of autoantigens in a complex mixture of liver homogenate or in bile duct epithelial cell homogenate *(2)*. In addition, we have used standard immunoprecipitation of crude liver homogenates coupled with matrix-assisted laser desorption/ionization time-of-flight mass spectrometry (MALDI-TOFMS) to analyze diseased vs. control liver for both known and unknown molecules. Although the following methods are somewhat specific with regard to the particular disease, primary biliary cirrhosis (PBC), studied by our laboratory, this technology provides a level of sensitivity in the femtomole range using crude cell lysates that we believe will be of wide application in the study of a number of disease states.

2. Materials

2.1. Isolation and Purification of Bile Duct Epithelial Cells

1. Type 1A collagenase used at 1 mg/mL.
2. Phosphate-buffered saline (PBS), pH 7.0.
3. Lymphoprep or other Ficoll-Hypaque solution for lymphocyte separation.
4. Monoclonal antibody HEA125 specific for bile duct epithelial cells (Research Diagnostics, Minneapolis, MN).
5. Anti-mouse Ig magnetic beads.
6. Magnet for use with beads.

2.2. Detection of Marker Protein(s) by Affinity Mass Spectrometry with Enzyme-Linked Signal Amplification

1. 3 M urea, pH 7.0.
2. 0.1% Tween in PBS.
3. Anti-mouse Ig conjugated to alkaline phosphatase.
4. 50 mM ammonium citrate, pH 5.5.
5. 50 mM ammonium bicarbonate, pH 7.8.
6. Synthetic phosphopeptide (Gly-Leu-phosphoSer-Pro-Ala-Arg).
7. α-Cyano-4-OH-cinnamic acid.

2.3. Immobilization of Monoclonal Antibodies onto Agarose Beads

1. Agarose beads containing Protein A, Protein G or both, or anti-mouse Ig.
2. Dimethylpimelimidate (DMP). It should be noted that DMP is moisture sensitive and should be stored refrigerated and desiccated. Warm vials to room temperature prior to opening. Dissolve in crosslinking buffer immediately before use.
3. Crosslinking buffer: 0.2 M triethanolamine, pH 8.2.
4. Blocking buffer: 0.1 M ethanolamine, pH 8.2.
5. Elution buffer: 0.05 M glycine-HCl, 0.15 M NaCl, pH 2.3.

2.4. Immunoprecipitation

1. Homogenization buffer for homogenizing tissue: PBS containing 1% Triton X-100 and 1 mM phenylmethysulfonylfluoride, pH 7.4.
2. Unbound agarose beads.
3. 50 mM ammonium citrate.
4. Sinapinic acid.

3. Methods

For the methodology described below, the examination of purified bile duct epithelial cells (*see* **Subheading 3.2.**) and crude liver homogenate (*see* **Sub-**

heading **3.4.**) are used as examples of sample material. This method will, however, work for any cell or tissue homogenate. The techniques described are a compilation of several methods.

3.1. Isolation and Purification of Bile Duct Epithelial Cells

1. Mince approximately 30 g of liver and incubate with 1 mg/mL type 1A collagenase with agitation for 4 h at 37°C.
2. Wash the resulting mixture 4 times with PBS, resuspend the cells in 30 mL of PBS, layer onto 15 mL of Lymphoprep for density gradient separation, and centrifuge at 6000g for 25 min.
3. Harvest the cells at the differential density interface and wash 3 times with PBS.
4. Incubate the washed cells with 5 mg/mL of the epithelial cell-specific monoclonal antibody, HEA125 (Research Diagnostics), in a 100–200-μL volume for 30 min at 37°C.
5. Wash 3 times with PBS, add 1 mL of washed sheep anti-mouse IgG-BioMag beads (Advanced Magnetics, Cambridge, MA) in 7 mL of PBS to the cells, and incubate at 37°C for 30 min followed by magnetic separation (*3,4*).

3.2 Detection of Marker Protein(s) by Affinity Mass Spectrometry with Enzyme-Linked Signal Amplification

This method, first described by Yip et al. (*2*), provides information only regarding relative amounts of the material recognized by the primary antibody used, which is especially useful when looking at diseased vs. control samples. The use of MALDI-TOF does not, in this case, provide any information as to the identity or molecular weight of an unknown molecule, for which a standard MALDI-TOF analysis would be required. In the following, AMS, coupled with enzyme-linked signal amplification, is illustrated by an analysis performed on isolated bile duct epithelial cells (*see* **Subheading 3.1.** and **Fig. 1**).

1. Wash the purified BDE cells with PBS (pH 7.0).
2. Add a 5-μL aliquot of primary monoclonal antibody, at an appropriate dilution, and allow to incubate at 25°C in a moist chamber for 30–60 min. For the controls, the cells or tissue are incubated with PBS only.
3. Wash the cells with 3 M urea in PBS (pH 7.0) followed by 0.1% Tween 20 in PBS (pH 7.0).
4. Add a 5- μL aliquot of goat anti-mouse IgG antibodies conjugated to bovine intestinal alkaline phosphatase (Calbiochem, San Diego, CA), 1 mg/mL, diluted 5000-fold in PBS, pH 7.0) and allow to incubate at 25°C in a moist chamber for 30–60 min.
5. Wash the cells with 3 M urea in PBS, followed by 0.1% Tween 20 in PBS, and then with 50 mM ammonium citrate (pH 5.5), and finally with 50 mM ammonium bicarbonate (pH 7.8).

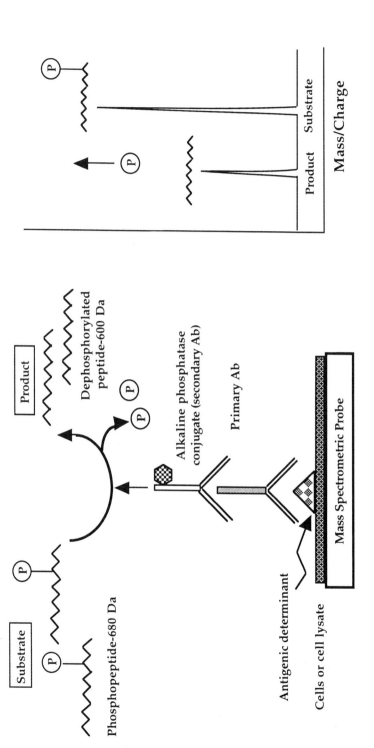

6. Add a 5-μL aliquot of synthetic phosphopeptide (Gly-Leu-phosphoSer-Pro-Ala-Arg) in 50 mM ammonium carbonate, 0.05 mM MgCl$_2$ (pH 9.5) and allow to incubate at 25°C in a moist chamber for at least 15 min.
7. Centrifuge briefly to remove the cells and mix a 0.5-μL aliquot of the supernatant with a 1-μL aliquot of α-cyano-4-OH-cinnamic acid (*see* **Note 1**) on the mass spectrometer probe and allow to air dry and crystallize.
8. Analyze the resulting sample preparation by MALDI-TOF (*see* **Note 2** and **Fig. 1**).

3.3. Immobilization of Antibodies on Agarose Beads

For the detection of known or unknown molecules, it is possible to perform immunoprecipitation using a mono- or polyclonal antibody that has been previously immobilized on a solid substrate such as agarose beads. As an example, mouse monclonal antibodies are used in the following protocol.

1. Incubate agarose beads containing Protein A and/or Protein G with the mouse mAbs in supernatant overnight at 4°C on a rocker platform. Add the antibody to the beads at the optimal concentration depending on the subclass of immunoglobulin and the affinity beads used.
2. Wash the beads in PBS 3 times to remove nonspecific bound protein.
3. Dissolve 13.2 mg of dimethylpimelimidate (DMP), just before use, in 2 mL of 0.2 M triethanolamine (pH 8.2) for crosslinking the antibody to the agarose beads (*see* **Note 3**). This amount is for a 2 mL volume of antibody-bound beads.
4. Incubate the DMP with the beads for 1 h at room temperature with gentle inversion.
5. Wash the beads with 5 mL of crosslinking buffer.
6. Following crosslinkage, block the remaining active sites by incubation with 2 mL of 0.1 M ethanolamine (pH 8.2) for 10 min with gentle inversion.
7. Wash the beads with 0.05 M glycine-HCl, 0.15 M NaCl (pH 2.3) to remove any unlinked antibody. Wash the beads with PBS prior to incubation with antigen.

Fig. 1. Schematic of detection of marker protein(s) by affinity mass spectrometry with enzyme-linked signal amplification. (**A**) The design of the laser desorption probe surface for the detection of antigenic determinants is illustrated where P is the phosphate group that is removed from the added substrate peptide (Gly-Leu-PhosphoSer-Pro-Ala-Arg; 680 Daltons). The presence of a specific marker analyte (antigen) is amplified by the coupled enzymatic conversion of substrate (680 Daltons) to dephosphorylated product (600 Daltons) directly *in situ*, i.e., on the probe surface (coupling is effected via the marker-specific monoclonal antibody). (**B**) This process is visualized as the mass spectrum in which the product peak (m/z 600) is the first peak and the substrate peak (background) is the second peak (m/z 680). The height of the product peak is directly correlated to the amount of product generated, which is directly proportional to the amount of analyte (antigen) present.

3.4. Immunoprecipitation

See **Note 4**.

A homogenate of cell suspensions is just one of the sources that can be used when performing immunoprecipitation analysis. The following is a reasonably standard method of immunoprecipitation, which is then followed by MALDI-TOF analysis.

1. Prepare the tissue by homogenization in PBS containing 1% Triton X-100 and 1 mM phenylmethylsulfonylfluoride (pH 7.4) at 4°C.
2. Incubate 100 µL of non-antibody-bound agarose beads with 500 µL of tissue homogenate for 1 h at 4°C with agitation to preabsorb nonspecific reactivity with these beads.
3. Incubate a 100-µL aliquot of preabsorbed homogenate with 50 µL of antibody-bound beads at 4°C for 18 h with agitation to perform antibody-specific immunoprecipitation.
4. Spin the beads in a microfuge for 5 min at 10,000g to remove the homogenate.
5. Wash the beads with 0.1% Tween 20 in PBS (pH 7.0), then with 50 mM ammonium citrate, and finally with water.
6. Add a 0.5-µL aliquot of the beads to the mass spectrometer probe element, followed by a 1-µL aliquot of sinapinic acid (*see* **Note 2**) and allow the mixture to air dry.
7. Analyze the resulting sample preparation by MALDI-TOF (*see* **Note 2**).

In this case, unlike **Subheading 3.2.**, MALDI-TOF provides, in a single spectrum, signals for all the proteins specifically captured by the chosen immobilized antibody. These captured proteins are detached by the laser used in MALDI-TOF so that the recorded spectrum shows peaks with m/z values corresponding to each of the captured proteins.

4. Notes

1. There are multiple choices where sample matrix for MALDI-TOF is concerned, including the two most common, sinapinic acid and α-cyano-4-OH-cinnamic acid. This chapter specifies those matrix materials used in our own studies.
2. Specifics regarding the actual mass spectrometric analysis of the sample have been omitted as mass spectrometers differ in their usage protocols, and settings vary with the individual molecules analyzed. Detection in the femtomole range should be possible with the above techniques.
3. The antibody is crosslinked to the agarose beads to prevent removal of the antibodies during antigen elution,
4. One problem that may occur when performing immunoprecipitation with crude tissue homogenates, which include particulate matter, is that this matter may become nonspecifically trapped within the beads. Therefore, it may be necessary to centrifuge the samples prior to immunoprecipitation. This, unfortunately, may also result in the loss of the molecule of interest, particularly if this is membrane

bound. In this case, it may be necessary to carry out fractionation of the various cell components or solubilization with different detergents to release antigens from the particulate material. Precleaning the homogenate with unbound Protein A or G beads will also help alleviate this problem.

References

1. Hutchens, T. W. and Yip, T. T. (1993) New desorption strategies for the mass spectrometric analysis of macromolecules. *Rapid Commun. Mass Spectrom.* **7,** 576–580.
2. Yip, T. T., Van de Water, J., Gershwin, M. E., Coppel, R. L., and Hutchens T. W. (1996) Cryptic antigenic determinants on the extracellular pyruvate dehydrogenase complex/mimeotope found in primary biliary cirrhosis. A probe by affinity mass spectrometry. *J. Biol. Chem.* **271,** 32,825–32,833.
3. Joplin, R. E., Johnson, G. D., Matthews, J. B., Hamburger, J., Lindsay, J. G., Hubscher, S. G., et al. (1994) Distribution of pyruvate dehydrogenase dihydrolipoamide acetyltransferase (PDC-E2) and another mitochondrial marker in salivary gland and biliary epithelium from patients with primary biliary cirrhosis. *Hepatology* **19,** 1375–1380.
4. Joplin, R., Strain, A. J., and Neuberger, J. M. (1989) Immunoisolation and culture of biliary epithelial cells from human liver. *In Vitro Cell. Dev. Biol.* **25,** 1189–1192.

27

Rapid Identification of Bacteria Based on Spectral Patterns Using MALDI-TOFMS

Jackson O. Lay Jr. and Ricky D. Holland

1. Introduction

The characterization of whole or "intact" bacteria, based on the mass spectral detection of genus, species, phenotypic, or genotypic biomarkers, desorbed and ionized directly without any isolation, purification, or concentration steps, has been accomplished using matrix-assisted laser/desorption ionization (MALDI) coupled with time-of-flight mass spectrometry (TOF-MS). Because MALDI-TOFMS plays such a key role in this method and also because these techniques may be unfamiliar to some readers, brief introductions to TOF-MS and MALDI and their advantages for bacterial characterization are presented in **Subheadings 1.1.** and **1.2.**

Another advantage of the use of MALDI for the characterization of bacteria that is less obvious than the purely instrumental ones but that might ultimately be just as important (although it has yet to be demonstrated experimentally) is the fact that the proteins derived from microorganisms might not need to be deliberately solubilized prior to mass spectral analysis. Indeed, the correspondence between the MALDI-TOF mass spectra from whole cells and from cellular supernatant supports the idea that many, if not all, of the proteins detected in spectra from whole cells are, in fact, soluble in either the solvents used to suspend the bacteria or the solvents used to mix bacteria directly with the MALDI matrix.

1.1. Time-of-Flight Mass Spectrometry

The separation of charged particles based on mass and flight time has been known since J.J. Thompson's famous experiments in 1897. Nevertheless, the

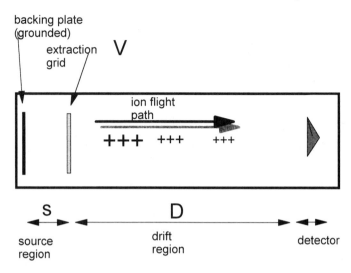

Fig. 1. Schematic diagram of a linear TOF mass spectrometer.

use of time-of-flight as a means for obtaining a mass spectrum was slow to develop. The first suggestion that a useful mass spectrometer might be developed using the time-of-flight principle is attributed to Stephens *(1)* 50 yr after Thompson's experiments. A few years later, in 1948, Cameron and Eggars *(2)* reported experimental data obtained using a TOF-MS instrument.

An attractive feature of both the early TOF instruments and their modern linear-TOF counterparts is the simplicity of their design. Ions formed in an ion source region (length = s; *see* **Fig. 1**), which is defined by a backing plate and an extraction grid, are subjected to an electric field ($E = V/s$) across this region, where V is the voltage applied to the extraction grid. The ions are consequently accelerated to a kinetic energy given by the product of the charge number (z, usually 1), the charge on the electron (e), and the applied potential (Es), so that **Eq. (1)** follows:

$$mv^2/2 = zeEs \quad (1)$$

where m is the mass of the ion and v is its velocity. Ions pass through the extraction grid with velocities that are inversely related to the square root of their masses. This relationship is derived simply by solving **Eq. (1)** for v, so that:

$$v = [2zeEs/m]^{0.5}. \quad (2)$$

The ions subsequently pass through the field-free drift region (length D) and impact the detector. In the absence of collisions, the ion velocities remain constant during flight so that the flight time measured the detector is given by the distance (D) divided by the velocity (v):

$$t = D[m/2zeEs]^{0.5} \quad (3)$$

For a typical instrument with an ion source region dimension ($s = 0.5$ cm), accelerating voltage ($V = 10$ kV), and drift region ($D = 1$ m), flight times are measured in microseconds. The TOF spectrum can easily be converted into a mass spectrum using these fixed (known) values for the accelerating voltage and drift region length as in **Eq. (4)**. However, because e, E, s, and D are constant, **Eq. (4)** is usually simplified to the form of **Eq. (5)** where the values of a and b are determined empirically using known mass standards.

$$m/z = 2eEs(t/D)^2 \qquad (4)$$

$$m/z = at^2 + b. \qquad (5)$$

In practice, Eqs. (1–5) do not completely describe the behavior observed with a TOF-MS system. Initial conditions for the ions involving time, space, and kinetic energy distribution all contribute to uncertainty in the mass spectrum because of a loss of mass (or time) resolution. However, more exacting consideration of the performance of TOF mass analyzers is beyond the scope of this chapter and unnecessary to an understanding of the MALDI-TOFMS behavior of bacterial components. The three principal advantages of the TOF mass analyzer for the characterization of bacteria can be deduced from the treatment given above. First, the only requirement for detection of either low or high mass ions is an ability to measure small differences in ion arrival times (t) and to continue to sample for long enough to detect the eventual arrival of higher mass ions. Second, the TOF-MS instrument does not scan the mass scale, so that all ions (positive or negative, as selected) eventually reach the detector, giving TOF-MS a much higher inherent sensitivity than conventional scanning mass spectrometers. Finally, the microsecond time scale for the ion flight times is suitable for the pulsed lasers used in the MALDI ionization process, described below. Flight times are synchronized with the firing of the laser pulse, and the values of a and b in **Eq. 5** are determined empirically. Many spectra can be obtained within a few seconds so that the MALDI-TOF-MS analysis of bacteria can be accomplished very quickly.

1.2. MALDI

MALDI involves the use of a large excess of a solid, light-absorbing material as a carrier or matrix in which the analyte is suspended. In the positive-ion mode, the matrix absorbs light from a laser (typically a Nd-YAG at 355 nm), the crystal lattice disintegrates, and protons are transferred from the gas-phase matrix molecules (typically weak acids) to the analyte, perhaps with enhancement of the pK of the matrix by laser excitation. The resulting protonated analyte molecule and any fragment ions can easily be detected by a TOF mass analyzer.

The development of modern, commercial TOF instruments can be attributed to the performance enhancement of laser desorption/ionization (LD) by matrix techniques, such as MALDI *(3)*, with the resulting ability to record useful mass spectra from proteins with masses considerably in excess of 10,000 Daltons *(4)*.

Prior to the advent of the matrix concept, the irradiation wavelength in LD-MS studies was changed to suit the expected analyte(s). Unfortunately, selection of an inappropriate wavelength resulted in a situation in which the signal for the analyte ions was either greatly enhanced or greatly diminished depending on the wavelength selected. Such uncontrolled behavior diminished the utility of this technique for the analysis of unknowns, because the best wavelength could not be known prior to analysis. Subsequently, Karas et al. *(5)* noted that an alternative, the use of a strongly absorbing matrix and a fixed laser wavelength, selected to suit the matrix rather that the analyte, offered an opportunity for controlled, reproducible energy deposition with an enhanced yield of analyte ions, regardless of their individual absorption characteristics.

The coupling of this soft ionization technique, capable of producing protonated molecules from nonvolatile analytes, to a TOF mass analyzer offering both high-sensitivity detection (all ions eventually reach the detector) and the detection of very high mass ions, led to the development of a new technique, MALDI-TOFMS, with the ability to provide reasonably accurate masses for proteins. Because this capability greatly exceeded the then available alternative means for the determination of protein molecular weights, the technique was widely adopted for the analysis of proteins as well as other high mass or highly nonpolar analytes. The advantages of MALDI-TOFMS for the characterization of proteins could reasonably be expected to apply to proteins desorbed from intact cells, and this is the basis for the application of MALDI-TOFMS to the characterization of bacterial cells.

1.3. Chemical Analysis and the Identification of Bacteria

1.3.1. Chemotaxonomy

A variety of methods have been investigated for the rapid identification of microbial species. For bacteria found in foods, it is important to distinguish pathogenic from similar nonpathogenic species. Desirable characteristics of a method for the identification of food-borne pathogens include the ability to detect minor differences between strains as well as rapidity of analysis, analytical ruggedness, simplicity of sample preparation, amenability to automation, and low cost. The polymerase chain reaction (PCR) has been widely adopted for the identification of microbes in similar clinical and environmental applications but has been less successfully applied to the identification of food-borne microorganisms because of matrix-specific problems encountered during the preparation of template DNAs *(6)*. Although PCR is a very powerful technique, the DNA probes may give false-negative results for mutant strains and cannot distinguish living and dead microorganisms. Furthermore, the tech-

Rapid Identification of Bacteria

nique requires extensive expertise in sample handling steps that are not easily automated.

Chemotaxonomy offers an alternative approach. In this case, classification of microbial strains is based on the analytical measurement of chemical constituents, which are either unique biomarkers or contributors to distinctive ratios of common chemical components. A well-known chemotaxonomic method (the MIDI system), which is based on gas chromatographic profiles of bacterial fatty acid methyl esters (FAME), has been developed. Even when autosamplers are used, however, this method still requires considerable sample preparation followed by a 45-min chromatographic separation. For some applications, a more rapid analysis step is needed.

1.3.2. Chemotaxonomy and Mass Spectrometry

In principle, chemotaxonomy using mass spectrometric rather than chromatographic or other instrumental measurements, could shorten analysis times enough to justify the greater expense of the mass spectrometry instrumentation. The use of mass spectrometry for the characterization of bacteria was investigated as early as 1975 by Anhalt and Fenselau (7). In a series of studies from 1975 to 1987, Fenselau and co-workers at The Johns Hopkins University compared fast-atom bombardment (FAB), laser desorption, and plasma desorption as ionization techniques for mass spectrometry (7–11). In these early studies, the principal analytes were phospholipids or other small molecules already known to be biomarkers for bacteria. More recently, the use of sample pyrolysis with pattern recognition has been the basis for the identification of intact bacteria using mass spectrometry (12). Because pyrolysis mass spectra from bacteria are so similar, co-analysis of authentic bacterial samples with unknowns has usually been required, followed by a complex evaluation of the data using computerized pattern recognition. Chemotaxonomy has also been successfully demonstrated using gas chromatography (GC)/MS techniques. For example, Fox et al. (13) have demonstrated the use of GC/MS for the differentiation of two closely related pathogenic organisms, *Bacillus anthracis* and *Bacillus cerus* (13), that are difficult to differentiate phenotypically or genotypically.

1.3.3. Bacterial Identification by MALDI-TOFMS of Whole Cells

Recently, Cain et al. (14) reported the use of MALDI-TOFMS to differentiate bacteria based on the analysis of proteins isolated from disrupted cells. Cells were harvested by centrifugation, washed twice with Tris buffer, resuspended, and disrupted by sonication. Proteins were isolated from a crude extract with methanol. This work was novel because the bacterial biomarkers detected were proteins rather than the smaller molecules examined in the earlier laser

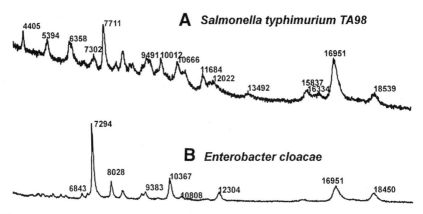

Fig. 2. MALDI-TOF mass spectra of *Salmonella typhimurium TA98* (**A**) and *Enterobacter cloacae* (**B**) obtained using whole cells.

desorption studies by Fenselau and coworkers. If the proteins selectively detected using this technique included either toxins or toxin-related biomarkers, the chemotaxonomic power already demonstrated by mass spectrometry for genus- and species-specific characterization in earlier studies might be extended to include differentiation of specific toxin-producing (or non-toxin-producing) strains. Thus, the analysis of whole cells, rather than proteins, by MALDI-TOFMS can be viewed as a logical extension of prior experiments that used laser desorption (but not MALDI) or that used MALDI with isolated proteins rather than with whole cells. In this way, a new method was developed, by Holland et al. *(15)*, using MALDI-TOFMS for the characterization of whole bacteria, rather than extracts, and based on ions formed directly from bacterial cells by laser desorption/ionization, using conditions much like those used for the analysis of isolated proteins. Indeed, the only significant difference between the procedures first used for characterizing bacteria and the procedures commonly used for pure proteins, was the observation that α-cyano-4-hydroxycinnamic acid appeared to be the optimum matrix for analysis of whole cells *(15)*.

Initially, five bacteria were analyzed by MALDI-TOFMS, and each gave a unique spectrum. Partial spectra of two of these bacteria are shown in **Fig. 2**. The spectra presented here differ somewhat from the spectra initially presented in ref. *15* because of improvements in the methodology since its introduction. These improvements are discussed in detail below, but, for this figure, include the addition of a mass calibration standard (m/z 16951 in the spectra), which allows more accurate mass assignments, and the use of 0.1% formic acid, which suppresses sodium and potassium adduct ion formation, giving the appearance of higher mass resolution. Whereas one bacterium, *Salmonella* *typhimirium TA98* (**Fig. 2A**), shows prominent ions near m/z 4405, 5396, 6358, 7711, and 10,012, as well as several smaller or unresolved peaks, the other, *Enterobacter cloacae* (**Fig. 2B**), shows ions at m/z 7294, 8028, 10,367, 12,304, and 18,450. Unlike pyrolysis mass spectra from whole bacteria, which usually show only relatively low-mass ions, MALDI-TOFMS gives signals in excess of m/z 10,000. Moreover, whereas differentiation of bacteria by pyrolysis mass spectrometry is often based on differences in intensity ratios rather than on the observation of unique ions, the MALDI-TOF mass spectra of bacteria give patterns that are clearly different with respect to both mass and intensity. The observation of such unique spectra suggested that MALDI-TOFMS might be used to identify bacteria. However, the proof of this concept, which is the identification of bacteria based on comparisons with either authentic samples or reference spectra, required demonstration that spectral differences between bacteria were reproducible and larger than experimental variations observed on reanalysis of samples of the same bacteria.

Evidence for this reproducibility came when the same five bacteria mentioned above were analyzed, again in ref. *15*, using standard MALDI-TOFMS conditions, to obtain reference spectra for comparison with a larger set of bacteria that were reanalyzed, using nominally identical conditions, after blind coding. Representative spectra from the nine bacteria were compared with the five reference spectra. Five spectra from the nine unknowns were judged to match reference spectra, whereas the other four did not. After the samples were decoded, the spectra from the five bacteria that matched reference spectra were found to represent the same species as the identified bacteria. In other words the five blind-coded spectra were correctly identified. The other four bacteria were strains that had not been previously analyzed. **Figures 3** and **4** show typical spectra from two of the unknowns and their matching reference spectra as they were presented in ref. *15*. In each case, including the spectra not shown, at least two unique ions were observed in both the reference and the unknown spectra that were not seen with the other bacteria. For example, with *Enterobacter cloacae* (**Fig. 3**), unique signals were observed near m/z 7700 and 9600, whereas the ion at m/z 7139 was not unique and difficult to see in the reference spectrum. With *Proteus mirabilis*, (**Fig. 4**), signals were observed near m/z 6200, 6400, 12,700, and 16,300 (weak). The correct identification of blind-coded bacteria, even in such a small study, provided evidence that MALDI-TOFMS from whole bacteria might be useful for the rapid determination of bacterial genera. Within a few months of the appearance of the work summarized above, similar results were reported in studies using whole bacteria in two other laboratories *(16,17)*.

Figures 3 and **4** represent typical, although perhaps not optimal, examples of the MALDI-TOF mass spectra of whole bacteria. The two prominent ions

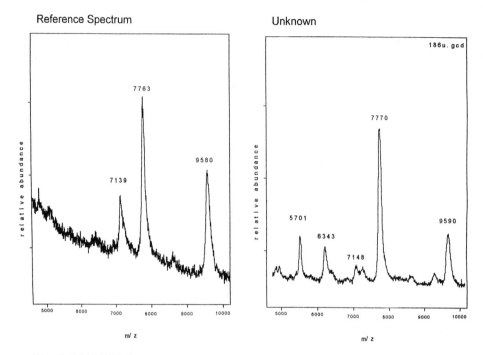

Fig. 3. MALDI-TOF mass spectra from a blind-coded sample and the corresponding reference spectrum of *Enterobacter cloacae* produced using nominally identical experimental conditions. (Reproduced from **ref. 15** with permission.)

observed in the first analysis used to obtain the reference spectrum in **Fig. 3** were also observed, in somewhat similar ratios, in the blind-coded spectrum. However, two additional ions were also observed near m/z 5700 and 6400, in the blind-coded sample, even though nominally identical conditions had been used to obtain both spectra. This example illustrates one of the most significant difficulties with this technique, the fact that, even under controlled conditions, the spectra of the bacteria change somewhat in replicate experiments. **Figure 4** shows the same sort of behavior for *Proteus mirabilis*. Although many of the same ions are present in both spectra, the ratio of ion intensity values, and the specific ions observed varies somewhat in the two spectra presented in the figure, as it also does in other examples, from experiment to experiment. Many of the methods used for obtaining spectra from bacteria are fraught with opportunities to introduce unwanted or variable signals into the spectra from bacteria. For example, the sample collection method used in the experiments described above retrieves both living and dead organisms and potentially also material from the culture medium as well. Moreover, these bacteria were analyzed without purification or pretreatment, whereas some investigators wash

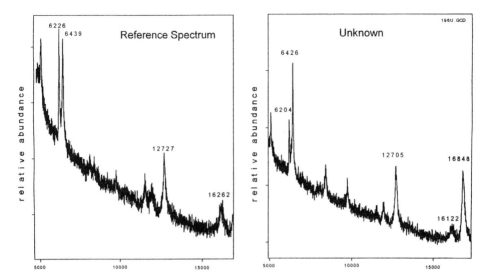

Fig. 4. MALDI-TOF mass spectra from a blind-coded sample and the corresponding reference spectrum of *Proteus mirabilis* produced using nominally identical experimental conditions. (Reproduced with permission from **ref. 15**.)

bacteria prior to analysis. Because of the variation between reference and unknown spectra and also because of the presence of apparent non-bacteria-specific ions, co-analysis of samples and reference bacteria can be used either to identify unknowns or to confirm identifications based on preliminary comparisons with reference spectra. Although the use of reference cultures rather than reference spectra is clearly more time consuming, variables associated with the mass spectrometry measurement can be minimized by analyzing the reference and unknown cultures side by side.

1.3.3.1. SPECIES SPECIFICITY

The methodology described above was also used to differentiate microorganisms at the species level; however, this was done based on co-analysis rather than by using reference spectra *(15)*. **Figure 5** shows the spectra obtained from three *Pseudomonas* species under nominally identical experimental conditions. Although some ions were observed with more than one bacterium (in one case with all three), enough unique ions were observed to allow all three strains to be differentiated unambiguously, especially when spectra were compared with reference cultures analyzed at the same time. As noted above, one ion near m/z 8700 was observed with all three *Pseudomonas* strains, but not with the other bacteria analyzed. This ion might be tentatively termed a "genus-specific biomarker" for *Pseudomonas*, pending further studies to determine whether

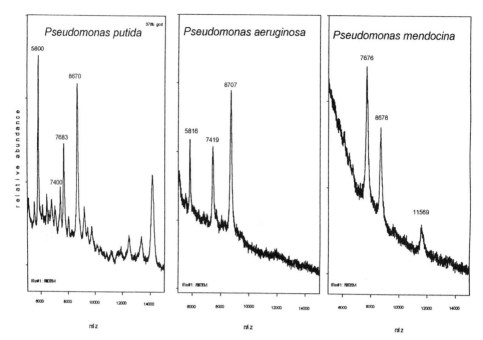

Fig. 5. Species specificity. MALDI-TOF mass spectra from three species of *Pseudomonas*. (Reproduced with permission from **ref. 15**.)

this ion is in fact only detected with this genus and not with other bacteria. The observation of such a common ion from a given genus might warrant isolation and identification of the protein if a protein biomarker for such a genus was needed. An example of the identification of a phenotypic biomarker from two bacteria exhibiting a common behavior is illustrated in **Subheading 3.6.**

2. Materials

1. α-Cyano-4-hydroxycinnamic acid matrix material.
2. Acetonitrile and trifluoroacetic acid (TFA).
3. Cytochrome c mass standard.
4. Tryptic soy agar culture medium.
5. C18 Protein/Peptide column (Vydac, Hesperia, CA).
6. LaserTec Research linear MALDI-TOFMS (Vestec, Houston, TX) with frequency tripled (355 nm) Nd-YAG laser or nitrogen laser.

3. Methods

3.1. Preparation of MALDI Matrix Solution

Prepare a saturated matrix solution by adding the matrix (α-cyano-4-hydroxycinnamic acid) to 2:1 H_2O + acetonitrile (containing 0.1% TFA) until a small amount of insoluble material is observed. Shake and centrifuge the solution and decant the clear liquid for use.

3.2. Preparation of Bacteria

1. Culture the bacteria on tryptic soy agar (*see* **Note 1**).
2. Remove bacteria from the agar using a culture loop, taking care to minimize the simultaneous removal of the agar. For typical colonies grown on culture plates, an amount that can easily be removed using a 2-mm culture loop in one to three attempts is used.
3. Suspend enough bacteria in solution to make a cloudy suspension. For example, suspend the amount removed using a 2-mm culture loop as described in **step 1** in 1 mL of the suspension solvent/solution. Use correspondingly smaller volumes of solvent for smaller numbers of bacteria. Typical solvents used to prepare the suspension are 0.1 N HCl, 0.1% aqueous TFA, or acetonitrile + water containing 0.1% TFA (*see* **Note 2**).

3.3. MALDI-TOFMS Analysis

1. Mix a small volume (typically 2 µL) of the bacterial suspension with the saturated matrix solution (typically 18 µL) for MALDI analysis (0.1% formic acid can be added to the matrix solution to suppress adverse effects on the MALDI spectrum of sodium chloride from the agar medium; *see* **Note 3**). Place approx 2 µL of this mixture on the MALDI target, allow to air dry, and insert into the mass spectrometer (*see* **Note 4**).
2. Record a mass spectrum in the positive-ion mode (*see* **Note 5**).

3.4. HPLC Separation of Bacterial Proteins

1. Shake and then centrifuge the cloudy bacterial suspension/solution used to produce spectra from whole cell bacteria as described above. Filter the product through a 0.5-mm filter to recover any soluble proteins.
2. Subject the clear supernatant to high-performance liquid chromatography (HPLC) separation. For example, for the fractions from which hdeAB proteins (*see* **Subheading 3.6.**) were identified, inject 100 µL of the filtered supernatant onto a Vydac C18 Protein/Peptide column. Separate using a linear gradient of 20–90% acetonitrile containing 0.1% TFA (mobile phase A) against H_2O (mobile phase B) over 30 min with a flow rate of 1 mL/min.
3. Collect individual fractions for analysis by MALDI-TOFMS. Mix 2 µL of the collected fraction with 18 µL of the matrix solution and analyze 2 µL of this mixture exactly as described in **Subheading 3.3.**

3.5 Effects of Experimental Parameters in General

As noted in **Subheading 1.3.3.**, the spectra of bacteria can vary, sometimes considerably, in replicate analyses *(18,19)*. This suggests that factors other than the specific bacteria being analyzed may significantly affect the spectra recorded using MALDI-TOFMS. The difference between the spectra of *Enterobacter cloacae*, shown in **Figs. 2** and **3**, illustrates this point. The data in the two figures were collected using the same species of bacteria, but under quite different conditions. **Fig. 2B**, produced about a year later than **Fig. 3**, was obtained using an internal mass reference and experimental conditions optimized for the analysis of bacteria based on incremental improvements over the course of time. The effects of a few parameters associated with both the instrumental measurement and the handling of the bacteria are summarized in **Subheadings 3.5.1.–3.5.5.** These include the effects on the bacterial spectra of the growth medium, the MALDI matrix, and the solvents in which bacteria are suspended. The effect of culture age is also quite significant, and this is discussed as well.

MALDI instruments typically incorporate different types of laser, either a nitrogen laser or a Nd-YAG laser. Although the wavelengths are close enough (355 nm for the Nd-YAG and 337 for the nitrogen laser) to allow good MALDI using either light source, the power of the two types of laser is quite different. For typical commercially available instruments, the Nd-YAG is about 100 times more powerful than the nitrogen laser, depending on how much the Nd-YAG laser beam is attenuated. In our laboratory, the results obtained for a given bacterial culture, using instruments equipped with the two different types of lasers, have been different, even when the TOF extraction voltage and the matrix used were the same *(18)*. Our experience has been that ions are typically observed spanning a larger range of *m/z* values using the more powerful laser and that, although it always more difficult to obtain spectra from cells of gram-positive than from gram-negative bacteria, we have never observed spectra from any gram-positive bacterium using a nitrogen laser. However, excellent spectra have been produced using nitrogen-laser-based instruments in other laboratories *(20,21)*. Thus, it is difficult, and perhaps unwise, to speculate too much on the importance of the laser, except as a caution that identifications based on comparisons of spectral patterns should be attempted only after it has been established that data generated on different instruments are comparable for identical bacteria. Such interlaboratory comparisons of spectra are possible, as has been recently demonstrated by researchers at the University of Alberta, in Canada and at the U.S. Army's Aberdeen Proving Ground facilities in Maryland *(20)*. Otherwise, until a larger number of laboratories have reported on this phenomenon, or until whole cells are studied using a single instrument equipped with both types of laser, neither the extent to which the type of laser contributes to the differences in spectra, nor the relative merits of the two types of laser, can be resolved. It is, however, clear that the specific instrument used does seem to have some effect on the spectral pattern, and this should be considered when building a library including spectra of bacteria produced using different MALDI-TOFMS instruments.

Rapid Identification of Bacteria

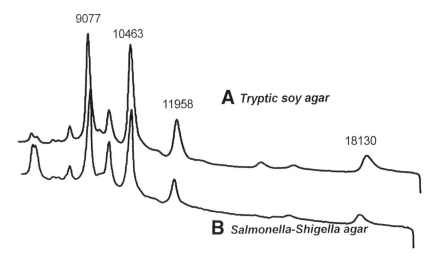

Fig. 6. MALDI-TOF mass spectra of *Shigella flexneri* grown using tryptic soy agar (**A**) and Salmonella-Shigella agar (**B**).

3.5.1. Effects of the Culture Medium

A parameter that might be expected to have a profound effect on the spectra of bacteria is the medium on which they are grown. Different media could result in changes in both the bacteria-specific (analyte) and the media-specific (background) ions. We have studied the MALDI mass spectral behavior of *Shigella flexneri* cultured using several different media *(18)*. Surprisingly, for the three most extensively used media there were only minor changes in the number or the intensity of the bacteria-specific peaks that were observed in the mass spectra. For example, **Fig. 6** shows partial mass spectra for *Shigella flexneri* grown on (**A**) tryptic soy agar and (**B**) on Salmonella-Shigella agar. As the figure shows, the characteristic ions between m/z 9000 and 18,000 are nearly identical using either medium. The ability to produce comparable spectra using more than one medium should be helpful for the construction of spectral libraries, because the same growth medium cannot always be used. Some bacteria simply cannot be easily cultured except by using very specific media. As expected, different media do produce different media-specific ions. For example, tryptic agar media does not produce peaks above m/z 3000, whereas blood agar shows peaks as high as m/z 15,000, which are presumed to be from proteins in the blood used to make the agar. Of course, by using background spectra obtained directly from the medium, medium-specific ions can be easily identified, or alternatively, they can be avoided altogether if adequate care is exercised in sampling the bacteria from the medium surface.

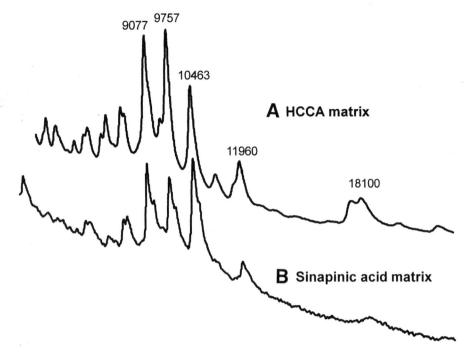

Fig. 7. The effects of matrix compounds. MALDI-TOF mass spectra of *Shigella flexneri* obtained using HCCA (**A**) and sinapinic acid (**B**).

3.5.2. MALDI Matrix Effects

The selection of the most appropriate matrix for analyzing bacteria may depend on a number of factors, including the specific instrument, the laser used for MALDI, the solvents used, or any of the steps used in manipulation of the bacteria. In the earliest reported experiments, α-cyano-4-hydroxycinnamic acid (HCCA) clearly gave much better spectra than any other matrix compound *(15,16)*. For example, **Fig. 7** shows two spectra obtained using the same bacterial culture with HCCA (**A**) and sinapinic acid (**B**) for the MALDI matrix. Typical of the results observed with other matrix compounds, fewer signals are produced with sinapinic acid than with HCCA. For example, **Fig. 7B** shows only a hint of signal above m/z 12,000 and although signals are observed below m/z 9000 with both matrix compounds, the spectrum produced using HCCA generally seems to show signals for a larger number of m/z values. In more recent studies in our own laboratory, using a TOF instrument incorporating a nitrogen laser, sinapinic acid seems to provide better spectra *(18)*. Other laboratories have reported that they have obtained spectra with both HCCA and sinapinic acid, using a nitrogen-laser-based instrument, but they report that HCCA provides higher sensitivity *(20)*. Hence, although other laboratories may well observe different results, we and others have reported either that HCCA is the best matrix for MALDI, especially using an Nd-YAG laser, or that HCCA is one of the matrix compounds that can be used to produce good-quality spectra from bacteria.

3.5.3. Suppression of Salt-Containing Adducts

Because the agar media can contain up to about 10,000 ppm sodium chloride, we often add formic acid (0.1%) to the matrix solution to improve both the mass assignments and the apparent resolution of the spectra *(18)*. The improvement in precision of the mass measurements using formic acid is attributed in part to the suppression of salt adducts that are incompletely resolved at low resolution. Formic acid has also been used for this purpose in the analysis of peptides or proteins by MALDI-TOFMS *(22)*. This suppression of salt-containing adduct ions causes the spectra to look as if the resolution has been increased because the sodium or potassium portion of the unresolved mass envelope is greatly reduced. This effect can easily be seen by a comparison of **Figs. 2** and **3**.

3.5.4. Solvent Effects

Typically, the analysis of bacteria by MALDI-TOFMS requires that the bacteria be added to a solution containing a matrix compound or suspended in some solvent just prior to mixing with the matrix in another solution. We have mixed the bacteria directly with the matrix solution in some experiments, and in others we have first suspended the bacteria in a separate solution, which was then mixed with the matrix solution. Although this was initially done entirely for convenience, we have found that spectra can be made more reproducible from run to run by first suspending the bacteria in a solvent and then mixing a portion of this suspension with a small volume of matrix solution. However, this use of solvents in the suspension step can result in differences in the spectra based on the specific solvent used *(18)*. Different components may be observed or the components may be similar but observed in different proportions. This can be explained by the effects of cell lysis as the bacteria reside in the suspension solvent or by differences in the solubility of proteins in the suspension solvent, and perhaps by the interactions between the solvent and any free proteins. To minimize potential loss of proteins to the solvent in which the bacteria are suspended, small (microliter) volumes can be used so that the bulk of the material, bacteria, solvent, and any suspended proteins, can be applied to the MALDI target rather than diluting the proteins or leaving them behind in a large volume of unused liquid. **Figure 8** shows spectra from *Shigella flexneri* suspended in two different solvents prior to MALDI-TOFMS analysis. The use of 0.1 N HCl (**Fig. 8B**) produces a greater number

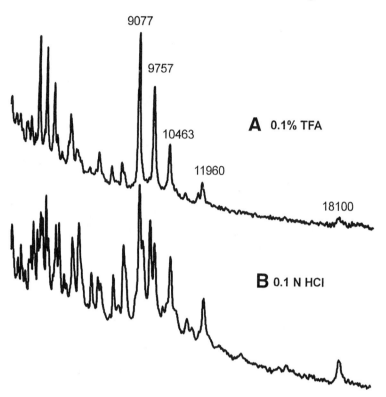

Fig. 8. MALDI-TOF mass spectra of *Shigella flexneri* obtained using 0.1% TFA (**A**) and 0.1 N HCl (**B**).

of spectral components than either 0.1% aqueous TFA (**Fig. 8A**) or mixed acetonitrile/water/0.1% TFA solutions (spectra not shown). Obviously, the spectral patterns produced from whole bacteria are affected by the pH of the suspension solvent as well as by its composition. Similar effects are reported in more detail in **ref. 20**.

3.5.5. Culture Age

Based on careful evaluation of all the effects described in the preceding sections, we have developed a standard set of conditions for the analysis of whole cells by MALDI-TOFMS, which are given in detail in **Subheading 3.3**. Although the optimum conditions will no doubt vary somewhat from laboratory to laboratory (see, for example, the conditions used in the interlaboratory validation reported in **ref. 20**), the conditions described in **Subheading 3.3.** should provide a reasonable starting point for the analysis of bacteria by MALDI-TOFMS. Spectra produced using these conditions, for example **Figs. 2** and **9**,

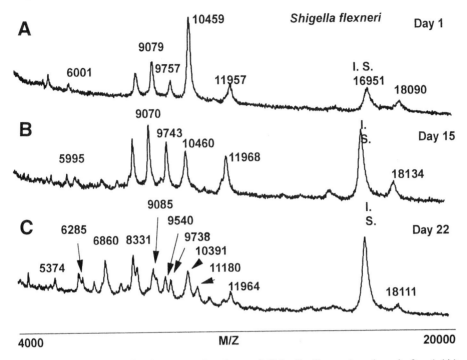

Fig. 9. The effects of culture age. A culture of *Shigella flexneri* analyzed after 1 (**A**), 15 (**B**), and (**C**) 22 d of storage in a refrigerator.

show the accumulated effects of small incremental changes in the experimental conditions on the quality of the spectra compared with spectra presented in some of the other figures.

One of the most important variables affecting MALDI-TOFMS spectra of bacteria is the age of the culture. Replicate experiments using different bacteria demonstrate this phenomenon and the kind of changes that may be seen *(18,19,23)*. **Figure 9** depicts these changes for *Shigella flexneri* over 22 d for cultures stored in a refrigerator. A numerical evaluation of the effects of culture age on spectra has been reported for four bacteria: *Escherichia coli*, *Shigella flexneri*, *Enterobacter cloacae*, and *Salmonella typhimurium* TA98 *(19,23)*. Some of these results are depicted in **Fig. 10** and **Tables 1** and **2**. Each bacterium was cultured for 24 h prior to analysis. Then a second set of the same bacteria was placed in the refrigerator and reanalyzed after 1, 8, 15, and 22 d. The first set of mass spectra (produced after 24 h) was used as a reference (major peaks only) for identification and compared with spectra produced in the subsequent set of analyses (1, 8, 15, and 22 d of ageing) using a BASIC computer program written for this purpose. The program calculated a value describing the percent of matches for peaks in the aged culture that matched

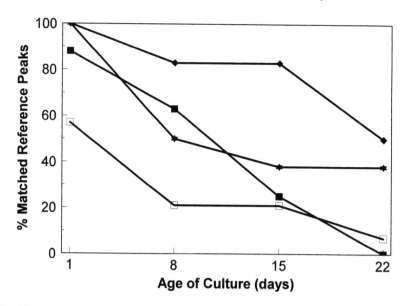

Fig. 10. Change in the percentage of matched reference ions (within 20 Daltons) as four different cultures age in a refrigerator for up to 22 d.

the peaks in the reference spectra within a predefined mass range tolerance (± 20 mass units). Under very carefully controlled conditions, bacteria typically showed a very good match (up to 100% match for the reference ions) after the initial growth period. After this 24-h period, the match between spectra taken on different days decreased, sometimes remarkably after 8 d of ageing.

As noted above, the experimental spectrum from each bacterial culture was compared with all four reference spectra. Typical of most of the bacteria we have studied, 24 h after inoculation, *E. coli*, *Shigella flexneri*, and *Enterobacter cloacae* gave mass spectra that were in good agreement with their own reference spectra (**Fig. 10**) but not with the reference spectra for the other three bacteria. As **Fig. 10** shows, between 88 and 100% of the reference ions were observed in the second (test) set of spectra for these three bacteria. Atypical results are also depicted in this figure, from *Salmonella typhimurium* TA98, which showed only a 57% match for the 24-h culture. This low match between otherwise identical reference and experimental spectra is attributed in part to the low resolution of the TOF used to make the measurements. Mass assignment and matching of peaks was compromised by doublets that were sometimes resolved and in other experiments merged. Although the match with the reference spectrum for this bacterium was only 57%, this was a much better match than with the reference spectra from the other bacteria (data not shown).

Table 1
Comparison of the Percentage of Peaks from Cultures of *Enterobacter cloacae* that Match Reference Peaks in the Spectral Fingerprint of Four Bacteria, for Cultures Stored in a Refrigerator for up to 22 d

	Days			
Bacterium	1	8	15	22
E. coli	0[a]	0	0	0
S. flexneri	0	0	0	0
E. cloacae	100	50	38	38
S. typhimurium	7	7	7	7

[a]Percentage of matched reference peaks.

Table 2
Comparison of the Percentage of Peaks from Cultures of *S. flexenri* that Match Reference Peaks in the Spectral Fingerprint of Four Bacteria for Cultures Stored in a Refrigerator for up to 22 d

	Days			
Bacterium	1	8	15	22
E. coli	38[a]	75	75	50
S. flexneri	100	83	83	50
E. cloacae	0	13	13	25
S. typhimurium	0	7	14	7

[a]Percentage of matched reference peaks.

All four bacteria showed a decrease in the number of correctly matched peaks as the cultures aged (**Fig. 10**). For example, *E. coli* matched 88, 63, and 25% of the reference ions after d 1, 8, and 15, respectively. No spectrum was produced on d 22. Typically though, as the number of peaks that matched the correct reference peaks decreased, the number of mismatched peaks, i.e., peaks that matched with another bacterium, did not change. This is illustrated in **Table 1** for *Enterobacter cloacae* where the maximum number of matches for reference peaks with the other bacteria remains constant at 0–7% over time. However, in one case, the spectra of two bacteria became more alike as the cultures aged. This effect was observed with *Shigella flexneri* and *E. coli*. These two bacteria are reported to be genetically similar and difficult to distinguish from each other using some traditional methods of identification. Thus, it is not surprising that when the spectra are compared, *Shigella flexneri* and *E. coli*

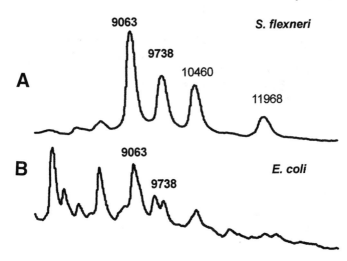

Fig. 11. A comparison of portions of the MALDI-TOF mass spectra near m/z 10,000 for *Shigella flexneri* and *E. coli* showing some of the common ions in this mass region.

(**Fig. 11**) show some of the same major peaks. Nevertheless, they can be distinguished using reference spectra based on their overall mass spectra. As **Table 2** shows, although a replicate spectrum from *E. coli* matched 34% of the reference peaks from *Shigella flexneri*, it matched 100% of the reference peaks from the original *E. coli* culture. As the cultures age, however, some of the peaks that are critical for differentiation of these two bacteria either disappear or merge with other peaks. Moreover, after 22 d, the spectra from these two bacteria cannot be distinguished because their spectra contain ions that match some of the reference ions for each bacterium. **Table 2** shows this trend for *Shigella flexneri* as the spectra change for cultures from d 1 through d 22. The 22-d-old *Shigella flexneri* culture matched 50% of the reference peaks for itself and also 50% for *E. coli*. Rather than representing a defect in the method, this behavior may, in fact, be an indication of the genetic similarity between the two bacteria. In other words, after storage for 22 d, the bacteria may be producing some of the same proteins. In comparison, the other unrelated bacteria, although they did not show a good match with their own reference spectra after 22 d, also did not seem to match each other either. This observation provides some preliminary evidence to support the hypothesis that this type of time-related spectral behavior might be indicative of some common genetic heritage. Other evidence supporting the notion that these spectral changes result from changes in the biochemistry of the bacteria, rather than changes (mutation) in the strain of bacteria, is the fact that when old cultures are regrown under identical conditions, almost identical spectra can be obtained (**Fig. 12**).

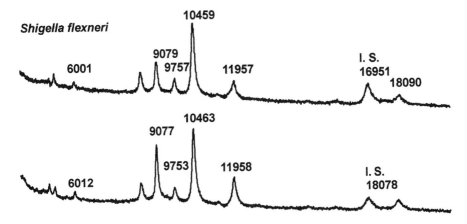

Fig. 12. MALDI-TOF mass spectra obtained from two 1-d-old cultures of *Shigella flexneri*, one of which was recultured from the other.

If the changes in the spectra resulted from mutation, the regrown cultures should exhibit patterns that differ from the initial spectra.

3.6. Identification of Proteins in MALDI-TOF Mass Spectra

The components responsible for the pattern produced from whole bacteria by MALDI-TOFMS have been assumed to be proteins. As noted in the preceding sections, the nature of the proteins observed from whole cells by MALDI depends on a number of experimental parameters including factors associated with the biochemistry of the bacteria. For this reason, identification of the proteins observed in spectra from whole cells could prove useful. In addition to providing a basis for understanding biology-based changes in spectra from cells, the use of characteristic biomarkers in MALDI-TOF mass spectra could be more meaningful if the functions associated with the apparent bacteria-specific proteins were understood. Moreover, because it is likely that the specificity of supposed bacteria-specific ions will, at least in the near future, be based on experimental demonstration of their absence from MALDI-produced spectra from all or almost all other bacteria, biological function may provide the best evidence regarding the bacterial specificity of the proteins observed in MALDI-TOF mass spectra. We also believe that a subgroup of these proteins could be used as biomarkers to differentiate bacteria based on either genotypic or phenotypic characteristics.

The feasibility of this approach has been demonstrated by the identification of two proteins common to the MALDI-TOF mass spectra of the closely related bacteria, *E. coli* and *Shigella flexneri (24,25)*. Some strains of these bacteria are pathogenic in humans, and because most are also resistant to digestion by

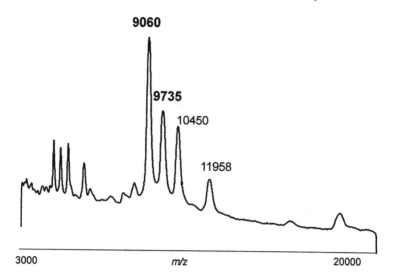

Fig. 13. MALDI-TOF mass spectrum obtained from a suspension of *Shigella flexneri*, showing many of the same ions as spectra from whole cells.

stomach acids, even a few cells can lead to infection. In principle, some of the peaks in the mass spectra of these two bacteria should be associated with acid resistance. The proteins with masses *m/z* 9060 and 9735, selected for identification, typically give the most prominent ions in the mass spectra from cells of *Shigella flexneri,* and they are also observed with *E. coli* (the same approach could also be used with other, less abundant, proteins). The spectrum obtained from a suspension of *Shigella flexneri*, shown in **Fig. 13**, is very similar to the whole-cell mass spectra shown in **Figs. 9** and **12**. The mass assignments for both components (*m/z* 9060 and 9735 in **Fig. 13**) showed a measured standard deviation of ±5 mass units based on triplicate analyses. **Figure 14** shows an HPLC chromatogram from the same supernatant of a centrifuged suspension of *Shigella flexneri*. Individual HPLC fractions were collected and analyzed, using MALDI-TOFMS to find the retention times for analytes with the same masses, presumed to be the two target proteins. They were found in fractions corresponding to the peaks marked 1 and 2 in the figure. The MALDI-TOF mass spectra from representative fractions are shown in **Fig. 15**. Mass spectral analysis of the other major HPLC peaks (data not shown) associated with the separation revealed that many other proteins observed in the MALDI spectrum from whole bacteria were present in the supernatant as well, but these proteins have not yet been identified.

To identify the proteins, the *m/z* 9060 and 9735 peaks were collected by HPLC some 20 times each, and the 20 protein-containing fractions correspond-

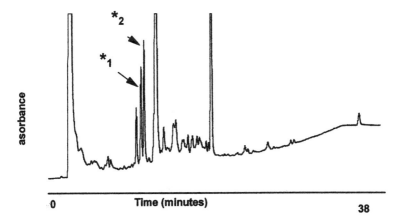

Fig. 14. HPLC trace of supernatant from a suspension of *Shigella flexneri* showing two peaks (1 and 2) having components with masses of 9735 and 9060 Daltons, respectively.

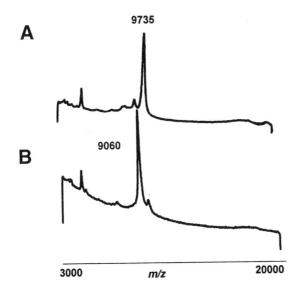

Fig. 15. MALDI-TOF mass spectra from HPLC fractions marked 1 (**A**) and 2 (**B**) showing separation of the two proteins giving masses of 9735 and 9060 Daltons, respectively.

ing to each mass were pooled separately. The two pooled samples were purified in a final HPLC step. **Figure 16** shows the HPLC chromatograms from these two samples, which were submitted to automated Edman analysis. Based on the first 10 amino acid residues from each protein, the identities were determined to be HdeB (*m/z* 9060) and HdeA (*m/z* 9735) from *E. coli*. For *m/z* 9060,

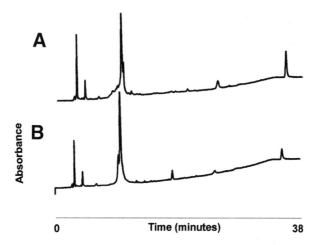

Fig. 16. HPLC traces for 20 combined fractions of *m/z* 9060 (**A**) and *m/z* 9735 (**B**).

the sequence Ala-Asn-Glu-Ser-Ala-Lys-Asp-Met-Thr-Cys was determined by Edman degradation. This sequence corresponds to a portion of the following sequence for HdeB from *E. coli*:

Met-Gly-Tyr-Lys-Met-Asn-Ile-Ser-Ser-Leu-Arg-Lys-Ala-Phe-Ile-Phe-Met-Gly-Ala-Val-Ala-Ala-Leu-Ser-Leu-Val-Asn-Ala-Gln-Ser-Ala-Leu-Ala-**Ala-Asn-Glu-Ser-Ala-Lys-Asp-Met-Thr-Cys-Gln-Glu-Phe-Ile-Asp-Leu-Asn-Pro-Lys-Ala-Met-Thr-Pro-Val-Ala-Trp-Trp-Met-Leu-His-Glu-Glu-Thr-Val-Tyr-Lys-Gly-Gly-Asp-Thr-Val-Thr-Leu-Asn-Glu-Thr-Asp-Leu-Thr-Gln-Ile-Pro-Lys-Val-Ile-Glu-Tyr-Cys-Lys-Lys-Asn-Pro-Gln-Lys-Asn-Leu-Tyr-Thr-Phe-Lys-Asn-Gln-Ala-Ser-Asn-Asp-Leu-Pro-Asn**

where the bold portion of the sequence depicts the smaller portion of these proteins actually collected from *Shigella flexneri*. The predicted mass for this sequence is 9063 Daltons, in good agreement with the measured mass (9060 Daltons). Similarly, for the peak corresponding to *m/z* 9735, the first 10 amino acids sequenced by the Edman method were again a part of the larger sequence for HdeA, where a smaller portion of the predicted protein, from *E. coli*, was isolated from *Shigella flexneri*. Again, the expected mass based on the protein sequence (not shown) is 9738, in good agreement with 9735 Daltons, the measured mass.

Tentatively, we attribute the fact that only a portion of the expected protein was detected, in each case, to proteolytic cleavage at the Ala-Ala bonds within the two precursors. This mass has been observed from whole cells, cellular suspensions, and the cellular supernatant. As noted above, the average masses of these proteins, calculated from the amino acid sequences were 9063 and 9738, respectively, 3 mass units higher than the values we measured. These data are consistent with close homology between the HdeA (and HdeB) proteins in *E. coli* and *Shigella flexneri*. In fact, these proteins are encoded by the σ^s-dependent genes, hdeAB, which are associated with an acid-resistant phenotype *(26)*. Although the function of the two proteins is not known, this acid-resistant phenotype is associated with the presence of the gene itself, which protects bacteria from acidic conditions in the stomach, allowing bacteria to infect the host with only a few cells. Although it was hoped that biologically relevant proteins would be detected using this approach, the finding that these two proteins were related to acid resistance was largely coincidental. In fact, it would have been difficult to find the proteins directly based on the difference between their predicted mass (for hdeA/hdeB) and their observed masses, presumed to be caused by proteolysis within the cell.

As noted above, the proteins isolated from *Shigella flexneri* were indistinguishable from the proteins predicted for *E. coli*, consistent with the genetic similarity, as well as the common acid-resistant nature of these two bacteria. To demonstrate that the two components in these bacteria were in fact from the same proteins, they were also collected from *E. coli*. Fractions with the same retention time as those collected with *Shigella flexneri* were collected and analyzed. The MALDI-TOF mass spectrum from the fractions isolated from *E. coli* also corresponded to HdeB and HdeA, with the same apparent Ala-Ala proteolysis, as expected.

This identification of these two proteins demonstrates the detection of a specific bacterial phenotype, traceable to specific masses in the MALDI-TOF mass spectra obtained initially from whole cells and later from cellular supernatant. We believe that this same approach could be applied to the detection of proteins specific for other properties, such as antibiotic resistance or resistance to heating. Clearly, mass spectrometry provides a method of detecting bacteria-specific properties that complements the well-known capabilities of PCR for identifying specific bacterial strains. Using *E. coli* and *Shigella flexneri*, we have demonstrated that the proteins associated with acid resistance are readily detectable in the MALDI-TOF mass spectra obtained from whole cells. Using a similar approach, we believe that proteins specific for other properties of public health interest, such as drug resistance or food safety, could also be detected. Following this, MALDI-TOFMS could be used to screen bacterial samples for a specific property, or the protein could be identified and then used as the basis for a non-mass-spectrometric field assay.

4. Notes

1. As noted in **Subheading 3.5.1.**, other media can also be used. For optimum reproducibility, bacteria cultures should be grown and stored under the same conditions for each experiment.

2. Examples of the effect of the use of different solvents are discussed in **Subheading 3.5.4.** with details given in **Figs. 8A** and **B** and **ref. 20**.
3. Because agar media can contain up to about 10,000 ppm sodium chloride, it is often useful to add formic acid (0.1%) to the matrix solution as a means of improving both the mass assignments and the apparent resolution of the spectra which are otherwise adversely affected by the presence of the salt (*see* **Subheading 3.5.3.**).
4. Similar results can be obtained from the supernatant of filtered or centrifuged bacteria. However, in this case, a smaller number of proteins may be observed. Finally, bacteria can be added directly to the matrix solution and 2 µL applied to the target. All these procedures can be used to obtain spectra from cells.
5. To improve mass accuracy, bacteria are analyzed using an internal mass calibrant as is well known in the analysis of purified peptides and proteins. For example, the protein cytochrome c can be added to the target as an internal mass calibration standard.

References

1. Stephens, W. E. (1946) Pulsed Mass Spectrometer with Time Dispersion. *Phys. Rev.* **69,** 691.
2. Cameron, A. E. and Eggers, D. F. (1948) Ion velocitron. *Rev. Sci. Instrum.* **19,** 605–607.
3. Karas, M., Bachmann, D., Bahr, U., and Hillenkamp, F. (1987) Matrix-assisted laser desorption of nonvolatile compounds. *Int. J. Mass Spectrom. Ion Processes* **78,** 53–68.
4. Karas, M. and Hillenkamp, F. (1988) Laser desorption ionization of proteins with molecular masses exceeding 10,000 daltons. *Anal. Chem.* **60,** 2299–2301.
5. Karas, M., Bachmann, D., and Hillenkamp, F. (1985) Influence of wavelength in high-irradiance ultraviolet laser desorption mass spectrometry of organic molecules. *Anal. Chem.* **57,** 2935–2939.
6. Hill, W. E. (1996) The polymergse chain reaction: applications for the detection of food borne pathogens. *Crit. Rev. Food Sci. Nutr.* **36,** 123–173.
7. Anhalt, J. P. and Fenselau, C. (1975) Identification of bacteria using mass spectrometry. *Anal. Chem.* **47,** 219–225.
8. Heller, D. N., Fenselau, C., Cotter, R. J., Demirev, P., Olthoff, J. K., Honovich, J., et al. (1987) Mass spectral analysis of complex lipids desorbed directly from lyophilized membranes and cells. *Biochem. Biophys. Res. Commun.* **142,** 194–199.
9. Ho, B. C., Fenselau, C., Hansen, G., Larsen, J. and Daniel, A. (1983) Dipalmitoylphosphatidycholine in amniotic fluid quantified by fast-atom-bombardment mass spectrometry. *A. Clin. Chem.* **29,** 1349–1353.
10. Fenselau, C. and Cotter, R. J. (1987) Chemical aspects of fast atom bombardment. *Chem. Rev.* **87,** 501–512.
11. Heller, D. N., Cotter, R. J., and Fenselau, C. (1987) Profiling of bacteria by fast atom bombardment mass spectrometry. *Anal. Chem.* **59,** 2806–2809.
12. DeLuca S., Sarver, E. W., Harrington, P. de B., and Voorhees, K. J. (1990) Direct Analysis of bacterial fatty acids by Curie-point pryolysis tandem mass spectrometry. *Anal. Chem.* **62,** 1465–1472.
13. Fox, A., Rogers, J. C., Fox, K. F., Schnitzer, G., Morgan, S. L., Brown, A., et al. (1990) Chemotaxonomic differentiation of legionella by detection and character-

ization of aminodeoxyhexoses and other unique sugars using gas chromatography-MS. *J. Clinic. Microbiol.* **March 28,** 546–552.
14. Cain, T. C., Lubman, D. M., and Weber, W. J. Jr. (1994) Differentiation of bacteria using protein profiles from matrix-assisted laser desorption/ionization TOF-MS. *Rapid Commun. Mass Spectrom.* **8,** 1026–1030.
15. Holland, R. D., Wilkes, J. G., Sutherland, J. B., Persons C. E., Voorhees, K. J., and Lay, J. O. Jr. (1996) Rapid identification of intact whole bacteria based on spectral patterns using MALDI-TOF-MS. *Rapid Commun. Mass Spectrom* **10,** 1227–1232.
16. Claydon, M. A., Davey, S. N., Edward-Jones V., and Gordon, D. B. (1996) The rapid identification of intact microorganisms using mass spectrometry. *Nature Biotechnol.* **14,** 1584–1586.
17. Krishnamurthy, T. and Ross, P. L. (1996) Rapid identification of bacteria by direct MALDI-TOF-MS analysis of whole cells. *Rapid Commun. Mass Spectrom.* **10,** 1992–1996.
18. Holland, R. D., Rafii, F., Holder, C. L., Sutherland, J. B., Voorhees, K. J. and Lay, J. O. Jr. (1997) Investigation into the experimental parameters that affect the MALDI-TOF-MS produced from whole bacteria. *Proceedings of the 46th ASMS Conference on Mass Spectrometry and Allied Topics*, Orlando, FL, p. 154.
19. Arnold, R. J., Karty, J. A., Ellington, A. D., and Reily, J. P. (1989) Monitoring the growth of bacteria culture by MALDI-TOF-MS of whole cells. *Analytical Chemistry* **71,** 1990–1996.
20. Wang, Z., Russon, L., Li, L., Roser, D. C., and Long, S. R. (1998) Investigation of spectral reproducibility in direct analysis of bacteria proteins by matrix-assisted laser desorption/ionization time-of-flight mass spectrometry. *Rapid Commun. Mass Spectrom.* **12,** 456–464.
21. Haag, A. M., Taylor, S. N., Johnston, K. H., and Cole, R. B. (1998) Rapid identification and speciation of *Haemophilus* bacteria by MALDI-TOF-MS. *J. Mass Spec.* **33,** 750–756.
22. Walker, K. L. Chiu, R. W. Monnig, C. A., and Wilkins, C. L. (1993) Off-line coupling of capillary electrophoresis and MALDI-TOF-MS. *Anal. Chem.* **67,** 4197–4204.
23. Holland, R. D., Burns, G., Parsons, C., Rafii, F., Sutherland, J. B. and Lay, J. O. Jr. (1997) Evaluating the effects of culture age on the rapid identification of whole bacteris using MALDI/TOF-MS. *Proceedings of the 45th ASMS Conference on Mass Spectrometry and Allied Topics*, Palm Springs, CA, p. 1353.
24. Holland, R. D., Rafii, F., Holder, C. L., Sutherland, J. B., Voorhees, K. J., and Lay, J. O. Jr. (1997) Identification of the proteins observed in MALDI TOF mass spectra of whole cells. *Proceedings of the 46th ASMS Conference on Mass Spectrometry and Allied Topics*, Orlando, FL, p. 194.
25. Holland, R. D., Duffy, C., Rafii, F., Sutherland, J. B., Heinz, T. M., Holder, C. L., Voorhees, K. J., and Lay, J. O. Jr. (1999) Identification of bacterial proeins observed in MALDI TOF mass spectra from whole cells. *Analytical Chemistry* **71,** 3226–3230.
26. Waterman, S. R. and Small, P. L. C. (1996) Identification of sigma S-dependent genes associated with the stationary-phase acid resistance phenotype of *Shigella flexneri. Mol. Microbiol.*, **21,** 925–940.

28

Appendices

In compiling the appendices, I have omitted the detailed discussion of mass and abundance values in mass spectrometry presented in the first volume of this series since I hope that this is, by now, all common knowledge. I have also omitted the tables of mass standards, proteolytic reagents, MALDI matrices, and contaminant effects in MALDI since these topics are more than adequately dealt with in this volume in the chapters themselves. Thus, there remains a glossary of abbreviations (Appendix I), a table of mass and abundance values for individual elements (**Table 1**), listings of mass values and abbreviations for amino acid and monosaccharide residues (Appendix II, **Tables 2** and **3**), a summary of peptide fragmentation (Appendix III), and, finally, an update for references on the use of mass spectrometry in protein and peptide analysis (Appendix IV).

Appendix I: Glossary of Abbreviations and Acronyms

The following terms and their preferred contracted forms, together with those to be found in Appendix II, are used in this volume specifically or are in general use. In some cases, alternative contracted forms are given in parentheses.

Å	angstrom (10^{-10} m)
Aβ	amyloid β-peptide
ACE	affinity capillary electrophoresis
ACN	acetonitrile
ACR	ancient conserved region
ADP	adenosine 5′-diphosphate
AIDS	acquired immunodeficiency syndrome
Aminopeptidase M	amino acid arylamidase aminopeptidase M
AMP	adenosine 5′-monophosphate
AMP-PCP	adenylyl methylene diphosphate
AMP-PNP	adenylyl imidodiphosphate

From: *Methods in Molecular Biology, vol. 146:*
Protein and Peptide Analysis: New Mass Spectrometric Applications
Edited by: J. R. Chapman © Humana Press Inc., Totowa, NJ

AMS	affinity mass spectrometry
ASCII	American standard for information exchange (computer encoding system)
ATP	adenosine 5'-triphosphate
ATPase	adenosine 5'-triphosphatase
ATT	6-aza-2-thiothymine
ave	average
BASIC	beginners' all-purpose symbolic instruction code (computing language)
BDE	bile duct epithelial
$[BH_2]^+$	positive ion corresponding to a base, e.g., adenine, with H and H^+ added
bHLH	basic helix-loop-helix
BLAST	homology-based search program for sequence analysis (*see* Chapter 4)
BO	bacterioopsin
BSA	bis(trimethylsilyl)acetamide
BSA	bovine serum albumin
BTX	benzyl-1-thio-β-*O*-xylopyranoside
CAM	carboxamidomethyl
CCSD	complex carbohydrate structure database (http://www.ccrc.uga.edu)
CD	celiac disease
Cdk	cyclin-dependent kinase
cDNA	complementary DNA
CE	capillary electrophoresis
CGM	fetal bovine serum
CHAPS	3-[(3-cholamidopropyl)dimethylammonio]-1-propanesulfonate
CID	collision-induced dissociation
CMC (Cmc)	critical micelle concentration
Cmc	carboxymethylcysteine
CMP	cytidine 5'-monophosphate
CPY	carboxypeptidase-Y
CRC	compact reaction column
CS	citrate synthase
CsgA	C-signaling A protein
Cymal-6	cyclohexyl-hexyl-β-D-maltoside
αCys	cysteine in the α-chain of hemoglobin
βCys	cysteine in the b-chain of hemoglobin
Cys-NO	*S*-nitrosocysteine

DAHC	diammonium hydrogen citrate
dAMP	deoxyadenosine monophosphate
dATP	deoxyadenosine triphosphate
DCM	dichloromethane
DDM	n-dodecyl-β-D-maltoside
DE	delayed extraction
$\Delta\varepsilon_n$	absorbance change at wavelength n
DHAP	2,6-dihydroxyacetophenone
DHB (2,5-DHB)	2,5-dihydroxybenzoic acid (gentisic acid)
DHC	diammonium hydrogen citrate
2D LC	two-dimensional liquid chromatography
DMF	dimethylformamide
DMP	dimethylpimelimidate
DMSO	dimethylsulfoxide
DNA DNAs	deoxyribonucleic acid
DP IV	dipeptidyl peptidase IV
DTAC	dodecyltrimethylammonium chloride
DTSSP	3,3'-dithiobis(sulfosuccinimidyl propionate)
DTT	1,4-dithiothreitol
e^-	electron
ε	extinction coefficient
E. coli	*Escherichia coli*
EBEB	E and B represent electrostatic and magnetic sectors, respectively, in an instrument configuration. EBEB is a four-sector instrument configuration
EDTA	ethylenediamine tetraacetic acid
EGF	epidermal growth factor
ELISA	enzyme-linked immunosorbent assay
E:S	enzyme-to-substrate ratio
ESI	electrospray ionization
ESI-MS	electrospray ionization mass spectrometry
ESI-MS/MS	electrospray ionization tandem mass spectrometry
EST	expressed sequence tag
ethidium bromide	2,7-diamino-10-ethyl-9-phenyl-phenanthridinium bromide
eV	electron volt
EX1 EX2	forms of kinetics for hydrogen exchange
FA	ferulic acid
FAB(MS)	fast-atom bombardment (mass spectrometry)
Factor Xa	restriction protease factor Xa
FAD	flavin-adenine nucleotide
FAME	fatty acid methyl ester

FASTA	homology-based search program for sequence analysis (*see* Chapter 4)
2F-DNPX	2,4-dinitrophenyl-2-deoxy-2-fluoro-β-D-xylopyranoside
femtomole	10^{-15} mole
FHV	Flock house virus
FMN	flavin mononucleotide
Fmoc	*N*-(9-fluorenyl)methoxycarbonyl
fmol	symbol for femtomole
FPLC	fast protein liquid chromatography
FSH	follicle-stimulating hormone
FTICR-MS	fourier transform ion cyclotron resonance mass spectrometry
FWHM	full-width at half-maximum intensity
GC/MS	on-line gas chromatography/mass spectrometry
GIP	glucose-dependent insulinotropic polypeptide
GLP-1	glucagon-like peptide-1
GMP	guanosine 5′-monophosphate
GSH	glutathione
GST	glutatione-*S*-transferase
GTP	guanosine 5′-triphosphate
GUI	graphic user interface
HbSNO	*S*-nitrosohemaglobin
HCCA (4-HCCA)	4-hydroxy-α-cyanocinnamic acid
hCG	human chorionic gonadotrophin
hCG-β	β-subunit of human chorionic gonadotrophin
HdeA	protein from *E. coli*
hdeAB	genes encoding HdeA and HdeB proteins
HdeB	protein from *E. coli*
HE	HEPES and EDTA-based buffer
HEPES	4-(2-hydroxyethyl)-1-piperazineethanesulfonic acid
HFIP	hexafluoro-2-propanol
HIV	human immunodeficiency virus
hLH	human lutenizing hormone
HOAc	acetic acid
HPA	3-hydroxypicolinic acid
HPLC	high-performance (pressure) liquid chromatography
HRV14	human rhinovirus 14
HX	hydrogen exchange
I	inhibitor concentration
IAA	iodoacetamide
ID	inner diameter
IDA	isotope dilution assay

IDE	insulin degrading enzyme
IEF	isoelectric focusing
IgG	immunoglobulin G
iNOS	recombinant inducible mouse macrophage NO-synthase
IOC	International Olympic Committee
Iodogen	1,3,4,6-tetrachloro-3α,6α-diphenylglycouril
IP/MS	immunoprecipitation/mass spectrometry
IPTG	isopropyl-β-D-thiogalactopyranoside
IR	infrared
ISV	ion-source voltage
k_{cat}	rate constant for catalyzed reaction
K_A	association constant
kb	kilobase
kDa	kilodalton (10^3 Daltons)
KER	kinetic energy release
KERD	kinetic energy release distribution
keV	kiloelectron volt (10^3 eV)
k_i	inhibitor (inactivator) rate constant
K_i	inhibitor (inactivator) dissociation (binding) constant
kJ	kilojoule (10^3 J)
K_m	Michaelis-Menten constant
kV	kilovolt (10^3 V)
LB	Luria Bertani culture medium
LC	liquid chromatography
LC/ESI-MS	on-line liquid chromatography/electrospray ionization mass spectrometry
LC/MS	on-line liquid chromatography/mass spectrometry
LC/LC/MS/MS	on-line two-dimensional liquid chromatography/tandem mass spectrometry
LC/MS/MS	on-line liquid chromatography/tandem mass spectrometry
LD	laser desorption
LD-MS	laser desorption mass spectrometry
$M^{+\bullet}$	molecular ion-radical—positively charged
$M^{-\bullet}$	molecular ion-radical—negatively charged
mA	milliamp (10^{-3} A)
mAbs	monoclonal antibodies
MALDI	matrix-assisted laser desorption/ionization
MALDI-FTICR-MS MALDI-FTMS	matrix-assisted laser desorption/ionization fourier transform mass spectrometry
MALDI-MS	matrix-assisted laser desorption/ionization mass spectrometry

MALDI-TOFMS	matrix-assisted laser desorption/ionization time-of-flight mass spectrometry
MalS	α-amylase
MCH	melanin concentrating hormone
ME	2-mercaptoethanol
MEM	minimum essential medium
MeOH	methanol
meV	milli-electron volt (10^{-3} eV)
mFA1	murine fetal antigen 1
µg	microgram (10^{-6} g)
mg	milligram (10^{-3} g)
$(M-H)^-$	molecular ion after loss of proton—negatively charged
$[M+H]^+$ or MH^+	singly protonated molecular ion—positively charged
$[M+nH]^{n+}$ or $[MH_n]^n$	multiply protonated (n-fold) molecular ion—positively charged
mi	monoisotopic
MIDI	chemotaxonomic method for bacteria
MIKE	mass-analyzed ion kinetic energy
µJ	microjoule (10^{-6} J)
mL	milliliter (10^{-3} L)
µL	microliter (10^{-6} L)
mM	millimolar (10^{-3} molar)
µM	micromolar (10^{-6} molar)
$[M+Na]^+$ $[MNa]^+$	monosodiated molecular ion—positively charged
mol	symbol for mole
MOPS	3-(N-morpholino)propane sulfonic acid
mRNA	messenger ribonucleic acid
MS	mass spectrometry
MS/MS	tandem mass spectrometry (mass spectrometry/mass spectrometry)
MS/MS/MS (MS^3)	a three-stage MS/MS experiment
M_r (mol wt)	relative molecular mass (molecular weight)
Muc 1a′	mucin-derived peptide
m/z	mass-to-charge ratio
Na$_2$EDTA	disodium salt of ethylenediamine tetraacetic acid.
NaAz	sodium azide
NAD	nicotinamide adenine dinucleotide
NADH (β-NADH)	reduced form of β-nicotiamide adenine dinucleotide
NADPH	reduced form of nicotinamide adenine dinucleotide phosphate
nano	10^{-9}
nanomole	10^{-9} mole

NC	nitrocellulose
Nd-YAG	neodymium-doped yttrium aluminium garnet (laser)
NeuAc	sialic acid (N-acetylneuraminic acid)
ng	nanogram (10^{-9} g)
NH_4OAc	ammonium acetate
NHS	N-hydroxysuccinimido-
v_i	inhibited relative reaction rate
nL	nanoliter (10^{-9} L)
nM	nanomolar (10^{-9} molar)
nm	nanometer (10^{-9} m)
nmol	nanomole
NMR	nuclear magnetic resonance
v_o	uninhibited relative reaction rate
NOG	n-octyl glucoside
Nonidet p-40	trade name for nonionic detergent
NtrC	nitrogen assimilation regulatory protein (from *E. coli*)
O.D.	optical density
OD	outer diameter
OGP	n-octyl-glucopyranoside
OMPA	outer membrane protein
OR	orifice energy
ORF	open reading frame
PAGE	polyacrylamide gel electrophoresis
PAPS	phosphoadenosine-phosphosulfate
PBC	primary biliary cirrhosis
PBS	phosphate-buffered saline
PCR	polymerase chain reaction
pD	a measure of deuterium ion activity, measured analogously to pH
PDB	Protein Data Bank (Brookhaven National Lab. http://pdb.pdb.bnl.gov/)
PDC-E2	pyruvate dehydrogenase dihydrolipoamide acetyltransferase
PEEK	polyetherether ketone
Pefabloc SC	4-(2-aminoethyl)-benzenesulfonyl fluoride hydrochloride
PEG	polyethylene glycol
PEP	prolyl endopeptidase
PFPA	pentafluoropropionic acid
pGEX	expression system
picomole	10^{-12} mole
pK_n	$pK_n = -\log K_n$ when K_n is the constant for a given solute dissociation step

pK_a	pK_a = – log K_a when K_a is the dissociation constant of an acid in water
pL	picoliter (10^{-12} liter)
pM	picomolar (10^{-12} molar)
PMF	peptide mass fingerprint
pmol	picomole
PMSF	phenylmethylsulfonyl fluoride
PNGase F	peptide N-glycosidase
pNPX	p-nitrophenyl-β-D-xylopyranoside
PROSITE	protein family and domain database at http:/www.expasy.ch/prosite/
ps	picosecond (10^{-12} s)
PSD	postsource decay
PVDF	polyvinylidene difluoride
Q	a quadrupole mass filter in an instrument configuration
QTOF (QqTOF)	QTOF is a hybrid instrument (Q + TOF); q is a quadrupole collision cell
Q1	first quadrupole (mass analyzer) in a triple quadrupole MS
Q2	second quadrupole (collision cell) in a triple quadrupole MS
Q3	third quadrupole (mass analyzer) in a triple quadrupole MS
RAAM	reagent-array analysis method
RE	reflector
RIA	radioimmunoassay
RNA	ribonucleic acid
RP	reversed phase
RP-HPLC	reversed-phase high-performance (pressure) liquid chromatography
RSNO	nitrosothiol
S	substrate concentration
S/N	signal-to-noise ratio
SA	sinapinic acid
SCE	strong-cation exchange
SDS	sodium dodecyl sulfate
SDS-PAGE	sodium dodecyl sulfate-polyacrylamide gel electrophoresis
SIM	selected-ion monitoring
SPE	solid-phase extraction
TBS	tris-buffered saline
TCA	trichloroacetic acid
TCEP	tris(2-carboxyethyl)phosphine hydrochloride
TFA	trifluoroacetic acid
THAP	2,4,6-trihydroxyacetophenone

TIC	total-ion current
TLCK	N-α-p-tosyl-L-lysine chloromethyl ketone
TOF	time-of-flight
TOF-MS	time-of-flight mass spectrometry
TPCK	L-1-tosylamide-2-phenylethylchloromethyl ketone
Tricine	N-[2-hydroxy-1,1-bis(hydroxymethyl)ethyl]glycine
Tris	tris(hydroxymethyl)aminomethane
Tris-HCl	tris(hydroxymethyl)aminomethane hydrochloride
Triton	trade name for polyethylene ether surface-active compounds
Triton X-100	polyethylene ether nonionic surfactant
tRNA	transfer ribonucleic acid
TSH	thyrotropin
Tween	trade name for polyoxyethylene sorbitan surface-active compounds
Tween 80	primarily monooleate-esterified version of Tween
u	atomic mass unit
UHQ	ultra-high quality
UV	ultraviolet
V_{max}	maximal rate for a process
WHO	World Health Organization
X-gal	5-bromo-4-chloro-3-indolyl-β-D-galactoside
Xyl	xylose
z	number of charges on an ion

Appendix II: Mass Values and Abbreviated Forms for Amino Acid and Monosaccharide Residues

The molecular weight of a protein or peptide (M_p), of known amino acid composition, can be calculated by adding the masses of each individual amino acid residue (M_{ai}), using data from Table 2, to the masses of the appropriate N- and C-terminal groups (M_N, M_C) as in Eq. *(1)*:

$$M_p = \Sigma\, n_i\, M_{ai} + M_N + M_C \qquad (1)$$

where n_i represents the number of occurrences of residue i. For a straightforward protein or peptide with unmodified carboxy and amino termini, then M_N = the mass of a hydrogen atom (monoisotopic = isotopic average = 1.008) and M_C = the mass of a hydroxyl group (monoisotopic = 17.003, isotopic average = 17.007). Note that the masses of the amino acid residues are equivalent to those of the corresponding amino acid less the mass of a water molecule.

In the same manner as for amino acids, the molecular weight of an oligosaccharide (M_{os}) of known monosaccharide composition may be calculated from Eq. *(2)* using the residue masses (M_{si}) given in Table 3 and adding the mass of

Table 1
Mass and Abundance Values for Individual Elements

Element	Isotopically averaged mass[a]	Isotope	Isotopic mass	Isotopic abundance (%)
H	1.00794	^{1}H	1.007835[b]	99.985
		^{2}H	2.01410	0.015
C	12.011	^{12}C	12.00000	98.90
		^{13}C	13.003355[b]	1.10
N	14.00674	^{14}N	14.00307	99.634
		^{15}N	15.00011	0.366
O	15.9994	^{16}O	15.99491	99.762
		^{17}O	16.99913	0.038
		^{18}O	17.99916	0.200
F	18.99840	^{19}F	18.99840	100
Na	22.98977	^{23}Na	22.98977	100
Si	28.0855	^{28}Si	27.97693	92.23
		^{29}Si	28.97649	4.67
		^{30}Si	29.97377	3.10
P	30.97376	^{31}P	30.97376	100
S	32.066	^{32}S	31.97207	95.02
		^{33}S	32.97146	0.75
		^{34}S	33.96787	4.21
		^{36}S	35.96708	0.02
Cl	35.4527	^{35}Cl	34.96885	75.77
		^{37}Cl	36.96590	24.23
K	39.0983	^{39}K	38.96371	93.258
		^{40}K	39.96400	0.012
		^{41}K	40.96183	6.730

[a]Equivalent to atomic weight.
[b]Not rounded off where result is equivocal.

a water molecule (M_{H2O}; monoisotopic = 18.011, isotopic average = 18.015) to represent the terminating groups:

$$M_{os} = \Sigma\, n_i\, M_{si} + M_{H2O} \tag{2}$$

where n_i represents the number of occurrences of residue i. Again, the masses of sugar residues are equivalent to those of the corresponding sugar less the mass of a water molecule.

Appendix III: Peptide Fragmentation

Peptides are generally based on a linear arrangement of repeating units (– NH – CHR – CO –), which differ only in the nature of the side chain, R. Fragmentation of peptide molecular ions (or, more usually, [M+H]$^+$ ions in the

Table 2
Amino Acid Residues

Amino acid	Three (one)-letter code	Monoisotopic mass	Average mass
Glycine	Gly (G)	57.021	57.052
Alanine	Ala (A)	71.037	71.079
Serine	Ser (S)	87.032	87.078
Proline	Pro (P)	97.053	97.117
Valine	Val (V)	99.068	99.133
Threonine	Thr (T)	101.048	101.105
Cysteine	Cys (C)	103.009	103.145
Isoleucine	Ile (I)	113.084	113.159
Leucine	Leu (L)	113.084	113.159
Asparagine	Asn (N)	114.043	114.104
Aspartic acid	Asp (D)	115.027	115.089
Glutamine	Gln (Q)	128.059	128.131
Lysine	Lys (K)	128.095	128.174
Glutamic acid	Glu (E)	129.043	129.116
Methionine	Met (M)	131.040	131.199
Histidine	His (H)	137.059	137.141
Phenylalanine	Phe (F)	147.068	147.177
Arginine	Arg (R)	156.101	156.188
Tyrosine	Tyr (Y)	163.063	163.176
Tryptophan	Trp (W)	186.079	186.213
Dehydroalanine	Dha	69.021	69.063
Homoserine lactone	Hsl	83.037	83.090
Homoserine	Hse	101.048	101.105
Pyroglutamic acid	Glp	111.032	111.100
Hydroxyproline	Hyp	113.048	113.116
Hydroxylysine	Hyl	144.090	144.174
Carbamidomethylcysteine		160.031	160.197
Carboxymethylcysteine		161.015	161.181
Pyridylethylcysteine		208.067	208.284

positive-ion mode) can, as a result, be described in terms of a limited number of fragmentation paths, each of which has been assigned a letter-based code. Thus, a, b, and c ions are formed by main-chain fragmentation, with the positive charge on the N terminus, whereas the d ion is produced by side-chain fragmentation. On the other hand, v, w, x, y, and z ions are formed with the positive charge on the C terminus, with the v and w ions being produced by side chain fragmentation. The numeric subscript, seen on the diagrams, denotes the position in the amino acid chain where fragmentation occurs.

Table 3
Monosaccharide Residues

Monosaccharide name	Abbreviation	Monoisotopic mass	Average mass
Deoxypentose	DeoxyPen	116.117	116.047
Pentose	Pen	132.116	132.042
Deoxyhexose[a]	DeoxyHex	146.143	146.058
Hexose[b]	Hex	162.142	162.053
Hexosamine	HexN	161.158	161.069
N-Acetylhexosamine[c]	HexNAc	203.195	203.079
Hexuronic acid	HexA	176.126	176.032
N-Acetylneuraminic acid	NeuAc	291.258	291.095
N-Glycolylneuraminic acid	NeuGc	307.257	307.090

[a]For example, Fucose (Fuc), Rhamnose (Rha).
[b]For example, Mannose (Man), Glucose (Glc), Galactose (Gal).
[c]For example, Glucosamine (GlcNAc), Galactosamine (GalNAc).

N-Terminal charge retention

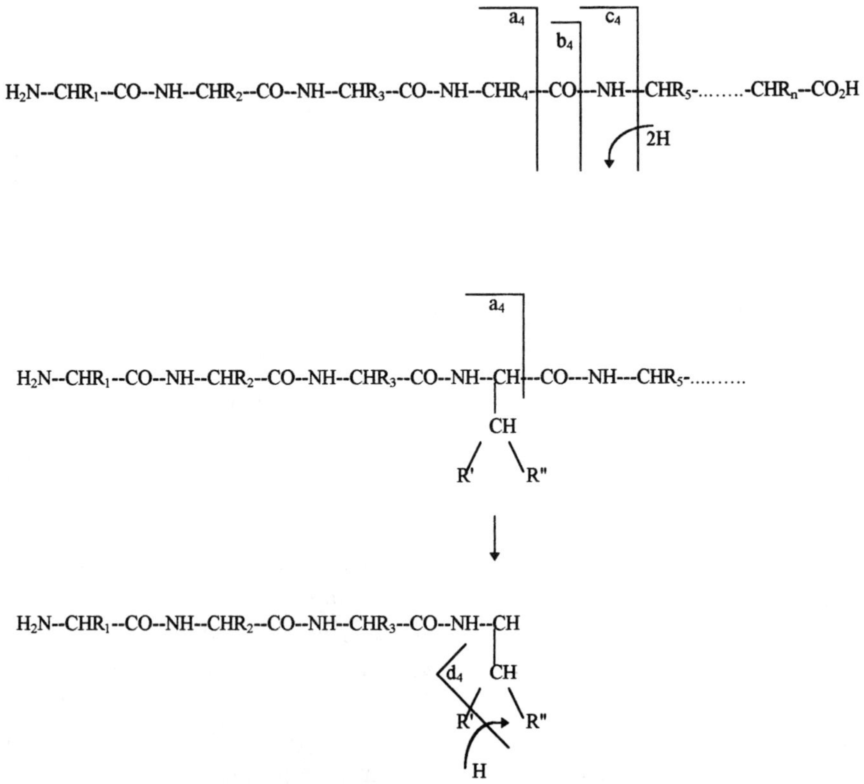

H_2N--CHR_1--CO--NH--CHR_2--CO--NH--CHR_3--CO--NH--CHR_4--CO--NH--CHR_5-.........-CHR_n--CO_2H

a_4 | b_4 | c_4

2H

H_2N--CHR_1--CO--NH--CHR_2--CO--NH--CHR_3--CO--NH--CH--CO---NH---CHR_5-.........

a_4

CH
R' R"

↓

H_2N--CHR_1--CO--NH--CHR_2--CO--NH--CHR_3--CO--NH--CH

d_4 CH
R' R"
H

C-Terminal charge retention

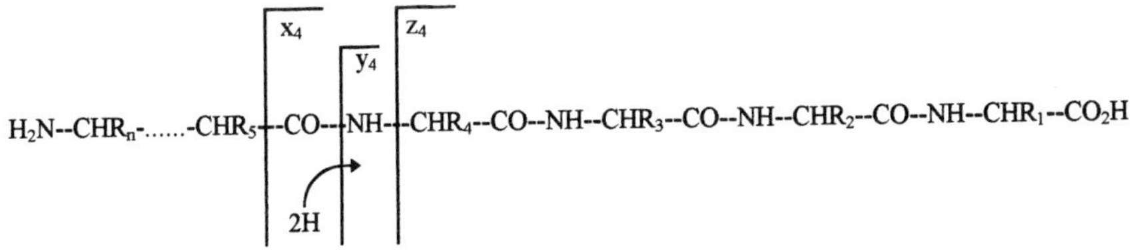

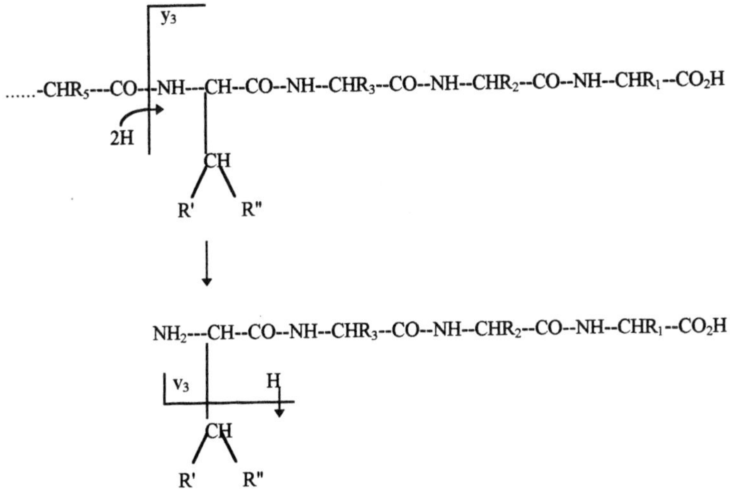

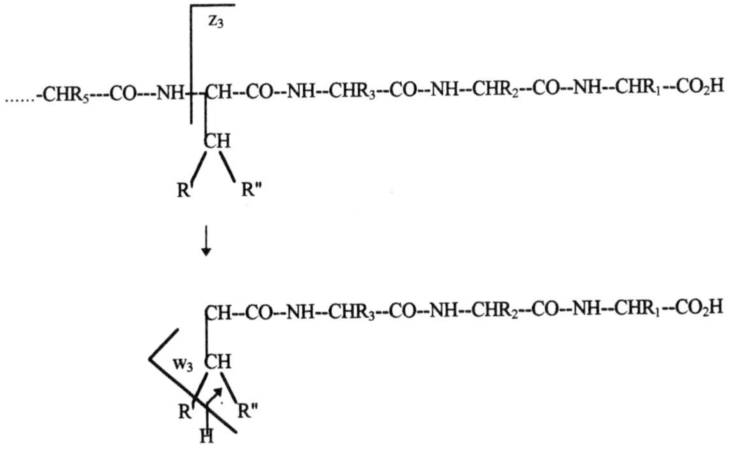

$$H_2N-CHR_n-......-CHR_5-CO-|NH-CHR_4-CO-NH-CHR_3-CO-NH-CHR_2-CO-NH-CHR_1-CO_2H$$

with x_4, y_4, z_4 cleavage sites; 2H transfer at y_4.

$$......-CHR_5-CO-|NH-CH-CO-NH-CHR_3-CO-NH-CHR_2-CO-NH-CHR_1-CO_2H$$

y_3 cleavage; 2H transfer; side chain CH with R' and R''.

↓

$$NH_2-CH-CO-NH-CHR_3-CO-NH-CHR_2-CO-NH-CHR_1-CO_2H$$

v_3 cleavage with H transfer; side chain CH with R' and R''.

$$......-CHR_5-CO-NH-|CH-CO-NH-CHR_3-CO-NH-CHR_2-CO-NH-CHR_1-CO_2H$$

z_3 cleavage; side chain CH with R' and R''.

↓

$$CH-CO-NH-CHR_3-CO-NH-CHR_2-CO-NH-CHR_1-CO_2H$$

w_3 cleavage; side chain CH with R', R'' and H.

Appendix IV: Current References to the Use of Mass Spectrometry in Protein and Peptide Analysis

> For out of old fields, as men say,
> Cometh all this new corn from year to year,
> And out of old books, in good faith,
> Cometh all this new science that men learn
> (Geoffrey Chaucer: *The Parlement of Foules*)

This final appendix provides an update to the main text by listing references from current literature in this fast-moving field. These references are mostly taken from the appropriate section of the mass spectrometry current awareness service which is regularly compiled for the *Journal of Mass Spectrometry*. Further references, taken directly from the *Journal of Mass Spectrometry, Rapid Communications in Mass Spectrometry*, and *European Mass Spectrometry*, have been added to those in the current awareness listings. Most references selected were published in 1997/8. Coverage generally extends to the end of 1998 but also covers the first four months of 1999 for the above three journals. The listed references have been categorized under the following headings: peptide sequencing, peptide analysis (qualitative and quantitative), peptide synthesis, protein sequencing and peptide mapping, protein analysis (qualitative and quantitative), protein configuration and folding, general analysis of glycosylated or phosphorylated proteins and peptides, protein interactions and kinetics, and method development in protein analysis.

A survey of these current awareness listings suggests the following principal sources for articles concerned with the use of mass spectrometry in protein and peptide analysis. This list of the "top 10" journals, which carry just over 60% of the published articles, is presented in order of decreasing frequency of citation. Note that publication is becoming more widespread since the corresponding list in the first volume of this series reported that the "top 10" journals carried over 80% of the listed papers.

1. *Rapid Communications in Mass Spectrometry*
2. *Analytical Chemistry*
3. *Electrophoresis*
4. *Journal of Mass Spectrometry*
5. *Analytical Biochemistry*
6. *International Journal of Mass Spectrometry and Ion Processes*
7. *Journal of the American Society for Mass Spectrometry*
8. *Journal of Chromatography (Parts A/B)*

9. *Protein Science*
10. *Biochemistry—USA*

Peptide Sequencing

Bahr, U., Karas, M., and Kellner, R. (1998) Differentiation of lysine/glutamine in peptide sequence analysis by electrospray ionization sequential mass spectrometry coupled with a quadrupole ion trap. *Rapid Commun. Mass Spectrom.* **12**, 1382.

Dreisewerd, K., Kingston, R., Geraerts, W. P. M., Li, K. W. (1997) Direct mass spectrometric peptide profiling and sequencing of nervous tissues to identify peptides involved in male copulatory behavior in *Lymnaea stagnalis*. *Int. J. Mass Spectrom. Ion Proc.* **169**, 291.

Fernandez-de-Cossio, J., Gonzalez, J., Betancourt, L., Besada, V., Padron, G., Shimonishi, Y., and Takao, T. (1998) Automated interpretation of high-energy collision-induced dissociation spectra of singly protonated peptides by 'SeqMS', software aid for *de novo* sequencing by tandem mass spectrometry. *Rapid Commun. Mass Spectrom.* **12**, 1867.

Jainhuknan, J. and Cassady, C. J. (1998) Negative ion post-source decay time-of-flight mass spectrometry of peptides containing acidic amino acid residues. *Anal. Chem.* **70**, 5122.

Jimenez, C. R., Li, K. W., Dreisewerd, K., Spijker, S., Kingston, R., Bateman, R. H., et al. (1998) Direct mass spectrometric peptide profiling and sequencing of single neurons reveals differential peptide patterns in a small neuronal network. *Biochemistry-USA* **37**, 2070.

Kratzer, R., Eckerskorn, C., Karas, M., and Lottspeich, F. (1998) Suppression effects in enzymatic peptide-ladder sequencing using ultraviolet matrix-assisted laser desorption/ionization mass spectrometry. *Electrophoresis* **19**, 1910.

Lehmann, E., Vetter, S., and Zenobi, R. (1998) Detection of specific zinc-finger peptide complexes with matrix-assisted laser desorption/ionization mass spectrometry. *J. Protein Chem.* **17**, 517.

Lin, T. and Glish, G. L. (1998) C-terminal peptide sequencing via multistage mass spectrometry *Anal. Chem.* **70**, 5162.

Lindh, I., Sjovall, J., Bergman, T., and Griffiths, W. J. (1998) Negative-ion electrospray tandem mass spectrometry of peptides derivatized with 4-aminonaphthalenesulphonic acid. *J. Mass Spectrom.* **33**, 988.

Miyagi, M., Nakao, M., Nakazawa, T., Kato, I., and Tsunasawa, S. (1998) A novel derivatization method with 5-bromo-nicotonic acid N-hydroxysuccinimide for determination of the amino acid sequences of peptides. *Rapid Commun. Mass Spectrom.* **12**, 603.

Naylor, S. and Tomlinson, A. J. (1998) Membrane preconcentration-capillary electrophoresis tandem mass spectrometry (mPC-CE-MS/MS) in the sequence analysis of biologically derived peptides. *Talanta* **45**, 603.

Reiber, D. C., Brown, R. S., Weinberger, S., Kenny, J., and Bailey, J. (1998) Unknown peptide sequencing using matrix-assisted laser desorption/ionization and in-source decay *Anal. Chem.* **70**, 1214.

Schurch, S., Scott, J. R., and Wilkins, C. L. (1997) Alternative labeling method for peptide ladder sequencing using matrix-assisted laser desorption-ionization Fourier transform mass spectrometry. *Int. J. Mass Spectrom. Ion Proc.* **169,** 141.

Settlage, R. E., Russo, P. S., Shabanowitz, J., and Hunt, D. F. (1998) A novel micro-ESI source for coupling capillary electrophoresis and mass spectrometry. Sequence determination of tumor peptides at the attomole level. *J. Microcolumn* **10,** 281.

Spengi, B., Luetzenkirchen, F., Metzger, S., Chaurand, P., Kaufmann, R., Jeffrey, W., et al. (1997) Peptide sequencing of charged derivatives by post-source decay MALDI mass spectrometry. *Int. J. Mass Spectrom. Ion Proc.* **169,** 127.

Stevenson, T. I., Loo, J. A., and Greis, K. D. (1998) Coupling capillary high-performance liquid chromatography to matrix-assisted laser desorption/ionization mass spectrometry and N-terminal sequencing of peptides via automated microblotting onto membrane substrates. *Anal. Biochem.* **262,** 99.

Stimson, E., Truong, O., Richter, W. J., Waterfield, M. D., and Burlingame, A. L. (1997) Enhancement of charge remote fragmentation in protonated peptides by high- energy CID MALDI-TOF-MS using "cold" matrices. *Int. J. Mass Spectrom. Ion Proc.* **169,** 231.

Strahler, J. R., Smelyanskiy, Y., Lavine, G., and Allison, J. (1997) Development of methods for the charge-derivatization of peptides in polyacrylamide gels and membranes for their direct analysis using matrix-assisted laser desorption-ionization mass spectrometry. *Int. J. Mass Spectrom. Ion Proc.* **169,** 111.

Yates, J. R., Morgan, S. F., Gatlin, C. L., Griffin, P. R., and Eng, J. K. (1998) Method to compare collision-induced dissociation spectra of peptides. Potential for library searching and subtractive analysis. *Anal. Chem.* **70,** 3557.

Peptide Analysis (Qualitative and Quantitative)

Ayyoub, M., Monsarrat, B., Mazarguil, H., and Gairin, J. E. (1998) Analysis of the degradation mechanisms of MHC class I-presented tumor antigenic peptides by high performance liquid chromatography/electrospray ionization mass spectrometry: application to the design of peptidase-resistant analogs. *Rapid Commun. Mass Spectrom.* **12,** 557.

Belder, D., Husmann, H., Schoppenthau, J., and Paulus, G. (1998) CE/MALDI-MS of peptides and low molecular weight drugs using non-volatile buffers *J. High Res. Chromatogr.* **21,** 59.

Boss, H. J., Watson, D. B., and Rush, R. S. (1998) Peptide capillary zone electrophoresis/mass spectrometry of recombinant human erythropoietin. An evaluation of the analytical method. *Electrophoresis* **19,** 2654.

Brown, R. S., Feng, J. H., and Reiber, D. C. (1997) Further studies of in-source fragmentation of peptides in matrix-assisted laser desorption-ionization. *Int. J. Mass Spectrom. Ion Proc.* **169,** 1.

Bulet, P., Uttenweiler-Joseph, S., Moniatte, M., Van Dorsselaer, A., and Hoffmann, J. A. (1998) Differential display of peptides induced during the immune response to *Drosophila.* A matrix-assisted laser desorption/ionization time-of-flight mass spectrometry study. *J. Protein Chem.* **17,** 528.

Cheong, W. J., Oh, C. S., and Yoo, J. S. (1998) Analyses of phthalates and peptides using a gradient micro-LC/MS system. *Bull. Kor. Chem. Soc.* **19**, 495.

Chiu, D. T. and Zare, R. N. (1998) Assaying for peptides in individual *Aplysia* neurons with mass spectrometry (Commentary). *Proc. Natl. Acad. Sci. USA* **95**, 3338.

Clarke, N. J., Tomlinson, A. J., Ohyagi, Y., Younkin, S., and Naylor, S. (1998) Detection and quantitation of cellularly-derived amyloid β-peptides by immunoprecipitation/HPLC/MS. *FEBS Lett.* **430**, 419.

Curcuruto, O., Rovatti, L., Pastorino, A. M., and Hamdan, M. (1999) Peptide nitration by peroxynitrite. Characterisation of the nitration sites by liquid chromatography/tandem mass spectrometry. *Rapid Commun. Mass Spectrom.* **13**, 156.

Desiderio, D. M. and Zhu, X. G. (1998) Quantitative analysis of methionine enkephalin and p-endorphin in the pituitary by liquid secondary ion mass spectrometry and tandem mass spectrometry. *J. Chromatogr. A.* **794**, 85.

Deterding, L. J., Barr, D. P., Mason, R. P., and Tomer, K. B. (1998) Characterization of cytochrome *c* free radical reactions with peptides by mass spectrometry. *J. Biol. Chem.* **273**, 12863.

Ducrocq, C., Dendane, M., Laprevote, O., Serani, L., Das, B. C., Bouchemal-Chibani, N., et al. (1998) Chemical modifications of the vasoconstrictor peptide angiotensin II by nitrogen oxides (NO, HNO_2, HOONO). Evaluation by mass spectrometry. *Eur. J. Biochem.* **253**, 146.

Duncan, M. W. and Poljak, A. (1998) Amino acid analysis of peptides and proteins on the femtomole scale by gas chromatography/mass spectrometry. *Anal. Chem.* **70**, 890.

Erdjument-Bromage, H., Lui, M., Lacomis, L., Grewal, A., Annan, R. S., McNulty, D. E., et al. (1998) Examination of micro-tip reversed-phase liquid chromatographic extraction of peptide pools for mass spectrometric analysis. *J. Chromatogr. A.* **826**, 167.

Escoubas, P., Whiteley, B. J., Kristensen, C. P., Celerier, M. L., Corzo, G., and Nakajima, T. (1998) Multidimensional peptide fingerprinting by high performance liquid chromatography, capillary zone electrophoresis and matrix-assisted laser desorption /ionization time-of-flight mass spectrometry for the identification of tarantula venom samples. *Rapid Commun. Mass Spectrom.* **12**, 1075.

Gobom, J., Nordhoff, E., Mirgorodskaya, E., Ekman, R., and Roepstorff, P. (1999) Sample purification and preparation technique based on nanoscale reversed phase columns for the sensitive analysis of complex peptide mixtures by matrix-assisted laser desorption/ionization mass spectrometry. *J. Mass Spectrom.* **34**, 105.

Green-Church, K. B., and Limbach, P. A. (1998) Matrix-assisted laser desorption/ionization mass spectrometry of hydrophobic peptides. *Anal. Chem.* **70**, 5322.

Harriman, S. P., Hill, J. A., Tannenbaum, S. R., and Wishnok, J. S. (1998) Detection and identification of carcinogen-peptide adducts by nanoelectrospray tandem mass spectrometry. *J. Am. Soc. Mass Spectrom.* **9**, 202.

Herring, C. J. and Qin, J. (1999) An on-line preconcentrator and the evaluation of electrospray interfaces for the capillary electrophoresis/mass spectrometry of peptides. *Rapid Commun. Mass Spectrom.* **13**, 1.

Hsieh, S., Dreisewerd, K., Van der Schors, R. C., Jimenez, C. R., Stahl-Zeng, J. R., Hillenkamp, F., et al. (1998) Separation and identification of peptides in single neurons by micro-column liquid chromatography/matrix-assisted laser desorption/ionization time-of-flight mass spectrometry and post-source decay analysis. *Anal. Chem.* **70**, 1847.

Janaky, T., Szabo, P., Kele, Z., Balaspiri, L., Varga, C., Galfi, M., et al. (1998) Identification of oxytocin and vasopressin from neurohypophyseal cell culture. *Rapid Commun. Mass Spectrom.* **12**, 1765.

Jimenez, C. R. and Burlingame, A. U. (1998) Ultra-micro analysis of peptide profiles in biological samples using MALDI mass spectrometry. *Exp. Nephrol.* **6**, 421.

Jones, M. D., Patterson, S. D., and Lu, H. S. (1998) Determination of disulfide bonds in highly bridged disulfide-linked peptides by matrix-assisted laser desorption/ionization mass spectrometry with post-source decay. *Anal. Chem.* **70**, 136.

Lavanant, H., Heck, A., Derrick, P. J., Mellon, F. A., Parr, A., Dodd, H. M., et al. (1998) Characterization of genetically modified nisin molecules by Fourier transform ion cyclotron resonance mass spectrometry. *Eur. Mass Spectrom.* **4**, 405.

Lindemayr, K., Brueckner, A., Koerner, A., Hahn, R., Jungbauer, A., Josic, D. J., et al. (1999) Matrix-assisted laser desorption/ionization time-of-flight and nanoelectrospray ionization ion trap mass spectrometric characterization of 1-cyano-2-substituted-benz[*f*]isoindole derivatives of peptides for fluorescence detection. *J. Mass Spectrom.* **34**, 427.

Molle, D., Morgan, F., Bouhallab, S., and Leonil, J. (1998) Selective detection of lactolated peptides in hydrolysates by liquid chromatography/electrospray tandem mass spectrometry. *Anal. Biochem.* **259**, 152.

Naylor, S., Ji, Q. C., Johnson, K. L., Tomlinson, A. J., Kieper, W. C., and Jameson, S. C. (1998) Enhanced sensitivity for sequence determination of major histocompatibility complex class, I., peptides by membrane preconcentration-capillary electrophoresis/microspray-tandem mass spectrometry. *Electrophoresis* **19**, 2207.

Nilsson, C., Westman, A., Blennow, K., and Ekman, R. (1998) Processing of neuropeptide Y and somatostatin in human cerebrospinal fluid as monitored by radioimmunoassay and mass spectrometry. *Peptides* **19**, 1137.

Nilsson, C. L. and Brodin, E. (1998) Substance P and related peptides in porcine cortex: whole tissue and nuclear localization. *J. Chromatogr. A.* **800**, 21.

Prokai, L., Kim, H. S., Zharikova, A., Roboz, J., Ma, L., Deng, L., et al. (1998) Electrospray ionization mass spectrometric and liquid chromatographic/mass spectrometric studies on the metabolism of synthetic dynorphin A peptides in brain tissue *in vitro* and *in vivo*. *J. Chromatogr. A.* **800**, 59.

Prokai, L. and Zharikova, A. D. (1998) Identification of synaptic metabolites of dynorphin A (1-8) by electrospray ionization and tandem mass spectrometry. *Rapid Commun. Mass Spectrom.* **12**, 1796.

Redeker, V., Toullec, J. Y., Vinh, J., Rossier, J., and Soyez, D. (1998) Combination of peptide profiling by matrix-assisted laser desorption/ionization time-of-flight mass spectrometry and immunodetection on single glands or cells. *Anal. Chem.* **70**, 1805.

Salih, B. and Zenobi, R. (1998) MALDI mass spectrometry of dye-peptide and dye-protein complexes. *Anal. Chem.* **70**, 1536.

Sandin, J., Nylander, I., and Silberring, J., (1998) Metabolism of β-endorphin in plasma studied by liquid chromatography/electrospray ionization mass spectrometry. *Regul. Pepti.* **73**, 67.

Schnolzer, M. and Lehmann, W. D. (1997) Identification of modified peptides by metastable fragmentation in MALDI mass spectrometry. *Int. J. Mass Spectrom. Ion Proc.* **169**, 263.

Skogsberg, U., McEwen, I., and Stenhagen, G. (1998) Liquid chromatography-electrospray mass spectrometry method to separate and detect *N-tert*-butoxycarbonyl peptides. *J. Chromatogr. A.* **808**, 253.

Uttenweiler-Joseph, S., Moniatte, M., Lagueux, M., Van Dorsselaer, A., Hoffmann, J. A., and Bulet, P. (1998) Differential display of peptides induced during the immune response of *Drosophila*. A matrix-assisted laser desorption/ionization time-of-flight mass spectrometry study. *Proc. Natl. Acad. Sci. USA* **95**, 11342.

Vaisar, T. and Urban, J. (1998) Low-energy collision-induced dissociation of protonated peptides. Importance of an oxazolone formation for a peptide bond cleavage. *Eur. Mass Spectrom.* **4**, 359.

Venkateshwaran, T. G., Stewart, J. T., Bishop, R. T., De Haseth, J. A., and Bartlett, M. G. (1998) Solution conformation of model peptides with the use of particle beam LC/FT-IR spectrometry and electrospray mass spectrometry. *J. Pharm. Biomed. Anal.* **17**, 57.

Worster, B. M., Yeoman, M. S., and Benjamin, P. R. (1998) Matrix-assisted laser desorption/ionization time-of-flight mass spectrometric analysis of the pattern of peptide expression in single neurons resulting from alternative mRNA splicing of the FMRFamide gene. *Eur. J. Neurosci.* **10**, 3498.

Yoshida, S., Tajima, M., and Takamatsu, T. (1998) Structural characterization of a potent vasodilatory recombinant peptide by a multiple mass spectrometric approach including LC/electrospray ionization MS (Japanese, English Abstract). *Bunseki Kagaku* **47**, 17.

Yuan, M., Namikoshi, M., Otsuki, A., Rinehart, K. L., Sivonen, K., and Watanabe, M. F. (1999) Low-energy collisionally-activated decomposition and structural characterization of cyclic heptapeptide microcystins by electrospray ionization mass spectrometry. *J. Mass Spectrom.* **34**, 33.

Yuan, M., Namikoshi, M., Otsuki, A., and Sivonen, K. (1998) Effect of amino acid side-chain on fragmentation of cyclic peptide ions. Differences of electrospray ionization/collision-induced decomposition mass spectra of toxic heptapeptide microcystins containing ADMAdda instead of Adda. *Eur. Mass Spectrom.* **4**, 287.

Zhang, H., Stoeckli, M., Andren, P. E., and Caprioli, R. M. (1999) Combining solid-phase preconcentration, capillary electrophoresis and off-line matrix-assisted laser desorption/ionization mass spectrometry. Intracerebral metabolic processing of peptide E *in vivo*. *J. Mass Spectrom.* **34**, 377.

Zigrovic, I., Versluis, C., Horvat, S., and Heerma, W. (1998) Mass spectrometric characterization of Amadori compounds related to the opioid peptide morphiceptin. *Rapid Commun. Mass Spectrom.* **12**, 181.

Peptide Synthesis

Aubagnac, J. L., Enjalbal, C., Subra, G., Bray, A. M., Combarieu, R., and Martinez, J. (1998) Application of time-of-flight secondary ion mass spectrometry to *in situ* monitoring of solid-phase peptide synthesis on the Multipin™ system. *J. Mass Spectrom.* **33**, 1094.

Cavelier, F., Enjalbal, C., El Haddadi, M., Martinez, J., Sanchez, P., Verducci, J., et al. (1998) Monitoring of peptide cyclization reactions by electrospray ionization mass spectrometry. *Rapid Commun. Mass Spectrom.* **12**, 1585.

D'Agostino, P. A., Hancock, J. R., Provost, L. R., Semchuk, P. D., and Hodges, R. S., (1998) Liquid chromatographic/high-resolution mass spectrometric and tandem mass spectrometric identification of synthetic peptides, using electrospray ionization. *J. Chromatogr. A.* **800**, 89.

Hakansson, K., Zubarev, R., and Hakansson, P. (1998) Combination of nozzle-skimmer fragmentation and partial acid hydrolysis in electrospray ionization time-of-flight mass spectrometry of synthetic peptides. *Rapid Commun. Mass Spectrom.* **12**, 705.

Thurmer, R., Meisenbach, M., Echner, H., Weiler, A., Al-Qawasmeh, R. A., Voelter, W., et al. (1998) Monitoring of liquid-phase peptide synthesis on soluble polymer supports via matrix-assisted laser desorption/ionization time-of-flight mass spectrometry. *Rapid Commun. Mass Spectrom.* **12**, 398.

Protein Sequencing and Peptide Mapping

Cao, P. and Moini, M. (1998) Capillary electrophoresis/electrospray ionization high mass accuracy time-of-flight mass spectrometry for protein identification using peptide mapping. *Rapid Commun. Mass Spectrom.* **12**, 864.

Chen, Y., Wall, D., and Lubman, D. M. (1998) Rapid identification and screening of proteins from whole cell lysates of human erythroleukemia cells in the liquid phase, using non-porous reversed phase high-performance liquid chromatography separations of proteins followed by matrix-assisted laser desorption/ionization mass spectrometry analysis and sequence database searching. *Rapid Commun. Mass Spectrom.* **12**, 1994.

Denzinger, T., Przybylski, M., Savoca, R., and Sonderegger, P. (1997) Mass spectrometric characterisation of primary structure, sequence heterogeneity, and intramolecular disulfide loops of the cell adhesion protein axonin-1 from chicken. *Eur. Mass Spectrom.* **3**, 379.

Gobom, J., Nordhoff, E., Ekman, R., and Roepstorff, P. (1997) Rapid micro-scale proteolysis of proteins for MALDI-MS peptide mapping using immobilized trypsin. *Int. J. Mass Spectrom. Ion Proc.* **169**, 153.

Hirayama, K., Yuji, R., Yamada, N., Kato, K., Arata, Y., and Shimada, I. (1998) Complete and rapid peptide and glycopeptide mapping of mouse monoclonal antibody by LC/MS/MS using ion trap mass spectrometry. *Anal. Chem.* **70**, 2718.

Jungblut, P. R., Otto, A., Favor, J., Lowe, M., Muller, E. C., Kastner, M., et al. (1998) Identification of mouse crystallins in 2D protein patterns by sequencing and mass spectrometry. Application to cataract mutants. *FEBS Lett.* **435**, 131.

Katta, V., Chow, D. T., and Rohde, M. F. (1998) Applications of in-source fragmentation of protein ions for direct sequence analysis by delayed extraction MALDI-TOF mass spectrometry. *Anal. Chem.* **70,** 4410.

Laidler, P., Cowan, D. A., Houghton, E., Kicman, A. T., and Marshall, D. E. (1998) Discrimination of mammalian growth hormones by peptide-mass mapping. *Rapid Commun. Mass Spectrom.* **12,** 975.

Marina, A., Garcia, M. A., Albar, J. P., Yague, J., Lopez de Castro, J. A., and Vazquez, J. (1999) High-sensitivity analysis and sequencing of peptides and proteins by quadrupole ion trap mass spectrometry. *J. Mass Spectrom.* **34,** 17.

Opiteck, G. J., Jorgenson, J. W., Moseley, M. A., and Anderegg, R. J. (1998) Two-dimensional micro-column HPLC coupled to a single quadrupole mass spectrometer for the elucidation of sequence tags and peptide mapping. *J. Microcolumn Sep.* **10,** 365.

Pfeifer, T., Ruecknagel, P., Kuellertz, G., and Schierhorn, A. (1999) A strategy for rapid and efficient sequencing of Lys-C peptides by matrix-assisted laser desorption/ionization time-of-flight mass spectrometry post-source decay. *Rapid Commun. Mass Spectrom.* **13,** 362.

Raftery, M. J. and Geczy, C. L. (1998) Identification of post-translational modifications and cDNA sequencing errors in the rat S100 proteins MRPS and 14 using electrospray ionization mass spectrometry. *Anal. Biochem.* **258,** 285.

Roepstorff, P. (1998) Protein sequencing or genome sequencing. Where does mass spectrometry fit into the picture? *J. Protein Chem.* **17,** 542.

Scheler, C., Lamer, S., Pan, Z. M., Li, X. P., Salnikow, J., and Jungblut, P. (1998) Peptide mass fingerprint sequence coverage from differently stained proteins on two-dimensional electrophoresis patterns by matrix-assisted laser desorption/ionization mass spectrometry (MALDl-MS). *Electrophoresis* **19,** 918.

Swiderek, K. M., Davis, M. T., and Lee, T. D. (1998) The identification of peptide modifications derived from gel-separated proteins using electrospray triple quadrupole and ion trap analyses. *Electrophoresis* **19,** 989.

Whittal, R. M., Keller, B. O., and, Li, L. (1998) Nanoliter chemistry combined with mass spectrometry for peptide mapping of proteins from single mammalian cell lysates. *Anal. Chem.* **70,** 5344.

Protein Analysis (Qualitative and Quantitative)

Amoresano, A., Andolfo, A., Siciliano, R. A., Cozzolinn, R., Minchiotti, L., Galliano, M., et al. (1998) Analysis of human serum albumin variants by mass spectrometric procedures. *BBA-Protein Struct. Mol. Enzym.* **1384,** 79.

Arnott, D., Henzel, W. J., and Stults, J. T. (1998) Rapid identification of comigrating gel-isolated proteins by ion trap mass spectrometry. *Electrophoresis* **19,** 968.

Asgeirsson, B., Haebel, S., Thorsteinsson, L., Helgason, E., Gudmundsson, K. O., Gudmundsson, G., et al. (1998) Hereditary cystatin, C., amyloid angiopathy. Monitoring the presence of the Leu-68-→Gln cystatin, C., variant in cerebrospinal fluids and monocyte cultures by MS. *Biochem. J.* **329,** 497.

Badock, V., Raida, M., Adermann, K., Forssmann, W. G., and Schrader, M. (1998) Distinction between the three disulfide isomers of guanylin 99-115 by low-energy collision-induced dissociation. *Rapid Commun. Mass Spectrom.* **12,** 1952.

Berger, S. J., Claude, A. C., and Melancon, L. (1998) Analysis of recombinant human ADP-ribosylation factors in reversed-phase high-performance liquid chromatography and electrospray mass spectrometry. *Anal. Biochem.* **264,** 53.

Bisse, E., Zorn, N., Eigel, A., Lizama, M., Huaman-Guillen, P., Marz, W., et al. (1998) Hemoglobin Rambam (β69[E13]Gly→Asp), a pitfall in the assessment of diabetic control. Characterization by electrospray mass spectrometry and HPLC. *Clin. Chem.* **44,** 2172.

Brennan, S. O. (1998) Electrospray ionization mass analysis of normal and genetic variants of human serum albumin. *Clin. Chem.* **44,** 2264.

Buzy, A., Millar, A. L., Legros, V., Wilkins, P. C., Dalton, H., and Jennings, K. R. (1998) The hydroxylase component of soluble methane mono-oxygenase from *Methylococcus capsulatus* (Bath) exists in several forms as shown by electrospray-ionisation mass spectrometry. *Eur. J. Biochem.* **254,** 602.

Camafeita, E. and Mendez, E. (1998) Screening of gluten avenins in foods by matrix-assisted laser desorption/ionization time-of-flight mass spectrometry. *J. Mass Spectrom.* **33,** 1023.

Campagnini, A., Foti, S., Jetschke-Schmachtel, M., Maccarrone, G., Przybylski, M., and Saletti, R. (1998) Characterization of cyanogen bromide fragments of reduced human sperm albumin by matrix-assisted laser desorption/ionization mass spectrometry (Letter). *J. Mass Spectrom.* **33,** 673.

Chassaigne, H. and Lobinski, R. (1998) Characterization of horse kidney metallothionein isoforms by electrospray MS, and reversed-phase HPLC-electrospray MS. *Analyst* **123,** 2125.

Chassaigne, H. and Lobinski, R. (1998) Polymorphism and identification of metallothionein isoforms by reversed-phase HPLC with on-line ion spray mass spectrometric detection. *Anal. Chem.* **70,** 2536.

Chassaigne, H. and Lobinski, R. (1998) Study of the polymorphism of metallothioneins by ion-spray mass spectrometry (French). *Analusis* **26,** M65.

Chong, B. E., Lubman, D. M., Rosenspire, A., and Miller, F. (1998) Protein profiles and identification of high performance liquid chromatography isolated protein of cancer cell lines using matrix-assisted laser desorption/ ionization time-of-flight mass spectrometry. *Rapid Commun. Mass Spectrom.* **12,** 1986.

Chua, E. K. M., Brennan, S. O., and George PMU (1998) Albumin Church Bay 560 Lys→Glu; a new mutation detected by electrospray ionisation mass spectrometry. *BBA-Protein Struct. Mol. Enzym.* **1**382, 305.

Clayton, P. T., Doig, M., Ghafari, S., Meaney, C., Taylor, C., Leonard, J. V., et al. (1998) Screening for medium chain acyl-CoA dehydrogenase deficiency using electrospray ionisation tandem mass spectrometry. *Arch. Dis. Child.* **79,** 109.

Cosenza, L., Sweeney, E., and Murphy, J. R. (1997) Disulfide bond assignment in human interleukin-7 by matrix-assisted laser desorption/ionization mass spectroscopy and site-directed cysteine to serine mutational analysis. *J. Biol. Chem.* **272,** 32995.

Courchesne, P. L., Jones, M. D., Robinson, J. H., Spahr, C. S., McCracken S. Bentley, D. L., et al. (1998) Optimization of capillary chromatography/ion trap mass spectrometry for identification of gel-separated proteins. *Electrophoresis* **19**, 956.

Crowley, J. R., Yarasheski, K., Leeuwenburgh, C., Turk, J., and Heinecke, J. W. (1998) Isotope dilution mass spectrometric quantification of 3-nitrotyrosine in proteins and tissues is facilitated by reduction to 3-aminotyrosine. *Anal. Biochem.* **259**, 127.

Dai, Y., Li, L., Roser, D. C., and Long, S. R. (1999) Detection and identification of low-mass peptides and proteins from solvent suspensions of *Escherichia coli* by high performance liquid chromatography fractionation and matrix-assisted laser desorption/ionization mass spectrometry. *Rapid Commun. Mass Spectrom.* **13**, 73.

Davis, M. T., and Lee, T. D. (1998) Rapid protein identification using a microscale electrospray LC/MS system on an ion trap mass spectrometer. *J. Am. Soc. Mass Spectrom.* **9**, 194.

De Sain-van der Velden, M. G. M., Rabelink, T. J., Gadellaa, M. M., Elzinga, H., Reijngoud, D. J., Kuipers, F., and Stellaard, F. (1998) In vivo determination of very-low-density lipoprotein-apolipoprotein B100 secretion rates in humans with a low dose of L-[1-^{13}C]valine and isotope ratio mass spectrometry. *Anal. Biochem.* **265**, 308.

Ducret, A., Van Oostveen, I., Eng, J. K., Yates, J. R., and Aebersold, R. (1998) High throughput protein characterization by automated reversed-phase chromatography/electrospray tandem mass spectrometry. *Protein Sci.* **7**, 706.

Duewell, N. S. and Honek, J. P. (1998) CNBr/formic acid reactions of methionine- and trifluoroethionine-containing lambda lysozyme: probing chemical and positional reactivity and formylation side-reactions by mass spectrometry. *J. Protein Chem.* **17**, 337.

Dufresne, C. P., Wood, T. D., and Hendrickson, C. L. (1998) High-resolution electrospray ionization Fourier transform mass spectrometry with infrared multiphoton dissociation of glucokinase from *Bacillus stearo-thermophilus*. *J. Am. Soc Mass Spectrom.* **9**, 1222.

Dukan, S., Turlin, E., Biville, F., Bolbach, G., Touati, D., Tabet, J. C., et al. (1998) Coupling 2D SDS-PAGE with CNBr cleavage and MALDI-TOFMS. A strategy applied to the identification of proteins induced by a hypochlorous acid stress in *Escherichia coli*. *Anal. Chem.* **70**, 4433.

Dworschak, R. G., Ens, W., Standing, K. G., Preston, K. R., Marchylo, B. A., Nightingale, M. J., et al. (1998) Analysis of wheat gluten proteins by matrix-assisted laser desorption/ionization mass spectrometry. *J. Mass Spectrom.* **33**, 429.

Easterling, M. L., Colangelo, C. M., Scott, R. A., and Amster, I. J. (1998) Monitoring protein expression in whole bacterial cells with MALDI time-of-flight mass spectrometry. *Anal. Chem.* **70**, 2704.

Erhard, M., Von Doehren, H., and Jungblut, P. R. (1999) Rapid identification of the new anabaenopeptin, G., from *Planktothrix agardhii* HUB 011 using matrix-assisted laser desorption/ionization time-of-flight mass spectrometry. *Rapid Commun. Mass Spectrom.* **13**, 337.

Fenyo D., Qin, J., and Chait, B. T. (1998) Protein identification using mass spectrometric information. *Electrophoresis* **19**, 998.

Fernandez, J., Gharahdaghi, F., and Mische, S. M. (1998) Routine identification of proteins from sodium dodecyl sulfate-polyacrylamide gal electrophoresis (SDS-PAGE) gels or polyvinyl difluoride membranes using matrix-assisted laser desorption/ionization time-of-flight mass spectrometry (MALDI-TOF-MS). *Electrophoresis* **19**, 1036.

Figeys, D., Zhang, Y., and Aebersold, R. (1998) Optimization of solid-phase microextraction capillary zone electrophoresis/mass spectrometry for high sensitivity protein identification. *Electrophoresis* **19**, 2338.

Fligge, T. A., Bruns, K., and Przybylski, M. (1998) Analytical development of electrospray and nanoelectrospray mass spectrometry in combination with liquid chromatography for the characterization of proteins. *J. Chromatogr. B.* **706**, 91.

Fukuo, T., Kubota, N., Kataoka, K., Nakai, M., Suzuki, S., and Arakawa, R. (1998) Matrix-assisted laser desorption/ionization and electrospray ionization mass spectrometry analysis of blue copper proteins, azurin and mavicyanin. *Rapid Commun. Mass Spectrom.* **12**, 1967.

Garzotti, M. and Hamdan, M. (1998) Liquid chromatography/tandem mass spectrometry of synthesis products associated with the viral protein $U_S 11$. *Rapid Commun. Mass Spectrom.* **12**, 843.

Gatlin, C. L., Kleemann, G. R., Hays, L. G., Link, A. J., and Yates, J. R. (1998) Protein identification at the low femtomole level from silver-stained gels using a new fritless electrospray interface for liquid chromatography/microspray and nanospray mass spectrometry. *Anal. Biochem.* **263**, 93.

Green, B. N., Kuchumov, A. R., Hankeln, T., Schmidt, E. R., Bergtrom, G., and Vinogradov, S. N. (1998) An electrospray ionization mass spectrometric study of extracellular hemoglobins from *Chironomus thummi thummi*. *BBA-Protein Struct. Mol. Enzym.* **1383**, 143.

Green, M. K., Vestling, M. M., Johnston, M. V., and Larsen, B. S. (1998) Distinguishing small molecular mass differences of proteins by mass spectrometry. *Anal. Biochem.* **200**, 204.

Griffiths, W. J., Gustafsson, M., Yang, Y., Curstedt, T., Sjovell, J., and Johansson, J. (1998) Analysis of variant forms of porcine surfactant polypeptide-C by nanoelectrospray mass spectrometry. *Rapid Commun. Mass Spectrom.* **12**, 1104.

Haebel, S., Albrecht, T., Sparbier, K., Walrlen, P., Korner, R., and Steup, M. (1998) Electrophoresis-related protein modification. Alkylation of carboxy residues revealed by mass spectrometry. *Electrophoresis* **19**, 679.

Hagmann, M. L., Kionka, C., Schreiner, M., and Schwer, C. (1998) Characterization of the $F(ab')_2$ fragment of a murine monoclonal antibody using capillary isoelectric focusing and electrospray ionization mass spectrometry. *J. Chromatogr. A.* **816**, 49.

Haynes, P., Miller, I., Aebersold, R., Gemeiner, M., Eberini, I., Lovati, M. R., et al. (1998) Proteins of rat serum: I. Establishing a reference two-dimensional electrophoresis map by immunodetection and microbore high performance liquid chromatography/electrospray mass spectrometry. *Electrophoresis* **19**, 1484.

Haynes, P. A., Fripp, N., and Aebersold, R. (1998) Identification of gel-separated proteins by liquid chromatography/electrospray tandem mass spectrometry: comparison of methods and their limitations. *Electrophoresis* **19,** 939.

Hendricker, A. D., Basile, F., and Voorhees, K. J. (1998) A study of protein oxidative products using a pyrolysis-membrane inlet quadrupole ion trap mass spectrometer with air as the buffer gas. *J. Anal. Appl. Pyrol.* **46,** 65.

Hines, W. H., Parker, K., Peltier, J., Patterson, D. H., Vestal, M. L., and Martin, S. A. (1998) Protein identification and protein characterization by high-performance time-of-flight mass spectrometry. *J. Protein Chem.* **17,** 525.

Jensen, C., Andersen, S. O., and Roepstorff, P. (1998) Primary structure of two major cuticular proteins from the migratory locust, *Locusta migratoria,* and their identification in polyacrylamide gels by mass spectrometry. *BBA-Protein Struct. Mol. Enzym.* **1429,** 151.

Jensen, O. N., Larsen, M. R., and Roepstorff, P. (1998) Mass spectrometric identification and micro-characterization of proteins from electrophoretic gels. Strategies and applications. *Proteins* **Suppl 2,** 74.

Kalkum, M., Przybylski, M., and Glocker, M. O. (1998) Structure characterization of functional histidine residues and carbethoxylated derivatives in peptides and proteins by mass spectrometry. *Bioconj. Chem.* **9,** 226.

Kasheverov, I., Utkin, Y., Weise, C., Franke, P., Hucho, F., and Tsetlin, V. (1998) Reversed-phase chromatography isolation and MALDI mass spectrometry of the acetylcholine receptor subunits. *Protein Express Purif* **12,** 226.

Kele, Z., Janaky, T., Meszaros, T., Feher, A., Dudits, D., and Szabo, P. T. (1998) Capillary chromatography/micro-electrospray mass spectrometry used for the identification of putative cyclin-dependent kinase inhibitory protein in *Medicago. Rapid Commun. Mass Spectrom.* **12,** 1564.

Keough, T., Lacey, M. P., Trakshel, G. M., and Asquith, T. N. (1997) The use of MALDI mass spectrometry to characterize synthetic protein conjugates. *Int. J. Mass Spectrom. Ion Proc.* **169,** 201.

Kertscher, U., Beyermann, M., Krause, E., Furkert S, Berger, H., Bienert, M., et al. (1998) The degradation of corticotrophin-releasing factor by enzymes of the rat brain studied by liquid chromatography/mass spectrometry. *Peptides* **19,** 649.

Knudsen, C. B., Bjornsdottir, I., Jons, O., and Hansen, S. H. (1998) Detection of metallothionein isoforms from three different species using on-line capillary electrophoresis/mass spectrometry. *Anal. Biochem.* **265,** 167.

Krell, T., Chakrewarthy, S., Pin, A. R., Elwell, A., and Coggins, J. R. (1998) Chemical modification monitored by electrospray mass spectrometry. A rapid and simple method for identifying and studying functional residues in enzymes. *J. Pept. Res.* **51,** 201.

Kulik, W., Meesterburrie, J. A. N., Jakobs, C., and, D. E., Meer, K. (1998) Determination of $\delta^{13}C$ values of valine in protein hydrolysate by gas chromatography/combustion isotope-ratio mass spectrometry. *J. Chromatogr. B.* **710,** 37.

Kuwata, H., Yip, T. T., Yip, C. L., Tomita, M., and Hutchens, T. W. (1998) Bactericidal domain of lactoferrin. Detection, quantitation, and characterization of lactoferricin in serum by SELDI affinity mass spectrometry. *Biochem. Biophys. Res. Commun.* **245,** 764.

Lampi, K. J., Ma, Z. X., Hanson, S. R. A., Azuma, M., Shih, M., Shearer, T. R., et al. (1998) Age-related changes in human lens crystallins identified by two-dimensional electrophoresis and mass spectrometry. *Exp Eye Res* **67,** 31.

Laprevote, O., Serani, L., Das, B. S., Halgand, F., Forest, E., and Dumas, R. (1999) Stepwise building of a 115-kDa macromolecular edifice monitored by electrospray mass spectrometry: the case of acetohydroxy acid isomeroreductase. *Eur. J. Biochem.* **259,** 356.

Lei, Q. P., Cui, X. Y., Kurtz, D. M., Amster, I. J., Chernushevich, I. V., and Standing, K. G. (1998) Electrospray mass spectrometry studies of non-heme iron-containing proteins. *Anal. Chem.* **70,** 1838.

Liu, D. L., Zhu, Z. H., Khao, S. K., Sun, D. Y., and Yang, Y. S. (1997) Matrix-assisted laser desorption/ionization mass spectrometry for determination of molecular weight of CaMs. *Chem. J. Chinese Univ-Chinese* **18,** 1966.

Liu, S. M. and Figliomeni, S. (1998) Gas chromatography/mass spectrometry analyses of [2,3,3-d3]serine, [2,3,3-d3]cysteine and [3-^{13}C]cysteine in plasma and skin protein. Measurement of trans-sulphuration in young sheep. *Rapid Commun. Mass Spectrom.* **12,** 1199.

Loo, J. A., Holler, T. P., Foltin, S. K., McConnell, P., Banotai, C. A., Horne, N. M., et al. (1998) Application of electrospray ionization mass spectrometry for studying human immunodeficiency virus protein complexes. *Protein* **Suppl 2,** 28.

Ma, Z. X., Hanson, S. R. A., Lampi, K. J., David, L. L., Smith, D. L., and Smith, J. B. (1998) Age-related changes in human lens crystallins identified by HPLC and mass spectrometry. *Exp. Eye Res.* **67,** 21.

Mamome, G., Malorni, A., Scaloni, A., Sannolo, N., Basile, A., Pocsfalvi, G., et al. (1998) Structural analysis and quantitative evaluation of the modifications produced in human hemoglobin by methyl bromide using mass spectrometry and Edman degradation. *Rapid Commun. Mass Spectrom.* **12,** 1783.

Mao, Y., Moose, R. J., Wagnon, K. B., Pierce, J. T., Debban, K. H., Smith, C. S., et al. (1998) Analysis οφ $α_2$-microglobulin in rat urine and kidneys by liquid chromatography/electrospray ionization mass spectrometry. *Chem. Res. Toxicol.* **11,** 953.

Marentes E. and Grusak, M. A. (1998) Mass determination of low-molecular-weight proteins in phloem sap using matrix-assisted laser desorption/ionization time-of-flight mass spectrometry. *J. Exp. Bot.* **49,** 903.

Marx, M. K., Mayer-Posner, F., Soulimane, T., and Buse, G. (1998) Matrix-assisted laser desorption/ionization mass spectrometry analysis and thiol-group determination of isoforms of bovine cytochrome *c* oxidase, a hydrophobic multisubunit membrane protein. *Anal. Biochem.* **256,** 192.

McComb, M. E., Oleschuk, R. D., Chow, A., Ens, W., Standing, K. G., Perreault, H., et al. (1998) Characterization of hemoglobin variants by MALDI-TOF-MS using a polyurethane membrane as the sample support. *Anal. Chem.* **70,** 5142.

Mckendrick, S. E., Frormann, S., Luo, C., Semchuck, P., Vederas, J. C., and Malcolm, B. A. (1998) Rapid mass spectrometric determination of preferred irreversible proteinase inhibitors in combinatorial libraries. *Int. J. Mass Spectrom.* **176,** 1 13.

Morgan, F., Bouhallab, S., Molle, D., Henry, G., Maubois, J. L., and Leonil, J. (1998) Lactolation of β-lactoglobulin monitored by electrospray ionisation mass spectrometry. *Int. Dairy J.* **8,** 95.

Mueller, D. R., Schindler, P., Coulot, M., Voshol, H., and Van Oostrum, J. (1999) Mass spectrometric characterization of stathmin isoforms separated by 2D-PAGE. *J. Mass Spectrom.* **34,** 336.

Nakanishi, T., Miyazaki, A., Kishikawa, M., Shimizu, A., Aoki, Y., and Kikuchi, M. (1998) Hb Sagami [ββ139(H17)Asn→Thr]: a new hemoglobin variant not detected by isoelectrofocusing and propan-2-ol test, was detected by electrospray ionization mass spectrometry (Letter). *J. Mass Spectrom.* **33,** 565.

Nolte, L., Van der Westhuizen, F. H., Pretorius, P. J., and Erasmus, E., (1998) Carnitine palmitoyltransferase, I., activity monitoring in fibroblasts and leukocytes using electrospray ionization mass spectrometry. *Anal. Biochem.* **256,** 178.

Ofori-Acquah, S. F., Green, B. N., Wild, B. J., Lalloz, M. R. A., and Layton, D. M. (1998) Quantification of (G)γ- and (A)γ-globins by electrospray ionisation mass spectrometry. *Int. J. Mol. Med.* **2,** 451.

Otto, A., Muller, E. C., Brockstedt, E., Schumann, M., Rickers, A., Bommert, K., and Wittmann-Leibold, B. (1998) High performance two-dimensional gel electrophoresis and nanoelectrospray mass spectrometry as a powerful tool to study apoptosis-associated processes in a Burkitt lymphoma cell line. *J. Protein Chem.* **17,** 564.

Peter, J., Unverzagt, C., Engel, W. D., Renauer, D., Seidel, C., and Hosel, W. (1998) Identification of carbohydrate deficient transferrin forms by MALDI-TOF mass spectrometry and lectin ELISA. *BBA-Gen. Subjects* 1**380,** 93.

Peterson, K. P., Pavlovich, J. G., Goldstein, D., Little, R., England, J., and Peterson, C. M. (1998) What is hemoglobin A1c? An analysis of glycated hemoglobins by electrospray ionization mass spectrometry. *Clin. Chem.* **44,** 1951.

Rabilloud, T., Kieffer, S., Procaccio, V., Louwagie, M., Courchesne, P. L., Patterson, S. D., et al. (1998) Two-dimensional electrophoresis of human placental mitochondria and protein identification by mass spectrometry: toward a human mitochondrial proteome. *Electrophoresis* **19,** 1006.

Reid, G. E., Rasmussen, R. K., Dorow, D. S., and Simpson, R. J. (1998) Capillary column chromatography improves sample preparation for mass spectrometric analysis. Complete characterization of human α-enolase from two-dimensional gels following *in situ* proteolytic digestion. *Electrophoresis* **19,** 946.

Reijngoud, D. J., Hellstern, O., Elzinga, H., De Sain-van der Velden, M. G., Okken, A., and Stellaard, F. (1998) Determination of low isotopic enrichment of L-[1-^{13}C]valine by gas chromatography/combustion/isotope-ratio mass spectrometry: A robust method for measuring protein fractional synthetic rates *in vivo*. *J. Mass Spectrom.* **33,** 621.

Reynolds, T. M., McMillan, F., Smith, A., Hutchinson, A., and Green, B. (1998) Haemoglobin Le, Lamentin (α20 (B1) His→Gln) in a British family. Identification by electrospray mass spectrometry. *J Clin Pathol* **51,** 467.

Rohde, E., Tomlinson, A. J., Johnson, D. H., and Naylor, S. (1998) Comparison of protein mixtures in aqueous humor by membrane preconcentration capillary electrophoresis/mass spectrometry. *Electrophoresis* **19,** 2361.

Schnaible, V., Michels, J., Zeth, K., Freigang, J., Welte, W., Buhler, S., et al. (1997) Approaches to the characterization of membrane channel proteins (porins) by UV MALDI-MS. *Int. J. Mass Spectrom. Ion Proc.* **169,** 165.

Schrimer, D. C., Yalcin, T., and, Li, L (1998) MALDI mass spectrometry combined with avidin-biotin chemistry for analysis of protein modifications. *Anal. Chem.* **70,** 1569.

Srinivasan, J. R., Kachman, M. T., Killeen, A. A., Akel, N., Siemieniak, D., and Lubman, D. M. (1998) Genotyping of apolipoprotein, E., by matrix-assisted laser desorption/ionization time-of-flight mass spectrometry. *Rapid Commun. Mass Spectrom.* **12,** 1045.

Stevens, R. C. and Davis, T. N. (1998) Mlc1p is a light chain for the unconventional myosin Myo2p in *Saccharomyces cerevisiae*. *J. Cell. Biol.* **142,** 711.

Stone, K. L., De Angelis, R., Lo Presti, M., Jones, J., Papov, V. V., and Williams, K. R. (1998) Use of liquid chromatography/electrospray ionization tandem mass spectrometry (LC/ESI-MS/MS) for routine identification of enzymatically digested proteins separated by sodium dodecyl sulfate polyacrylamide gel electrophoresis. *Electrophoresis* **19,** 1046.

Sun, M. Z., Ding, L., Ji, Y. P., Zhao, D. Q., Liu, S. Y., and, N. I., J. Z. (1999) Matrix-assisted laser desorption/ionization time-of-flight mass spectrometric analysis of phospholipase A_2 and fibrinolytic enzyme, two enzymes obtained from Chinese *Agkistrodon blomhoffi ussurensis* venom. *Rapid Commun. Mass Spectrom.* **13,** 150.

Tamura, A., Azuma, T., Ikeda, K., Ozawa, K., and Masujima, T. (1998) Determination of insulin content in pancreatic β cell line MIN6 cells by matrix-assisted laser desorption/ionization time-of-flight mass spectrometry. *Biol Pharm Bull* **21,** 1240.

Theberge, R., Connors, L., Skinner, M., Skare, J., and Costello, C. E. (1999) Characterization of transthyretin mutants from serum using immunoprecipitation, HPLC/ electrospray ionization and matrix-assisted laser desorption/ionization mass spectrometry. *Anal. Chem.* **71,** 452.

Tolic, L. P., Harms, A. C., Anderson, G. A., Smith, R. D., Willie, A., and Jorns, M. S. (1998) Electrospray ionization mass spectrometry characterization of heterotetrameric sarcosine oxidase. *J. Am. Soc. Mass Spectrom.* **9,** 510.

Valentine, S. J., Counterman, A. E., Hoaglund, C. S., Reilly, J. P., and Clemmer, D. E. (1998) Gas-phase separations of protease digests. *J. Am. Soc. Mass Spectrom.* **9,** 1213.

Van Adrichem, J. H. M., Bornsen, K. O., Conzelmann, H., Gass, M. A. S., Eppenberger, H., Kresbach, G. M., et al. (1998) Investigation of protein patterns in mammalian cells and culture supernatants by matrix-assisted laser desorption/ ionization mass spectrometry. *Anal. Chem.* **70,** 923.

Wang, Z., Russon, L., L. I., L., Roser, D. C., and Randolph-Long, S. (1998) Investigation of spectral reproducibility in direct analysis of bacteria proteins by matrix-assisted laser desorption/ionization time-of-flight mass spectrometry. *Rapid Commun. Mass Spectrom.* **12,** 456–464.

Weiss, K. C., Yip, T. T., Hutchens, T. W., and Bisson, L. F. (1998) Rapid and sensitive fingerprinting of wine proteins by matrix-assisted laser desorption/ionization time-of-flight (MALDI-TOF) mass spectrometry. *Amer. J. Enol. Viticult.* **49,** 231.

Westman, A., Nilsson, C. L., and Ekman, R. (1998) Matrix-assisted laser desorption/ionization time-of-flight spectrometry analysis of proteins in human cerebrospinal fluid. *Rapid Commun. Mass Spectrom.* **12**, 1092.

Whitelegge, J. P., Gundersen, C. B., and Faull, K. F. (1998) Electrospray-ionization mass spectrometry of intact intrinsic membrane proteins. *Protein Sci.* **7**, 1423.

Wolf, B. P., Sumner, L. W., Shields, S. J., Nielsen, K., Gray, K. A., and Russell, D. H. (1998) Characterization of proteins utilized in the desulfurization of petroleum products by matrix-assisted laser desorption/ionization time-of-flight mass spectrometry. *Anal. Biochem.* **260**, 117.

Wu, J. and Watson, J. T. (1998) Optimization of the cleavage reaction for cyanylated cysteinyl proteins for efficient and simplified mass mapping. *Anal. Biochem.* **258**, 268.

Wu, X. H. D., Marakov, K. I., Martinez, P. A., McVerry, P. H., and Malinzak, D. A. (1997) Analysis of lipidation of recombinant Lyme vaccine protein (rOspA) by electrospray mass spectroscopy. *J. Pharm. Biomed. Anal.* **16**, 613.

Wykes, L. J., Jahoor, F., and Reeds, P. J. (1998) Gluconeogenesis measured with [U-^{13}C]glucose and mass isotopomer analysis of ApoB-100 amino acids in pigs. *Am. J. Physiol.* **274**, E365.

Xiang BS. Ferretti, J., and Fales, H. M. (1998) Use of mass spectrometry to ensure purity of recombinant proteins. A cautionary note. *Anal. Chem.* **70**, 2188.

Xu, Y. Z., Zhang, Y. B., Sun, Y. L., Wang, Y. H., and Huang, Z. X. (1997) Molecular weight determination of cytochrome Tb5 and Lb5 and their mutant proteins by electrospray ionization mass spectrometry technique. *Acta Chim Sin* **55**, 1116.

Yang, L. Y., Lee, C. S., Hofstadler, S. A., Pasa-Tolic, L., and Smith, R. D. (1998) Capillary isoelectric focusing/electrospray ionization Fourier transform ion cyclotron resonance mass spectrometry for protein characterization. *Anal. Chem.* **70**, 3235.

Zaia, J., Fabris, D., Wei, D., Karpel, R. L., and Fenselau, C. (1998) Monitoring metal ion flux in reactions of metallothionein and drug-modified metallothionein by electrospray mass spectrometry. *Protein Sci.* **7**, 2398.

Zappacosta, F., Di Luccia, A., Ledda, L., and Addeo, F. (1998) Identification of C-terminally truncated forms of β-lactoglobulin in whey from Romagnola cows' milk by two-dimensional electrophoresis coupled to mass spectrometry. *J. Dairy Res.* **65**, 243.

Zhang, X. Y., De Meester, I., Lambeir, A.-M., Dillen, L., Van Dongen, W., Esmans, E. L., et al. (1999) Study of the enztymatic degradation of vasostatin I and II, and their precursor, Cromogranin A, by dipeptidyl peptidase IV, using high-performance liquid chromatography and electrospray mass spectrometry. *J. Mass Spectrom.* **34**, 255.

Zhou, G. H., Luo, G. A., Zhou, Y., Zhou, K. Y., Zhang, X. D., and Huang, L. Q. (1998) Application of capillary electrophoresis, liquid chromatography, electrospray mass spectrometry and matrix-assisted laser desorption/ionization time-of-flight mass spectrometry to the characterization of recombinant human erythropoietin. *Electrophoresis* **19**, 2348.

Zugaro, L. M., Reid, G. E., Ji, H., Eddes, J. S., Murphy, A. C., Burgess, A. W., and Simpson, R. J. (1998) Characterization of rat brain stathmin isoforms by two-dimensional gel electrophoresis matrix-assisted laser desorption/ionization and electrospray ionization ion trap mass spectrometry. *Electrophoresis* **19,** 867.

Protein Configuration and Folding

Fligge, T. A., Przybylski, M., Quinn, J. P., and Marshall, A. G. (1998) Evaluation of heat-induced conformational changes in proteins by nanoelectrospray Fourier transform ion cyclotron resonance mass spectrometry. *Eur. Mass Spectrom.* **4,** 401.

Happersberger, H. P., and Glocker, M. O. (1998) A mass spectrometric approach to the characterization of protein folding reactions. *Eur. Mass Spectrom.* **4,** 209.

Happersberger, H. P., Stapleton, J., Cowgill, C., and Glocker, M. O. (1998) Characterization of the folding pathway of recombinant human macrophage-colony stimulating-factor β (rhM-CSF β) by bis-cysteinyl modification and mass spectrometry. *Proteins* **Suppl 2,** 50.

Jensen, P. K., Harrata, A. K., and Lee, C. S. (1998) Monitoring protein refolding induced by disulfide formation using capillary isoelectric focusing/electrospray ionization mass spectrometry. *Anal. Chem.* **70,** 2044.

Konermann, L. and Douglas, D. J. (1998) Equilibrium unfolding of proteins monitored by electrospray ionization mass spectrometry. Distinguishing two-state from multi-state transitions. *Rapid Commun. Mass Spectrom.* **12,** 435.

Konermann, L. and Douglas, D. J. (1998) Unfolding of proteins monitored by electrospray ionization mass spectrometry. A comparison of positive and negative ion modes. *J. Am. Soc. Mass Spectrom.* **9,** 1248.

Li, A. Q., Fenselau, C., and Kaltashov, I. A. (1998) Stability of secondary structural elements in a solvent-free environment. II: The β-pleated sheets. *Proteins* **Suppl 2,** 22.

Resing, K. A. and Ahn, N. G. (1998) Deuterium exchange mass spectrometry as a probe of protein kinase activation. Analysis of wild-type and constitutively active mutants of MAP kinase kinase-1. *Biochemistry-USA* **37,** 463.

Smith, D. L. (1998) Local structure and dynamics in proteins characterized by hydrogen exchange and mass spectrometry. *Biochemistry* **63,** 285.

Tsui, V., Garcia, C., Cavagnero, S., Siuzdak, G., Dyson, H. J., and Wright, P. E. (1999) Quench-flow experiments combined with mass spectrometry show apomyoglobin folds through an obligatory intermediate *Protein Sci.* **8,** 45.

Veenstra, T. D., Johnson, K. L., Tomlinson, A. J., Craig, T. A., Kumar, R., and Naylor, S. (1998) Zinc-induced conformational changes in the DNA-binding domain of the vitamin, D., receptor determined by electrospray ionization mass spectrometry. *J. Am. Soc. Mass Spectrom.* **9,** 8.

Veenstra, T. D., Johnson, K. L., Tomlinson, A. J., Kumar, R., and Naylor, S. (1998) Correlation of fluorescence and circular dichroism spectroscopy with electrospray ionization mass spectrometry in the determination of tertiary conformational changes in calcium-binding proteins. *Rapid Commun. Mass Spectrom.* **12,** 613.

Veenstra, T. D., Johnson, K. L., Tomlinson, A. J., Naylor, S., and Kumar, R. (1997) Electrospray ionisation mass spectrometry temperature effects on metal ion: protein stoichiometries and metal-induced conformational changes in calmodulin. *Eur. Mass Spectrom.* **3**, 453.

Wang, F., Li, W. Q., Emmett, M. R., Hendrickson, C. L., Marshall, A. G., Zhang, Y. L., et al. (1998) Conformational and dynamic changes of *Yersinia* protein tyrosine phosphatase induced by ligand binding and active site mutation and revealed by H/D exchange and electrospray ionization Fourier transform ion cyclotron resonance mass spectrometry. *Biochemistry-USA* **37**, 15289.

Wang, F., Scapin, G., Blanchard, J. S., and Angeletti, R. H. (1998) Substrate binding and conformational changes of *Clostridium glutamicum* diaminopimelate dehydrogenase revealed by hydrogen/deuterium exchange and electrospray mass spectrometry. *Protein Sci.* **7**, 293.

Waring, A. J., Mobley, P. W., and Gordon, L. M. (1998) Conformational mapping of a viral fusion peptide in structure-promoting solvents using circular dichroism and electrospray mass spectrometry. *Proteins* **Suppl 2**, 38.

Yang, H. J. H., Li, X. L. C., Amft, M., and Grotemeyer, J. (1998) Protein conformational changes determined by matrix-assisted laser desorption mass spectrometry. *Anal. Biochem.* **258**, 118.

Glycosylated or Phosphorylated Proteins and Peptides—General Analysis

Alving, K., Paulsen, H., and Peter-Katalinic, J. (1999) Characterization of *O*-glycosylation sites in MUC2 glycopeptides by nanoelectrospray QTOF mass spectrometry. *J. Mass Spectrom.* **34**, 395.

Bateman, K. P., White, R. L., Yaguchi, M., and Thibault, P. (1998) Characterization of protein glycoforms by capillary-zone electrophoresis/nanoelectrospray mass spectrometry. *J. Chromatogr. A.* **794**, 327.

Beranova-Giorgianni, S., Desiderio, D. M., and Pabst, M. J. (1998) Structures of biologically active muramyl peptides from peptidoglycan of *Streptococcus sanguis*. *J. Mass Spectrom.* **33**, 1182.

Cleverley, K. E., Belts, J. C., Blackstock, W. P., Gallo, J. M., and Anderton, B. H. (1998) Identification of novel *in vitro* PKA phosphorylation sites on the low and middle molecular mass neurofilament subunits by mass spectrometry. *Biochemistry-USA* **37**, 3917.

Cramer, R., Richter, W. J., Stimson, E., and Burlingame, A. L. (1998) Analysis of phospho- and glycopolypeptides with infrared matrix-assisted laser desorption and ionization. *Anal. Chem.* **70**, 4939.

Dage, J. L., Sun, H. J., and Halsall, H. B. (1998) Determination of diethylpyrocarbonate-modified amino acid residues in a,-acid glycoprotein by high-performance liquid chromatography/electrospray ionization mass spectrometry and matrix-assisted laser desorption/ionization time-of-flight mass spectrometry. *Anal. Biochem.* **257**, 176.

De Corte, V., Demol, H., Goethals, M., Van Damme, J., Gettemans, J., and Vandekerckhove, J. (1999) Identification of Tyr438 as the major *in vitro* c-Src phosphorylation site in human gelsolin. A mass spectrometric approach. *Protein Sci.* **8**, 234.

De Gnore, J. P. and Qin, J. (1998) Fragmentation of phosphopeptides in an ion trap mass spectrometer. *J. Am. Soc. Mass Spectrom.* **9,** 1175.

Denzinger, T., Diekmann, H., Bruns, K., Laessing, U., Stuermer, C. A., and Przybylski, M. (1999) Isolation, primary structure characterization and identification of the glycosylation pattern of recombinant neurolin, a neuronal cell adhesion protein. *J. Mass Spectrom.* **34,** 435.

Fouques, D., Ralet, M. C., Molle, D., Leonil, J., and Meunier, J. C. (1998) Enzymatic phosphorylation of soy globulins by the protein kinase CK2. Determination of the phosphorylation sites of β-conglycinin α subunit by mass spectrometry. *Nahrung* **42,** 148.

Graham, M. E., Dickson, P. W., Dunkley, P. R., and Von Nagy-Felsobuki, E. I. (1998) Characterization of the phosphorylation of rat tyrosine hydroxylase using electrospray mass spectrometry (Letter). *Rapid Commun. Mass Spectrom.* **12,** 746.

Hanger, D. P., Betts, J. C., Loviny, T. L. F., Blackstock, W. P., and Anderton, B. H. (1998) New phosphorylation sites identified in hyperphosphorylated τ (paired helical filament-τ) from Alzheimer's disease brain using nanoelectrospray mass spectrometry. *J Neurochem* **71,** 2465.

Hiki, Y., Tanaka, A., Kokubo, T., Iwase, H., Nishikido, J., Hotta, K., et al. (1998) Analyses of IgA1 hinge glycopeptides in IgA nephropathy by matrix-assisted laser desorption/ionization time-of-flight mass spectrometry. *J Amer Soc Nephrol* **9,** 577.

Immler, D., Gremm, D., Kirsch, D., Spengler, B., Presek, P., and Meyer, H. E. (1998) Identification of phosphorylated proteins from thrombin-activated human platelets isolated by two-dimensional gel electrophoresis by electrospray ionization tandem mass spectrometry (ESI-MS/MS) and liquid chromatography-electrospray ionization mass spectrometry (LC/ESI-MS). *Electrophoresis* **19,** 1015.

Jaffe, H., Veeranna, Shetty, K. T., and Pant, H. C. (1998) Characterization of the phosphorylation sites of human high molecular weight neurofilament protein by electrospray ionization tandem mass spectrometry and database searching. *Biochemistry-USA* **37,** 3931.

Juhasz, P. and Martin, S. A. (1997) The utility of nonspecific proteases in the characterization of glycoproteins by high-resolution time-of-flight mass spectrometry. *Int. J. Mass Spectrom. Ion Proc.* 169 217.

Kinumi, T., Tobin, S. L., Matsumoto, H., Jackson, K. W., and Ohashi, M. (1997) The phosphorylation site and desmethionyl N-terminus of *Drosophila* phosrestin, I., *in vivo* determined by mass spectrometric analysis of proteins separated by two-dimensional gel electrophoresis. *Eur. Mass Spectrom.* **3,** 367.

Korsmeyer, K. K., Guan, S. H., Yang, Z. C., Falick, A. M., Ziegler, D. M., and Cashman, J. R. (1998) *N*-glycosylation of pig flavin-containing mono-oxygenase form 1. Determination of the site of protein modification by mass spectrometry. *Chem. Res. Toxicol.* **11,** 1145.

Lapolla, A., Fedele, D., Plebani, M., Garbeglio, M., Seraglia, R., D'Alpaos, M., et al. (1999) Direct evaluation of glycated and glyco-oxidized globins by matrix-assisted laser desorption/ionization mass spectrometry. *Rapid Commun. Mass Spectrom.* **13,** 8.

Matsumoto, H., Kahn, E. S., and Komori, N. (1998) Non-radioactive phosphopeptide assay by matrix-assisted laser desorption/ionization time-of-flight mass spectrometry: application to calcium/calmodulin-dependent protein kinase II. *Anal. Biochem.* **260**, 188.

Neubauer, G. and Mann, M. (1999) Mapping of phosphorylation sites of gel-isolated proteins by nanoelectrospray tandem mass spectrometry: potentials and limitations. *Anal. Chem.* **71**, 235.

Tholey, A., Reed, J., and Lehmann, W. D. (1999) Electrospray tandem mass spectrometry studies of phosphopeptides and phosphopeptide analogues. *J. Mass Spectrom.* **34**, 117.

Tsarbopoulos, A., Bahr, U., Pramanik, B. N., and Karas, M. (1997) Glycoprotein analysis by delayed extraction and post-source decay MALDI- TOF-MS. *Int. J. Mass Spectrom. Ion Proc.* **169**, 251.

Udiavar, S., Apffel, A., Chakel l, Swedberg, S., Hancock, W. S., and Pungor, E. (1998) The use of multidimensional liquid-phase separations and mass spectrometry for the detailed characterization of post-translational modification in glycoproteins. *Anal. Chem.* **70**, 3572.

Wei, J., Yang, L. Y., Harrata, A. K., and Lee, C. S. (1998) High resolution analysis of protein phosphorylation using capillary isoelectric focusing electrospray ionization mass spectrometry. *Electrophoresis* **19**, 2356.

Wolfender, J.-L., Chu, F., Ball, H., Wolfender, F., Fainzilber, F., Baldwin, M. A., et al. (1999) Identification of tyrosine sulfation in *Conus pennaceus* conotoxins α-PnIA and α-PnIB. Further investigation of labile sulfo- and phosphopeptides by electrospray, matrix-assisted laser desorption/ionization (MALDI) and atmospheric pressure MALDI mass spectrometry. *J. Mass Spectrom.* **34**, 447.

Yamauchi, E., Kiyonami, R., Kanai, M., and Taniguchi, H. (1998) The C-terminal conserved domain of MARCKS is phosphorylated *in vivo* by proline-directed protein kinase. Application of ion trap mass spectrometry to the determination of protein phosphorylation sites. *J. Biol. Chem.* **273**, 4367.

Zeng, C. and Biemann, K. (1999) Determination of *N*-linked glycosylation of yeast external invertase by matrix-assisted laser desorption/ionization time-of-flight mass spectrometry. *J. Mass Spectrom.* **34**, 311.

Zhang, X. Y., Dillen, L., Bauer, S. H. J., Van Dongen, W., Liang, F., Przybylski, M., et al. (1997) Mass spectrometric identification of phosphorylated vasostatin II, a chromogranin A-derived protein fragment (1-113). *BBA-Protein Struct. Mol. Enzym.* **1343**, 287.

Protein Interactions and Kinetics

Ayed, A., Krutchinsky, A. N., Ens, W., Standing, K. G., and Duckworth, H. W. (1998) Quantitative evaluation of protein-protein and ligand-protein equilibria of a large allosteric enzyme by electrospray ionization time-of-flight mass spectrometry. *Rapid Commun. Mass Spectrom.* **12**, 339.

Bennett, J. S., Bell, D. W., Buchholz, B. A., Kwok, E. S. C., Vogel, J. S., and Morton, T. H. (1998) Accelerator mass spectrometry for assaying irreversible covalent modification of an enzyme by acetoacetic ester *Int. J. Mass Spectrom.* **180**, 185.

Bruce, J. E., Smith, V. F., Liu, C. L., Randall, L. L., and Smith, R. D. (1998) The observation of chaperone-ligand noncovalent complexes with electrospray ionization mass spectrometry. *Protein Sci.* **7,** 1180.

Buhler, S., Michels, J., Wendt, S., Ruck, A., Brdiczka, D., Welte-West, et al. (1998) Mass spectrometric mapping of ion channel proteins (porins) and identification of their supramolecular membrane assembly. *Proteins* **Suppl 2,** 63.

Chazin, W. and Veenstra, T. D. (1999) Determination of the metal-binding cooperativity of wild-type and mutant calbindin D_{9K} by electrospray ionization mass spectrometry. *Rapid Commun. Mass Spectrom.* **13,** 548.

Connor, D. A., Falick, A. M., Young, M. C., and Shetlar, M. D. (1998) Probing the binding region of the single-stranded DNA-binding domain of rat DNA polymerase β using nanosecond-pulse laser-induced cross-linking and mass spectrometry. *Photochem Photobiol* **68,** 299.

Constanzer, M. L., Chavez-Eng, C. M., and Matuszewski, B. K. (1998) Determination of a novel selective inhibitor of type 1 5α-reductase in human plasma by liquid chromatography with atmospheric pressure chemical ionization tandem mass spectrometry. *J. Chromatogr. B.* **713,** 371.

Counterman, A. E., Valentine, S. J., Srebalus, C. A., Henderson, S. C., Hoaglund, C. S., and Clemmer, D. E. (1998) High-order structure and dissociation of gaseous peptide aggregates that are hidden in mass spectra. *J. Am. Soc. Mass Spectrom.* **9,** 743.

De Gnore, J. P., Konig, S., Barrett, W. C., Chock, P. B., and Fales, H. M. (1998) Identification of the oxidation states of the active site cysteine in a recombinant protein tyrosine phosphatase by electrospray mass spectrometry using on-line desalting. *Rapid Commun. Mass Spectrom.* **12,** 1457.

Furuya, H., Yoshino, K., Shimizu, T., Mantoku, T., Takeda, T., Nomura, K., et al. (1998) Mass spectrometric analysis of phosphoserine residues conserved in the catalytic domain of membrane-bound guanylyl cyclase from the sea urchin spermatozoa. *Zool. Sci.* **15,** 507.

Gervasoni, P., Staudenmann, W., James, P., and Pluckthun, A. (1998) Identification of the binding surface on β-lactamase for GroEL by limited proteolysis and MALDI mass spectrometry. *Biochemistry-USA* **37,** 11660.

Glocker, M. O., Nock, S., Sprinzl, M., and Przybylski, M. (1998) Characterization of surface topology and binding area in complexes of the elongation factor proteins EF-Ts and EF-TuGDP from *Thermus thermophilus*. A study by protein chemical modification and mass spectrometry. *Chem.-Eur* **4,** 707.

Gourlaouen, N., Bolbach, G., Florentin, D., Gonnet, F., and Marquet, A. (1998) Characterization of protein-hapten conjugates by mass spectrometry (French, English Abstract). *CR Acad. Sci. Ser. II C.* **1,** 35.

Green, B. N., Sannes-Lowery, K. A., Loo, J. A., Saiterlee, J. D., Kuchumov, A. R., Walz, D. A., et al. (1998) Electrospray ionization mass spectrometric study of the multiple intracellular monomeric and polymeric hemoglobins of *Glycera dibranchiata*. *J. Protein Chem* **17,** 85.

Helin, J., Caldentey, J., Kalkkinen, N., and Bamford, D. H. (1999) Analysis of the multimeric state of proteins by matrix-assisted laser desorption/ionization mass spectrometry after cross-linking with glutaraldehyde. *Rapid Commun. Mass Spectrom.* **13,** 185.

Jedrzejewski, P. T., Girod, A., Tholey, A., Konig, N., Thullner, S., Kinzel, V., et al. (1998) A conserved deamidation site at Asn 2 in the catalytic subunit of mammalian cAMP-dependent protein kinase detected by capillary LC-MS and tandem mass spectrometry. *Protein Sci.* **7,** 457.

Jeyarajah, S., Parker, C. E., Sumner, M. T., and Tomer, K. B. (1998) Matrix-assisted laser desorption/ionization mass spectrometry mapping of human immunodeficiency virus-gp120 epitopes recognized by a limited polyclonal antibody. *J. Am. Soc. Mass Spectrom.* **9,** 157.

Jorgensen, T. J. D., Roepstorff, P., and Heck AIR (1998) Direct determination of solution binding constants to noncovalent complexes between bacterial cell wall peptide analogues and vancomycin group antibiotics by electrospray ionization mass spectrometry. *Anal. Chem.* **70,** 4427.

Leopold, I., Gunter, D., and Neumann, D. (1998) Application of high performance liquid chromatography/inductively coupled plasma mass spectrometry to the investigation of phytochelatin complexes and their role in heavy metal detoxification in plants. *Analusis* **26,** M28.

Lopaticki, S., Morrow, C. J., and Gorman, J. J. (1998) Characterization of pathotype-specific epitopes of Newcastle disease virus fusion glycoproteins by matrix-assisted laser desorption/ionization time-of-flight mass spectrometry and post-source decay sequencing. *J. Mass Spectrom.* **33,** 950.

Lyubarskaya, Y. V., Carr, S. A., Dunnington, D., Prichett, W. P., Fisher, S. M., Appelbaum, E. R., et al. (1998) Screening for high-affinity ligands to the Src SH_2 domain using capillary isoelectric focusing-electrospray ionization ion trap mass spectrometry. *Anal. Chem.* **70,** 4761.

Mak, M., Mezo, G., and Skribanek Z. Hudecz, F. (1998) Stability of Asp-Pro bond under high- and low-energy collision-induced dissociation conditions in the immunodominant epitope region of *herpes simplex* virion glycoprotein D. *Rapid Commun. Mass Spectrom.* **12,** 837.

Millar, A. L., Jackson, N. A. C., Dalton, H., Jennings, K. R., Levi, M., Wahren, B., et al. (1998) Rapid analysis of epitope-paratope interactions between HIV-1 and a 17-amino-acid neutralizing microantibody by electrospray ionization mass spectrometry. *Eur. J. Biochem.* **256,** 164.

Nettleton, E. J., Sunde, M., Lai, Z. H., Kelly, J. W., Dobson, C. M., and Robinson, C. V. (1998) Protein subunit interactions and structural integrity of amyloidogenic transthyretins. Evidence from electrospray mass spectrometry. *J. Mol. Biol.* **281,** 553.

Northrop, D. B. and Simpson, F. B. (1998) Kinetics of enzymes with isomechanisms. Britton-induced transport catalyzed by bovine carbonic anhydrase II, measured by rapid-flow mass spectrometry. *Arch. Biochem. Biophys.* **352,** 288.

Potier N. Donald, L. J., Chernushevich, I., Ayed, A., Ens, W., Arrowsmith, C. H., et al. (1998) Study of a noncovalent *trp* repressor-DNA operator complex by electrospray ionization time-of-flight mass spectrometry. *Protein Sci.* **7,** 1388.

Qiu, Y. C., Benet, L. Z., and Burlingame, A. L. (1998) Identification of the hepatic protein targets of reactive metabolites of acetaminophen *in vivo* in mice using two-dimensional gel electrophoresis and mass spectrometry. *J. Biol. Chem.* **273**, 17940.

Raftery, M. J. and Geczy, C. U. (1998) Identification of noncovalent dimeric complexes of the recombinant murine S100 protein CP10 by electrospray ionization mass spectrometry and chemical cross-linking. *J. Am. Soc. Mass Spectrom.* **9**, 533.

Rostom, A. A., Sunde, M., Richardson, S. J., Schreiber, G., Jarvis, S., Bateman, R., et al. (1998) Dissection of multi-protein complexes using mass spectrometry. Subunit interactions in transthyretin and retinol-binding protein complexes. *Proteins* **Suppl 2**, 3.

Salih, B., Masselon, C., and Zenobi, R. (1998) Matrix-assisted laser desorption/ionization mass spectrometry of noncovalent protein-transition metal ion complexes. *J. Mass Spectrom.* **33**, 994.

Siegel, M. M., Tabei, K., Bebernitz, G. A., and Baum, E. Z. (1998) Rapid methods for screening low molecular mass compounds non-covalently bound to proteins using size exclusion and mass spectrometry applied to inhibitors of human cytomegalovirus protease. *J. Mass Spectrom.* **33**, 264.

Sun, Y., Bauer, M. D., and Lu, W. (1998) Identification of the active site serine of penicillin-binding protein 2a from methicillin-resistant *Staphylococcus aureus* by electrospray mass spectrometry. *J. Mass Spectrom.* **33**, 1009.

Tsatsos, P. H., Reynolds, K., Nickels, E. F., He, D. Y., Yu, C. A., and Gennis, R. B. (1998) Using matrix-assisted laser desorption/ionization mass spectrometry to map the quinol binding site of cytochrome bo_3 from *Escherichia coli*. *Biochemistry-USA* **37**, 9884.

Vis, H., Heinemann, U., Dobson, C. M., and Robinson, C. V. (1998) Detection of a monomeric intermediate associated with dimerization of protein, H. U., by mass spectrometry. *J. Amer. Chem. Soc.* **120**, 6427.

Volz, J., Bosch, F. U., Wunderlin, M., Schuhmacher, M., Melchers, K., Bensch, K., Steinhilber, W., et al. (1998) Molecular characterization of metal-binding polypeptide domains by electrospray ionization mass spectrometry and metal chelate affinity chromatography. *J. Chromatogr. A.* **800**, 29.

Walker, A. K., Wu, Y. L., Timmons, R. B., Kinsel, G. R., and Nelson, K. D. (1999) Effects of protein-surface interactions on protein ion signals in MALDI mass spectrometry. *Anal. Chem.* **71**, 268.

Weller, V. A. and Distefano, M. D. (1998) Measurement of the α-secondary kinetic isotope effect for a prenyltransferase by MALDI mass spectrometry. *J. Amer. Chem. Soc.* **120**, 7975.

Wong, A. W., He, S. M., Grubb, J. H., Sly, W. S., and Withers, S. G. (1998) Identification of Glu-540 as the catalytic nucleophile on human β-glucuronidase using electrospray mass spectrometry. *J. Biol. Chem.* **273**, 34057.

Wood, T. D., Guan, Z. Q., Borders, C. L., Chen, L. H., Kenyon, G. L., and McLafferty, F. W. (1998) Creatine kinase; essential arginine residues at the nucleotide binding site identified by chemical modification and high-resolution tandem mass spectrometry. *Proc. Natl. Acad. Sci. USA* **95**, 3362.

Wu, X. H., Wang, H. F., Liu, Y. F., Lu, X. Y., Wang, J. J., and Li, K. (1997) Histone adduction with nicotine. A bio-AMS study. *Radiocarbon* **39**, 293.

Yu, L., Gaskell, S. J., and Brookman, J. L. (1998) Epitope mapping of monoclonal antibodies by mass spectrometry: identifica tion of protein antigens in complex biological systems. *J. Am. Soc. Mass Spectrom.* **9**, 208.

Zechel, D. L., Konermann, L., Withers, S. G., and Douglas, D. J. (1998) Pre-steady state kinetic analysis of an enzymatic reaction monitored by time-resolved electrospray ionization mass spectrometry. *Biochemistry-USA* **37**, 7664.

Protein Analysis—Method Development

Akashi, S., Takio, K., Matsui, H., Tate, S., and Kainosho, M. (1998) Collision-induced dissociation spectra obtained by Fourier transform ion cyclotron resonance mass spectrometry using a $^{13}C,^{15}N$-doubly depleted protein (Letter). *Anal. Chem.* **70**, 3333.

Carbeck, J. D., Severs, J. C., Gao, J. M., Wu, Q. Y., Smith, R. D., and Whitesides, G. M. (1998) Correlation between the charge of proteins in solution and in the gas phase investigated by protein charge ladders, capillary electrophoresis and electrospray ionization mass spectrometry. *J. Phys. Chem. B.* **102**, 10596.

Devreese, B. and Van Beeumen, J. (1998) The important contribution of high precision mass spectrometric methods in the study of the structure-function relationship of proteins and enzymes. *Analyst* **123**, 2457.

Fournier, I., Beavis, R. C., Blais, J. C., Tabet, J. C., and Bolbach, G. (1997) Hysteresis effects observed in MALDI using oriented, protein-doped matrix crystals. *Int. J. Mass Spectrom. Ion Proc.* **169**, 19.

Garzotti, M., Rovatti, L., and Hamdan, M., (1998) On-line liquid chromatography/electrospray tandem mass spectrometry to investigate acrylamide adducts with cysteine residues. Implication for polyacrylamide gel electrophoresis separations of proteins. *Rapid Commun. Mass Spectrom.* **12**, 484.

Heck, A. J. R. and Derrick, P. J., (1998) Selective fragmentation of single isotopic ions of proteins up to 17 kDa using 9.4 Tesla Fourier transform ion cyclotron resonance. *Eur. Mass Spectrom.* **4**, 181.

Hendricker, A. D. and Voorhees, K. J. (1998) Amino acid and oligopeptide analysis using Curie-point pyrolysis mass spectrometry with *in-situ* thermal hydrolysis and methylation. Mechanistic considerations. *J. Anal. Appl. Pyrol.* **48**, 17.

Hirschler, J., Halgand, F., Forest, E., and Fontecilla-Camps, J. C. (1998) Contaminant inclusion into protein crystals analyzed by electrospray mass spectrometry and X-ray crystallography. *Protein Sci.* **7**, 185.

Li, F., Dong, M., Miller, L. J., and Naylor, S. (1999) Efficient removal of sodium dodecyl sulfate (SDS) enhancss analysis of proteins by SDS-polyacrylamide gel electrophoresis coupled with matrix-assisted laser desorption/ionization time-of-flight mass spectrometry (Letter). *Rapid Commun. Mass Spectrom.* **13**, 464.

Marie, G., Serani, L., Laprevote, O., and Das, B. C. (1998) Influences of pH and in-source collisional energy on the cationization of insulin. *Rapid Commun. Mass Spectrom.* **12**, 1182.

McLuckey, S. A., Stephenson, J. L., and Asano, K. G. (1998) Ion/ion proton-transfer kinetics: implications for analysis of ions derived from electrospray of protein mixtures. *Anal. Chem.* **70**, 1198.

Meunier, C., Jamin, M., and, D. E., Pauw, E. (1998) On the origin of the abundance distribution of apomyoglobin multiply charged ions in electrospray mass spectrometry. *Rapid Commun. Mass Spectrom.* **12,** 239.

Osborn, B. L. and Abramson, F. P. (1998) Pharmacokinetics and metabolism studies using uniformly stable isotope proteins with HPLC/CRIMS detection. *Biopharm. Drug Dispos.* **19,** 439.

Puchades, M., Westman, A., Blennow, K., and Davidsson, P. (1999) Removal of sodium dodecyl sulfate from protein samples prior to matrix-assisted laser desorption/ionization mass spectrometry analysis. *Rapid Commun. Mass Spectrom.* **13,** 344.

Reid, G. E., Simpson, R. J., and O'Hair, R. A. J. (1998) A mass spectrometric and *ab initio* study of the pathways for dehydration of simple glycine and cysteine-containing peptide [M+H}' ions. *J. Am. Soc. Mass Spectrom.* **9,** 945.

Resemann, A., Mayer-Posner, F. J., Lapolla, A., Fedele, D., D'Alpaos, M., and Traldi, P. (1998) Further considerations on the use of matrix-assisted laser desorption/ionization mass spectrometry in the analysis o(glycated globins (Letter). *Rapid Commun. Mass Spectrom.* **12,** 805.

Schurenberg, M., Dreisewerd, K., and Hillenkamp, F. (1999) Laser desorption/ionization mass spectrometry of peptides and proteins with particle suspension matrixes. *Anal. Chem.* **71,** 221.

Shi, S. D. H., Hendrickson, C. L., and Marshall, A. G. (1998) Counting individual sulfur atoms in a protein by ultrahigh-resolution Fourier transform ion cyclotron resonance mass spectrometry: experimental resolution of isotopic fine structure in proteins. *Proc. Natl. Acad. Sci. USA* **95,** 11532.

Staudemann, W., Hatt, P. D., Hoving, S., Lehmann, A., Kertesz, M., and Jarnes, P. (1998) Sample handling for proteome analysis. *Electrophoresis* **19,** 901.

Stephenson, J. L. and McLuckey, S. A. (1998) Ion/ion reactions for oligopeptide mixture analysis. Application to mixtures comprised of 0.5-100 kDa components. *J. Am. Soc. Mass Spectrom.* **9,** 585.

Stockigt, D. and Haebel, S. (1998) Identification of collision-induced dissociation fragments from a protonated glutathione conjugate by isotope-specific MS^3 experiments in an ion trap. *Rapid Commun. Mass Spectrom.* **12,** 273.

Tsugita, A., Kamo, M., Miyazaki, K., Takayama, M., Kawakarni, T., Shen, R. Q., et al. (1998) Additional possible tools for identification of proteins on one- or two-dimensional electrophoresis. *Electrophoresis* **19,** 928.

Vachet, R. W., Ray, K. L., and Glish, G. L. (1998) Origin of product ions in the MS/MS spectra of peptides in a quadrupole ion trap. *J. Am. Soc. Mass Spectrom.* **9,** 341.

Zhang, W. Z., Niu, S. F., and Chait, B. T. (1998) Exploring infrared wavelength matrix-assisted laser desorption/ionization of proteins with delayed extraction time-of-flight mass spectrometry. *J. Am. Soc. Mass Spectrom.* **9,** 879.

Zhang, X., Jai Nhuknan, J., and Cassady, C. J. (1997) Collision-induced dissociation and post-source decay of model dodecapeptide ions containing lysine and glycine. *Int. J. Mass Spectrom. Ion Proc.* **171,** 135.

Index

"If you don't find it in the index, look very carefully through the entire catalogue."
Sears, Roebuck Catalogue, 1897.

A

Abbreviations and acronyms, 489–497
Adenosine 3',5'-cyclic
   monophosphate, 369
   FAB spectrum of, 374, 375
Adenosine 5'-triphosphate, FAB
   spectrum of, 374, 375
   regenerating systems,
      efficiency of, 378
Adenylyl cyclase, alternative substrates
   used with, 378
   FAB spectrum of, 375, 376
   quantitation using FAB, 376
Affinity capillary electrophoresis, 197
Affinity isolation, 430
   with Ni-NTA agarose gel, 297,
      302, 303
   with streptavidin coated
      Dynabeads, 119
Affinity mass spectrometry, 453
Agarose beads, antibody bound, 454,
   457, 458
   non-antibody bound, 454, 458
   Protein A, 454, 457
   Protein G, 454, 457
   Protein G-Plus/Protein A, 440,
      442–444, 446
Agarose gel, 300
Algal, ^{12}C enriched, 297
Alzheimer's disease, 439
Amino acids, abbreviated forms,
   497, 499
   equal masses of, 52, 53
   residue masses, 497, 499

Aminolysis, used to identify
   glycoproteins, 213, 216
Aminopeptidase M, 188, 196
Ammonium chloride, ^{15}N-labeled,
   226, 227
Amyloid β-peptide, characterization of,
   439, 440
   fragment [12-28] as internal standard
      for quantitation, 448
   quantitation by MALDI, 445
Angiotensin, as mass calibrant, 187, 321
Antibody, anti-Aβ specific, 439
   anti-Ab[17-24] specific, 440, 446
   anti-Ab[5-11] specific, 446
   anti-mouse IgG conjugated to
      alkaline phosphatase, 454, 455
   anti-p24 specific, 187, 188, 191
   bile duct epithelial cell specific,
      454, 455
   immobilized for epitope excision, 185
Antigens, detection and characterization
   of, 453
Arg-C endoprotease, 226, 229
Asp-N endoprotease, 30, 33, 171, 226,
   229, 283
Avenins, preparation of standards,
   356, 357
   quantitation in food, 362
6-Aza-2-thiothymine MALDI matrix, 168

B

Bacteria, chemotaxonomy and mass
   spectrometry of, 464–466
   identification from MALDI spectra
      of whole cells, 465–471

MALDI spectra from whole, effect of experimental parameters on, 471–481
reproducibility of, 467–469
species specificity of, 469, 470
preparation for MALDI analysis, 471
proteins from, solubility of, 461
analysis of, 471, 481–485
Bacterioopsin, 426
analysis of protease fragments, 428
ESI analysis of, 428
solvent adduct formation with, 428
Beads, magnetic, 454, 455
Benzyl 1-thio-β-D-xylopyranoside, 212
Bile duct epithelial cells, isolation and purification of, 454, 455
BLAST, 52, 53, 80
Bombesin, 135
structure of gas-phase ions from, 144
Bovine serum albumin, as mass calibrant, 187, 358, 359

C

Carbonic anhydrase, as mass calibrant, 187
Carboxypeptidase B, 297, 303
Carboxypeptidase Y, 188, 194, 226, 229
Cdk2 regulatory protein, preparation of, 227
Celiac disease, 355
CHAPS, 440, 443, 448
Chemiluminescence, enhanced, 31
Chromatography, anion-exchange, 226, 227
Ni^{2+}-affinity, 226, 227
perfusion, 99, 103
strong-cation exchange, 19–21
two-dimensional, 18, 19, 23
α-Chymotrypsin, 349
CIDentify program, 42, 52–56, 58
CIDentify result compiler program, 42, 56–58
Cirrhosis, primary biliary, 453

Citrate synthase, electrospray analysis of, 244, 245
multimers, 244
association constant for, 245
NADH complex, dissociation constant for, 246
NADH inhibition of, 240, 242, 245, 246
preparation of, 240, 241
Coomassie blue, 8, 297
CP10 chemotractant protein, 28
cDNA for, 36, 37
mass spectrometry of, 31–35
preparation of, 30, 31
C-peptide, 293
peptide mapping data from, 306, 309
preparation from proinsulin, 302, 303
quantitation of, 294
Cyclic AMP-dependent protein kinase, activity determined using FAB, 379–381
effect of alternative substrates on, 379–381
Cyclohexyl-hexyl-β-D-maltoside detergent, 426
removal during sample preparation, 433–435
Cysteine, identification of nitrosylation of, 158–161
S-nitroso, preparation of solution, 151
pyridylethylation of, 51, 339
Cytidylyl cyclase, quantitation using FAB, 378, 379
Cytochrome c, as MALDI test compound, 389
as mass calibrant, 470
on-target tryptic digest of, 174, 179
Cytotoxins, 326, 328
analogs of, 326, 328

D

Databases, of complex carbohydrate structure (CCSD), 280
of DNA sequences, 30, 57

Index

of glycosidases, 203
of protein families and domains (PROSITE), 65
of protein structure, 1, 2, 22, 23, 30, 41, 57, 139, 324
of venoms, 324
Dehydroalanine, 339, 350
Deoxyhexose, 281, 500
Desalting techniques, 7, 9, 116, 117, 172, 177–179, 349
Detergents, preventing protein agglomeration, 391
preventing surface adsorption of proteins, 391–393
Deuterium oxide, 99–101, 104
Deuterium, control of exchange, 100, 101
correction for back-exchange, 104, 105
incorporation measured from average mass, 105, 106
location of exchange sites, 96, 97
Diabetes, 293
Diammonium hydrogen citrate, 252, 388, 389
in oligonucleotide analysis, 390
in sample pretreatment for MALDI, 394, 395
in suppression of alkali ion adducts, 254
2′,6′-Dihydroxyacetophenone MALDI matrix, 252, 254, 388, 389, 390
2,5-Dihydroxybenzoic acid MALDI matrix, 116, 227, 228, 276, 277, 283, 287, 298, 408, 416
3,5-Dimethoxy-4-hydroxycinnamic acid (sinapinic acid) MALDI matrix, 116, 227, 229, 277, 356, 358, 389, 408, 412, 416, 454, 458, 474
Dimethylpimelimidate cross-linker, 454, 457
Dinitrophenolate leaving group, 207
2,4-Dinitrophenyl 2-deoxy-2-fluoro-β-D-xylospyranoside, 212, 214

Dipeptidyl peptidase IV (DP-IV), 252, 257–266
inhibition of, 264, 265
serum equivalent activity of, 265, 266
Dissociation, collision-induced, 42, 159, 211, 348, 351, 370, 373
Disulfide bridges, characterization of, 168, 174–177
formation of, 302, 303
MALDI-induced cleavage of, 168, 171
3,3′-Dithiobis(sulfosuccinimidyl propionate) cross-linker, 115, 123, 124
Dithiothreitol 1,4- (DTT), 5, 7, 20, 115, 227, 276, 241, 345, 409, 410, 412, 418
n-Dodecyl-β-D-maltoside detergent, 426, 427
used in ESI analysis, 430, 435
Dodecyltrimethylammonium chloride, 426, 427, 430
Domains, protein, 69–71, 73, 74
Drop dialysis as sample pretreatment for MALDI, 394–396
Drug binding, effects on viral capsid dynamics, 234
Dynabeads, 440, 443, 444, 446
Dynorphin A fragments, 135
structure of gas-phase ions from, 142–144

E

Edman degradation, 33, 47, 63, 173, 211, 483, 484
Electrospray ionization (ESI), 27, 103, 214, 215, 239, 240, 243, 317, 318, see also Nanospray
fragmentation induced in, 159, 348, 351
ion source operation, 31, 32, 136, 153, 213, 298, 320, 321, 427
ion source temperature, 428, 435
mass accuracy in, 324, 326–329, 426, 428, 435

mass calibration in, 12, 32, 242, 298, 321
micro-scale, 346, 349, 350, 352
with MS/MS, 43, 215, 216
with TOFMS, 240
Electrostatic repulsion, in gas-phase peptide ions, 135, 141, 142, 144, 145
Elements, mass and abundance values for, 498
β-Elimination, 339, 346, 349, 353
Enterobacter cloacae, spectra from whole cells, 467, 477
Enzyme-linked immunosorbent assay (ELISA), 355, 356, 361, 362, 363, 365, 366
Enzyme-linked signal amplification, 453
Enzymes, concentration of buffer for assay of, 372
FAB spectra of incubates, 373, 374
Enzyme-to-substrate ratio, 33, 155, 157, 171, 172, 179, 181, 200, 228, 229, 412, 413
Epitope, excision, 185
extraction, 185
HPVHAGPIAPG, structural relevance of, 197, 198
mapping, 191, 197
Escherichia coli, 477–481
protein identification in, 481–485
Exoglycosidase digestions, parallel, 281
sequential with MALDI analysis, 276, 281, 283

F

FASTA, 52, 53, 80
Fast-atom bombardment (FAB), 370
quantitation using, 371, 374, 381
Ferulic acid MALDI matrix, *see* 4-hydroxy-3-methoxycinnamic acid
Filter, cutoff, in sample pretreatment for MALDI, 388, 396, 397
Flock house virus, 228, 229, 232–235

proteolytic digestion of, 229
N-(9-Fluorenyl)methoxycarbonyl derivatives, 297
Fluorosugars, 206–210
Formic acid, as solvent, 427, 428, 430, 433, 441, 442
in suppression of adduct ion formation, 466, 471, 475
Fragment ions in TOFMS, metastable, 278
prompt, 277, 278
reduction of, 278
Fructose, as co-matrix in MALDI, 298

G

β-Galactosidase, 276, 286, 388, 389, 390
MALDI analysis of, 389, 390
sequential digestions with, 286
Genes, ancient conserved regions in, 68, 69
domains and modules in, 69–71
evolution of, 72–74
orthologous and paralogous, 69
Gliadins, preparation of standards, 356, 357
quantitation in food, 362
rapid screening in food, 360
γ-Globulin, as MALDI test compound, 389, 396, 398
Glu-C endoprotease, 150, 157, 224, 226, 229
Glucagon-like peptide-1-[7-36] amide (GLP-1$_{7-36}$), 252
kinetics of hydrolysis by DP IV, 257–266
quantitation by MALDI, 257, 258
Glucoamylase, 276
analysis of glycosylation in, 279
β-Glucosaminidase, N-acetyl, 276, 286
Glucose-dependent insulinotropic polypeptide/gastric inhibitory polypeptide (GIP$_{1-42}$), 252
kinetics of hydrolysis by DP IV, 257–266

Index

quantitation by MALDI, 257, 258
Glucose-specific IIBC component of
  E. coli phosphotransferase, ESI
  analysis of, 430
Glutathione-agarose beads, 28, 31
Gluten-free foods, analysis of, 362
  extraction of, 358, 361, 362
Gluten-rich foods, preparation for
  MALDI analysis, 357
Glutens, analysis in foods, 355, 363
  MALDI mass profiles of, 358, 359
Glycans, direct mass spectrometric
  sequencing by post-source decay
  analysis, 275
  N- and O-linked differentiated,
    280, 281
  N-linked, 273–275
  O-linked, 275
  structural characterization of,
    281–286
Glycerol as FAB matrix, 371, 373,
  374, 384
Glycoforms, quantitation by
  MALDI, 280
Glycopeptides, β-elimination from,
  339, 346, 349, 353
  carbohydrate microheterogeneity in
    mass spectra, 280
  identified by aminolysis, 211, 213,
    216
  MALDI matrix for analysis of, 416
  mucin-type, localization of
    glycosylated sites in, 286
  non-specific hydrolysis of, 286, 287
  sialylated, analysis of, 278, 280
  suppression of component signal in
    mixture analysis, 278
Glycoproteins, analysis of glycosylation
  in, 280, 281, 286–289, 339, 345,
  346, 348–350
  carbohydrate content from MALDI
    spectrum, 278, 279
  MALDI matrix for analysis of, 416

structural diversity of, 273–275
Glycosidase intermediate, difluorosugar
  strategy for trapping, 209, 210
  2-fluorosugar strategy for trapping,
    206–208, 214
  5-fluorosugar strategy for trapping,
    208, 209
  peptic digest of, 210, 213, 214–216
Glycosidases, active sites of, 204, 205
  inverting mechanism for, 204
  labeled, proteolytic digestion of,
    210, 211, 213, 214–216
  location of active site by neutral-loss
    MS/MS, 211, 212, 215, 216
  retaining mechanism for, 204, 205
  used in glycan analysis, 282
Glycosylation, analysis of, current
  references to use of mass
  spectrometry in, 519–521
  consensus sequences for, 273, 275
  recognised using collision-induced
    dissociation, 348, 351
  site-specific characterization in
    glycoproteins, 279–286
H-Gly-Pro-4-nitroanilide, 252, 265,
  266, 268, 270
H-Gly-Pro-Pro-4-nitroanilide, 252,
  266, 268
Guanosine 3',5'-cyclic
  monophosphate, 369

H

HdeA protein identified in *E. coli*,
  483, 484
HdeB protein identified in *E. coli*, 483-
  484
Hemoglobin, identification of reactive
  cysteine in, 159, 160
  S-nitroso, 147, 148
  chemical synthesis of, 149,
    151, 152
  enzymatic synthesis of, 149, 152
  peptide mapping of, 150, 157, 158

separation and analysis of globin chains in, 153–155
Hexafluoro-2-propanol, 426, 428
Hexahistidine tag, 296, 300, 302, 430
  removal of, 302
Hexosamine, N-acetyl, 280, 281, 500
Hexose, 280, 281
HIV-1$_{IIIB}$ p26 AIDS-related protein, 187
  affinity-binding to immobilized antibody, 191
  preparation of, 189, 190
  proteolytic footprinting of affinity-bound, 191–198
Human chorionic gonadotrophin (hCG), α- and β-subunits, separation of, 346
  β-subunit of (hCG-b), 338
  glycosylation states determined in, 346, 348–350
  isolation of, 339–341, 346
  N-linked carbohydrates of, 342
  O-linked carbohydrates of, 343
  primary sequence of, 340
  reduction and pyridylethylation of, 341, 345, 346, 347, 351
  residues 109–145 isolated from, 339, 345, 349
  tryptic digest of, 343, 347
Human rhinovirus 14, 228, 229, 232, 233
  proteolytic digestion of, 229
Hydrogen bonding, in gas-phase peptide ions, 135, 142–145
Hydrogen exchange, calculation of rates, 107, 108
  coupled with mass spectrometry, 95–97
  EX1 kinetics, 98, 99
    isotope distribution from, 106
  EX2 kinetics, 99
    isotope distribution from, 106
  phosphate buffers in, 99–101
Hydrophobic interactions, in peptide structures, 135, 144, 145

4-Hydroxy-α-cyanocinnamic acid (HCCA) MALDI matrix, 116, 168, 175, 187, 276, 277, 281, 408, 411–413, 415-416, 440, 442, 444, 454, 457, 466, 470, 474, 475
  adducts formed with peptides, 176
  purification of, 446, 447
  used for analysis of glycopeptides, 416
  with nitrocellulose as matrix additive, 411
4-Hydroxy-3-methoxy-cinnamic acid (ferulic acid) MALDI matrix, 388, 389
3-Hydroxypicolinic acid MALDI matrix, 277, 278
N-Hydroxysuccinimidobiotin, 115, 119–123

I

Immunoprecipitation, 439, 443, 444, 454, 458
  non-specific binding reduced in, 448
  used with mass spectrometry, 439, 440
Insulin, 293, 297
  as mass calibrant, 187, 321, 441, 448
  biosynthetic analog of, 306
  isotopic patterns of, 295, 306–310
  peptide mapping data from, 306, 308
  preparation from proinsulin, 302, 303
  quantitation of, 294
Iodine, attachment to a protein, 118
Iodoacetamide 5, 7, 8, 19, 20, 170, 175, 409, 410
  in-gel alkylation with, 410
  on-target alkylation with, 175, 176
Iodoacetic acid, carboxymethylation with, 345, 348, 349
Iodogen iodination reagent, 115, 118
Isopropanol as MALDI solvent, 388–391, 441, 442
Isotope dilution assay, 293, 294
Isotopic labeling, deuterium in pro-insulin fusion protein, 301, 305

negative, with ^{12}C in pro-insulin fusion protein, 294–296, 301, 305, 309–311
^{15}N in pro-insulin fusion protein, 301, 305
^{18}O in peptide C-terminal carboxyl groups, 5, 10

J

Journals, mass spectrometry, 502, 503

K

K_{cat} value, calculation of, 265
Kinetic analysis using MALDI, 260–266
Kinetic energy release (KER), 133, 134
  distribution of, 136–139
  measurement of, 134, 136
K_m value, calculation of, 265

L

Leupeptin, protease inhibitor, 188, 189, 440, 442
Liquid chromatography, 19, 30, 212, 226, 227, 297, 321, 322, 409, 427, 471
  coupled to mass spectrometry, 21, 22, 150, 152–155, 157, 158, 210, 213, 319, 324–329, 343–345, 347, 348, 428, 430–435
  following hydrogen exchange, 102, 103
  for desalting samples, 170, 177–179, 407
  for sample recovery, 179
  microbore, 321, 343–345
  microscale columns for, 19, 170, 177–179, 407, 413, 414, 419, 420
  nanoscale columns for, 407, 414, 415, 419, 420
  of bacterial proteins, 471, 482–484
  of globins, 153–155
  of hydrophobic proteins, 427, 428, 430–435
  of insulin-related peptides, 303
  of protease digests, 20, 34, 35, 102, 103, 157–168, 172, 306, 345, 349
  of venom peptides, 322, 323, 324–329
  split flow, 153
Lutefisk97 program, 41–52
Lys-C endoprotease, 19, 20, 188, 191–194, 226, 229, 283
Lysine, cross-linking via, 123, 124
  modification of, 118–123

M

Mannose transporter complex from *E. coli,* 426, 427
  analysis of components, 430–433
Mass analyzed ion kinetic energy (MIKE) spectrometry, 133, 134, 136, 370, 373
Mass analyzed ion kinetic energy spectra, deconvolution of, 136, 137
Mass spectra, deconvolution of, 106
Mass spectrometer, Fourier transform ion cyclotron resonance, 298, 309–311
  hybrid quadrupole/time-of-flight 4
  ion trap, 42, 43, 50, 51, 117, 121, 123, 314
  magnetic sector, 104, 136, 372
  quadrupole, 30, 103, 150, 314, 322, 426
    mass range of, 240, 298, 322
  time-of-flight (TOF), 116, 168, 188, 189, 240, 242, 252, 409, 461–463, 470
    reflectron instrument, 278, 311
    with delayed extraction, 227, 277, 409
  triple quadrupole 2, 18, 42, 212, 213, 345
Mass spectrometry current awareness service, 502

Matrix-assisted laser desorption/
  ionization (MALDI), 463, 464
  adverse effects in, from hygroscopic
    salts, 391
  from phosphate buffer, 391
  from SDS, 391, 419
  mass accuracy in, 223, 231, 425
  mass calibration in, 32, 116, 117,
    187–189, 298, 321, 358, 359,
    441, 448, 466, 470
  matrix, see also individual matrix
    entries
    crystallization of, 394
    for glycopeptide and glycoprotein
      analysis, 277
    quantitation by, 255, 256, 395, 397–
      399, 443–445
      drop dialysis as sample
        preparation for, 395, 396
      internal standard for, 443, 444
      reproducibility of, 397
      standard curve for, 445
      without internal standards, 258
    sample clean-up, on-target, 406, 407
    small-scale, 170, 177–179, 394–
      397, 407, 413–415, 420
    sample digestion, on-target, 169,
      170, 170–174, 179
      small-scale, 415
    sample preparation, choice of buffer
      for, 391
      choice of matrix for, 277, 389,
        408, 415, 416, see also
        individual matrix entries
      choice of solvent for, 389–391
      detergent used in, 391–396
      dried-droplet method, 116, 410
      in the presence of salts, 389,
        394, 418
      preformed matrix layer for,
        411, 412
      sandwich method, 412, 417, 418
      sensitivity of methods, 419
      spin-dry method, 411, 412
      thick-layer method with
        nitrocellulose, 444, 445
      thin-layer method, 411, 412
    sample reaction, on-target, 170,
      174–176, 412, 413, 418, 419
    sample recovery from matrix, 179
    signal loss and remedial actions for,
      399, 400
    signal suppression by salts, 396
Max protein–DNA complex, structural
  studies of, 225, 226
Melanine-concentrating hormone, 171
Melittin, 135
  structure of gas-phase ions from,
    140–142
β-Mercaptoethanol, 118, 170, 302, 388
Modules, protein, 69–71
Molecular weight, change from
  monoisotopic to average, 50
Monosaccharides, abbreviated forms,
  497, 498, 500
  residue masses, 497, 498, 500
Morpholine, N-ethyl, 188, 195
  N-methyl, 188, 196
Motifs, protein, 65, 66, 71
Muc1a' mucin-derived peptide,
  glycosylation analysis of, 287
Murine fetal antigen 1, glycosylation
  analysis of, 283
Myoglobin, as mass calibrant, 32,
  298, 321

N

Nanospray ionization, 9, 10, 117, 241–243
  with MS/MS, 1, 43
Neuraminic acid, N-acetyl, 281, 282, 500
Neuraminidase, 276, 283
Neurotoxins, 326, 328
β-Nicotinamide adenine dinucleotide
  reduced form (b-NADH),
  preparation of, 242
Nitric oxide, interaction with proteins, 147

Index

Nitrocellulose as matrix additive in MALDI, 411, 419
p-Nitrophenyl-β-D-xylopyranoside, 212
Nitrosothiols, S-nitrosylation of proteins with, 147
Nitrosyl peptides, hydrolysis of the S-NO bond in, 159
Nitrosylated α- and β-globins identified by mass shift, 153–155
Nitrosylated globins, quantitative analysis of, 153
Nitrosylated peptides, electrospray mass spectrometry of, 157, 158
Nitrosylated proteins, monitoring of, 160–162
NtrC nitrogen assimilation regulatory protein, 171, 172

O

n-Octyl-β-D-glucopyranoside detergent, use in MALDI, 356, 358, 413, 419, 440, 441, 448
Oligonucleotides, MALDI analysis of, 390
Oligosaccharides, complex-type, 274, 275
   high-mannose type, 274
   hybrid-type, 274, 275
Outer membrane protein from *E. coli*, analysis of, 433–435
   preparation of, 426, 427
Ovalbumin, as MALDI test compound, 389

P

p21-B regulatory protein, 226
   binding site for Cdk2 on, 232
   ^{15}N-labeled, 227
   preparation and purification of, 227
   proteolytic digestion of, 228
p21-B/Cdk2 complex, 227, 228
   proteolytic digestion of, 228
ParR DNA-binding protein, 119
   dimers, 115, 119–121
      biotinylated, 119–121
      cross-linked, 124
   monomers, 119–124
      location of interaction site, 124
Pefablok SC, protease inhibitor, 188, 190
Pentafluoropropionic acid, 286, 287
Pepsin, 96, 150, 210
Pepstatin, protease inhibitor, 188, 189, 440, 442
Peptic digest, 102, 155, 159, 213
Peptide N-glycosidase (PNGaseF), used to differentiate N- and O-linked glycans, 280, 281, 282, 283
Peptides, α-helix, gas-phase ion stability of, 135, 140–142, 144, 145
   aminolysis of, 213, 216
   analysis, current references to use of mass spectrometry in, 504–507
   β-hairpin, 135, 142
   β-pleated sheet, gas-phase ion stability of, 135, 142, 144, 145
   fragmentation, 498–501
      b-type ions from, 45, 46, 50, 51, 499, 500
      y-type ions from 2, 5, 43, 45, 46, 50, 51, 52, 499, 501
   hydrophobic, ESI analysis of, 426
   MALDI analysis of, 425
   MALDI spectra of, using n-octyl glucopyranoside detergent, 419
   mapping, current references to use of mass spectrometry in, 508, 509
   processing program for, 117
   used to identify glycosylation sites, 280, 281
   used to identify protein primary structure 1, 117, 124, 150, 157, 223, 224, 280, 281, 306, 308, 309, 405, 415, 416
   molecular modeling of ions from, 139, 140
   sequence tag in 2, 51, 52, 63

sequencing, current references to use of mass spectrometry in, 503, 504
tandem mass spectrometry in 1–4, 9, 10, 12, 17, 18, 20–22, 41–44, 216, 217
synthesis, current references to mass spectrometric analysis used in, 508
three-strand β-sheet former, 135
pH, influence in MALDI sample preparation, 417
Phenylmethylsulfonyl fluoride, 188, 226, 228, 440, 454, 458
Z-Phe-Pro-(isonicotinic acid methyl ester) methyl ketone, as prolyl endopeptidase inhibitor, 253
hydrolytic stability of, 253–256
rate constant for hydrolysis of, 256
Phosphodiesterase inhibitors, efficiencies determined by FAB, 378
Phospholipase D, 388
analysis by MALDI, 392–396
as internal standard for quantitation by MALDI, 397
Phosphopeptide Gly-Leu-phosphoSer-Pro-Ala-Arg, 454, 457
Phosphorylation, analysis of, current references to use of mass spectrometry in, 519–521
Phosphotransferase system from *E. coli*, preparation of glucose-specific IIBC component of, 426, 430
Polyacrylamide gel electrophoresis (PAGE), 30, 31, 151, 230, 297, 301, 409, 426
Polyalanine peptide, 135, 144
Polymerase chain reaction (PCR), 298–300
Polyvinylidene difluoride membrane, 31
Proinsulin, 293, 294, 298–300, 302, 303
Prolyl endopeptidase, 252

Protease inhibitors, 188, 189, 190, 440, 442, 443, 446
Protein A, binding specificity of, 447
Protein analysis worksheet, 227, 229
Protein complexes, analysis of by ESI, 18, 226, 239, 240
cross-linked sites identified in, 114, 123, 124
denaturing with urea, 19
dissociation of, 19
isotopic labeling of subunit, 231
quaternary structure of, 224, 225, 230, 231
Protein G, binding specificity of, 447
Protein G-Plus, binding specificity of, 447
Protein G-Plus/Protein A, binding specificity of, 447
Protein–DNA interaction, analysis of, 225, 226
Proteins, *see also* individual protein entries
analysis, current references to use of mass spectrometry in, 509–518, 523–526
ancient conserved regions in, 68, 69
bacterial, as biomarkers, 481
associated with acid-resistance, 485
configuration and folding, current references to mass spectrometric analysis of, 518, 519
cross-linking, used in structural analysis, 113–115, 123, 124
deuterium labeled, 100, 104, 105
pepsin digestion of, 96, 97, 102, 103
evolution of, 72–74
high molecular weight, MALDI matrix for, 389
higher-order structural properties of, 113–115
hydrogen exchange models for, 97–99

Index

hydrophobic, ESI analysis of, 426–435
  MALDI analysis of, 425
  use of detergent in analysis of, 391, 393, 427, 430
  interactions and kinetics, current references to mass spectrometric analysis of, 521–525
  lower molecular weight, MALDI matrix for, 389
  MALDI spectra of, using HEPES or MOPS buffer, 391
    using n-octyl glucopyranoside detergent, 419
  marker, detection by affinity mass spectrometry, 453–457
  mass mapping, *see* peptides, mapping
  mixture analysis, 18, 419
  sequencing, current references to use of mass spectrometry in, 508, 509
  surface labelling, used in structural analysis, 113–115, 117–126
  unfolding, dynamics of, 98, 99
Proteolytic digestion, *see also* individual enzymes and substrates
  in-gel 5, 7, 8, 409, 410
  limited, 224, 225, 232
  of affinity-bound material, 191–196
  of protein complex, 228, 231, 232
  on-target, for MALDI, 169–174, 412, 413
  time-resolved, 232–235
Proteus mirabilis, spectra from whole cells, 467, 468
Pseudomonas species, spectra from whole cells, 469
Pyridylethylation, 51, 339, 346

Q

Quantitation, *see* entries under individual analytes, FAB and MALDI

R

Reaction columns, compact, 188–190
Reduction and alkylation, combined on-target, 175
Reduction, in-gel, 7, 410
  on-target, 170, 174–176

S

Salmonella typhimurium, spectra from whole cells, 466, 467, 477
Sepharose beads, CNBr-activated, 188, 190
Sepharose, glutathione (GSH-sepharose) beads, 188, 189
Sep-Pak cartridges, 297, 298, 303
Serine, conversion to dehydroalanine by β-elimination, 339, 350
Shigella flexneri, protein identification in, 481–485
  spectra from whole cells, 473, 475, 477
Sialic acid, 274, 280, 283
Sinapinic acid MALDI matrix, *see* 3,5-dimethoxy-4-hydroxycinnamic acid
Snake toxins, classification by mass spectrometry, 317, 320, 330
Sodium dodecyl sulfate (SDS), 427
  adverse effects on MALDI mass spectra, 391, 419
Substance P, as mass calibrant, 242
Sulfitolysis, oxidative, 302

T

Tandem mass spectrometry (MS/MS), 17, 18, 21, 22, 114, 117, 210
  charge state of precursor ions in, 43, 48
  mass accuracy in, 44, 48
  multiple stage, 43
  neutral-loss scan, 211, 212, 215, 216
  scanning modes, 213, 215, 216
N-α-p-Tosyl-L-lysine chloromethyl ketone (TLCK), 227, 229, 440, 442

N-Tosyl-L-phenylalanine chloromethyl
    ketone (TPCK), 440, 442
2,4,6-Trihydoxyacetophenone MALDI
    matrix, 277, 278, 408, 411, 416
Trinitrophenolate leaving group, 209
Tris(2-carboxyethyl)phosphine
    hydrochloride, 170, 175, 181
Triton X-100 detergent, use in MALDI,
    388, 391, 392, 393, 394, 396
Triton-X detergent, use in
    immunoprecipitation, 454, 458
Trypsin 5, 8, 19, 20, 115, 117, 169,
    170–174, 223, 224, 226, 228, 229,
    343, 346, 349, 410, 412, 413
  MALDI sample plate derivatized
    with, 229
  poroszyme-immobilised, 115
  TPCK treated, 188, 194, 297, 303
Tryptic digestion, on-target, 173, 174, 176
Tween-20 detergent, 454, 455, 458
Tween-80 detergent, 388

   use in MALDI, 391, 392
Tyrosine, modification of, 118

V

VENOMS database, 322
Venoms, composition of, 318
  ESI analysis of, 322–329
4-Vinylpyridine, 341, 347, 351, 409
Viral structure, investigated using mass
    spectrometry, 232–236

W

Waterbugs, 242
WIN 52084 antiviral agent, 234
Words, peptide, 75–83

X

β-Xylosidase, 212
  identification of active site of, 214–216
  inactivated, 212, 214
  peptic digest of, 213–216